1st International Conference on Design, Energy, Materials and Manufacture (ICDEMM 2018)

IOP Conference Series: Materials Science and Engineering
Volume 539

Bali, Indonesia
24 – 25 October 2018

ISBN: 978-1-5108-9085-5
ISSN: 1757-8981

Printed from e-media with permission by:

Curran Associates, Inc.
57 Morehouse Lane
Red Hook, NY 12571

CURRAN ASSOCIATES INC. proceedings.com

Some format issues inherent in the e-media version may also appear in this print version.

Printed by Curran Associates, Inc. (2019)

For permission requests, please contact the Institute of Physics at the address below.

Institute of Physics
Dirac House, Temple Back
Bristol BS1 6BE UK

Phone: 44 1 17 929 7481
Fax: 44 1 17 920 0979

techtracking@iop.org

Additional copies of this publication are available from:

Curran Associates, Inc.
57 Morehouse Lane
Red Hook, NY 12571 USA
Phone: 845-758-0400
Fax: 845-758-2633
Email: curran@proceedings.com
Web: www.proceedings.com

Table of contents

Volume 539

International Conference on Design, Energy, Materials and Manufacture

24–25 October 2018, Bali, Indonesia

Accepted papers received: 18 April 2019

Preface

International Conference on Design, Energy, Materials and Manufacture

Peer review statement

Papers

Weighbridge vibration analysis using 3 accelerometer sensors and single excitation

Syafri and Darman.....1

Wing design of uav serindit v-1

K Anuar, M Akbar and Herisiswanto.....9

Multi-step ahead prediction of vapor compression air conditioning system behaviour using neural networks

S Sholahudin, K Ohno, S Yamaguchi and K Saito.....21

Hardness improvement of aluminum alloy 2024 t3 after artificial aging treatment

I M Astika.....29

Assessment of thermal and transport properties of ionic liquids as suitable absorbent for absorption cooling applications

H M Ariyadi, S Yamaguchi and K Saito.....34

Study of stinging nettle fibers as a reinforcing of composite materials based on its growing region

I G P A Suryawan, N P G Suardana, I N S Winaya, I W B Suyasa and I P Lokantara.....42

Investigation and analysis on the performance of hydraulic ram pump at various design its snifter valve

M Sucipta and M Suarda.....49

Investigation on flow pattern in a hydraulic ram pump at various design and setting of its waste valve

M Suarda, M Sucipta and I G K Dwijana.....57

The effect of using poly (methyl methacrylate) (pmma) in the preparation of free standing 3d graphene

P Pransisco.....67

Proximate and morphology properties of swat bamboo activated carbon carburized under different carbonization temperature

I M Astika, D N K P Negara, C I P K Kencanawati, T G T Nindhia and F Hidajat.....73

Activated carbon characteristics of tabah bamboo that physically activated under different activation time

D N K P Negara, T G T Nindhia, I W Surata, M Sucipta and F Hidajat.....79

Hardness and wear rate of basalt/alumina/shellfish powder reinforced phenolic resin matrix hybrid composite brake lining pads

I K A Atmika, I D G A Subagia, I W Surata and I N Sutantra.....85

Experimental study on combustion characteristics of an indirect injection diesel engine fuelled by biodiesel blends

A Ghurri and S K Keun.....91

Wear of nickel-tungsten layer on carbon steel with varying contact surface temperature

I M Widiyarta, I M Parwata and I P Lokantara.....100

Impact Strenght and fluid absorption at the brake lining of hybrid composite

I G A K Suriadi and I K A Atmika.....106

Properties of thermal conductivity, density and hardness of aluminium matrices with reinforcement of sicw / al_2o_3 hybrid after sintering process

K Suarsana, I M Astika and P W Sunu.....111

Pyrolysis study of coconut leaf's biomass using thermogravimetric analysis

I M Rajendra, I N S Winaya, A Ghurri and I K G Wirawan.....121

The metal gasket sealing performance of bolted flanged with fem analysis

I M G Karohika and I N G Antara.....129

Influence of dies type for gasket forming shape

I M G Karohika, I N G Antara and I M D Budiana.....137

Sea water absorption on corrosion and hardness properties of lining brake with hybrid particulates composites

I D G A Subagia and N P G Suardana.....144

Utilization plastic waste using pyrolysis fixed bed

I B A Swamardika, I N S Winaya and R S Hartati.....152

Observations of mechanical and manual forging on bronze as a gamelan material

I G N Priambadi and I K G Sugita.....160

Energy analysis of solar farm planning in weh island, aceh, indonesia

S Rahmawati, M Iqbal and I D Sara.....167

The effect of diameter downcomer in air entrainment process from vertical plunging water jet with downcomer

Warjito, Budiarso and R L Jaya.....178

Two water wheels performance in series for empowerment of irrigation

Asral, F Huda and M Akbar.....185

Experimental study of an organic rankine cycle system using r134a as working fluid with helical evaporator and condenser

A Martin, Romy, D Agustina and A M Ibra.....194

Design and manufacturing of organic rankine cycle (orc) system using working fluid r-134a with helical evaporator and condenser

A Martin, Romy, D Agustina and A M Ibra.....201

The effect of fiber volume fraction on tensile and impact properties of *eleusine indica* grass reinforced polypropylene bio composite

I P Lokantara, N P G Suardana, I W Surata and I N S Winaya.....210

Smoldering combustion propagation of subtropic peat; case study on pelalawan peat

A Martin, H Orion, A Sandhyavitri, P Palamba and Y S Nugroho.....215

Analysis of transient elastohydrodynamic lubrication of point contact subjected to sinusoidal dynamic loads

D R P Cupu, A Stratmann and G Jacobs.....223

The characterization tensile and thermal properties of hibiscus tiliaceus cellulose fibers

N H Sari and Y A Padang.....229

Performance of vertical axis Savonius wind turbines related to the fin number on the blade

Ridwan, I Setyawan and Setiyono.....235

Application of stiffened plate technique to reduce crack propagation in spring plate of girth gear connector in rotary cement kiln

H Dahlan, M Rusli and R Muchtar.....242

The using of sound signal and simple microphone to detect damages in induction motor

F Huda, A Anggriawan and M Rusli.....252

Remaining strength assessment of dented pipe based on depth and strain based criteria

M Akbar, F Huda and D K Dewi.....260

Thermal resistance analysis of central processing unit cooling system based on cascade straight heat pipe

W N Septiadi, W A W Ula, I G A A D Wulandari, I A Tnunay and M R Murti.....274

Experimental study of thermoacoustic cooling with parallel-plate stack in different distances

Amirin, Triyono and M Yulianto.....281

Central processing unit based on cascade straight heat pipe

W N Septiadi, I G A A D Wulandari, W A W Ula, I A Tnunay, I K G Wirawan, M R Murti and N Putra.....301

Combustion characteristics of gas fuel from basic materials *arak bali*

I G K Sukadana and I K G Wirawan.....309

Solidification characteristic of cu-20%sn bronze alloys casting process by using sand and metal molds

I K G Sugita, I G N Priambadi, C I P K Kencanawati and K Astawa.....315

Experimental investigation on the effect of resonator shapes on the temperature characteristic of thermoacoustic cooling device

D Agustina, Sabri and S Purnama.....322

Optimization on strength spot welding joint trough finite element modelling indentation approach

I N Budiarsa, I M Astika, I N G Antara and I M G Karohika.....328

THE EFFECT OF COMPOSITION OF PLASTIC WASTE LOW DENSITY POLYETHYLENE (LDPE) WITH SAND TO PRESSURE STRENGTH AND DENSITY OF SAND/LDPE COMPOSITES

I M Susila, N P G Suardana, C I P K Kencanawati, I N A Thanaya and I W B Adnyana.....334

IOP Conf. Series: Materials Science and Engineering **539** (2019) 011001 doi:10.1088/1757-899X/539/1/011001

PREFACE

We welcome to you the *1st International Conference on Design, Energy, Materials and Manufacture* (ICDEMM) held October 24 - 25, 2018 in Bali, Indonesia. It is a great pleasure for the Faculty of Engineering University of Riau and University of Udayana to be the joint host organisation, in the spirit of strengthening of cooperation and mutual growth to be the world class university.

The aim of ICDEMM with the theme of "Engineering for Green Technology and Sustainable Development", is to provide an excellent international forum for engineers, scientists, researchers, academics, practitioners and professionals from all over the world to connect, and to share knowledge and learn about advances in green and sustainable development.

The conference run over 2 days with a wide array of quality technical presentations by a number of excellent key note and invited speaker presentations, authors of accepted papers, and the chance for the many participants to meet face-to-face whilst experiencing one of Indonesia's most vibrant tourist destination. The conference also include a number of social activities such as a conference dinner and a number of opportunities for delegates and accompanying persons to enjoy Bali and its surrounds.

Our deepest gratitude to all our speakers, participants, contributors, partners and professional association, who have given this conference their support. We would also like to thank all members of Organizing Committee, Steering Committee, Scientific Committee and Editorial for all of their support and advice. We also owe our success to the full support of the Rector of University of Riau and University of Udayana; the Dean of Faculty of Engineering University of Riau and University of Udayana.

We hope that this program will further stimulate research in Mechanical Engineering. We feel honoured and privileged to serve the best recent developments in the field of Mechanical Engineering to you through this exciting program. Thank you and we hope to see you again at the ICDEMM 2020.

Feblil Huda PhD
University of Riau
Pekanbaru, Indonesia

Dr.Eng I Made Gatot Karohika
University of Udayana
Bali, Indonesia

Peer review statement

All papers published in this volume of *IOP Conference Series: Materials Science and Engineering* have been peer reviewed through processes administered by the proceedings Editors. Reviews were conducted by expert referees to the professional and scientific standards expected of a proceedings journal published by IOP Publishing.

IOP Conf. Series: Materials Science and Engineering **539** (2019) 012001 doi:10.1088/1757-899X/539/1/012001

Weighbridge Vibration Analysis Using 3 Accelerometer Sensors and Single Excitation

Syafri[1], Darman[1]
[1]Department of Mechanical Engineering, Universitas Riau, Kampus Bina Widya Panam,Jln Soebrantas km 12,5 Pekanbaru, Riau, Indonesia, 28293

Email: Prie_00m022@yahoo.com

Abstract. Weighbridge was a measuring instrument that was widely used by the Ministry of Transportation to scale the weight of loaded vehicles on the roadways. Due to being traversed by vehicles with various kind of weight and sizes, weighbridges often experience resonance due to the vibrations of vehicles above it. This will potentially cause damages to weighbridges. To anticipate it, researches regarding vibration analysis on weighbridge were needed to decrease resonance on weighbridges. The Vibration Analysis was done by Modal Analysis Method using three accelerometers and one shaker as the measurement sensor. In testing, weighbridge was modeled into a cylindrical beam shape with a length of 724 mm and a diameter of 16 mm. As the supports, Fix-fix Support was used at both ends of the beam. Next, Accelerometer measurement sensor mounted in three positions with a distance of 181 mm, 362 mm, 543 mm from beam supports. Then, Excitation force with four frequency variations (1 Hz, 3 Hz, 5 Hz, 10 Hz) was applied to the four measuring points located between the accelerometer sensor. The measurements result in the form of FRF graph can be obtained by QuickDAQ. From the FRF graph, the five lowest natural frequency of the weighbridge was obtained that equal to 89.9 Hz, 91 Hz, 69.9 Hz, 75 Hz, and 96.9 Hz.

1. Introduction

Weighbridge was a structure used to scale the weight of trucks on the roadways. This was to make sure for trucks not to be overloaded so that the asphalt lifetime will be sustained according to design. Weighbridge was included as a structural design that receives resonances caused by vibration from trucks traversing above it. If the vibration frequency from the truck is close or equal to the weighbridge's natural frequency, this might lead to the increased vibration amplitude of the weighbridge, that potentially do some damages to it.

The negative impact that might happen from the resonances could be minimalized by doing tests and analyzing the natural frequency of the weighbridge. After obtaining the weighbridge's natural frequency, next, we can do some actions so that the natural frequency will not get close or even equal with disturbance frequency, example by adding some mass or increasing rigidity to the structure design and etc. The same research ever being done before in which the disturbance signal was coming from earthquakes. And not from vehicles. [1]

In this research, measuring the vibration was done by putting a cylinder beam with fix support underneath the weighbridge, so that the beam could receive the vibration. Next, we use *accelerometer* and *shaker* as the input sensor dan the test output. After testing the natural frequency of the beam, we can use it to know the vibration characteristic that happens on the weighbridge.

2. Literature Review

2.1. Weighbridge

Weighbridge was a device to scale the weight of loaded vehicles that can be assembled permanently or portable [2]. The bridge construction was shown In Figure 1, where the device was consist of the load cell, bridge structure, RCC pillar, and ground level.

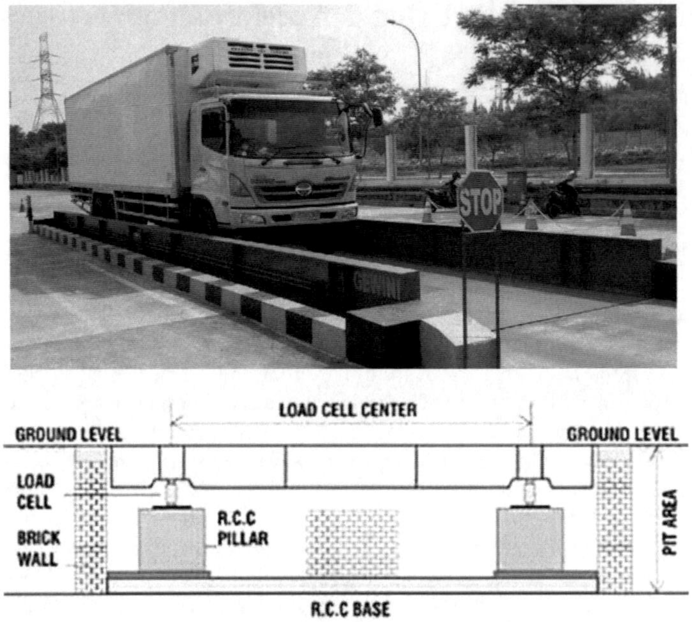

Figure 1. Weighbridge Construction

2.2. One DOF Free Vibration Analysis

Figure 2 shows the schematic structure of one DOF free vibration that consist of a load (m) and supported by a spring with rigidity (k) and excitation force ($x, \dot{x}, \ddot{x}$). Using motion differential equation, natural frequency can be determined by the following equation [3].

$$m\ddot{x}+kx=0 \quad (1)$$

And assuming the result as:

$$x=A\sin\omega t \quad (2)$$

Natural Frequency can be obtained by substituting Equation 2 to Equation 1 that expressed to be Equation 3

$$w_n=\sqrt{\frac{k}{m}} \quad (3)$$

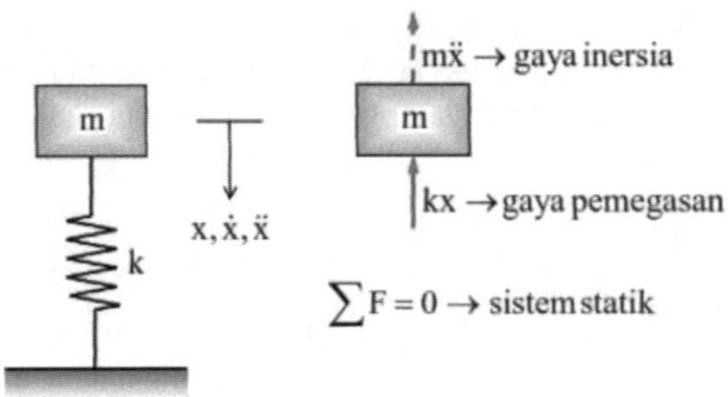

Figure 2. Spring-mass system and free-body diagram.

IOP Conf. Series: Materials Science and Engineering **539** (2019) 012001 doi:10.1088/1757-899X/539/1/012001

2.3. Modal Analysis

Modal Analysis was a process to determine the dynamical characteristic or a modal parameter of a structure. This modal parameter consists of Natural Frequency (ω), mode shapes (ϕ), and damping ratio (ξ). Some of the common methods to be used were by a test using a shaker as the excitation force like shown in Figure 3. [4].

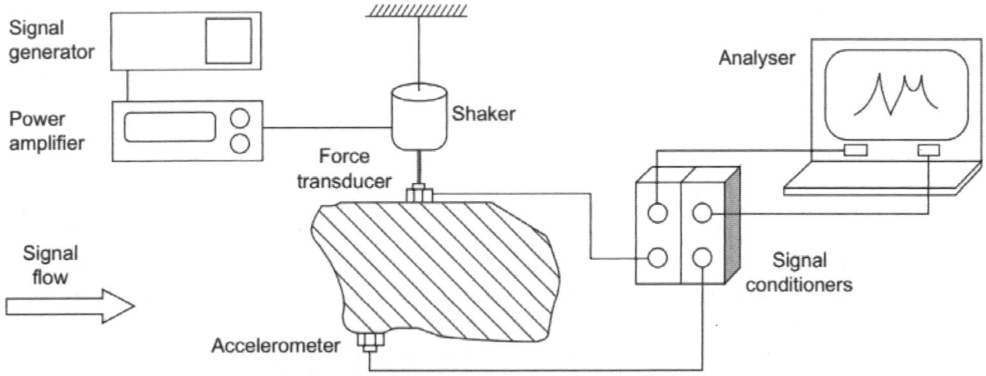

Figure 3. Measurement set-up with shaker excitation

Schematically, devices of the loaded impact method consist of

1. Shaker, a device that can produce an excitation force to the structure.
2. Accelerometer, a device that collects responses from the structure due to excitation force.
3. FFT analyzer, a device to be used for FRF computation.
4. *PC* and modal testing Software to identify the modal parameter and show it in the form of graphs.

In the modal analysis test, accelerometer's sensors were placed on the determined spots of the structure. Next, the excitation force was given by shaking the specimen with the shaker on the determined spots. Due excitation force, there will responses from the structure that will be collected by the accelerometer. Excitation force and the measurement responses are graphically shown in Figure 4.

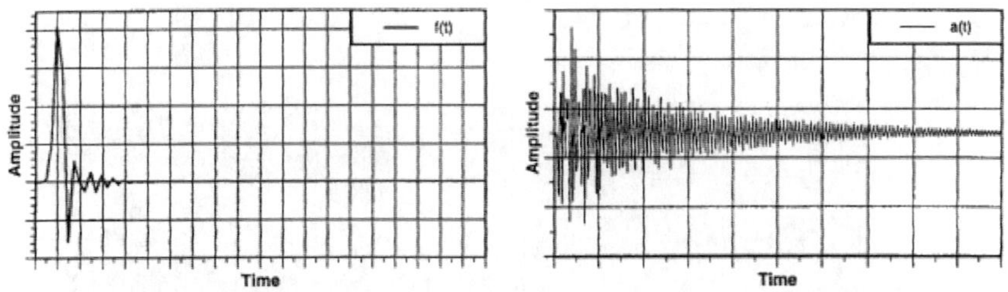

Figure 4. (a). Excitation force graph (b). Responses graph

Excitation forces and measured responses will be sent to the FFT analyzer. Input in the time domain will be converted to the frequency domain. By this, measured FRF is obtained like shown in Figure 5. From the graph, we can obtain the natural frequency of the structure. If we connect a point on the imaginer part of measured FRF into another measuring point, we can obtain the structural vibration modus.

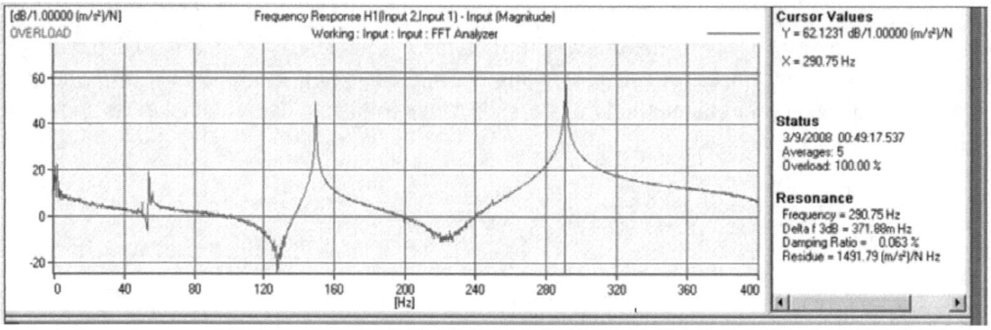

Figure 5. FRF Graph [5]

3. Methodology

Modal analysis to measure the personal frequency of the weighbridge structure was ideally performed directly on the measuring object. However, due to the various problem, the measurement is not yet possible to be implemented, So as an alternative measurement was done on the beam cylinder that was modeled under the weighbridge. After that, the beam cylinder was given an excitation force to model the outer vibration causing resonance at the weighbridge. At the capital analysis test setup, it was then used 2 units of the laptop, data acquisition, Arduino Uno, relay, 3 units accelerometer CMCP770A, solenoid valve CAMSCO TAS-15, and beam cylinder with aluminum material as in Figure 5 below. In this case, the accelerometer and shaker (solenoid valve CAMSCO TAS-15) function as the input and output test sensors shown by Figure 6.

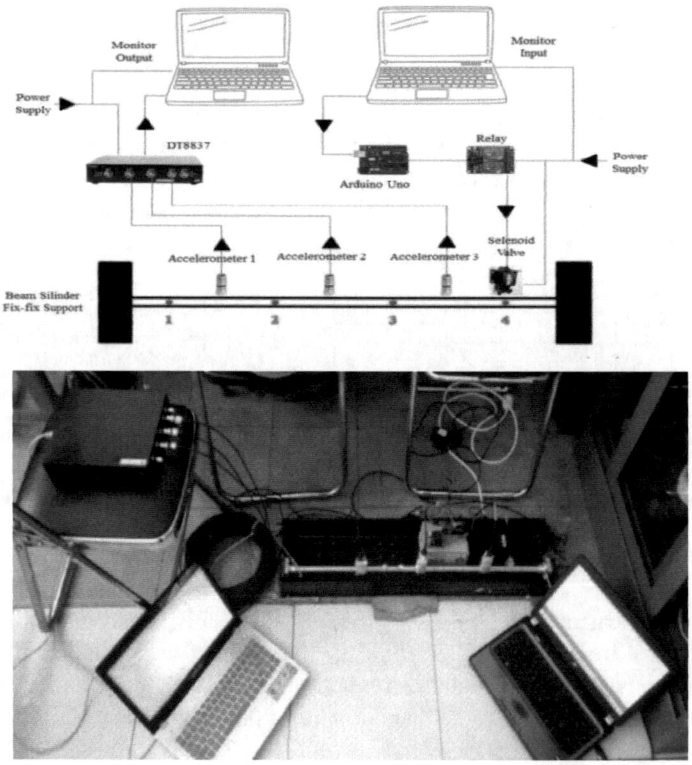

Figure 6. Experiment set up

IOP Conf. Series: Materials Science and Engineering **539** (2019) 012001 doi:10.1088/1757-899X/539/1/012001

The cylindrical beam used in this experiment has a length of 724 mm and a diameter of 16 mm. Then the cylindrical beam was supported by fix-fix support at both ends. The next beam cylinder was divided into 4 test points, namely points 1,2,3 and 4, where the point will be used as a reference point for the installation of shaker (solenoid valve CAMSCO TAS-15) which serves as an excitation style. Then the three accelerometer output sensors were placed on beams with a distance of 181 mm, 362 mm, 543 mm respectively from the left beam as shown in Figure 7.

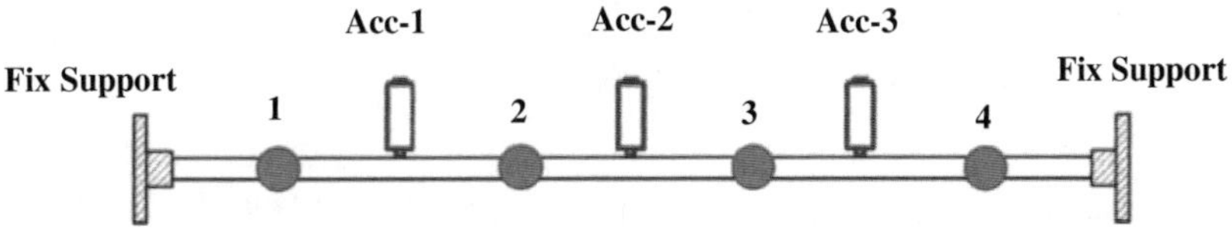

Figure 7. Sensor installation on the measuring object

Information
Acc1=*Accelerometer*1(Distance181mm)
Acc2=*Accelerometer*2(Distance362mm)
Acc3=*Accelerometer*3(Distance543mm)
1,2,3,4 = Point of excitation force

Modal analysis was done by giving the excitation force at point 1 four times with each different frequency that was 1 Hz, 3 Hz, 5 Hz, and 10 Hz. Then, the structural response due to the excitation force excitation was captured by three accelerometer sensors displayed in FRF graphic form. Next three FRF graphs from each accelerometer were compared to get the best results. The same test was continued at the next test point 2, 3 and 4 as shown in Figure 8.

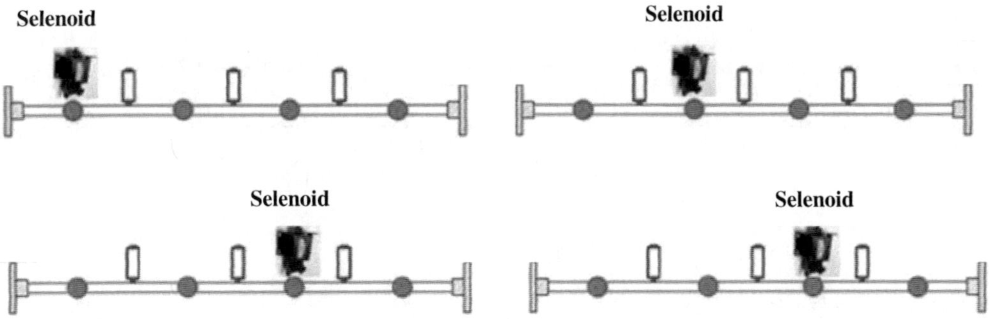

Figure 8. Modal analysis on beam cylinder

4. Result And Discussion

Based on the tests result, some obtained FRF graphs derived from four test points with four variations of excitation force frequency. The graph was subsequently grouped by the excitation force frequency of each test starting from the lowest frequency of 1 Hz as shown in Figure 9. In the test with the excitation force of 1 Hz, the response captured by the accelerometer tends not well, where the peak of the five lowest frequencies of the structure on the graph FRF was not clearly visible. This condition occurs because the frequency of excitation force given was relatively low, so the noise coming from outside the system was captured by the accelerometer sensor.

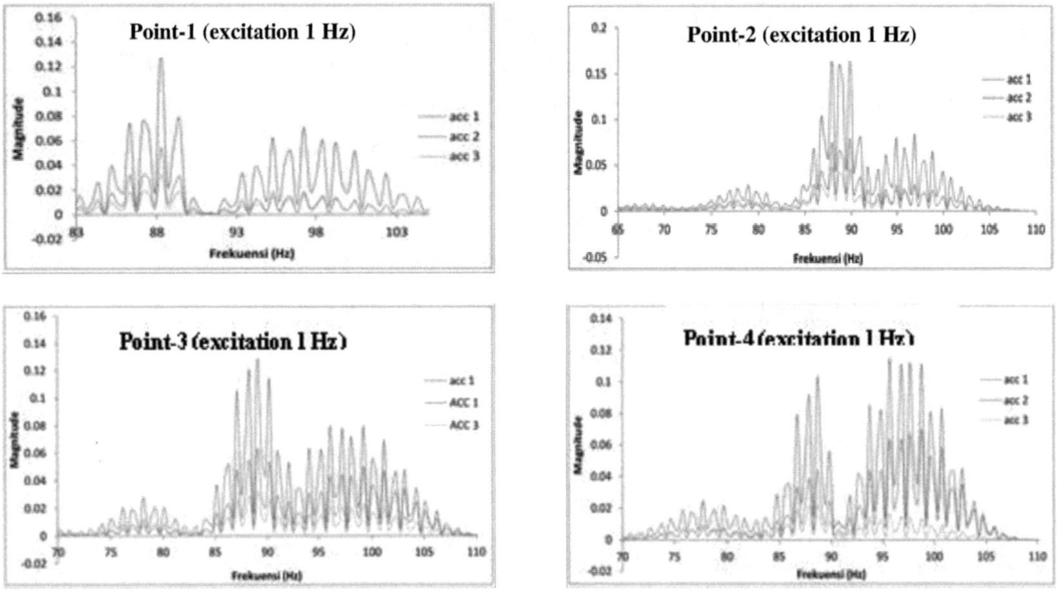

Figure 9. FRF Graph with 1 Hz excitation force

In tests with excitation force frequency 3 Hz, the results obtained tend to be better, which was the top five lowest natural frequencies of structures can be seen clearly on the FRF graph. This was because, at the 3 Hz excitation force, the difference in magnitude of each personal frequency was relatively lower so that the peak of the five lowest personal frequencies can be displayed as well as in Figure 10

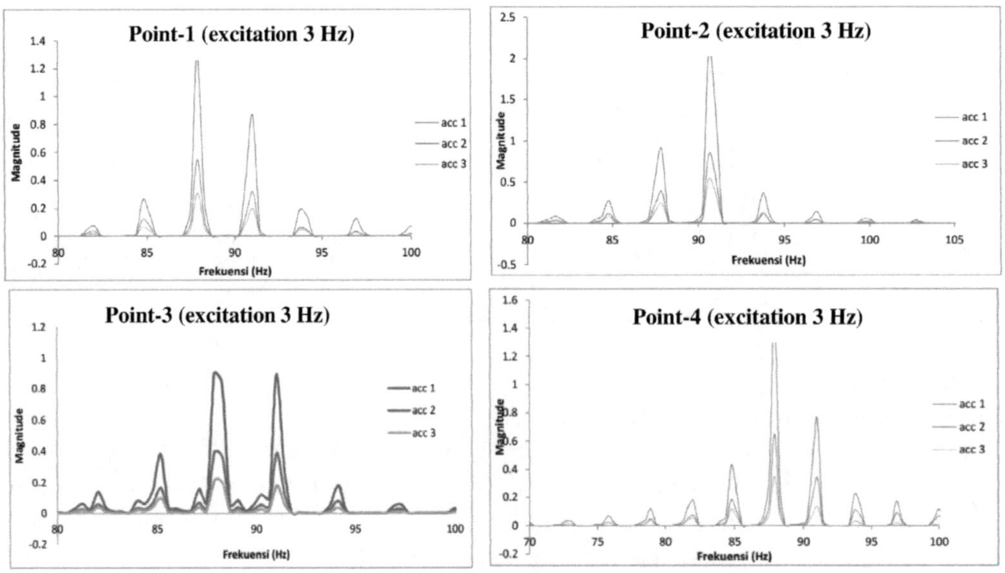

Figure 10. FRF Graph with 3 Hz excitation force

While in the test with the 5 Hz and 10 Hz excitation force, the FRF graphs that were obtained have a magnitude difference that tends to be greater than the previous test, so that some of the lowest natural frequency peak structures cannot be presented well as shown in Figure 11 and Figure 12.

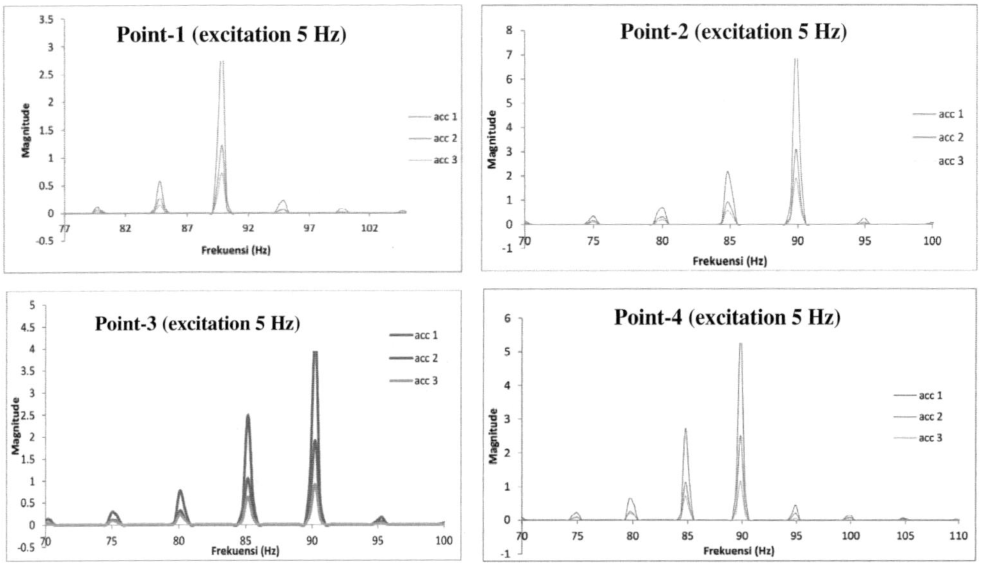

Figure 11. FRF Graph with 5 Hz excitation force

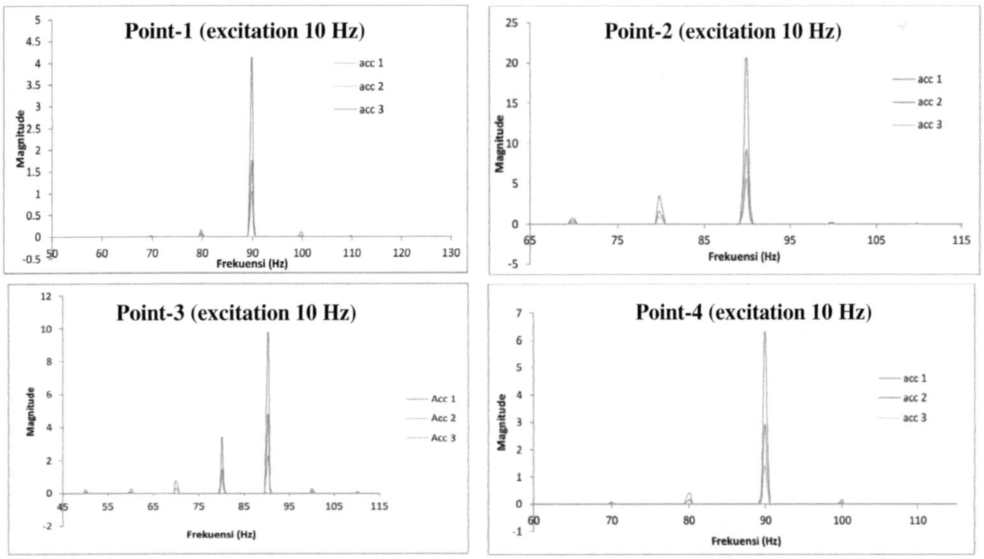

Figure 12. FRF Graph with10 Hz excitation force

Based on FRF graph, it was obtained five lowest beam cylinder frequencies. It was 82 Hz, 85 Hz, 87.5 Hz, 91 Hz, and 94 Hz. The best natural frequency values were displayed by the accelerometer no 2 which was located at the center of the beam cylinder 362 mm from the beam support. It was due to

IOP Conf. Series: Materials Science and Engineering **539** (2019) 012001 doi:10.1088/1757-899X/539/1/012001

accelerometer position from the beam support, so that accelerometer response to the excitation force becomes better. Furthermore, in order to avoid the resonance that can harm the structure, the excitation force acting on the structure must be controlled in such a way that its frequency was not close to or equal to the natural frequency of the structure. Otherwise, the resonance amplitude can be minimized by increasing the stiffness of the structure, or leaving the additional masses so that the natural frequency of the structure becomes changed.

5. Conclusion

Based on modal analysis with one excitation force and three accelerometer sensors, we got the five lowest frequency of weighbridge structure 82 Hz, 85 Hz, 87,5 Hz, 91 Hz, and 94 Hz. The best response of the sensor was obtained on accelerometer no 2 with a 3 Hz test excitation force.

References

[1] Demulawa, Meilan, Resonance analysis of weighbridge toward earthquake (case study weighbridge of Bantul Yogyakarta), Final Project, UGM, 2012
[2] Sulistiani, Ani, Weighbridge, instrumentation electronic paper, UGM 2013
[3] Thomson, William, Theory of Vibration with Application, Department of Mechanical and Environmental Engineering, University of California, 1993.
[4] Fu, He, Modal Analysis, Butterworth-Heinemann Linacre House, Jordan Hill, Oxford OX2 8DP, 2001
[5] Syafri, Bur, Free free Vibration analysis, SNTTM XI, UGM, 2012

Wing design of uav serindit v-1

K Anuar[1], M Akbar[1] and Herisiswanto[1]
[1] Department of Mechanical Engineering, Universitas Riau, Indonesia

Email: kaspul.anuar@lecturer.unri.ac.id

Abstract. UAVs are controlled aircraft without the presence of pilots in them. In its use, UAVs have been used for both military and civilian purposes. In the field of plantations, UAVs are used for mapping and monitoring (survey) of an area such as areas of industrial plantations, oil palm plantations and areas of forest fire disaster in a timely and precise manner. To develop a UAV capable of carrying out the mission of mapping the coastal estate's area, in this research is done wing design UAV which is named serindit V-1. The design process begins with the design concept, then proceed with a preliminary design that aims to determine the dimensions and initial configuration of the wing UAV serindit V-1. Furthermore, the wing UAV serindit V-1 is simulated on XFLR5 software using Clark's airfoil, GOE 611 and S7055. Simulations were performed to see the characteristics of aerodynamic at takeoff speed (12 m / s), cruising speed (20 m / s) and maximum speed (30 m / s). Simulation result at the take-off speed and cruising speed show the lift coefficients produced by the three airfoils meet the requirements. Then, the highest value of the lift coefficient ratio to the air drag coefficient at cruising speed, produced by airfoil S7055 at a 3-degree angle of attack. Furthermore, the graph of the pitching moment coefficient on all angles of attack produced by the three airfoils shows a negative gradient. This shows the condition of wing stability in the pitching direction.

1. Introduction

Unmanned Aerial Vehicle (UAV) is a controlled aircraft with no pilot in it [1]. In control, it can be done manually via a radio wave transmitter or independently (autopilot) using a small computer device. Initially, UAV is only used for military purposes. One of its uses is as a target plane shooting in war exercises. However, as the development of UAV technology in the world, the use of UAV began widely used for civil purposes [2]. UAVs are utilized in the mission of mapping and monitoring of an area such as natural forest areas, industrial plantation areas, palm oil plantations and timely areas of forest fire disaster and precision. Excess UAV in performing this mission compared to the image displayed satellite is not the occurrence of image distortion due to the influence of cloud cover. With the ability of UAVs to fly under clouds, it allows UAVs to carry out mapping and surveillance missions in a detailed and clear area.

Utilization of UAV in the realm of Riau Province, Indonesia, began to be used both government agencies such as the Regional Disaster Management Agency (BPBD) Riau as well as private parties engaged in plantation industry and palm oil plantations. BPBD Riau utilizing UAV in the monitoring mission of forest fire disaster. While the private sector utilizes UAV not only on monitoring missions of forest fires but also began to be used for mapping missions and monitoring of industrial timber plantations. This breakthrough is able to reduce operational costs in terms of mapping and monitoring of industrial forest areas that previously used helicopters.

The use of UAV for civilian purposes in Riau Province has the potential to grow because Riau Province is the province with the largest oil palm plantation area in Indonesia with a total of 2,381,895 hectares [3]. Then Riau Province also has a large industrial timber plantation area (HTI) with a total of 1,673,060 hectares.

2. Fundamental Theory

2.1 Lift Force On Aircraft
While the airplane on the air, there are four forces that work, the lift, gravity, thrust, and air dry [1]. The lift force acting on the aircraft is produced due to the difference in pressure on the upper surface and the underside of the wing. The magnitude of the raised force is influenced by the wing surface area, the air type, the aircraft velocity relative to the air, and the lift coefficient. Mathematically, the correlation can be seen in the following equation [1]

$$F_l = \frac{\rho . c_l . A . V^2}{2} \quad (1)$$

2.2 Wing Typical Design
The UAV used for mapping is designed to have maximum flight range. To get a UAV that has a maximum flight range with efficient power usage, the value of the lift-drag ratio is a factor highly considered by a UAV designer [4]. Then the next factor that needs to be evaluated is the value of Reynold number. In relatively small UAVs, the airflow that operates along the UAV wing generally works on low Reynolds value. The low Reynolds number value causes the maximum lift coefficient generated by the wing is difficult to achieve [4]. The following is shown in equation 2 to determine the Reynold number [5]

$$Re = \frac{\rho . V . c}{\mu} \quad (2)$$

2.3 Basic Design of Wings Geometry and Tail of UAV
One factor to consider in determining the geometry of the wings is wing loading. In a planned UAV flying at relatively low speeds, the required wing loading value is 4kg / m^2 - 5kg / m^2 [6]. Wingloading can be determined using the following equations [6]:

$$Wing_l = \frac{m}{A} \quad (3)$$

$$A = s \ x \ c \quad (4)$$

In designing a horizontal tail, the initial dimension of a span and its surface area can be determined using the basic rules of aircraft design. The size of the horizontal tailed span is generally a quarter of the wingspan, while the size of the horizontal tail area is one-fifth of the wing's area [4]. The dimensions and position of the horizontal tail will affect the pitching stability of an aircraft. In order for a horizontal tail UAV capable of making attitude UAV stable in the direction of pitching, then the value of horizontal tail volume coefficient (Vh) usually ranges from $0.3 \leq Vh \leq 0.6$ [7]. The value of the horizontal tail volume coefficient can be determined by using equation as follows [8]:

$$V_h = \frac{A_h . l_h}{A . c} \quad (5)$$

Furthermore horizontal tail volume coefficient also affects the position of the neutral point and the position of the center of gravity (CG) of UAV. To determine the position of the neutral point and the center of gravity of the UAV can use equations as follows [7]:

$$\frac{x_{np}}{c} = \frac{1}{4} + \frac{1+\left(\frac{2}{AR}\right)}{1+\left(\frac{2}{AR_h}\right)}\left(1 - \frac{4}{AR+2}\right) V_h \quad (6)$$

$$SM = \frac{Xnp - Xcg}{c} \quad (7)$$

The vertical tail volume coefficient (Vv) of the UAV has an important role in maintaining the stability of the yawing direction. The effectiveness of the vertical tail function can be achieved if the value of the vertical tail volume coefficient (Vv) is in the range of values between 0.02 to 0.05 [7]. The value of the vertical tail volume coefficient can be determined by using equation as follows [8].

$$V_v = \frac{A_v\, l_v}{A\, c} \tag{8}$$

3. Methodology

3.1 Design Concept UAV Serindit V-1

The design of a UAV concept is based on the UAV / design requirement object's mission. In this research, UAV serindit V-1 is designed to be suitable for mapping missions and coastal area cultivation. In accordance with its mission, the UAV serindit V-1 is designed to refer to a type of UAV glider with high-wing configured wings. The following table 1 shows the parameters of UAV Serindit V-1 as defined in this research.

Table 1. Parameter of UAV Serindit V-1

Parameter	Requirement
MTOW	1 kg – 1,5 kg
Wing Loading	5 kg/m^2
Wingspan	1,5 m
Wide Area	0,3 m^2 – 0,35 m^2
Wing Chords	0,2 m – 0,25 m
Stall Speed	10 m/s - 12 m/s
Cruise Speed	20 m/s – 23 m/s
Maximum Speed	26 m/s – 30 m/s
Flight Time	10 minutes – 15 minutes
Material	Composite and foam

3.2 Preliminary design of UAV Serindit V-1

In the initial design phase, the configuration of the wing's UAV serindit V-1 selected high wing type with a sweep angle of 5^0. The high wing configuration is selected because the lateral stability is better than other configurations. Furthermore, the initial dimensions of the wing and tail of the UAV serindit V-1 are calculated using the basic rules of aircraft design. All of the initial dimensions of the wings and the tail of the UAVseridnit V-1 will be used as parameters in the simulation process of XFLR5 software. Using equations (3) through (8), the calculation result of the initial dimension of the wing and the tail of the UAV serindit V-1 is shown in table 2 below.

Table 2. Initial dimensions of the wing and tail of the UAV Serindit V-1

Parameter	Area	Chord	Span	Volume Coefficient
Wing Dimension	0.3 m^2	0.2 m	1.5 m	-
Horizontal Tail Dimension	0.06 m^2	0.16 m	0.375 m	0.55
Vertical Tail Dimension	0.03 m^2	0.16 m	0.188 m	0.035

After determining the configuration and initial dimensions of the wing's UAV serindit V-1, then the wing design of UAV Serindit V-1 drawn on solid work software. Here is shown figure 1 wing design of UAV Serindit V-1.

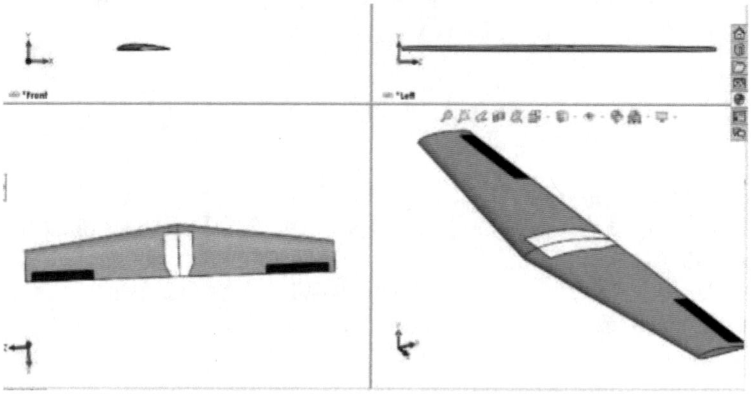

Figure 1. Wing Design of UAV Serindit V-1

3.3 Centre Of Gravity's (CoG) Design of UAV Serindit V-1

The position of the neutral point and the CG point of the UAV Serindit V-1 in the longitudinal direction is determined by equations (4), (5) and horizontal tail values of 0.55. Where the position of the neutral point and the CG point of UAV serindit V-1 based on the calculation respectively 9.35 cm and 6.35 cm.

3.4 Minimum and Maximum of Reynolds Number

One of the important parameters to determine its value is the Reynolds number. This number will be one of the parameters to be included in the simulation process in XFLR5 software. In determining the Reynolds number, the data relating to the air condition is adjusted to the coastal estate's area. UAV is planned to fly at an altitude of 100 meters above sea level. Where the air density parameter (ρ) is set at 1.177 kg / m3, the temperature (T) is 28^0 C and the air viscosity (μ) is 1.846 x 10^{-5} kg/m.s. Using equation (2) and air condition at an altitude of 100 m above sea level, the result of the Reynolds number calculation is shown in table 3 below.

Table 3. The result of calculation of Reynolds number on wings of UAV Serindit V-1

Flight Mode	Speed (m/s)	Reynolds Number
Take Off	12	$1.53x10^5$
Cruise Speed	20	$2.55x10^5$
Maximum Speed	30	$3.826\ x10^5$

3.5 Lift Coefficient Design

The calculation of the minimum lift coefficient is required by the vehicle of the UAV Serindit V-1 at take-off and at cruise speed. Using equation (1), the calculated value of the minimum lift coefficient required by the UAV Serindit V-1 at take-off and at cruise speed is shown in table 4 below.

Table 4. The value of the minimum lift coefficient (Cl) needs

Flight Mode	Speed (m/s)	Angle of Attack	Cl
Take off	12	8^0 - 20^0	0.5788
Cruise	20	0^0 - 3^0	0.208
Maximum	30	0^0 - 3^0	0.09

3.6 Wing Airfoil design of UAV Serindit V-1

The wing design of UAV Serindit V-1 is done with the selection of three flat bottom airfoils namely Clark y, GOE 611 and S7055 [9]. This airfoil is suitable for UAV glider type because it has the ability to stay floating when the motor drive is turned off. Here is shown figure 2 type airfoil used on wing design of UAV Serindit V-1 [9]

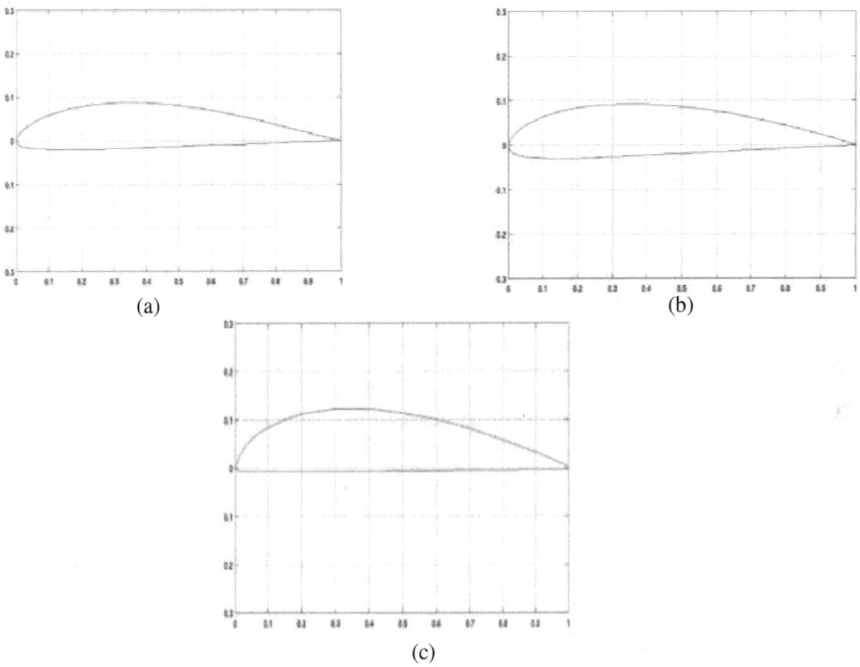

(a) (b) (c)

Figure 2. (a) airfoil Clark Y, (b) airfoil S7055, (c) airfoil GOE 611

4. Result And Discussion

4.1. Comparing simulation results and experiment result

Software xflr 5 is a software that is widely used to simulate aerodynamic characteristics of a UAV. In Figures 3 and 4, there is a relation between experiment result of lift coefficient vs alpha of airfoil S7055 [10] with the simulation result of software xflr 5. In figure 3, show that experimental result of lift coefficient vs alpha at Reynolds number equal to 60,000, has a trend value almost equal to the xflr5 result of lift coefficient vs alpha at the same Reynolds number.

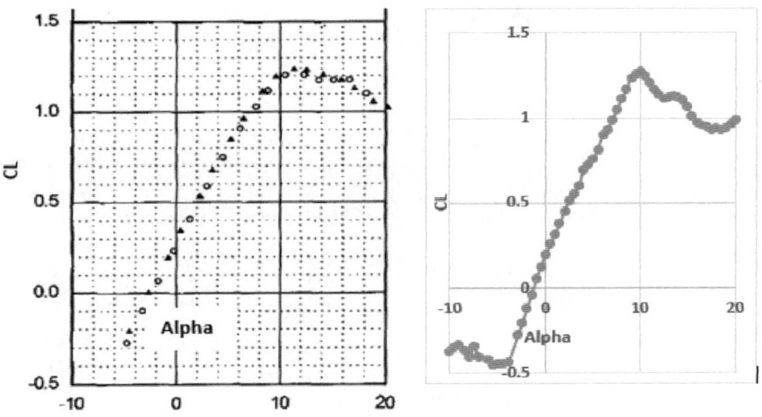

Figure 3. Lift Coefficient Vs Aoa (a) experiment [10] (b) XFLR5

Experiment result of lift coefficient vs alpha on airflow with Reynolds number of 300,000, also has a trend value similar to XFLR 5 simulation results. This graph is shown in Figure 4, where the left is a graph of airfoil S7055 obtained through experiment [10], whereas the right is a graph obtained through XFLR5 simulation.

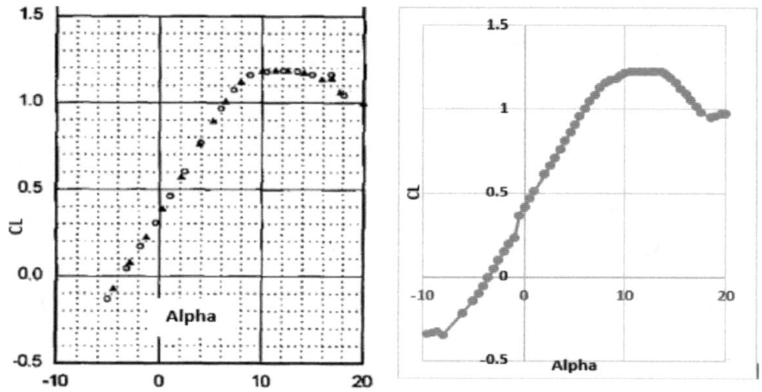

Figure 4. Lift Coefficient Vs Aoa at Reynolds number 300,000 (a) experiment [9] (b) XFLR5

4.2 Wing simulation result of UAV Serindit V-1 at takeoff speed (12 m/s)

To simulate the three airfoils with XFLR5 software, Operational speeds at takeoff and cruise speed are designed 12 m/s and 20 m/s respectively. Other conditions required are airflow conditions at an altitude of 100 meters above sea level such as air density (ρ) is 1.177 kg / m^3, temperature (T) is 28^0 C and air viscosity (μ) is 1.846 x 10^{-5} kg/m.s The following shows graphs of simulation results of UAV Serindit V-1 at take-off speed (12 m/s).

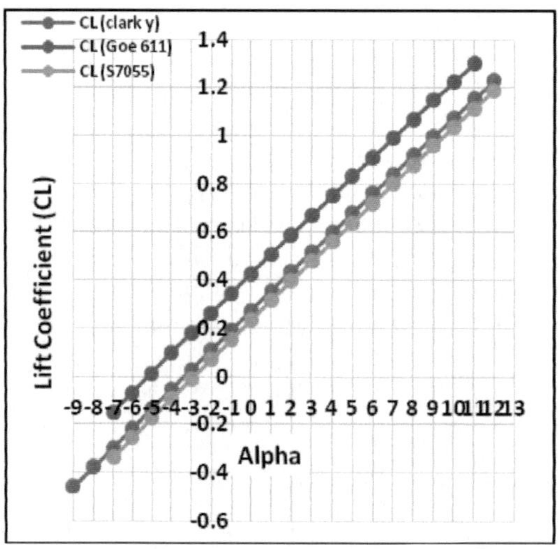

Figure 5. Lift coefficient vs Alpha graph at speed 12 m/s

The wing simulation result of UAV Serindit V-1 in Figure 5 shows the three types of wing airfoil UAV Serindit V-1 capable of producing lift force coefficient (Cl) more than 0.5788 at an 8-degree angle of attack. This shows that the lift coefficient of the three airfoils at an angle of attack of 8 degrees sufficient to lift the plane at take-off. From the graph also seen, the lift coefficient trends generated by the GOE 611 airfoil are greater than Clark airfoil or S7055 airfoil.

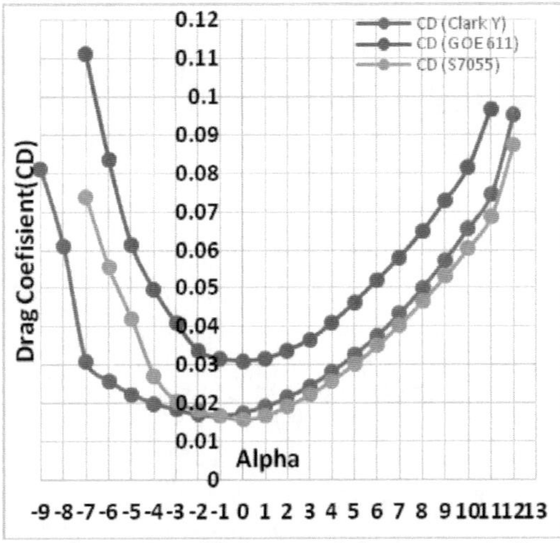

Figure 6. Drag coefficient vs alpha at speed 12 m / s

Figure 6 shows the coefficient of drag force coefficient at an angle of attack of 8 degrees occurring on the Clark Y, Goe 611 and S7055 airfoils respectively valued at 0.049, 0.064 and 0.046. When viewed as a whole at a positive angle of attack, the greatest airborne drag coefficient trends occur on the Goe 611 airfoil followed by Clark Y airfoil and S7055 airfoil.

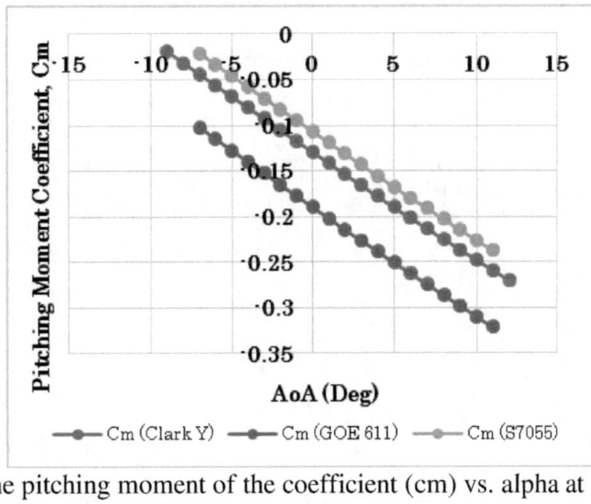

Figure 7. The pitching moment of the coefficient (cm) vs. alpha at speed of 12 m/s

The simulation results of the pitching moment coefficients of the three types airfoils as shown in Figure 7 show the pitching moment coefficients at all angles of the attack (AoA) having a negative gradient. This trend indicates good stability of the pitching direction of the wing UAV Serindit V-1 in all angle of attack conditions.

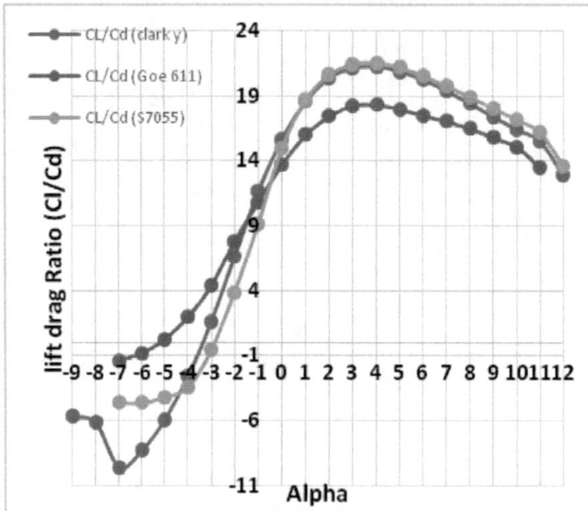

Figure 8. Lift drag ratio (Cl/Cd) vs alpha at a speed of 12 m/s

Figure 8 shows the graph of the lift drag coefficient ratio (glide ratio) at all angle of attack. The graph shows that the largest glide ratio at the positive attack angle produced by the S7055 airfoil, followed by the Clark Y airfoil and GOE 611 airfoil. The glide ratio value of the UAV will affect the duration and flight range of a UAV.

4.3 Wing Simulation Results of UAV Serindit V-1 at Cruise Speed (20 m/s)
The following graph shows the wing simulation results of the UAV Serindit V-1 at cruise speed (20 m/s).

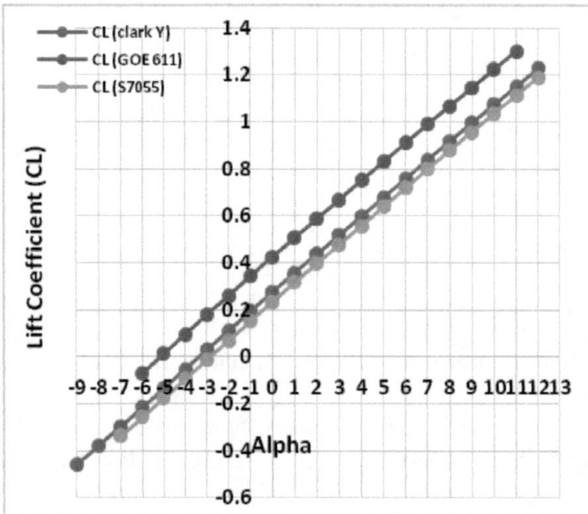

Figure 9. Lift coefficient vs Alpha at speed 20 m/s

The wing simulation results of UAV Serindit V-1 at the cruise speed (20 m/s) as shown in Fig. 9 shows the three types of airfoil UAV Serindit V-1 capable to produce lift coefficient of more than 0.208 at an angle of attack of 0 degrees. This shows that the lift coefficient of the three airfoils at the angle of attack of 0 degree meets the requirements to maintain its flying height. From the graph also seen, the trend value of lift coefficient produced by three airfoils at cruise speed equal to trend lift coefficient which resulted at take-off speed.

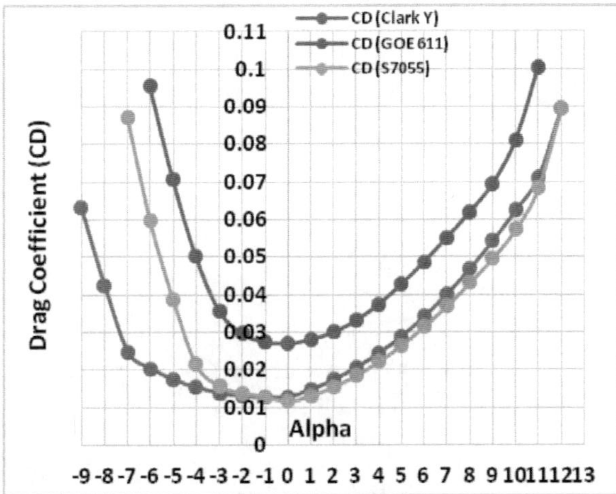

Figure 10. Drag coefficient vs. alpha at speed 20 m/s

Then for the air drag coefficient produced by Clark Y, Goe 611 and S7055 airfoils at 0-degree angles of attack are 0.0128, 0.027 and 0.0116 respectively. From Figure 10 seen in all angles of attack, the trends of drag coefficient at cruise speeds (20 m/s) are smaller than at takeoff speeds (12 m/s).

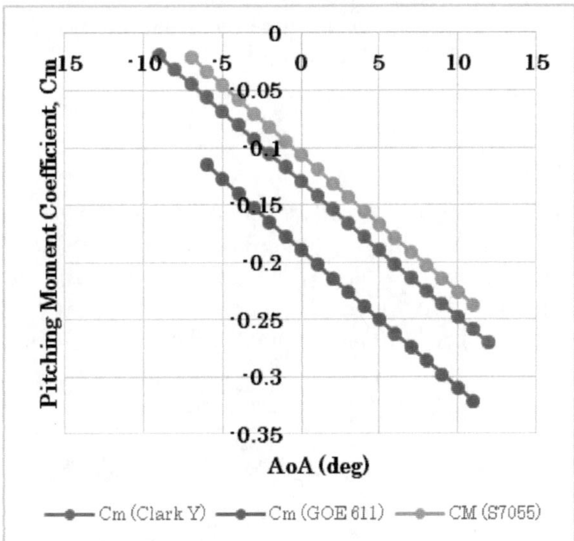

Figure 11. Pitching moment of the coefficient (Cm) vs. AoA at speed 20 m/s

Furthermore, as shown in Fig. 11, the simulation results of the pitching moment coefficients of the three types of airfoils at cruise speeds (20 m/s) have the same trend as the simulation result at the take-off speed (12 m / s). It saw that graph of pitching moment coefficients at all angles of attack having a negative gradient. This trend indicates that the good stability of the longitudinal wings of the UAV Serindit V-1 at all angle of attack conditions.

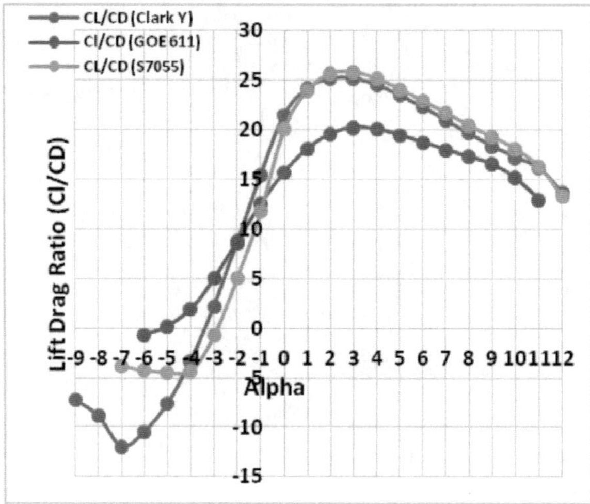

Figure 12. Lift drag coefficient ratio (Cl/Cd) vs. alpha at speed 20 m/s

Figure 12 shows the simulation results of the lift-drag coefficients ratio (Glide ratio) at cruise speed (20 m / s). In the graph, see the largest glide ratio at a 0-degree angle of attack is generated by the Clark Y airfoil, followed by the S7055 airfoil and GOE 611 airfoil. However, overall, the largest glide ratio occurs at a 3-degree angle of attack where the value of the glide ratio of S7055 airfoil, Clark Y and Goe 611 airfoil are 25. 803, 25.206 and 20.245 respectively. The largest glide ratio value at cruise speed is an important consideration in choosing the type of wing airfoil that requires high endurance flight.

4.4 Wing simulation result of UAV Serindit V-1 at maximum speed (30 m/s)

The result of simulation of UAV Serindit V-1 wing at maximum speed (30 m/s) shown the drag coefficient produced by Clark Y, Goe 611 and S7055 airfoil at 0-degree angle of attack are 0.0113, 0.0251 and 0.0102 respectively. From Figure 13 seen at all angles of attack, the trend of drag coefficient generated at maximum speed (30 m/s) is smaller than at the cruise speed.

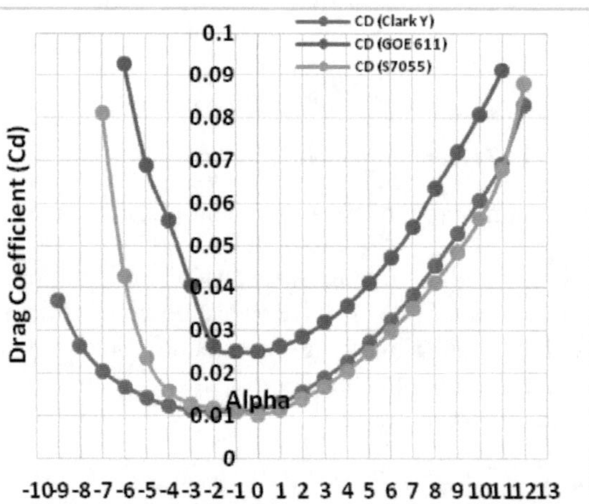

Figure 13. Drag coefficient vs. alpha at speed 30 m/s

The simulation results of the lift drag coefficient (Glide ratio) at maximum speed (30 m /s) shown in figure 14. In the graph show that the highest lift drag coefficient ratio at a 0-degree angle of attack is produced by Clark Y airfoil, followed by S7055 and GOE 611 airfoil. However, overall, the largest glide ratio value occurs at a 2-degree angle of attacks where the S7055, Clark Y and Goe 611 glide ratio values are 28.81, 28.13 and 20.51, respectively.

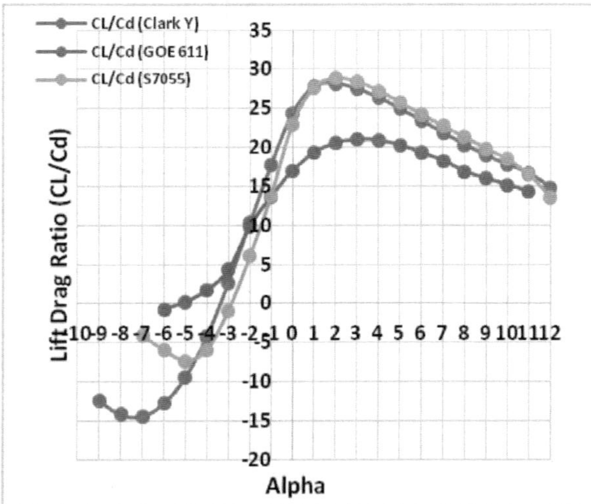

Figure 14. Lift drag coefficient ratio (Cl/Cd) vs. alpha at speed 30 m/s.

5. Conclusion

In this research, the initial geometry of the wing UAV serindit V-1 was determined using the basic method of aircraft design. Furthermore, the wing UAV Serindit V-1 is simulated on XFLR5 software to see its aerodynamic characteristics at take-off speed, cruise speed, and maximum speed. The simulation of the wing UAV Serindit V-1 was performed on the airfoils of Clark Y, GOE 611 and S7055. The following explains the conclusions of this research:

1. The lift coefficient produced by the three airfoils at the at take-off speed and at the cruise speed meets the requirements because the lift coefficient value exceeds the minimum lift coefficient required.
2. The highest value of the lift drag coefficient ratio at cruise speed, produced by S7055 airfoil at a 3-degree angle of attack.
3. The pitching moment coefficient at all angles of attack produced by the three airfoils is having a negative gradient. This shows the condition of wing stability in the pitching direction is a meet requirement.

References

[1] Austin, Reg. 2010. Unmanned Aircraft System. USA: John Wiley & Sons, Inc.
[2] Fahlstrom, paul. 2012. Introduction to UAV system. USA: John Wiley & Sons, Inc.
[3] Badan Pusat Statistik Provinsi Riau. 2015. Luas Areal Perkebunan Menurut Jenis Tanaman 2011 - 2015 (Ha). From https://riau.bps.go.id/linkTabelStatis/view/id/310.
[4] Rakesh and Sunil. 2015. Design of solar powered RC Aircraft. The international journal of engineering and science (IJES)
[5] Basaran, Bilek. 2015. Unmanned Aerial Vehicle Design And Production. Turkey. Ankara International Aerospace Conference (AIAC).
[6] Lennon, Andy. 2005. Model Aircraft Design. USA: Air Age Me.
[7] MIT. 2006. Basic Aircraft design Rule. The USA. MIT OPEN COURSEWARE
[8] Driessen F, Leussen M., Senders A., Vliet T., Vlugt S. (2013), *An autonomous quadcopter with environment mapper*, Huginn, Eindhoven University of Technology
[9] UIUC Applied Aerodynamic Group. 2017. UIUC Airfoil Coordinates Database. from http://m-selig.ae.illinois.edu/ads/coord_database.html

[10] Selig, Broeren and Guiglielmo. 1995 Low-Speed Airfoil test. University Of Illinois News Bureau. Soartech publication.

IOP Conf. Series: Materials Science and Engineering **539** (2019) 012003 doi:10.1088/1757-899X/539/1/012003

Multi-step ahead prediction of vapor compression air conditioning system behaviour using neural networks

S Sholahudin [1], K Ohno[1], S Yamaguchi[1] and K Saito[1]
Faculty of Science and Engineering, Waseda University, Shinjuku-ku, Tokyo, Japan

Email: sholahudin.mesin@gmail.com

Abstract. Cooling capacity and super heat temperature control for air conditioning (AC) system operation is necessary to ensure that the system operates efficiently. In this paper, multi-step-ahead prediction of AC system behaviour is presented using backpropagation neural network model as the first effort to develop the effective control strategy. Several step-ahead cooling capacity and superheat temperature performance are predicted under modulation of compressor speed and expansion valve opening. The prediction is proposed to capture the dynamic behaviour of system that can be applied in predictive control purpose. The configuration of ANN model is developed based on nonlinear autoregressive network with exogenous input (NARX) structure. Input and output data for training and validation of ANN model are generated by AC simulator. The ANN model is optimized by investigating the effect of number of neuron and time delay input on prediction accuracy. The results show that the ANN model developed in present study has good accuracy in predicting several step-ahead of cooling capacity and superheat temperature. Accordingly, this ANN model is applicable for predictive control in future study.

1. Introduction

In recent years, the increase of population and economic growth lead energy consumption increases especially in building sector. In daily life, it can be estimated that people may spend around 80% of their time inside the building. It forces the heating ventilating and air conditioning (HVAC) system to work every time to provide the thermal comfort for occupants. Accordingly, this system takes a large portion among the total energy consumption in building. Many researchers have paid much attention on building energy research especially on HVAC system. Developing a proper control strategy for cooling capacity and superheat temperature is one of the solutions to increase the efficiency of AC system operation. Since the system operates frequently, it can contribute in reducing building energy consumption significantly.

Several classical control strategy including ON/OFF and PI have been developed for AC system for residential building application. However they have limited operational condition and high energy consumption [1]. Additionally, it is hard to tune the proper gain for PI control that can cover wide operation condition. Alternatively, intelligent control strategy can be considered to improve the system efficiency.

Modelling dynamic behaviour is the important step to develop control strategy. Before being applied in actual system, the control performance should be tested and evaluated in the model that represents real system. It is not easy and very complex. The complexity comes from the nonlinear correlation between input and output of the system [2]. In the other hand, multi-input multi output (MIMO) characteristic of the system makes the system more difficult to model.

The modelling of AC system behaviour have been introduced using physical model which is established based on energy and mass balance, and thermodynamic and heat transfer theory [3]. Even though this method is believed to have high accuracy, however it requires many detail parameters in regard to the system characteristic. Even in some cases, they should be extracted from system's manufacturer. As the alternative solution, the dynamic behaviour of the system can be modelled using artificial neural network (ANN) approach. The system identification using ANN is simple and faster compared to physical model if input-output data is available.

Several ANN models have been successfully applied in AC system to model temperature and humidity [4]; building heating load [5]; chilled water temperature and superheat temperature [6]. In this paper, ANN model with Bayesian regularization is developed to predict superheat temperature and cooling capacity under modulation of expansion valve and compressor speed in unsteady condition. The motivation of present study is to develop ANN model that can capture dynamic response of superheat temperature and cooling capacity with different compressor speed and valve opening that can be applied for predictive control.

2. Methodology

2.1. System description

The schematic diagram of general vapour compression AC system used in present study is shown in Fig. 1. The system consists of several subsystems i.e. compressor, condenser, expansion valve, and evaporator which are connected to each other. The work of vapour compression cycle (VCC) can be briefly explained through the diagram shown in Fig. 1. The heat Q_E comes from conditioned space at room temperature T_R is absorbed by evaporator via refrigerant and released to the environment at ambient temperature T_A through condenser in amount of Q_C. The refrigerant R410A is selected as the working fluid to transfer the heat from low to high temperature condition. The notation T_{SH} represents the temperature difference between refrigerant at evaporator outlet with its saturation vapour at the same pressure. Number 1-4 indicates the refrigerant state in the cycle.

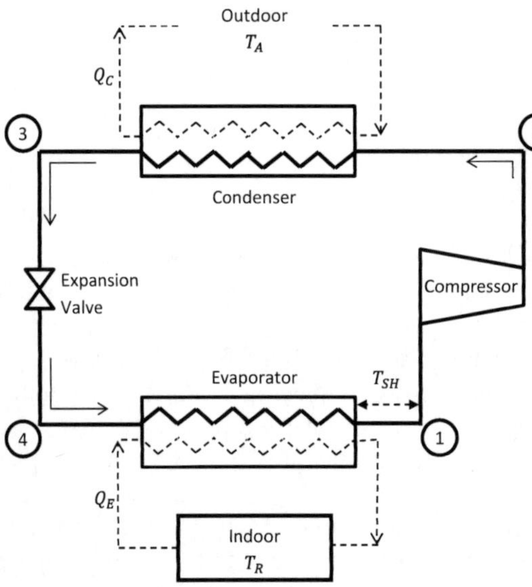

Figure 1. Schematic diagram of vapour compression cycle

2.2. Data sets fabrication

The feature of data sets has a significant influence on the performance of ANN model. The system behaviour that can be identified by ANN highly depends on the trend of data used in the training. It is important to note that when preparing data sets for identification in nonlinear system, the signal excitation scenario should cover all amplitudes. So the data feature that will be applied for learning can represent the dynamic of system. In this paper, N-samples-constant is proposed to meet the various amplitude signals [7]. To provide training data for dynamic system, the system is simulated by randomly varying compressor speed *N* and valve opening *V* to produce dynamic response of cooling capacity Q_E and superheat temperature T_{SH} with the condition provided in Table 1. The simulation is conducted using AC system simulator developed using physical-mathematical model and has been validated with actual system [8]. The input and output signal for identification is presented in Fig. 2.

Table 1. Simulation condition

Simulation condition	Setting	Unit
Condenser air inlet temperature	35	°C
Condenser air velocity	0.28	m/s
Evaporator air velocity	0.1	m/s
Cooling load sensible	1.89	kW
Volume	167.75	m^3
SHR	1	-
Initial room temperature	19.36	°C

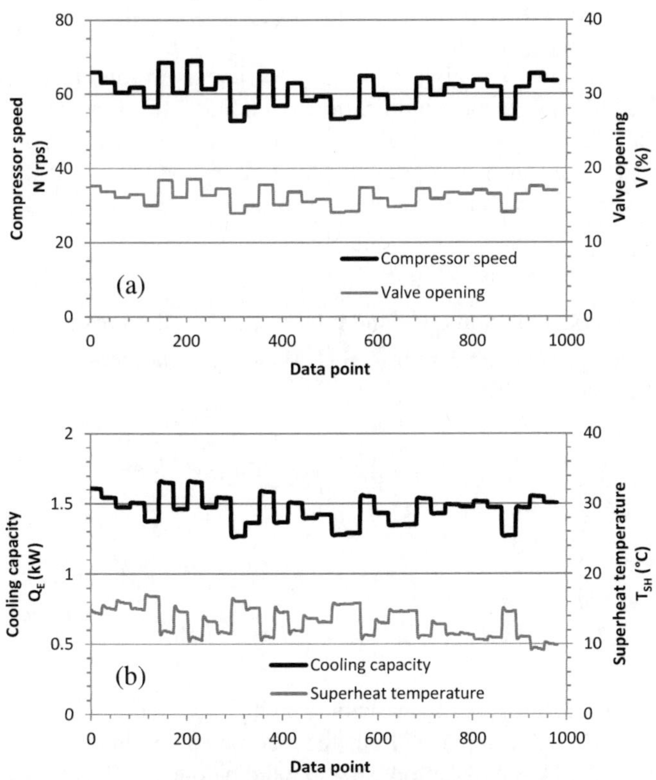

Figure 2. (a) Input and (b) output data for training

2.3 Dynamic neural network

Multilayer perceptron (MLP) ANN model has been applied in many applications to solve the approximation problem. In the simple form, the structure of MLP network is depicted in Fig. 3. This network is arranged by input layer, hidden layer and output layer. The relation between input and output is connected by some mathematical function via hidden neuron.

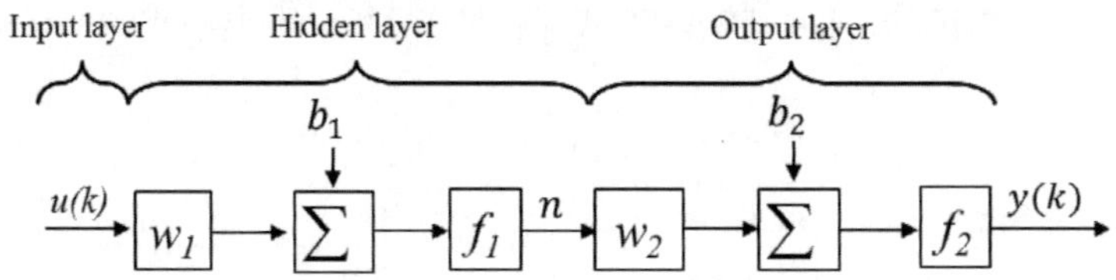

Figure 3. Multilayer perceptron network

The predicted output resulted from MLP network can be described through the following equation:

$$y(k) = f_2(w_2 n(k) + b_2) \tag{1}$$

$$n(k) = f_1(w_1 u(k) + b_1) \tag{2}$$

The network is trained to adjust the weight and bias coefficients to minimize the prediction error and weight coefficient. Weight coefficient minimization is intended to improve the generalization of the network. In this case, the network is trained to achieve high prediction accuracy with high generalization capability. Hence the Bayesian's theorem combined with Levenberg-Marquardt algorithm is employed to minimize objective function which can be written as follows:

$$J = \beta E_D + \alpha E_W \tag{3}$$

where E_D and E_W represents prediction error and total weight coefficients respectively. Notation β and α shows the regularization parameters determined by Bayesian's theorem. The ratio (α/β) controls the network performance.

$$\text{RMSE} = \left(\frac{1}{N}\sum_{j=1}^{N}(y_j - t_j)^2\right)^{1/2} \tag{4}$$

$$\text{CV} = \frac{RMSE}{\frac{1}{N}\sum_{j=1}^{N} y_j} x\ 100 \tag{5}$$

Generally there are two ANN model type namely static and dynamic networks. Static networks calculate the output directly using input without feedback element and input delays. While dynamic networks include the previous inputs or outputs to predict the next step output. In this paper the dynamic ANN model is considered to predict several step ahead dynamic system behaviour.

Fig. 4 shows dynamic ANN structure where the previous compressor speed *N* and valve opening *V* are used to predict the next cooling capacity Q_E and superheat temperature $T_{SH.}$ The number of time delay for input and output is indicated by *du* and *dy* respectively. The dynamic network can be used to predict one-step or multi-step ahead output prediction. The multi-step-ahead prediction is performed by feeding back the ANN output as the input for the next prediction; whereas one-step-ahead prediction takes the true output obtain from simulation as the input for the next prediction.

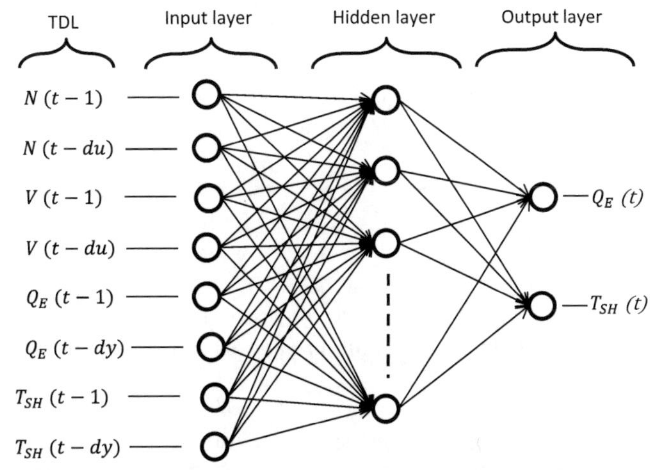

Figure 4. Configuration of NARX networks

3. Results and discussion

In this work, the ANN model is developed by investigating number of tapped delay lines (TDL) and neuron in hidden layer. The proper number of TDL is first analyzed by testing the performance of ANN using various number of TDL from 2 to 10 TDL. It should be noted that the initial weight and bias coefficients are randomly set for training, the prediction accuracy obtained from the same structure can be slightly different due to different convergence point at training process. To ensure that the network has a good performance at every structure, the network was trained 10 times with 20000 maximum iterations, learning rate of 0.01, and 10 neurons. The lowest error is then selected to be the best network. The prediction results for cooling capacity and superheat temperature under different number of TDLs is demonstrated in Fig. 5. It can be observed that the prediction error tends to decrease as number of TDL increases. This is due to the effect of the history of input sequence, the more input are used, the network has more information to fit the data with high accuration. The minimum error for cooling capacity is shown at 5 TDLs and superheat temperature at 7 TDLs. In this case, there are two options for the optimum number of TDL that can be considered for the prediction. Since the gap of accuracy between 5 and 7 TDL in cooling capacity prediction results is not significant, we select 7 TDLs as the optimum number of TDLs in present study.

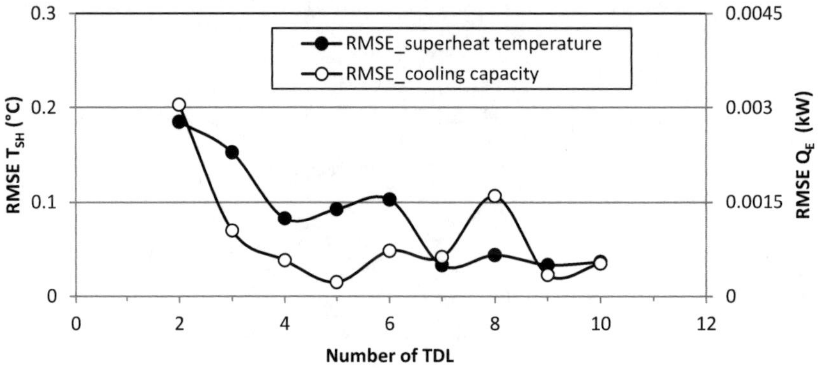

Figure 5. Prediction results under different number of TDL (a) cooling capacity (b) superheat temperature

Furthermore, to improve the prediction accuracy, the effect of neuron is also investigated using optimized input delay (7 TDLs). Number of neuron is varied from 2 to 20 neurons with the increment of 2 neurons. The ANN performance with various number neurons is shown in Fig. 6. As number of neuron increases the prediction accuracy for either cooling capacity or superheat temperature is getting better. The significant difference is shown by 2 and 4 neurons where the RMSE value goes down drastically. When only few neurons were applied, the network doesn't have sufficient weight coefficient to fit the data, therefore it has no capacity to perform accurate prediction. Otherwise, if too many neurons are used in the training, the network complexity increases and produces over fitting. Based on the results shown in Fig. 6, the optimum configuration for the dynamic ANN in present study can be proposed with 7 TDLs and 4 neurons.

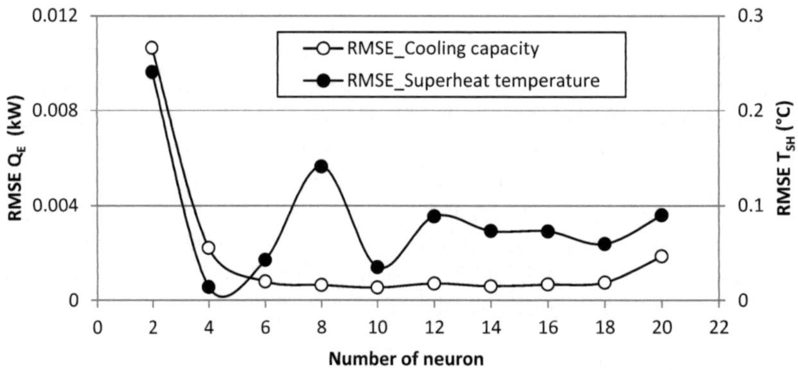

Figure 6. Performance of ANN under different number of neurons

The prediction output generated by optimized ANN configuration is shown in Fig. 7. The network was tested using constant input signal with the compressor speed and valve opening is 53.25 rps and 14.09 % respectively. The results show a good agreement between the predicted and corresponding values for both cooling capacity and superheat temperature with CV for superheat temperature is 0.066% and cooling capacity is 0.2%.

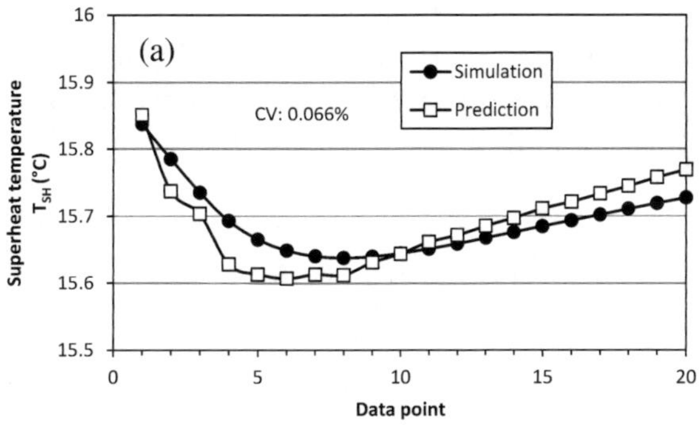

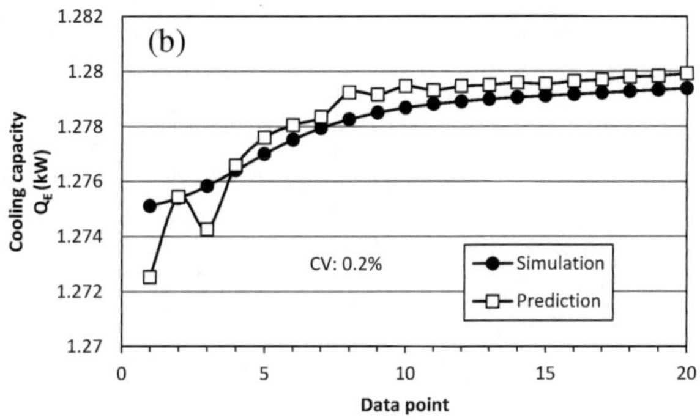

Figure 7. Prediction result (a) cooling capacity and (b) superheat temperature for 20 step ahead prediction

4. Conclusion

Dynamic ANN model has been developed to predict behavior of AC system with two inputs and two outputs in unsteady condition. The optimized network with 7 TDLs and 4 neurons shows satisfied accuracy for 20 steps-ahead prediction. Accordingly, the developed ANN model in present study can be implemented for predictive control in vapor compression AC system application to control cooling capacity and superheat temperature to achieve better efficiency compared to classical control.

Acknowledgements

The author would like to thank LPDP that has provided scholarship to fund doctoral course and supported this research.

References

[1] Ng, B. C., et al., Dynamic modelling of an automotive variable speed air conditioning system using nonlinear autoregressive exogenous neural networks. Applied Thermal Engineering 2014:73:1255-1269.
[2] Loannou PA, Pitsillides A. Modeling and control of complex systems. London, New York: CRC Press, Tayor & Francis Group; 2008.
[3] Jain, N. and A. Alleyne, Exergy-based optimal control of a vapor compression system. Energy Conversion and Management, 2015. 92: p. 353-365.
[4] Li, N., et al., Dynamic modeling and control of a direct expansion air conditioning system using artificial neural network. Applied Energy, 2012. 91(1): p. 290-300.

[5] Sholahudin, S. and H. Han, Simplified dynamic neural network model to predict heating load of a building using Taguchi method. Energy, 2016. 115: p. 1672-1678.
[6] Ekren, O., et al., Comparison of different controllers for variable speed compressor and electronic expansion valve. International Journal of Refrigeration, 2010. 33:p. 1161-1168.
[7] Norgaard, M., et al., Neural Networks for Modelling and Control of Dynamic Systems, Springer, 2001.
[8] Kuniyasu Matsumoto, Keisuke Ohno, Kiyoshi Saito, Numerical analysis of VRF Compression Heat-pump System under Unsteady-State Condition, Trans. of the JSRAE vol. 33 no. 2 (2016) 155-164.

IOP Conf. Series: Materials Science and Engineering **539** (2019) 012004 doi:10.1088/1757-899X/539/1/012004

Hardness improvement of aluminum alloy 2024 t3 after artificial aging treatment

I M Astika[1]
[1]Mechanical Engineering Study Program, Faculty of Engineering Udayana University, Indonesia

Email: imdastika@yahoo.com

Abstract. The needs of having strong and light material for many construction is increasing. Aluminum is one of materials that fulfill this requirement with other advantages such as corrosion resistant and easily formed. Generally, hardness of aluminum and its alloys is lower than ferrous metal such as iron and steel. Aluminum hardness improvement can be conducted by heat treatment namely artificial aging. Artificial aging in aluminum alloy type 2024 T3 in this study was conducted in 2 phases that is solution treatment in 500 ^{0}C for 1 hour and continued with artificial aging process in 180 ^{0}C by holding time 2, 4 and 6 hours. From the conducted research, it was found that there was hardness improvement on aluminum type 2024 T3 in line with the increasing of aging time. The gained microstructure was getting smooth which means the precipitation hardening process was getting perfect.

1. Introduction

Aluminum is one of non-ferrous materials which is corrosion resistant, light and strong. It is mostly applied on construction of airplane, vehicles or building. Compared to ferrous metal, aluminum and its alloys are less strong. Aluminum strength improvement can be done by heat treatment process in its fabrication process. One of heat treatment processes on aluminum and its alloys is artificial aging, especially in aluminum alloy with composition of Cu 2.5 – 5.0%.

Artificial aging with variation of time and temperature aging as well as variation of temperature solution treatment could improve the hardness of any kind of aluminum types as reported by previous researchers. Ekhlas Ahmed [1] conducted a test on Aluminum Alloy 2024 T6 with heating temperature of 500 ^{0}C for 1 hour and artificial aging on temperature of 190 ^{0}C with holding time variation 1, 2 and 5 hours and resulted the best precipitation phase in 5 hours of aging time. Naveed Afzal, et al [2] studied aluminum alloy 2024 with aging time of 2 hours and temperature variation of 105, 135, 165 and 195 ^{0}C. The study showed that the maximum tensile strength and hardness was gained with aging temperature of 135 and 195 ^{0}C. Gowrishankar M. C. et al [3] used AA 6061 material with aging temperature variation of 100, 150 and 200 ^{0}C and aging time of 3-10 hours. This study showed that better mechanical quality on lower temperature. Aytekin Polat et al [4] with AA 6061 material varied temperature and time aging. Temperature solution treatment of 550 ^{0}C for 2 hours. Artificial aging temperature variation of 160, 180 and 200 ^{0}C with holding time variation of 2, 5, 5, 10, 20, 40, 60 and 80 hours. The result showed that the mechanical quality was decreasing as the increasing of aging temperature but it was improving by the increasing of aging time. DAP Reis et al [5] varied solution treatment temperature on 495, 505 and 515 ^{0}C, and artificial aging on 190 and 208 ^{0}C. It was reported that the highest hardness was gained in combination 505 ^{0}C and 208 ^{0}C with aging time of 2 hours. MS Khan et al [6] used AA 2014 material with solution treatment temperature of 502 ^{0}C for 2 hours. The result showed improvement of hardness from 85 HV to 136 HV on artificial temperature of 180 ^{0}C and increased into 145 HV on aging temperature of 233 ^{0}C. MFIA Imam et al [7] got the highest hardness of aluminum alloy material on solution treatment temperature of 500 ^{0}C and artificial aging temperature of 165 ^{0}C.

Hardness of aluminum also could be improved by coating technique of the surface of the material and continued with artificial aging process. Jianzhong Zhou, et al [8] studied Aluminum Alloy 2024

IOP Conf. Series: Materials Science and Engineering **539** (2019) 012004 doi:10.1088/1757-899X/539/1/012004

T3 with Cryogenic Treatment (CT) with time variation of 2, 4, 6, 8 and 12 hours on 77 ^{0}K by using liquid nitrogen. The finding showed that CT heat treatment could improve the hardness of the material. M.H. Goodarzy et al [9] conducted ECAP process on AA 2024 T8 with natural aging on 70 and 100 ^{0}C. Heat treatment was conducted on 500 ^{0}C for 40 minutes. The result showed that on time aging 70 ^{0}C, the hardness increased.

Artificial aging process on composition variation of aluminum alloy could also improve aluminum hardness. Aondona P. Ihom et al [10] studied composite with composition of Al/0.05% particle glass. Treatment solution on 500 ^{0}C for 45 minutes. Variation of aging temperature 150 – 210 ^{0}C. The result showed that the tensile strength and hardness increased in line with the improvement of aging temperature. J.K. Odusote et al [11] proposed that his research used variation of Al-Cu-Zn with heat treatment temperature of 460 ^{0}C for 2 hours. Artificial aging on 160 ^{0}C for 5 hours. The best result was gained on composition 81.2 Al-1.56 Cu-8.33 Zn.

2. Method

Method of this research was heat treatment in form of solution treatment and artificial aging. Solution treatment process by heating material until 500 ^{0}C for 1 hour. The next process was artificial aging on 180 ^{0}C with holding time variation 2, 4 and 6 hours. Schematic heat treatment process is shown in Figure 2 [12, 13]. Composition of aluminum alloy type 2024 T3 used in this research are:

Cu	Mg	Mn	Fe	Zn	Ti	Cr	Si	Al
4.40%	1.50%	0.60%	0.50%	0.25%	0.15%	0.10%	0.05%	96.90%

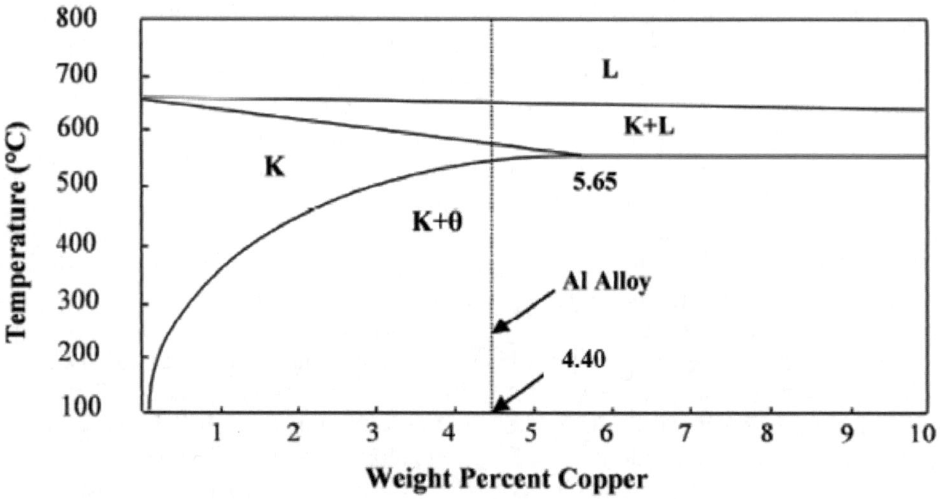

Figure 1. Phase diagram of the aluminum alloy

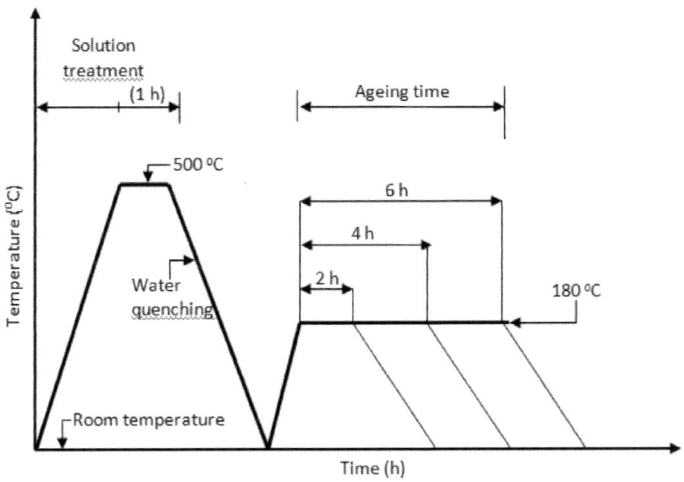

Figure 2. Schematic of aging time and solution treatment on aluminum alloy type 2024 T3 [12, 13]

2.1 Hardness Test

Hardness test was conducted to get the value of material hardness. It was done Vickers' hardness test with hardness test machine of Wolpert with pyramid-shaped diamond indenter with square pedestal and peak angle between two opposite sides of 136^0. The given load depended on the tested metal hardness and in this research, the load was 5 kg. The value of Vickers's hardness is expressed with HVN or HV.

$$HV = \frac{2P\,sin\left(\frac{\theta}{2}\right)}{d^2} = \frac{1,854P}{d^2}$$

in which: P = the applied load (kg)
d = average diagonal length (mm)
Θ = angle between the opposite diamond surface

3. Result and Discussion

Table 1. The result of hardness test of aluminum alloy 2024 T3

No	Hardness (HVN)			
		Aging time		
	Without heat treatment	2 hrs	4 hrs	6 hrs
1	56.8	61.2	66.5	75.7
2	57.9	62.8	67.2	69.6
3	54.7	59.4	64.8	71.5
4	54.6	58.8	64.3	77.9
5	53.5	58.6	63.9	73.6
6	56.6	60.9	66.2	71.5
7	57.4	62.6	67.2	67.7
8	62.6	66.7	72.2	67.1
9	67.7	70.5	73.3	69.6
10	64.2	68.9	72.3	69.6
Average	58.6	63.04	67.79	71.38

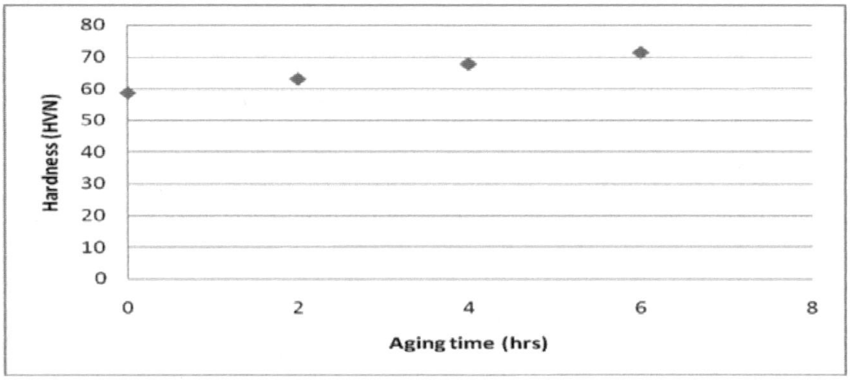

Figure 3. Graphic of the relation between hardness and aging time

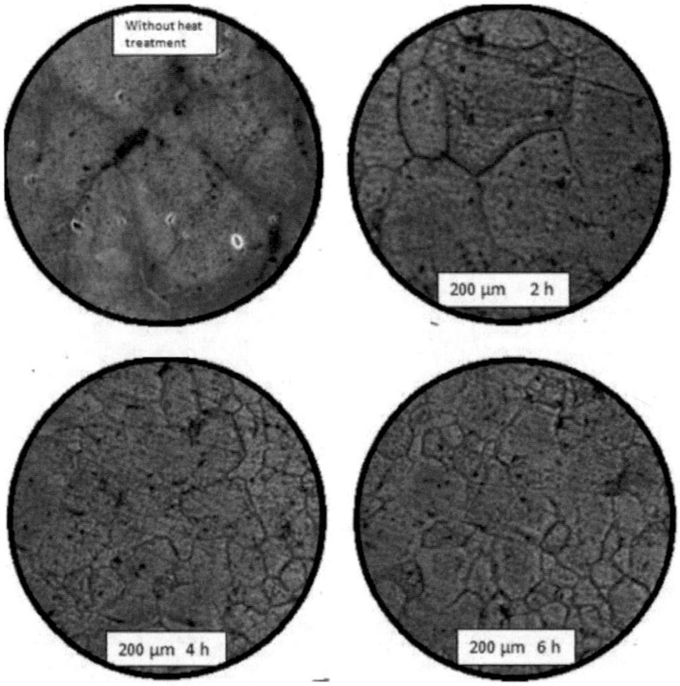

Figure 4. Microstructure of AA 2024 T3 with a variation of aging time.

The result of hardness test of aluminum alloy 2024 T3 after heat treatment of artificial aging shown on Table 1 and Figure 3. On graphic it can be seen the increasing of hardness in line with the addition of aging time. This increase happened because the improvement of precipitation to be more perfect shown by the smaller and more delicate precipitation size which is shown in Figure 4. The smaller and more delicate particle that was formed caused the hardness of material was getting higher.

4. Conclusion

The research showed that aging time could improve the hardness of aluminum alloy 2024 T3. The result of the micro photo showed the formed-particle was getting delicate with more aging time which caused the hardness of the material was increasing.

Acknowledgment
The authors are grateful to the Udayana University through the Institute of Research and Community Service and Faculty of Engineering which has funded this research.

References

[1] Ekhlas Ahmed 2013, **The Effect Of Heat Treatments On Mechanical Properties Of Aluminum Alloy Aa 2024-T6**, The Iraqi Journal For Mechanical And Material Engineering, Vol.13, No3, 2013, pp 402-409

[2] Naveed Afzal, Tariq Shah, and R. Ahmad 2013, **Microstructural Features and Mechanical Properties of Artificially Aged AA2024,** *ISSN 0556-171X. Ïðîáëåìû ïðî÷íîñòè, 2013, 6*, pp 69-81

[3] Gowrishankar M. C., Shravan, Rakesh, Rahul, Achutha Kini, S. S. Sharma 2014, **Effect of Artificial Aging on Strength and Wear Behaviour of Solutionized Aluminium 6061 Alloy,** Proceedings of the 3rd World Conference on Applied Sciences, Engineering and Technology 27-29 September 2014, Kathmandu, Nepal, ISBN 13: 978-81-930222-0-7, pp 388-393

[4] Aytekin Polat, Mustafa Avsar, Fahrettin Ozturk 2015, **Effects Of The Artificial-Aging Temperature And Time On The Mechanical Properties And Springback Behavior Of Aa6061,** Materiali in tehnologije / Materials and technology 49 (2015) 4, 487–493

[5] D.A.P. Reis, A.A. Couto, N.I. Domingues Jr, A.C.O. Hirschmann, S. Zepka, C. Moura Neto, 2012, **Effect of Artificial Aging on the Mechanical Properties of an Aerospace Aluminum Alloy 2024,** *Defect and Diffusion Forum Vols. 326-328 (2012) pp 193-198 © (2012) Trans Tech Publications, Switzerland doi:10.4028/www.scientific.net/DDF.326-328.193*

[6] M. S. Khan, A. Salam, N. Ikram 2015, **Effect of Cold Work on the Age Hardening Behaviour of Al 2014 Al alloy**, *Technical Journal, University of Engineering and Technology (UET) Taxila, Pakistan Vol. 20 No. I-2015, p 76-80*

[7] M. F. I. A. Imam, M. S. Rahman, M. Z. H. Khan 2015, **Influence Of Heat Treatment On Fatigue And Fracture Behavior Of Aluminium Alloy** , Journal Of Engineering Science And Technology Vol. 10, No. 6 (2015) 730 – 742,© School Of Engineering, Taylor's University

[8] Jianzhong Zhou, Suqiang Xu, Shu Huang, Xiankai Meng, Jie Sheng, Haifeng Zhang, Jing Li, Yunhui Sun and Emmanuel Agyenim Boateng 2016**, Tensile Properties and Microstructures of a 2024-T351 Aluminum Alloy Subjected to Cryogenic Treatment,** Metals **2016**, 6, 279; doi:10.3390/ met6110279.

[9] M.H. Goodarzy, H. Arabi, M.A. Boutorabi, S.H. Seyedein, S.H. Hasani Najafabadi 2014, **The effects of room temperature ECAP and subsequent aging on mechanical properties of 2024 Al alloy**, Journal of Alloys and Compounds 585 (2014) 753–759

[10] Aondona P. Ihom, Nyior G. Bem, Emmanuel E. Anbua, Joy N. Ogbodo, 2012**, The Effect of Ageing Time on Some Mechanical Properties of Aluminum/0.5% Glass Reinforced Particulate Composite,** *Journal of Minerals and Materials Characterization and Engineering*, 2012, 11, 919-923 Published Online September 2012 (http://www.SciRP.org/journal/jmmce)

[11] J.K. Odusote, A.A. Adeleke and P.A. Ajayi 2015 **Mechanical properties and microstructure of precipitation-hardened Al-Cu-Zn alloys ,** International Journal of Automotive and Mechanical Engineering (IJAME) ISSN: 2229-8649 (Print); ISSN: 2180-1606 (Online) Volume 12, pp. 3033-3042, July-December 2015 ©Universiti Malaysia Pahang DOI: http://dx.doi.org/10.15282/ijame.12.2015.17.0252

[12] I Made Astika, I Gusti Komang Dwijana dan I Gusti Ketut Sukadana 2017, **Pengaruh Waktu Aging Terhadap Kekuatan Tarik Aluminium Tipe 2024 T3,** Prosiding KNEP VIII Teknik Mesin Universitas Udayana, 20-21 Juli 2017 Denpasar Bali, ISSN 2338-414X, p 59-61

[13] I Made Astika and I G K Dwijana 2017 **Effect of Aging Time on the Fatigue Life of Aluminum Alloy 2024 T3,** Journal of Materials Engineering and Processing Technology, Volume 1 No. 2, ISSN 2549-6891 p 6-9

IOP Conf. Series: Materials Science and Engineering **539** (2019) 012005 doi:10.1088/1757-899X/539/1/012005

Assessment of thermal and transport properties of ionic liquids as suitable absorbent for absorption cooling applications

H M Ariyadi[1], S Yamaguchi[1] and K Saito[1]

[1]Department of Applied Mechanics and Aerospace Engineering, Waseda University, 3-4-1 Okubo, Shinjuku-ku, Tokyo 169-8555, Japan

Email: hifni.ariyadi@aoni.waseda.jp

Abstract. Ionic liquids, salts which have liquid phase at temperature below 100°C, have been widely introduced in engineering applications, including absorption cooling systems. The application of ionic liquids in absorption cooling systems is intended to remove the disadvantages of conventional working fluids such as corrosion and crystallization. In terms of thermodynamic performances, theoretical investigation based on solubility behavior of ionic liquids in natural refrigerants show a competitive performance in comparison with conventional working fluid. Nevertheless, heat transfer performance, which is also an important key in absorption cooling systems, particularly in terms of system design and size, needs to be deeply explored and investigated. This study aims to assess the thermal and transport properties of ionic liquids as absorbent in relation to the heat and mass transfer characteristics of these working fluids. The thermal and transport properties of ionic liquids proposed as absorbent for absorption cooling systems are collected, and heat and mass transfer characteristic of these ionic liquids based on their thermal and transport properties are investigated and analyzed. Finally, the most suitable ionic liquids for absorption machines, both in terms of thermodynamics and heat and mass transfer performances can be proposed.

1. Introduction

Absorption cooling systems are recently getting more attention and consideration with the rapid economic growth and increase of energy demand. This is due to the fact that these systems can be driven with wide variety of renewable energy such as solar energy, geothermal, and even waste heat to produce cold thus carries a primary energy saving, and emission reduction. Moreover, absorption systems contain non-CFC and therefore, environmentally friendly and becomes a competitive alternative to conventional vapor compression cooling systems. Similar to vapor compression cycle, absorption cycle is based on the cooling and heating process associated with phase changes of evaporation and condensation of refrigerant fluid at different temperature and pressure levels.

One of the most common working fluid pair in absorption cooling systems is water/LiBr. In this working pair, water works as refrigerant and aqueous LiBr solution works as absorbent. The water/LiBr absorption cycle is used mainly for cooling and heating purpose as it cannot produce cold below 0°C. Being natural fluids, both are emission free and ozone friendly resulting with zero global warming potential (GWP) and ozone depletion potential (ODP). However, the applications of absorption cycle working with water/LiBr are limited by corrosion and crystallization problems. Many attempts have been addressed to eliminate abovementioned drawbacks, such as proposing multicomponent salts solution absorbent [1], adding surfactants and additives [2], and proposing novel absorbent such as ionic liquids (IL) [3]. The latter application in absorption cooling systems can eliminate the abovementioned drawbacks as ionic liquids have liquid phase in wide range of temperature, negligible vapor pressure, non-corrosive, non-toxic, highly soluble with most of natural refrigerant, and thermally stable [3]. In addition, the thermophysical properties of ionic liquids can be adjusted by combining the cation and anion pair [4].

Studies on the use of water/ionic liquid working pairs as working fluid for absorption cooling applications are growing in number. Recent studies on the cycle performance of absorption cooling systems using water/ionic liquid working pairs have been published in many literatures, proposing some ionic liquids as new absorbent for water refrigerant such as 1,3-dimethylimidazolium dimethylphosphate ([dmim] [DMP]), 1-ethyl-3-methyl- imidazolium dimethylphosphate ([emim] [DMP]), 1-ethyl-3-methylimidazolium ethylsulfate ([emim] [$EtSO_4$]), etc. Based on theoretical thermodynamic calculation using vapor-liquid equilibrium data, the coefficient of performance and solution circulation ratio of single-stage absorption cooling systems using water/ionic liquid fluids is quite competitive in comparison with conventional water/LiBr working fluid [5] considering the above-mentioned advantages of ionic liquids over conventional aqueous LiBr solution.

In fact, the performances and efficiencies of an absorption system are not only determined by coefficient of performance and solution circulation ratio, but also determined by system design and size, initial cost and the operating cost of the machine, which are strongly dependent on the working fluid properties. Selecting and proposing a new working fluid for this system therefore should be carried out properly, considering not only solubility property of refrigerant in the absorbent, but also transport properties of the working fluid mixtures which influence the heat and mass transfer performance, such as viscosity, thermal conductivity, and diffusion coefficient.

Although studies on the cycle performance of absorption cooling systems using water/ionic liquid working pairs have been performed by many researchers, the assessment of thermal and transport properties of these new working pair remains incomplete. This study aims to assess the thermal and transport properties of ionic liquids as absorbent in order to obtain the suitable ionic liquids with optimum heat and mass transfer performance, which is an important parameter in absorption cooling systems, particularly in terms of system design and size.

2. Methodology

In this paper, two ionic liquids, namely 1-ethyl-3-methyl imidazolium ethyl sulfate [emim] [$EtSO_4$] and 1.3-dimethyl imidazolium dimethyl phosphate [dmim] [DMP] are selected as absorbent for water refrigerant. Thermal and transport properties and vapor-liquid equlibria of water/ [dmim] [DMP] mixtures necessary for the simulation are obtained from NIST database [6]. In addition, to obtain the diffusion coefficient of water in ionic liquid, a general correlation is used, as suggested by Krannich et al. [7]. The thermal and transport properties of coolant water and water vapor are calculated using correlation of IAPWS95 Revision 2016 [8].

The assessment of the properties is done in two parts. The first part is modelling the vapor liquid equilibria to build the pressure-temperature-composition (*PTX*) diagram. The *PTX* charts then can be used to estimates the specific operative conditions of the absorber based on operative conditions of the systems. From the first parts, the values of most important properties of selected working fluids can be obtained, compared, and analysed. And finally, the heat and mass transfer characteristics of the working fluids in an absorber can be evaluated using simple model.

Heat and mass transfer characteristics of ionic liquid based working fluid in the absorber is evaluated using a model constructed from energy and mass balance equations, derived from the absorption process takes place in an infinitesimal area of the tube as schematically shown in Fig. 1. Detail of mathematical description of the model can be found in [9].

The absorbent-refrigerant solution enters from the top of the tube, downwards through the inner wall as a falling film. In addition, the water vapour enters from the top of the vertical tube and be absorbed by the ionic liquid solution. The solution leaves the absorber through the bottom of the tube as less concentrated in salt than at the entrance.

The cooling water flows from the bottom of the absorber through tube annulus space to remove the heat of absorption. The heat is transferred from the solution to the cooling water while mass transfer process occurs in the vapor-solution surface.

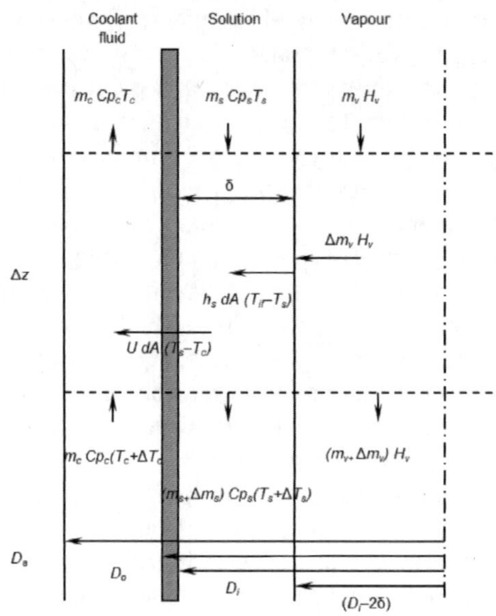

Figure 1. The schematic diagram of water vapor absorption process in vertical tube falling film absorber.

Assuming that the vapor-solution interface is in saturated condition and the bulk solution is perfectly mixed along the tube, the water vapor absorbed by the solution can be calculated using following equation

$$\frac{dm_s}{dz} = k_s(\pi\ Di)(c_{H2O,\text{int}} - c_{H2O}) \tag{1}$$

Moreover, with assumptions that the heat transfer only occurs in horizontal direction, and there is no heat loss in the absorber, from the energy balance at the solution side, the solution temperature at a control volume can be written as follows

$$\frac{dT_S}{dz} = \frac{H_V - Cp_S T_S}{m_S Cp_S}\frac{dm_S}{dz} - \frac{U\pi Di(T_S - T_C)}{m_S Cp_S} \tag{2}$$

Similarly, the cooling water temperature at a control volume can be written as follows

$$\frac{dT_c}{dz} = \frac{U\pi Di(T_s - T_c)}{m_c Cp_c} \tag{3}$$

3. Results and Discussions

3.1. Properties Comparison

In this subsection, thermos-physical and chemical properties of water/ionic liquid mixtures is discussed. The comparison is analysed under typical operative conditions of absorber, both in inlet and outlet. This consideration is taken because absorber plays as the most important component in the system which affects the overall performance of the systems.

Solubility. In absorption technologies, solubility of refrigerant in the absorbent is one of the most important properties that may reflects the performance of the systems. The coefficient of performance of an absorption cooling cycle can be roughly estimated from the solubility behaviour in a Dühring diagram plot. To be a good working pair for absorption application, the refrigerant must be well soluble in the absorbent under the conditions in which absorption takes place. High solubility and affinity of refrigerant in the absorbent reduce the amount of absorbent needed to circulate the refrigerant at certain cooling capacity, thus decreasing pumping power and reducing heat exchangers size.

Solubility of water in [dmim] [DMP] and [emim] [EtSO$_4$] have been measured and well correlated using Non-Random Two-Liquids (NRTL) model by He et al [10] and Wang et al. [11]. *PTX* diagrams of water in [dmim] [DMP] and [emim] [EtSO$_4$] mixtures are presented in Fig 2 (a) and (b), respectively. Although they concluded that both solutions show significant negative deviation to ideal Raoult's law, in term of mass concentration these two ionic liquids show considerably lower solubility than LiBr salt.

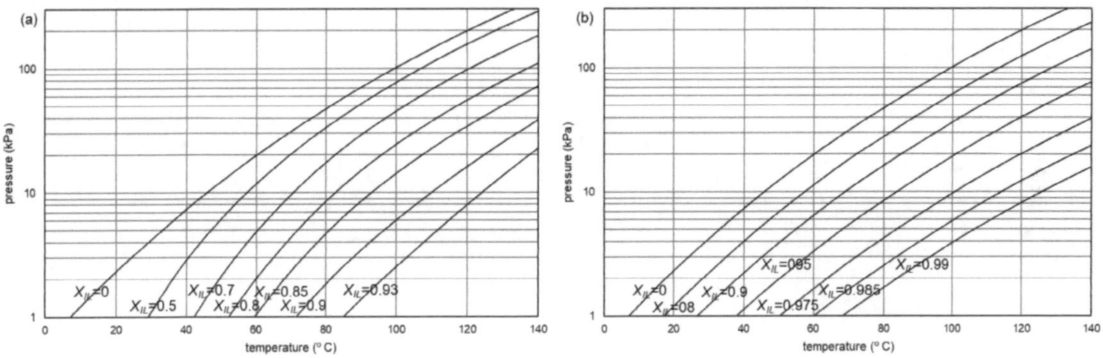

Figure 2. *PTX* diagram of water/[dmim][DMP] (a) and water/[emim][EtSO$_4$] (b)

PTX diagrams in Fig 2 show that at the same operation conditions working fluid mixtures with wide concentration line distribution have more refrigerant mass concentration in absorber-generator loop in comparison with working fluid mixtures with narrow concentration line. Therefore, the working fluid mixtures with wide concentration line distribution have lower solution flow rates than the fluid mixtures with narrow concentration line distribution, as it has higher concentration difference between strong and weak solution.

Apart of concentration distribution line, the solubility of water in [dmim][DMP] is much better than in [emim][EtSO$_4$], with the fact that the amount of water in water/[dmim][DMP] solution at a certain temperature and pressure is much higher than in water/[emim][EtSO$_4$]. An example of the concentration (both weak and strong solution) of studied ionic liquid in comparison with LiBr solution at different typical operative conditions (T_{abs}=T_{cond}=30~40 °C, T_{evap}=5~10 °C, T_{gen}=80~90 °C) is summarized in Table 1.

Table 1. Saturation absorbent concentration at various operative conditions

T_{evap} (°C)	Solution concentration leaving the absorber (% mass of absorbent)					
	T_{abs}=30 °C			T_{abs}=40 °C		
	[dmim][DMP]	[emim][EtSO$_4$]	LiBr	[dmim][DMP]	[emim][EtSO$_4$]	LiBr
5	54.02	93.01	52.57	68.83	96.14	57.81
8	50.76	91.38	50.57	66.31	95.27	56.08
10	48.46	90.08	49.13	64.55	94.58	54.90
T_{gen}	Solution concentration leaving the generator (% mass of absorbent)					

(°C)	T_{cond}=30 °C			T_{cond}=40 °C		
	[dmim][DMP]	[emim][EtSO$_4$]	LiBr	[dmim][DMP]	[emim][EtSO$_4$]	LiBr
80	85.65	97.50	62.94	81.32	95.65	57.62
85	87.54	97.99	65.20	83.89	96.50	59.91
90	89.06	98.37	67.45	85.95	97.17	62.15

3.1.1. Dynamic viscosity
One of the most common issue in ionic liquids that may reduce their potential as absorbent is their high viscosity. High viscosity of absorbent can cause significant pressure drops in the solution circulation loop, which can result in increase of pumping power or larger system size. He et al. [10] and Bhattacharjee et al. [12] have determined and correlated the viscosity of water/[dmim][DMP] and water/[emim][EtSO$_4$] solutions, respectively.

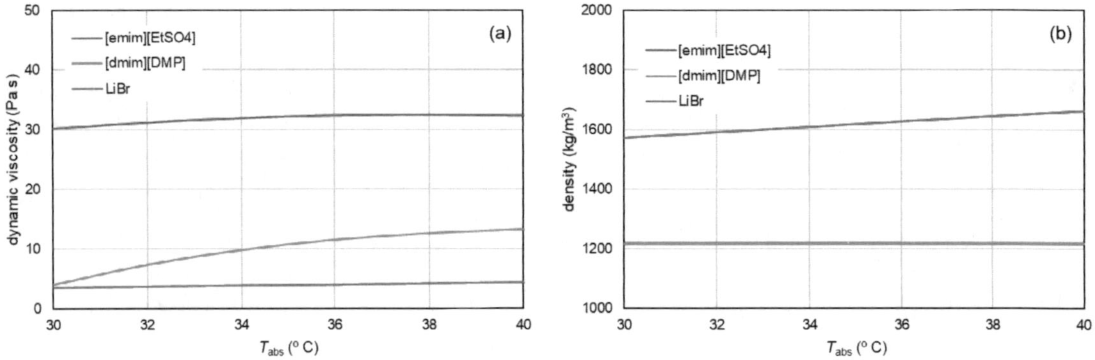

Figure 3. Viscosity (a) and density (b) of water/ionic liquid mixtures at saturation condition (T_{evap}=5 °C) and different absorber temperature

Fig 3 (a) shows viscosity of water/ionic liquid mixtures at saturation condition (T_{evap}=5 °C,) and different absorber temperature. As it can be observed from Fig. 3 (a), the dynamic viscosity of water/[emim][EtSO$_4$] is the highest in comparison with water/[dmim][DMP] and conventional fluid water/LiBr. An interesting point is that at temperature lower than 30 °C, the viscosity of water/[dmim][DMP] is lower than water/LiBr.

3.1.2. Density
The density of water/ [dmim][DMP] and water/[emim][EtSO$_4$] have been measured by He et al. [10] and Bhattacharjee et al. [12]. The density of water/ionic liquid mixtures at saturation condition (T_{evap}=5 °C) and different absorber temperature is shown in Fig 3 (b) with the density of water/[emim][EtSO$_4$] is almost similar with water/[dmim][DMP], and the density conventional fluid water/LiBr is much higher than other two mixtures. With lower density, the pumping power necessary to transport the solution from the low-pressure system to the high-pressure system can be reduce.

3.1.3. Heat capacity
Heat capacity is also one of the most important properties used to evaluate the thermal load and heat transfer characteristic of each components and thermal performance (COP) of the whole systems. Several experimental data of heat capacity of the studied working fluids have been published in literatures [10, 13].

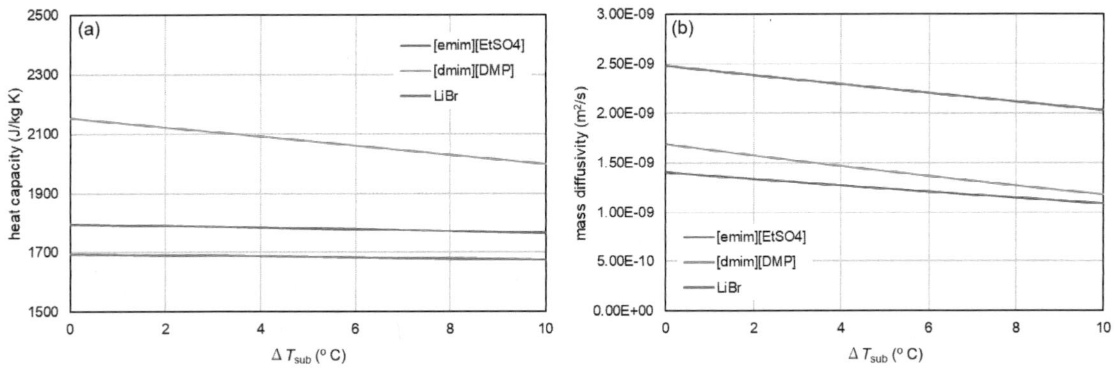

Figure 4. Heat capacity (a) and mass diffusivity (b) of water/ionic liquid mixtures at saturation condition leaving from the generator (T_{gen}=90 °C, T_{cond}=30 °C) and different subcooled absorber temperature (T_{evap}=5 °C)

Fig 4 (a) shows the heat capacity of studied solutions at typical entrance absorber conditions at various degree of subcooling. The solution concentration is determined from a condition than the solution is saturated when leaving the generator at temperature of 90 °C and pressure of 4.25 kPa (corresponds to T_{cond}=30 °C). This condition is used to plot the heat capacity over absorber temperature as the heat transfer phenomena in the absorption process starts at the entrance of absorber. The correlation based on experimental data shows that the heat capacity of aqueous ionic liquid solutions is higher than that of water/LiBr solution, being that of water/ [dmim][DMP] is the highest (see Fig 4 (a)).

3.1.4. Mass diffusivity

Mass diffusivity is an important property to evaluate mass transfer characteristic particularly in the absorber. Although this property is important, to the best of our knowledge there is no available data or correlation of mass diffusivity of water in studied ionic liquids and thus, to estimate the mass diffusivity of water in ionic liquids a method suggested by Krannich et al. [7] is applied.

Using similar condition to Fig 4 (a), the mass diffusivity of water in studied solutions at typical entrance absorber conditions at various degree of subcooling is then plotted in Fig 4 (b). In this figure, it shows that the mass diffusivity of studied ionic liquids is much lower in comparison with conventional aqueous LiBr solution, being [emim] [$EtSO_4$] is the lowest among three studied working fluids. This low mass diffusivity will significantly affect to the mass transfer behaviour, as will be discussed later.

3.2. Heat and Mass Transfer Characteristics

In order to get more understanding on the relation of properties assessment of ionic liquid based absorbents on the performance of the absorption process, heat and mass transfer characteristics of absorption process of water vapor in ionic liquid solutions is theoretically studied, using absorber configuration and operation conditions similar to the one studied by Ariyadi et al [14]. Mass flux and mass flux in water/[dmim][DMP] and water/[emim][$EtSO_4$] working fluids in comparison with conventional water/LiBr working fluid are shown in Fig. 5 (a) and (b), respectively, within solution Reynolds number varies up to 400. Fig. 4 shows that the mass flux of all studied working fluids increases with the increase of solution Reynolds number until reaches its peak then decreases with the increase of solution Reynolds number. This is because at low solution Reynolds number (i.e. Res < 20 in the case of [dmim] [DMP]) absorption process tends to be more dominant than heat transfer process, thus the absorption mass flux increases with the increase of solution Reynolds number. At higher solution Reynolds number heat transfer process is more dominant than absorption process, thus the absorption mass flux decreases with the increase of solution Reynolds number. In general, the mass

IOP Conf. Series: Materials Science and Engineering **539** (2019) 012005 doi:10.1088/1757-899X/539/1/012005

flux and its optimum values trends follow an order of water/LiBr > water/[dmim][DMP] > water/[emim][$EtSO_4$], because the mass flux is affected by diffusivity and the solubility of water in the absorbent. The low solution Reynolds number, where the optimum mass flux is reached, is affected by the high viscosity of water/ionic liquid solutions.

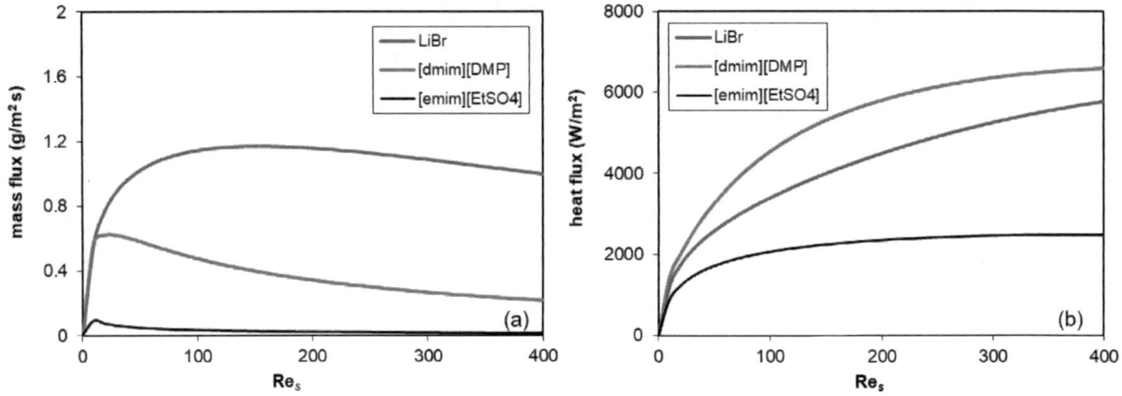

Figure 5. Mass flux (a) and heat flux (b) vs solution Reynolds number of water/LiBr and water/ionic liquid solutions.

Although mass flux of water/[dmim][DMP] solution is lower than that of water/LiBr solution, at the same absorber configuration and operation conditions, the heat flux of that water/[dmim][DMP] working fluid is slightly higher than that of water/LiBr working fluid due to higher heat capacity of water/[dmim][DMP] solution (see Fig. 5 (b)). The heat flux of that water/[emim][$EtSO_4$] working fluid remain the lowest as its low heat capacity and low heat of absorption.

4. Conclusions

Thermophysical and transport properties of ionic liquids absorbent for water refrigerant has been assessed at typical operative conditions both in inlet and exit of an absorber, together with theoretical studies on the heat and mass transfer characteristics of absorption process in a vertical tube absorber. The performances of a refrigerant/absorbent working pairs do not only depend on the vapor-liquid equilibrium or solubility, but is also affected by its thermophysical and transport properties. Among studied ionic liquids, in general, [dmim][DMP] provides better properties than [emim][$EtSO_4$]. Although, [dmim][DMP] still has undesirable properties that can make significance decrease in the system performance, such as low solubility and diffusivity, and high viscosity, this ionic liquid provides better properties, such as high heat capacity, low density, and no solidification risk in comparison with conventional water/LiBr solution which can make this ionic liquid as better alternative absorbent to LiBr solution. In theory, the thermophysical properties of ionic liquids can be adjusted by combining the cation and anion pair. However, in practice, tuning ionic liquids to have good properties for absorption application is not an easy task. Considering the high potential application of ionic liquids as an absorbent in absorption application, further study on other ionic liquids which have better properties for absorption cooling applications remains necessary.

Nomenclature

c	concentration
Cp	constant pressure heat capacity (J/kg K)
Di	inside diameter (m)
H	enthalpy (J/kg)
k	mass transfer coefficient (m/s)

Subscript

abs	absorber
c	cooling water
cond	condenser
evap	evaporator

L	tube length (m)	gen	generator
m	mass flowrate (kg/s)	H2O	water
P	pressure (kPa)	int	vapor-solution interface
Re	Reynolds number	s	solution
T	temperature (°C)	sub	subcooled
U	overall heat transfer coefficient (W/m^2 K)	v	vapor
X	solution concentration (% mass of absorbent)		

References

[1] Bourouis, M., Vallès, M., Medrano, M., Coronas, A., *Int. J. Therm. Sci.,* 44, 2005, pp. 491–498.
[2] Kim, J.K., Jung, J.Y., Kang, Y.T., *Int. J. Refrigeration,* 29, 2006, pp. 22–29.
[3] Kim, Y. J., Kim, S., Joshi, Y. K., Fedorov, A. G., Kohl, P. A., *Energy* 44, 2012, pp. 1005–1016.
[4] Popp, S., Bösmann, A., Wölfel, R., Wasserscheid, P., , *ACS Sustainable Chem. Eng.* 3, 2015, pp. 750–757.
[5] Yokozeki, A., and Shiflett, M. B., *Ind. Eng. Chem. Res.* 49, 2010, pp. 9496–9503.
[6] NIST, ILThermo: Ionic Liquids Database. URL: ilthermo.boulder.nist.gov/
[7] Krannich, M., Heym, F., Jess, A., *J. Chem. Eng. Data*, 61, 2016, pp. 1162–1176.
[8] The International Association for the Properties of Water and Steam (IAPWS), Revised Release on the IAPWS Formulation 1995 for the Thermodynamic Properties of Ordinary Water Substance for General and Scientific Use, Dresden, Germany, 2016.
[9] Tsai, B. B. and Perez-Blanco, H., *Int. J. Heat Mass Transfer,* 41 (15), 1998, pp. 240–2416.
[10] He, Z., Zhao, Z., Zhang, X., Feng, H., *Fluid Phase Equilibria*, 298, 2010, pp. 83–91
[11] Wang, J. F., Li, C. X., Wang, Z. H., *J. Chem. Eng. Data*, 52, 2007, pp. 1307-1312
[12] Bhattacharjee, A., Varanda, C., Freire, M. G., Matted., S., Santos, L. M. N. B. F., Marrucho, I. M., Coutinho, J. A. P., *J. Chem. Eng. Data*, 57 2012, pp. 3473–3482
[13] García-Miaja, G., Troncoso, J., Romaní, *J. Chem. Thermodynamics*, 41, 2009, pp. 161–166

IOP Conf. Series: Materials Science and Engineering **539** (2019) 012006 doi:10.1088/1757-899X/539/1/012006

Study of stinging nettle fibers as a reinforcing of composite materials based on its growing region

I G P A Suryawan[1,3], N P G Suardana[1], I N S Winaya[1], I W B Suyasa[2] and I P Lokantara[1]

[1]Department of Mechanical Engineering, Udayana University, Kampus Bukit Jimbaran, Bali Indonesia

[2]Doctoral Program of Enviromental Science, Udayana University, Denpasar, Bali Indonesia.

[3] Doctoral Program of Engineering Science, Udayana University, Denpasar, Bali Indonesia.

Email: suryaagus77@gmail.com

Abstract. Stinging nettle is a perennial herbaceous plant belonging to the Urticaceae family, In Bali Indonesia called jelatang. This species is considered a weed in intensive agriculture as its fast vegetative growth and high densities enable increased spread and soil coverage. In Tabanan Bali Indonesia several places to grow are in Denbantas altitude 200-500 meters and in Bedugul altitude 1000 -1500 meters above sea level. Nettle fibers are natural fibers that can be utilized as reinforcement for composite materials. Natural fibers have been used worldwide as a potential replacement for glass fiber reinforced composites, over the last few years has been applied in the automotive sector, aircraft interior and interior of the building. Properties of the natural fiber is lighter, easier to handle, not abrasive, and low cost. The objective of this study was to compare the composition and the tensile strength of the fiber nettle based growing areas from Bedugul and Denbantas. The method used is a fiber tensile test equipment, fourier transformer infrared red (FTIR), thermo-gravimetric analysis (TGA) and scanning electron microscope (SEM). From the research of nettle fibers that stinging nettle from Denbantas has a better tensile strength than Bedugul nettle. The TGA test the visible fiber from Denbantas is superior when used as a composite reinforcement.

1. Introduction

Stinging nettle plants thrive in Tabanan-Bali, Indonesia, of which its stem, leave and fiber are used by the people. Journals which discuss the use of nettle plants are; multipurpose of nettle plants for medicine, biomass and material [1]. Nettle plants are studied as antioxidant, antimicrobial, antiulcer and analgesic [2]. Nettle plants are studied as utility of its in layer diets as a natural yellow colorant for egg yolk [3]. Nettle plants are studied as adding its haylage to a total mixed ration on performance and rumen function of lactating dairy cows [4]. Nettle fiber is a natural fiber which is very strong with young's modulus 87 (±28) GPa, tensile strength 1594 (±640) MPa, strain fracture 2.11 (±0.81) % [5]. Method and the way of taking the fiber from the nettle stem highly influence the mechanical characteristic of the fiber, of which have been studied the method; decortication, chemical retting, water retting, microbiological retting, enzimatical treatment and chelating agent [6-7].

Nettle plants are covered by fine hairs especially at the leave and stem. If it is touched, these plants will release chemical substance, stinging and triggering inflammation which causes redness, itching and irritation of the skin. Wild nettle plants are considered as weeds (disturbing plants) in farm industry, which easily grow and seize food from the parent plant.

The average height of one dried sample of nettle stem taken from Bedugul was 54.2 cm and 49.4 cm from Denbantas. The average weight of one dried sample of nettle stem from bedugul was 16.45 gram and give fibers was 2.46 gram but sample Denbantas average weight was 11.31 gram and give fibers was 1.28 gram.

IOP Conf. Series: Materials Science and Engineering **539** (2019) 012006 doi:10.1088/1757-899X/539/1/012006

2. Method

2.1. Material

Stinging Nettle is harvested in Denbantas and Bedugul of Tabanan-Bali, Indonesia. Denbantas is located in altitude 200 – 500 meter above sea level, 008^0 27' 40" S – 115^0 08' 18" E with average humidity 83.12%, and rainfall 215.43 mm/month. Bedugul is located in altitude 1000 -1500 meter above sea level 008^0 15' 40" S – 115^0 09' 49" E, with rainfall 248.85 mm/month calculated in 5 years. Average temperatures in Bedugul are 22 ^{0}C at daylight and 16 ^{0}C nighttime, while average temperatures in Denbantas are 30 ^{0}C at daylight and 21 ^{0}C at nighttime.

The harvested stem of nettle plant was cut and dried for 3-5 days before it was rooted in a water for 4 days in the end of January 2017. Fiber of nettle plant was dried in a room temperature and kept in weeks. The fiber was extracted manually without damaging the fiber.

2.2. *Measurements*

The surface of fibers was coated with gold and then examined using a JOEL JSM-651OLA analytical scanning electron microscope (SEM) made in Japan.

Infrared spectra of fiber samples (4000-400 cm-1) were recorded using a Fourier Transformer Infrared (FTIR) Spectrophotometer (IRPrestige-21) Shimadzu.

Instrument of TGA701 LECO made in USA was uses for thermogravimetric analysis of fibers samples from room temperature to 600 ^{0}C at a heating rate of 10 ^{0}C/min under nitrogen gas flowing at 60 mL/min. About 1 g of sample was used in each test.

3. Result and Discussion

Characterization of nettle plants is a search of characteristic and basic nature of nettle fiber so the typical value can be discovered. In Fig. 1 and Fig. 2, it can be seen the nettle fiber from Bedugul and Denbantas respectively taken from one stem.

Table 1 – Table 4 are be authority by Badan Meteorologi, Klimatologi dan Geofisika Tuban Bali Indonesia. Physical environment such as rainfall, temperature, humidity and sunlight intensity highly influence the growth of nettle plants.

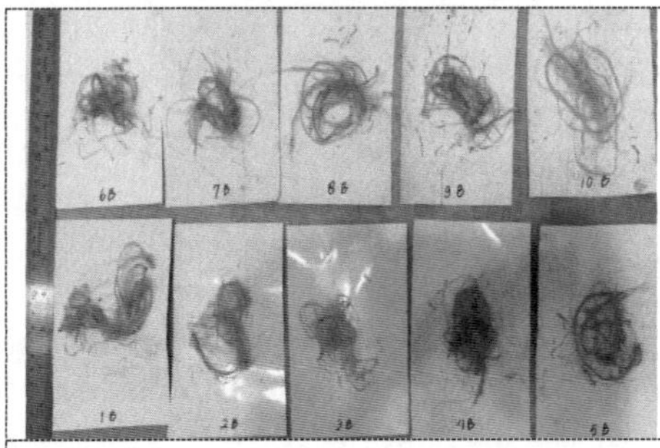

Figure 1. Fibers from Bedugul

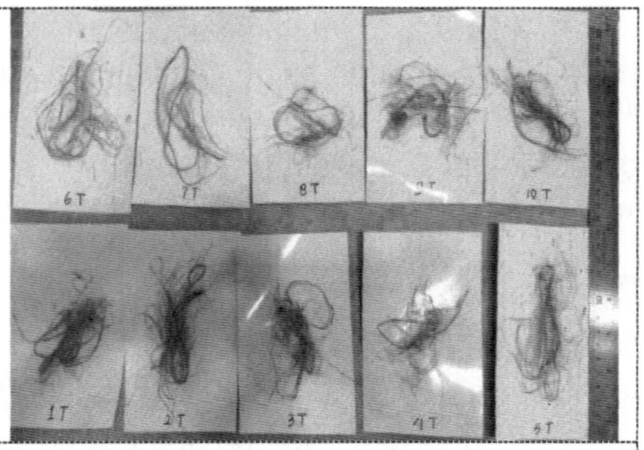

Figure 2. Fibers from Denbantas

Table 1. Monthly Rainfall in Bedugul Tabanan (millimeter).

Year	Jan	Feb	Mar	Apr	May	Jun	Jul	Aug	Sep	Oct	Nov	Des	Mean
2012	1,056.5	622.5	1,510.5	434.5	189.5	5.5	-	2.0	-	61.5	313.5	661.5	404.79
2013	542.0	766.0	244.0	293.5	177.0	111	62.0	10.0	6.0	133.0	307.1	587.0	269.93
2014	449.0	207.8	158.0	322.5	91.0	10.5	108.5	18.0	-	37.5	312.0	252.0	163.90
2015	210.0	254.0	351.0	355.0	41.0	22.0	8.0	-	-	11.0	110.0	106.0	122.33
2016	307.0	557.0	302.0	184.0	221.0	150	145.0	58.0	39.0	389.0	340.0	707.5	283.29

IOP Conf. Series: Materials Science and Engineering **539** (2019) 012006 doi:10.1088/1757-899X/539/1/012006

Table 2. Monthly Rainfall in Denbantas Tabanan (millimeter).

Year	Jan	Feb	Mar	Apr	May	Jun	Jul	Aug	Sep	Oct	Nov	Des	Average
2012	538.0	307.5	405.5	185.0	200.0	11.0	182.5	20.0	-	225.0	136.0	330.0	211.71
2013	516.5	200.5	186.0	265.5	418.0	161.5	124.0	40.0	34.0	19.0	333.0	555.0	237.75
2014	396.0	103.0	171.0	218.0	121.0	6.0	177.0	15.0	5.0	27.0	248.0	494.0	165.08
2015	356.0	243.0	218.0	90.0	162.0	174.0	68.0	10.0	2.0	11.0	289.0	132.0	146.25
2016	201.5	551.0	286.0	272.0	168.0	186.0	318.0	171.0	400.0	226.0	476.0	541.0	316.38

Table 3. The Average Humidity in Denbantas Tabanan (%).

Year	Jan	Feb	Mar	Apr	May	Jun	Jul	Aug	Sep	Oct	Nov	Des	Average
2012	85	86	83	83	84	83	83	81	80	81	82	84	82.9
2013	85	84	84	85	87	88	85	83	81	77	82	85	83.8
2014	84	83	83	84	84	84	82	81	78	79	79	86	82.3
2015	82	83	84	84	85	85	84	83	81	80	78	81	82.5
2016	79	88	82	85	85	84	86	83	82	85	85	85	84.1

Table 4. Duration of Sunlight in Denbantas Tabanan (%).

Year	Jan	Feb	Mar	Apr	May	Jun	Jul	Aug	Sep	Oct	Nov	Des	Average
2012	44	63	53	76	70	74	69	76	90	84	73	57	69.1
2013	52	57	70	54	63	58	72	77	81	92	64	47	65.6
2014	53	59	78	80	85	76	70	83	95	97	84	48	75.7
2015	58	66	66	60	79	82	82	79	89	93	85	61	75.0
2016	69	48	70	66	74	71	66	76	81	64	60	45	65.8

Table 1 shows the average of monthly rainfall in Bedugul within 5 years was 215.43 mm/month. Meanwhile, Table 2 shows the average of monthly rainfall in Denbantas within 5 years was 248.85 mm/month. Table 3 shows the average of humidity within 5 years in Denbantas was 83.12%. Meanwhile Table 4 shows the average of sunlight duration within 5 years in Denbantas was 70.23%.

3.1 TGA Test

TGA (Thermogravimetric analysis) is a heat analysis method in which the change in term of physical and chemical nature from the sample is measured in high precision. The measurement is conducted by the constant temperature escalation upon time function by calculating the mass loss.

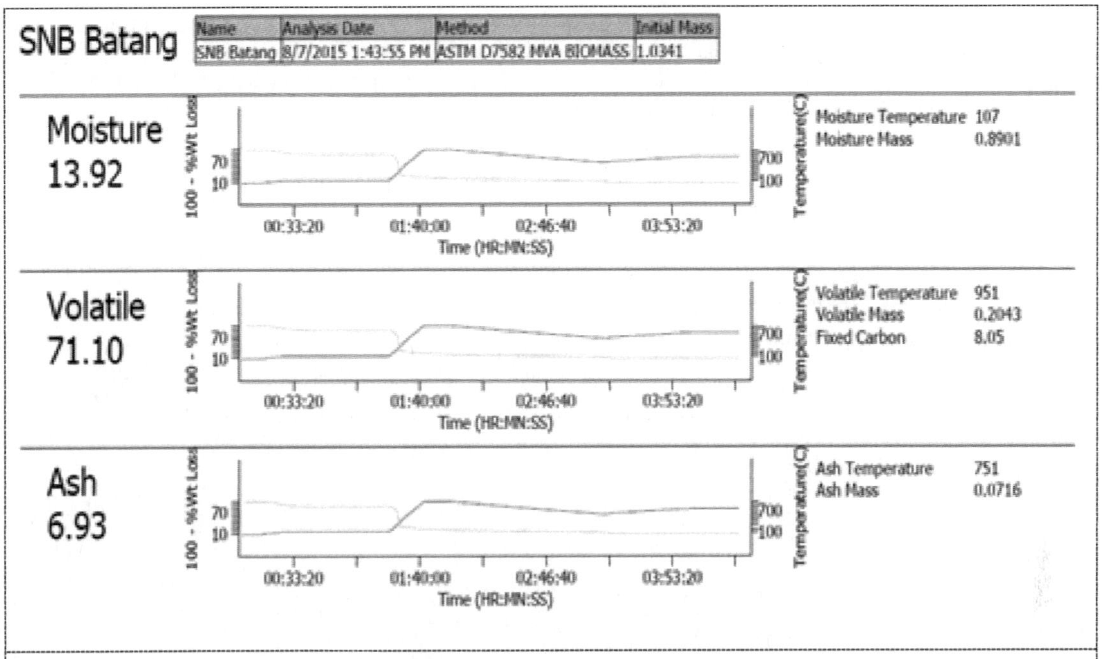

Figure 3. TGA fibers from Bedugul.

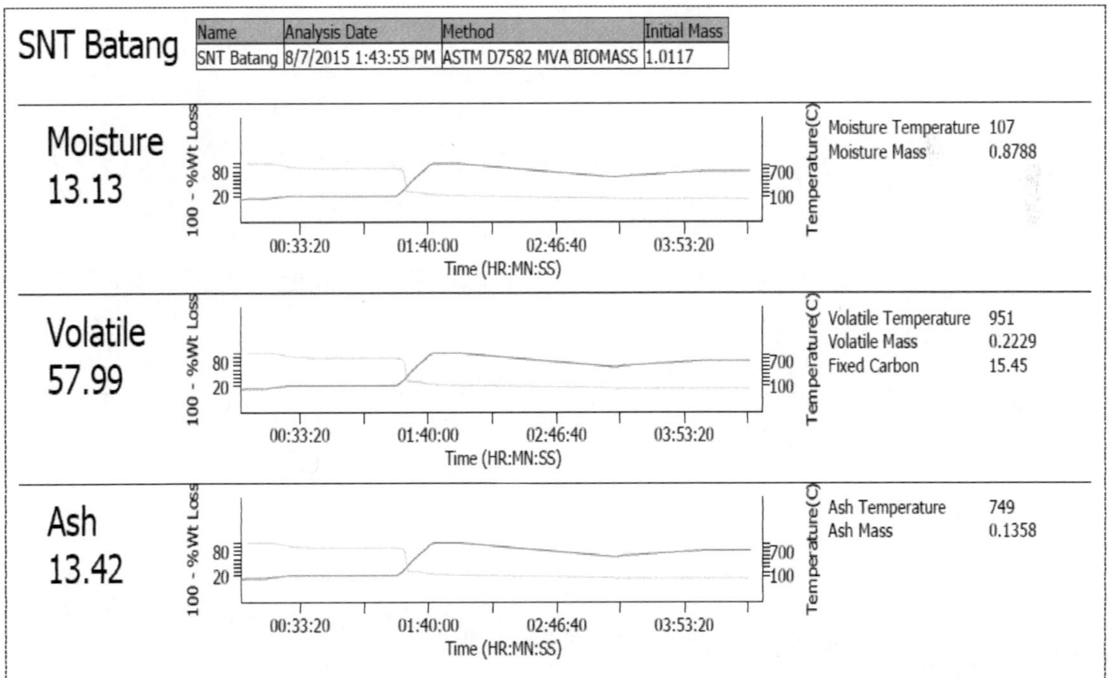

Figure 4. TGA fibers from Denbantas.

The drying of nettle stem from Bedugul took 4-5 days, while the Denbantas nettle stem took 3-4 days. The result of nettle stem TGA showed almost the same moisture values, because it was similarly conditioned that was dried under the sun until the nettle stem had the same weight after and before the drying.

IOP Conf. Series: Materials Science and Engineering **539** (2019) 012006 doi:10.1088/1757-899X/539/1/012006

Volatile value of Bedugul nettle stem was higher than from Denbantas. It showed Bedugul nettle stem had more volatile substance. Denbantas fibers has a higher fixed carbon that is 15, this is potential fiber.

Ash value of Denbantas nettle stem was higher than from Bedugul. It means that Denbantas nettle stem had more organic and anorganic materials. TGA showed the fiber of Denbantas nettle stem had better physical and chemical characteristic if it is used to strengthen composite material.

3.2 FTIR Test

FTIR (Fourier Transform Infra Red) is an infrared radiation through the sample. Some infrared radiation was absorbed by the sample and some of them passed the sample. The resulting spectrum is absorption and transmission molecule, creating traces of sample molecule, there will be two unique marks, absorbing sample molecule and sample molecule which was passed by infrared radiation.

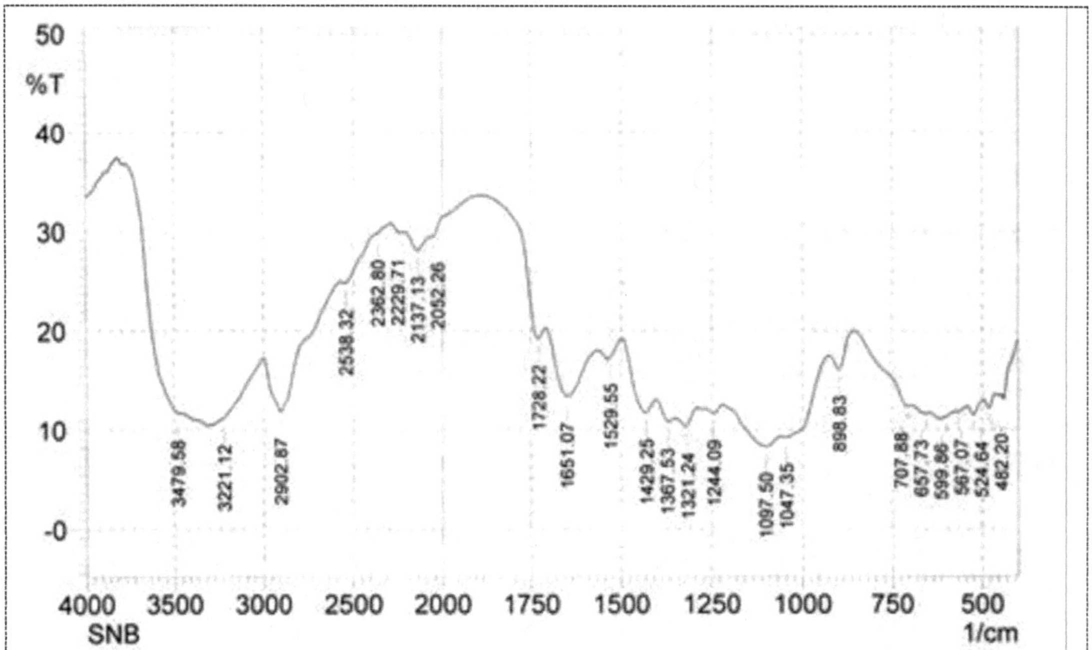

Figure 5. FTIR fibers from Bedugul.

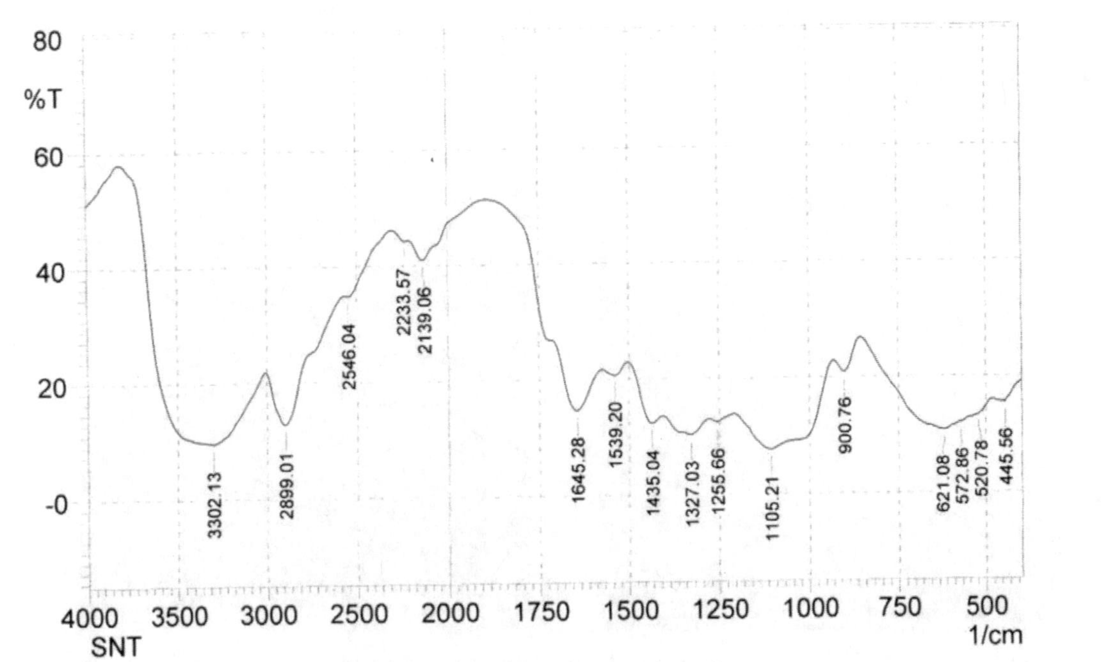

Figure 6. FTIR fibers from Denbantas.

FTIR test on the two samples of Bedugul and Denbantas nettle fiber, the absorption pattern and infrared radiation channel showed the same pattern, and it could be assured that both samples of the fiber came from the same element and compound.

Analysis on Denbantas nettle fiber was based on the biggest area so the occurred clusters were in wave number 3302.13 cm^{-1} which showed cluster NH_2, for wave number 2899.01 cm^{-1} showed cluster NH_3, for wave number 1645,28 cm^{-1} showed cluster CO and for wave number 1105.21 cm^{-1} showed cluster CH_3. These clusters were appropriate with chemical compounds for cellulose fiber.

3.3 Tensile Strength Test

Fiber tensile strength is the power of fiber to withstand the load up to break for each unit area. Single nettle fibers were glued onto a slotted paper holder (10 mm gap equal to L0 the nominal length of the fiber). Tensile tests were carried out on a MTS Tensile Manuals. Tesile test each number (code) is 30 samples.

Table 5. Tensile strength of fibers.

Code	Average (Mpa)	Standar deviasi
DB	886.0	275
4B	193.0	80
DD	900.7	247
4D	525.4	411

Code of fiber retting with decortication on the nettle stems which is harvested in Bedugul is symbolyzed by DB. Code of fiber ratting of nettle stems harvested in Bedugul with a soaked in water for 4 days is symbolyzed by 4B. Code of fiber retting with decortication on the nettle stem which is harvested in Denbantas is symbolyzed by DD. Code of fiber ratting from nettle stems harvested in Denbantas with soaked in water for 4 days is symbolyzed by 4D.

Tensile test result of Bedugul fiber on Table 5 shows the fiber taking with decortication method had averagely higher tensile strength with 866.0 MPa. Meanwhile, Denbantas fiber on Table 5 shows fiber taking with decortication method had averagely higher tensile strength with 900.7 MPa than soaked in water. If it is compared to the growth area, Denbantas fiber had higher tensile power.

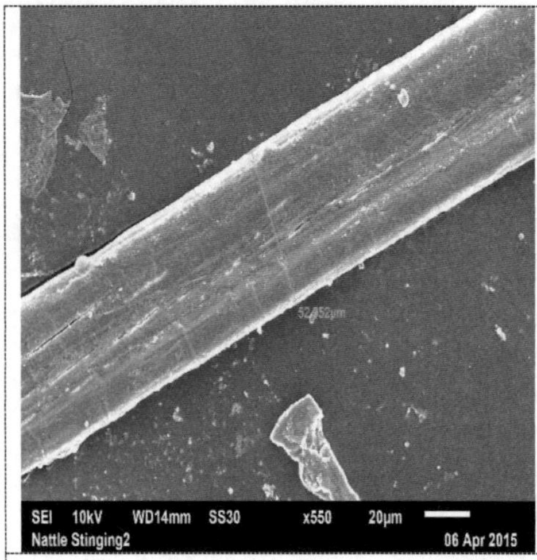

Figure 7. SEM fiber stinging nettle.

Figure 8. Plant stinging nettle Denbantas.

4. Conclusion

The potency of stinging nettle to be used as reinforcement of composite materials is highly recommended as it can be seen from the result of the characterization. From the research of nettle fibers that stinging nettle from Denbantas Tabanan has a better tensile strength, from the TGA test the visible fiber from Denbantas is superior when used as a composite reinforcement.

Acknowledgments

The authors thank the Ministry of Research, Tech., and Higher Education of the Republic of Indonesia and LPPM (Lembaga Penelitian dan Pengabdian Masyarakat) University of Udayana for supporting this research and paper through The Grant.

References

[1] N. D. Virgilio 2015 *Industrial Crops and Products* 68, 42-49
[2] Gülçin , O.I. Küfrevioglu, M. Oktay and M.E. Büyükokuroglu 2004 *Ethnopharmacology* 90205–15
[3] Y. Loetscher, M. Kreuzer and R.E. Messikommer 2013 *Animal Feed Science and Technology* 186 158– 68
[4] D.J. Humphries and C.K. Reynold 2014 *Animal Feed Science and Technology* 189 72–81
[5] E. Bodros, C. Baley 2008 *Materials Letters* 62 2143–45
[6] L. Bacci, S. Baronti, S. Predieri, N.D. Virgilio 2009 *Industrial Crops and Products* 29 480–84
[7] L. Bacci, S.D. Lonardo, L. Albanese, G. Mastromei and B. Perito 2010 *Textile Research Journal* 81 827–37
[8] H. Fischer, E. Werwein, and N. Graupner 2012 *Composite Materials* 46 3077–87
[9] M. Akgul 2013 *Composites: Part B* 45 925–29
[10] I G P Agus Suryawan et al *2017 IOP Conf. Ser.: Mater. Sci. Eng.* 201 012001
[11] C I P K Kencanawati et al *2017 IOP Conf. Ser.: Mater. Sci. Eng.* 201 012002

IOP Conf. Series: Materials Science and Engineering **539** (2019) 012007 doi:10.1088/1757-899X/539/1/012007

Investigation and analysis on the performance of hydraulic ram pump at various design its snifter valve

M Sucipta[1] and M Suarda[1]
[1]Mechanical Engineering Department, Faculty of Engineering, Udayana University, Indonesia

Email: m.sucipta@unud.ac.id

Abstract. Hydraulic ram pumps are equipped with air vessel to reduce the pulsation of the pumping water flow and the acceleration head loss. In their operation, the amount of air available in the air vessel will be decreased continuously flowing out with delivery water. In order to replace the air volume in the air vessel, the hydraulic ram pump is equipped with a snifter-valve. However, in its implementation, this valve is a hole or orifice with a diameter of 2 mm at a distance of about 2 cm below the delivery valve on the body of the hydraulic ram pump. On the other hand, this will reduce the water hammer pressure that occurs, furthermore, it reduces the performance of the hydraulic ram pump. Based on its working principle, a snifter-valve is an automatic anti-vacuum valve or a non-return valve or check valve. Therefore, the work has been carried out to investigate and analyze the performance of a hydraulic ram pump without a hole or any snifter-valve, with a hole of 1 mm diameter, and equipped with a check-valve of 1 mm orifice diameter. The results showed that the snifter-valve with check-valve of 1 mm orifice diameter gives the best hydraulic ram pump performance for the hydraulic ram pump model that has been investigated.

1. Introduction

The hydraulic ram pump is a device that utilizes available energy in the water stream to pump some of the water to a higher elevation without using electric or fuel [1,2]. The hydraulic ram pump was first designed in England at the end of the 18th century by John Whitehurst, originally called the "pulsation engine". The next generation design was done by Joseph Michel Montgolfier and patented in 1797. Hydraulic ram pumps in the history of its development are used as irrigation equipment to provide water at the locations where the electrical energy is unavailable. The main advantage of the hydraulic ram pump is its simple structure, using renewable and reliable sources of energy.

The components of a hydraulic ram pump are water supply tank, drive pipes, pump body, waste or impulse valve, delivery valve, snifter valve, air chamber, and delivery pipe [3]. The working mechanism of the waste valve results in the occurrence of a water hammer when the waste valve closes suddenly. This water hammer pressure energy is utilized in the hydraulic ram pump cycle [4]. The waste valve primarily affects the magnitude of the impulse force generated [5]. At the operation of the hydraulic ram pump, within a certain period of time the air volume in the air chamber will gradually decrease out with the pumped water, therefore it needs to take the air supply into the pump body through a snifter-valve.

In its implementation, this snifter-valve valve is generally only a hole with $1 \div 2$ mm diameter at a distance of about 2 cm below the height of delivery valve of the hydraulic ram pump body, as shown in Figure 1. This will result in a decrease of water hammer pressure due to a leakage of flow through the hole, and then it will also be able to reduce the performance of the hydraulic ram pump. Based on the working concept of the snifter-valve is an automatic anti-vacuum valve which will close during the compression phase and open when pressure back takes place in drive pipe at the recoil phase. Therefore, the concept of this valve should be a one-way valve (non-return valve or check valve). Hence, a non-return valve design has been investigated in this research. A ball shape design is chosen considering the ball is easier to move open and close than the form of a disc.

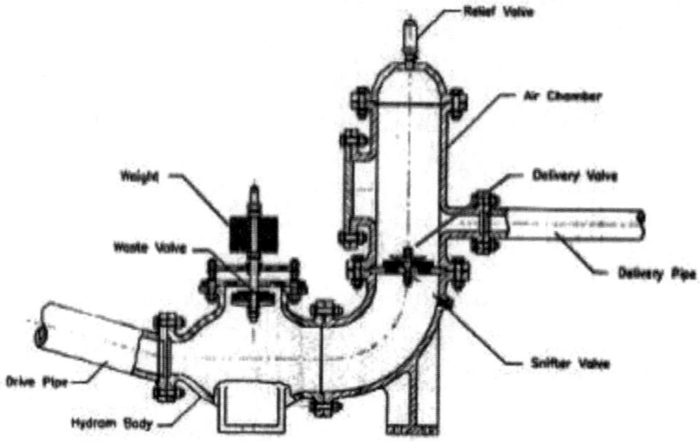

Figure 1. The main components of the hydraulic ram pump [6].

Figure 2. Orifice/hole of snifter-valve on the hydraulic ram pump.

In the second millennium, this hydraulic ram pump has been widely used by the communities. Suarda and Sukadana have implemented hydraulic ram pumps with membrane press valves for pumping water for community service (see Figure 2) in Belantih village, Kintamani sub-district, Bangli district, Bali province [7].

In addition, the hydraulic ram pump is constructed using a flange and bolt-nut connection system to simplify the assembly and disassembly, especially when repairing the valves. In line with the construction design, Yang proposed a novel design of hydraulic ram pumps with the note that it is advisable not to use a conical enlargement on the valve portion of the waste but cambered diffuser with an angle between 25° and 90°, to avoid the asymmetric pressure distribution on the disk valve waste [8]. Given the hydraulic ram pump, there are only two moving parts, namely waste valve and delivery valve. Their design determines the performance of the hydraulic ram pump. Therefore, Suarda has examined the forces that occur on both valves with algebraic analysis [9]. In general, methods that can be used in the main considerations of planning and application in various ways have been proposed by Balguda [10]. However, according to Nambiar, the waste valve is a key component of the hydraulic ram pumps, hence, it is requiring attention in the development and further optimization in improving the total efficiency of the hydraulic ram pump system [11].

Grygo develops mathematical equations of the performance stream of a water ram with regard to the height of water delivery [12]. The main aim of this work is to define the mathematical relationship that allows determination of the impact of the height of water on the performance of water ram. the obtained mathematical dependence (regression equation) shows that by increasing the height of water delivery it will reduce the performance. Then, Harith et al study how to determine the effect of improved design to significant effect on flow analysis and simulation study [13]. They confirmed that by adding a control mechanism to the new design component delivery and waste valve have enhanced about 20% more efficient than the current design. Moreover, they study on waste and delivery valve design modification to the pump performance [14]. The results of the investigation show that the projection of the velocity vector for every details case and pressure contour.

In addition, Mondol built a design, manufacture and test a hydraulic ram [15]. Then, he observed that if the water in the tank decreased the time taken between strokes increased and the pump ceased to work. Shortening the stroke increased the frequency of strokes. Thus, a certain amount of water has to be maintained in order to pump water via the ram pump satisfactorily. Furthermore, Guo et al develop a method for the optimal design and performance analysis of a hydraulic ram pump with numerical simulation and model experiment [16]. They argue that when the delivery head is less than 50 m, the efficiency of the new product ranges from 50% to 70% while the delivery flow is the largest.

Therefore, it is necessary to conduct research to investigate the performance of the hydraulic ram pumps without and/or equipped with snifter-valve. This research is very important because of the lack of research and development of the hydraulic ram pump design, especially the reference in determining the dimensions of its valves. With a more appropriate snifter-valve design, it is expected to produce more optimal hydraulic ram pump performance. Furthermore, with the availability of the design is expected to develop the manufacture and utilization of hydraulic ram pumps in the community.

2. Research Methods

2.1. Experimental Setup

Considering that in its implementation in community, the snifter-valve of hydraulic ram pump is only a small hole with a diameter of about 1 ÷ 2 mm, hence, the research was conducted to compare the performance of the hydraulic ram pump system which is unequipped with a snifter-valve, equipped with snifter-valve which is only a 1 mm diameter hole and a snifter-valve design/model of non-return valve with 1 mm orifice diameter, as shown in Figure 3. The investigation will be conducted experimentally on the hydraulic ram pump model that has been installed in the laboratory of Energy Conversion of Mechanical Engineering Department of Udayana University.

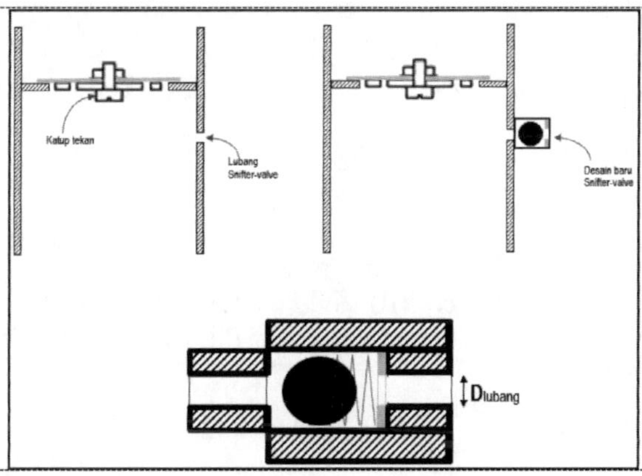

Figure 3. Snifter-valve design on the hydraulic ram pump.

In this experiment, the work was performed on the hydraulic ram pump installation system as shown in Figure 4. This research was conducted on the hydraulic ram model with the following specifications:

1) The supply water level is a drive head, Zs = 1.82 m
2) Delivery head Zd = 2.02 m, the addition of variation of the delivery head is adjusted by arranging the control valve opening on the delivery pipe.
3) Drive pipe diameter Ds= 36 mm with length Ls = 12.2 m.
4) The diameter of delivery pipe Ds=12 mm with a length of 1.36 m.
5) The diameter of the waste valve disk is 47 mm, and the stroke length is 6.0 mm.
6) Snifter-valve model: no hole, 1 mm diameter hole, and a non-return valve with 1 mm orifice diameter.
7) Waste valve mass (moving parts) 490 grams
8) The volume of the air chamber is 0.0083 m^3

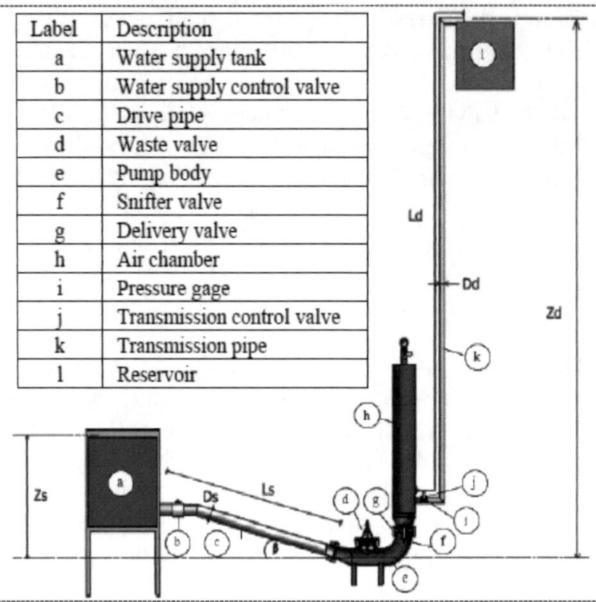

Label	Description
a	Water supply tank
b	Water supply control valve
c	Drive pipe
d	Waste valve
e	Pump body
f	Snifter valve
g	Delivery valve
h	Air chamber
i	Pressure gage
j	Transmission control valve
k	Transmission pipe
l	Reservoir

Figure 4. Installation scheme of the hydraulic ram pump system.

2.2. Experimental Procedure

The V-notch weir discharge equation used is the Kindvater-Shen equation [17], i.e.:

$$Q = \tag{1}$$

Where Q is the water flow rate (cfs), C is the flow coefficient, θ is the V-notch Weir angle, h is the hight of the water flow through V-notch Weir (ft), and k is the V-notch Weir flow correction factor.

The general method used to calculate the efficiency of the hydraulic ram pump installation is the Rankine method [18]:

$$\eta_R = \frac{Q_d H_d}{(Q_d + Q_w) H_s} 100\% \tag{2}$$

Where η_R is the total efficiency of the hydraulic ram system (%), Qw is the water outflow from the waste valve (m^3/s), Qd is the pumping water flow rate (m^3/s), Hs is the head of the driving water supply (m), and Hd is the pumping head (m).

Investigating the snifter-valve in the pump hydraulic ram system in the laboratory is carried out through the testing stages as follows:

1) Preparation of pumps and valve models, with water supply height Zs = 1.82 meters, the diameter of the drive pipe Ds = 1¼ inch and the length of Ls = 9 meters.
2) Setup hydraulic ram pump without hole or snifter-valve.
3) Setup of pumping height Zd = 2.02 meters
4) Flowing the water into the drive tank to full and be kept in an always-over-flow condition so that the height of the driving head is constant
5) Open the water supply control valve, to drain the water from the supply tank to the pump body.
6) Start the pump work by opening and closing the waste valve so that the pump can work and let the pump run for a while
7) Setup head of pumping pressure (Pd) of 0.5 bars
8) While the pump is working and it has been stable do data recording such as height of the water flow on V-notch weir (hvw) coming out from the waste valve, time for collecting of 1 liter of water volume coming out of the delivery pipes (Td), pressure on the delivery pipe manometer (Pd), and the frequency of the pump work cycle (F).
9) Repeat step 8 for 3 (three) times

IOP Conf. Series: Materials Science and Engineering **539** (2019) 012007 doi:10.1088/1757-899X/539/1/012007

10) Repeat steps 7 through 8 for pumping pressure heads (Pd): 1.0, 1.5, 2.0, 2.5, and 3.0 bars
11) Repeat steps 3 through 10 for a snifter-valve design/ model of 1 mm diameter hole, and the non-return snifter valve with a 1 mm orifice diameter.

3. Results and Discussions

Based on the data of the results of the experiment, then the performance of hydraulic ram pumps on the variation of the design of the snifter-valve hydraulic ram pumps tested at various ratios of pumping head to the static head of the hydraulic ram pump water supply can be illustrated as in Figure 5 to Figure 8.

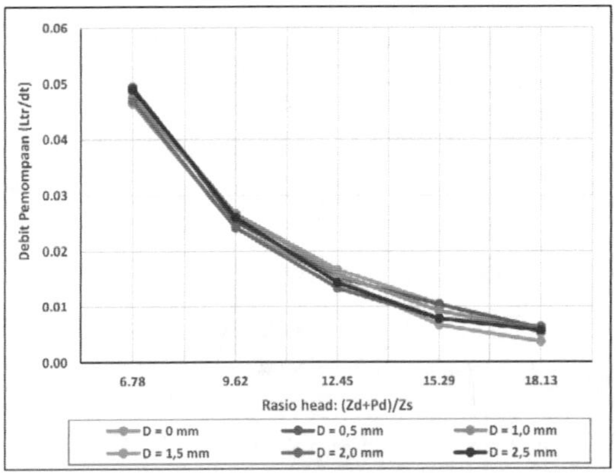

Figure 5. Pumping discharge at various of snifter-valve hole diameter.

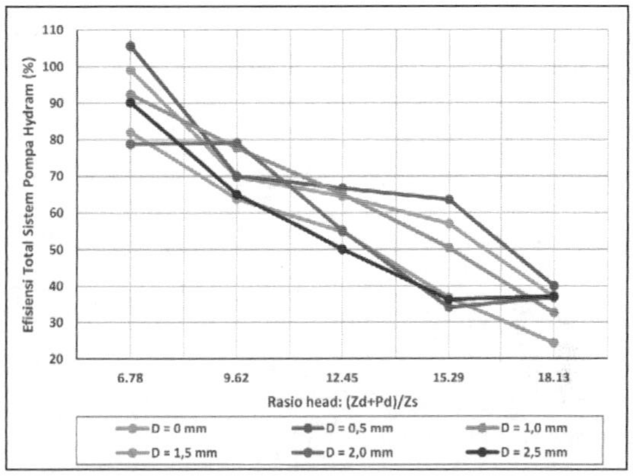

Figure 6. The total efficiency at various of snifter-valve hole diameter.

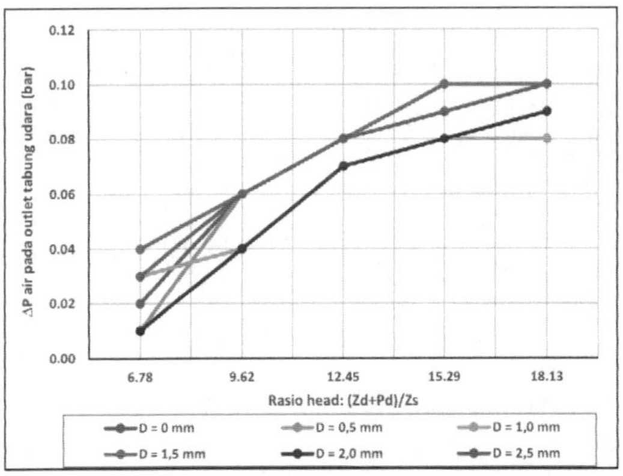

Figure 7. Pressure fluctuation of water at air vessel outlet.

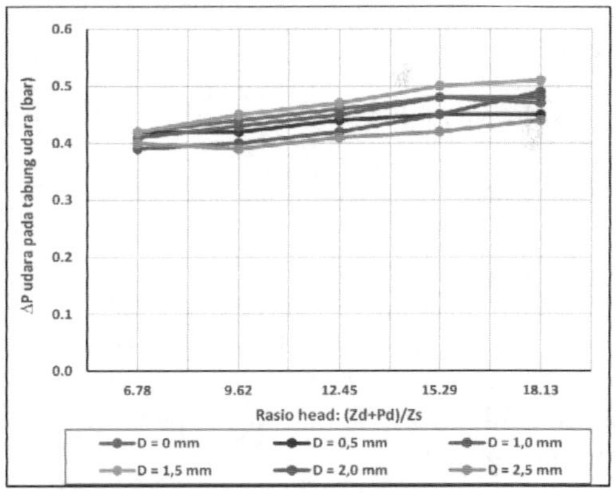

Figure 8. Pressure fluctuation of air in the top of air vessel.

Figure 5 shows that in general, the higher the pumping pump hydraulic pumping pump's pumping head is getting smaller, and the larger the snifter-valve pumping diameter of the pumping pump is relatively smaller. However, snifter-valve with a 0 mm (without hole) provides the highest pumping discharge. However, it should be noted that for pump operation within a certain period of time the amount of air in the press tube will be discharged and discharged thus disrupting the pump operation even stop operating. For that 1mm diameter, snifter-valve is the best alternative choice. Furthermore, Figure 6 shows that the higher the pumping head the smaller the total efficiency of the hydram pump generated. In addition, the smaller the Snifter-valve hole diameter the higher the total efficiency produced. Considering the function of the snifter-valve, then, snifter-valve with hole diameter 0.5 mm to 1.0 mm is the best alternative choice.

The larger the pumping head causes the slower the pumping cycle occurs, and the larger the diameter of the Snifter-valve hole results in a decrease in the frequency of the hydram pump cycle, as shown in Figure 7. This is due to the reduced back-pressure occurring in the moving pipe as a result of the greater pressure leakage water through the snifter-valve. Then the higher the pumping head results in the greater pressure fluctuation that occurs in the tube outlet tube to the delivery pipe, due to the higher pressure heads that occur in the compressive tube, as shown in 8. In addition, the small diameter hole-shaped snifters relatively result in fluctuations greater pressure.

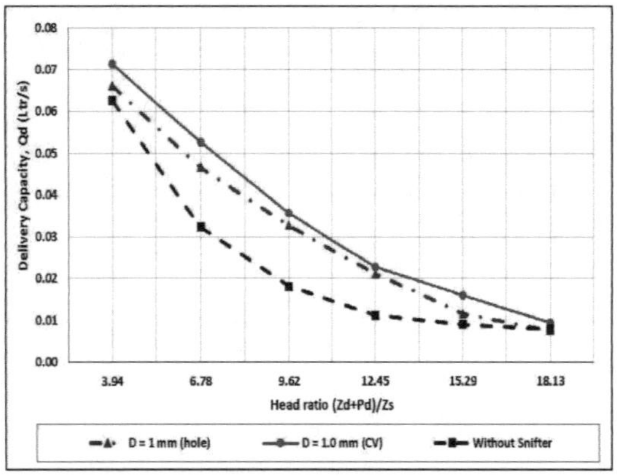

Figure 9. Pumping discharge.

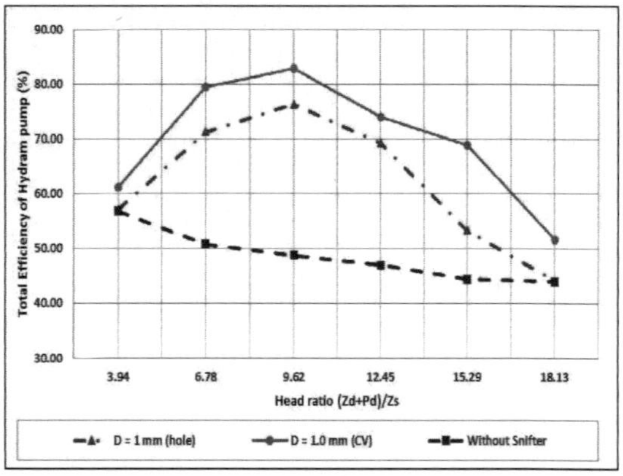

Figure 10. The total efficiency.

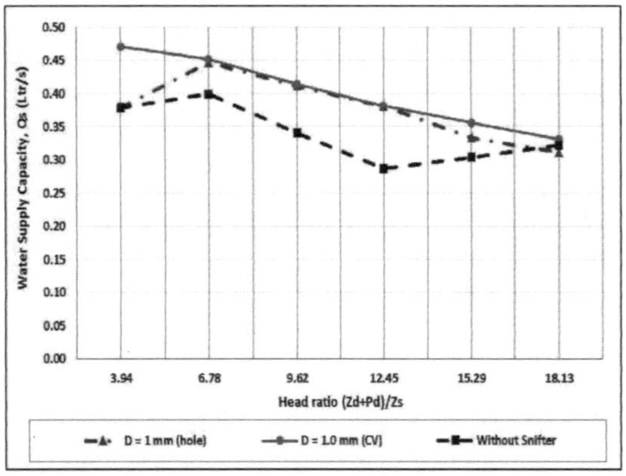

Figure 11. Water supply to drive the hydraulic ram pump.

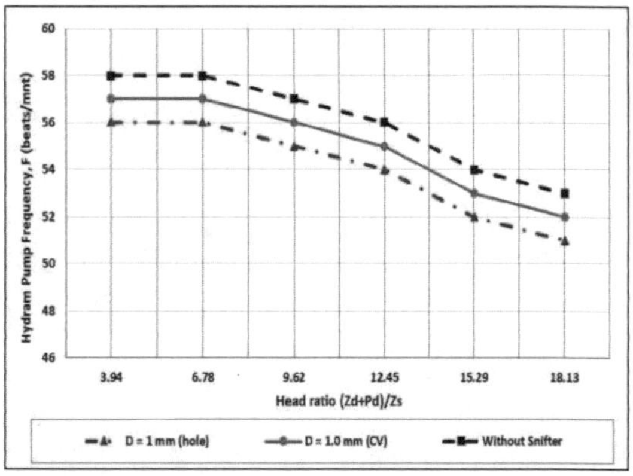

Figure 12. The frequency of the hydraulic ram pump cycle.

Figure 9 shows the effect of the snifter-valve design on the pumping discharge of the hydraulic ram pump. In general, the higher delivery head then pumping capacity of the hydraulic ram pump is getting smaller. This is because of the higher pumping head requires the greater of pumping power. The non-return snifter-valves (CV) with 1 mm orifice diameter results in greater pumping discharge compared to snifter-valve which is only a hole of the same diameter of 1 mm. It is due to leakage of flow through the hole during the compression phase. Furthermore, hydraulic ram pumps which are without equipped with snifter-valve produce the smallest pumping discharges, because most of the water hammer pressure occurs to cause higher back pressure on the drive pipe.

Figure 10 shows the effect of the snifter-valve design on the total efficiency of the hydraulic ram pump system. The hydraulic ram pump which is without equipped with snifter-valve has the lowest total efficiency, due to the minimal pulsation in the pump body. A 1 mm orifice diameter of the non-return snifter-valve provides greater total efficiency compared to the 1 mm diameter of the only hole of snifter-valve, because of there is no loss of pressure through the orifice of the snifter-valve at the time of compression phase. Furthermore, the highest efficiency of the tested hydraulic ram pump system is achieved at a pumping head ratio of 9.62. This means that if the head or high water source of the hydraulic ram pump is 2 meters then the highest efficiency is achieved at the head or pumping height of about 20 meters.

Moreover, Figure 11 shows that the hydraulic ram pump which is without equipped with snifter-valve requires the least discharging of the pump driving water flow, since in addition to the absence of waste water flow through the snifter-valve's hole, also due to the high pressure back on the drive pipe resulting in obstacles water flow on the drive pipe during acceleration phase.

The hydraulic ram pump cycle of the hydraulic ram pump which is without equipped with a snifter valve is the highest, due to the highest pressure back on the drive pipe resulted in delayed acceleration of water flow in the drive pipe, as shown in Figure 12 Conversely, the hydraulic ram pump equipped with a hole of snifter-valve produce the lowest pumping cycle due to the lowest pressure back that occurs in the drive pipe. While the hydraulic ram pump equipped with non-return snifter-valve model produces a pumping cycle that occurs between the without any snifter-valve and the hole snifter-valve model. In addition, the higher the pumping head will result in decreasing frequency of the hydraulic ram pump cycle.

As a result, a snifter-valve design with non-return valve system (CV) provides significantly improved pump performance. This is because it can optimize the utilization of water hammer pressure that occurs during the compression phase and the external air suction process during the recoil phase. However, further study is needed over a longer period of time, for instance, one week or more to determine the reduction or addition of air volume in the air chamber of the hydraulic ram pumps.

4. Conclusion

The snifter-valve model with non-return valve system has significantly improved the performance of the hydraulic ram pump compared to the only hole of snifter-valve and/or without any snifter-valve. However, further work is necessary to study the performance of hydraulic ram pump which is equipped with non-return snifter-valve model on the variation of orifice diameter in order to get the design reference of the snifter-valve sizing on various dimensions of hydraulic ram pump.

Acknowledgment

The authors would like to express their gratitude to the Faculty of Engineering of Udayana University for the fund support on the Unggulan Udayana research scheme for the fiscal year 2018.

References

[1] E. Celik, "Interactive Ram Pump Display Project", *in International Mechanical Engineering Congress & Exposition - IMECE2012*, November 9-15, 2012, Proceedings of the ASME 2012, (The American Society of Mechanical Engineers, Houston, Texas, 2012), pp 1-10.

[2] M. Inthachot, S. Saehaeng, J.F.J. Max, J. Müllerc and W. Spreer, "Hydraulic Ram Pumps for Irrigation in Northern Thailand", *in Agriculture and Agricultural Science Procedia* Vol. 5 (2015), (Elsevier B.V., 2015), pp.107-114.

[3] P. Diwan, A. Patel and L. Sahu, "Design and Fabrication of Hydraulic Ram with Methods of Improving Efficiency", *in International Journal of Current Engineering and Scientific Research (IJCESR)*, Vol. 3, No. 4, (Technical Research Organization India, 2016), pp. 5-13.

[4] A. Pathak, A. Deo, S. Khune, S. Mehroliya and M.M. Pawar, 2016. "Design of Hydraulic Ram Pump", *in International Journal for Innovative Research in Science & Technology*, Vol 2, No. 10, (IJIRST, 2016), pp. 290-293.

[5] A. Deo, A. Pathak, S. Khune and M.M. Pawar, "Design Methodology for Hydraulic Ram Pump", *in International Journal for Innovative Research in Science & Technology*, Vol 5, No. 4, (IJIRST, 2016), pp. 4737- 4745.

[6] A.A. Tessema, "Hydraulic Ram Pump System Design and Application", *in ESME 5th Conference on Manufacturing and Process Industry.* September 2000, (ESME, Addis Ababa, Ethiopia, 2000).J. van der Geer, J.A.J. Hanraads, R.A. Lupton, The art of writing a scientific article, J. Sci. Commun. 163 (2000) 51-59.

[7] M. Suarda, dan I G.K. Sukadana, "Perancangan dan Pengujian Unjuk Kerja Katup Tekan Pompa Hydraulic ram Model Plat, Membran, Bola dan Setengah-Bola", *Prosiding: Seminar Nasional Tahunan Teknik Mesin (SNTTM – XII)*. Hal. 387 - 394. Bandar Lampung, 22-23 Oktober 2013.

[8] K.L. Yang, at. al., "Design and Hydraulic Performance od a Novel Hydraulic ram", *11th International Conference on Hydroinformatics*, New York City, 8 January 2014, paper no 108.

[9] M. Suarda, "Forces Analysis on a Spherical Shaped Delivery Valve of Hydraulic ram Pump", *in Applied Mechanics and Materials*, Vol 776, (Trans Tech Publications, Switzerland, 2015), pp 377-383.

[10] R.D. Balguda, et al., "Designing of Hydraulic Ram Pump", *in International Journal of Engineering and Computer Science*, Vol 4, No 5, (IJECS Publication, Madhya Pradesh, India, 2015), pp. 11966-11971.

[11] P. Nambiar, A. Shetty, A. Thatte, S. Lonkar and V. Jokhi, 2015. "Hydraulic Ram Pump Maximizing efficiency", *in International Conference on Technologies for Sustainable Development (ICTSD)*, 4-6 Feb. 2015, Mumbai – India, (IEEE Explore Digital Library, 2015).

[12] Da. Grygo, "Effect of the Height of the Delivery Water on Performance of Water Ram", *Technical Sciences*, 2016, 19(2), pp. 139–149.

[13] M. N. Harith, R. A. Bakar, D. Ramasamy and M. Quanjin, "A Significant Effect on Flow Analysis & Simulation Study of Improve Design Hydraulic Pump", *in 4th International Conference on Mechanical Engineering Research (ICMER2017)*, IOP Conf. Series: Materials Science and Engineering 257 (2017) 012076., IOP Publishing.

[14] M. N. Harith, R. A. Bakar, D. Ramasamy, K. Kardigama and M. Quanjin, "A Study of Waste and Delivery Valve Design Modification to the Pump Performance", *in iCITES 2018*, IOP Conf. Series: Materials Science and Engineering 342 (2018) 012090, IOP Publishing.

[15] S.S. Mondol, "Design, manufacture and test a hydraulic ram", *ResearchGate Artikel*, May 2017.

[16] X. Guo, J. Li, K. Yang, H. Fu, T. Wang, Y. Guo, Q. Xia and Wei Huang, "Optimal Design and Performance Analysis of Hydraulic Ram Pump System", *Proceeding of Institution of Mechanical Engineering* 2018, Part A: J Power and Energy 0(0), pp. 1-15.

[17] LMNO Engineering, "V-Notch Weir Discharge Calculator and Equations", *(http://www.lmnoeng.com/weirs/vweir.php, 2015)*.

[18] T. Taye, "Hydraulic Ram Pump", *in Journal of the ESME*. Vol. II, July 1998, (ESME, Addis Ababa, Ethiopia, 1998).

IOP Conf. Series: Materials Science and Engineering **539** (2019) 012008 doi:10.1088/1757-899X/539/1/012008

Investigation on flow pattern in a hydraulic ram pump at various design and setting of its waste valve

M Suarda[1], M Sucipta[1] and I G K Dwijana[1]
[1]Mechanical Engineering Department, Faculty of Engineering, Udayana University, Indonesia

Email: made.suarda@unud.ac.id

Abstract. A hydraulic ram pumps are simple structures, consisting of two moving parts, the waste valve and the press valve. The waste valve mainly affects the magnitude of the impulse force generated. However, interaction of all elements of hydraulic ram pump system influence each other makes its operation becomes complex. Moreover, a detailed description of this hydraulic ram pump has not been well understood. In order to get the hydraulic ram pump designs, it needs to recognize comprehensively the water flow phenomenon that take place in the hydraulic ram pump working cycles. Therefore, visualization of the water flow characteristic in the hydraulic ram pump is done by constructing a hydraulic ram pump model using a transparent material that is acrylic. The characteristics of the waste valve are affected by parameters such as the orifice diameter of the valve, the diameter of the valve disk, the step length, and the valve weight and the discharge and water head of the hydraulic ram pump drive. Besides all these parameters affect each other. Therefore, in this experiment, experimental testing was conducted in the laboratory on hydraulic ram pump waste valve model to investigate the effect of waste valve orifice and disk diameter on hydraulic ram pump. The results showed that the diameter of orifice and disk of the waste valve have significantly influencing the flow pattern in the hydraulic ram pump body and its time movement each stage of the cycle. The greater the diameter of the orifice and the disk of the waste valve the velocity of water that has vortex more and more, so the speed of water flow in the body of the pump also decreases. Furthermore, the larger the diameter of the waste valve disk the shorter the interval of the hydraulic ram pump valve movements, so the frequency of the hydraulic ram pump valve movement increases.

1. Introduction

A hydraulic ram pump is a water pump powered by water with a height difference, to pump some of the falling water to a higher elevation than it originated [1]. As such, the hydraulic ram represents a renewable energy device [2]. It operates around two simple valves. Therefore, it is theoretically adaptable for manufacture in developing countries, and maintenance at village level.

In addition, the hydraulic pump system is very simple, consisting of two types of piping (drive and delivery pipe), pumping body, two valves (waste and delivery valve), and air vessel [3]. Therefore, ideally, the different combinations of head and discharge of the source water flow, the valve stroke and the weight of the valve, the ratio of length to the diameter of the driving pipe, and the volume of the air vessel are parameters of the hydraulic ram pump design [4]. The design of the waste valve and the delivery valve are essential in hydraulic ram performance [5]. Moreover, the valve is a key component of the hydraulic ram pump, which requires further focus and optimization to improve the overall efficiency of the pump [6].

Basically, the both valves of hydraulic ram pump, impulse valve and delivery valve [7], are non-return valves, as in Figure 1. The impulse valve is a simple valve that is closed by the drag force induced by a high through flow. The geometry of the valve is such that these drag forces increase rapidly as the valve moves towards its closed position, then inducing a rapid "sudden" closure. In addition, the sudden closure of the impulse valve induces a pressure rise in the drive pipe that is proportional to the velocity of the fluid in the pipe immediately before the valve closure. This pressure is maintained while pressure waves propagate along the drive pipe. During this period in which the drive pipe sustains a high pressure, a small discharge occurs through a non-return valve (delivery valve) into a vessel containing air at a pressure approximating to the delivery pressure of the pump. This discharge continues until such time as the pressure in the drive pipe subsides, at which stage, the non-return valve closes, and the discharge ceases [8,9].

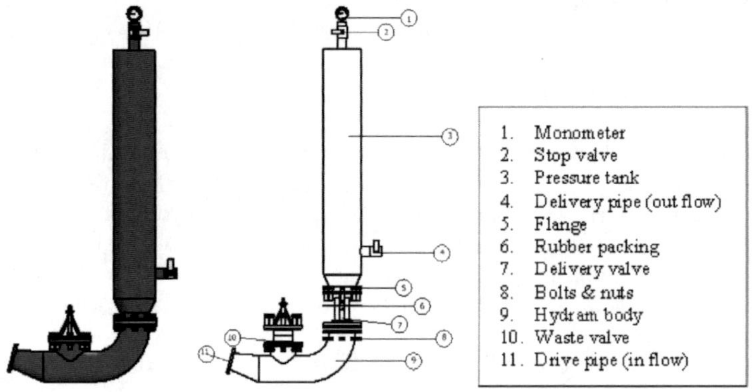

Figure 1. Components of a hydraulic ram pump.

The hydraulic ram pump has been designed using a conventional plat plate shaped delivery valve. Due to it has large dimension and works on high head, according to Pascal's law then its delivery valve receives a large water hammer force that proportional to the area of the plate valve, so it is very easily damaged or less reliable. Then, various shaped valve model for Hydraulic ram pump have been made and tested [10]. They are plate, membrane, spherical and half-spherical shaped. The test results showed that the spherical shaped delivery valve model produces the best volumetric efficiency, followed by membrane, half-spherical shaped, and the plate model gives the lowest efficiency.

The flow behind a sphere and a disk has been investigated experimentally with flow visualization and Particle Image Velocimetry [11]. They have observed in the two cases how the steady axisymmetric flow with toroid recirculation zone behind the body bifurcates to a steady flow with a planar symmetry containing two longitudinal counter-rotating vortices. A further instability leads to oscillation of these vortices, and for larger Reynolds numbers, to unsteady flow with regular hairpin shedding [12, 13, 14]. A significant research effort has been made into studying the flow past a stationary sphere over a wide range of Reynolds numbers both experimentally [15] and numerically [16]. Furthermore, Numerical solutions for the flow past stationary and rotating spheres at varies Reynolds number are calculated and performed using the finite volume solver CDP [17. The effect of rotating the sphere in either the stream wise or transverse direction on the wake structures and hydrodynamic forces are analysed.

Grygo develops mathematical equations of the performance stream of a water ram with regard to the height of water delivery [18]. The main aim of this work is to define the mathematical relationship that allows determination of the impact of the height of water on the performance of water ram. The obtained mathematical dependence (regression equation) shows that by increasing the height of water delivery it will reduce the performance. Then, Harith et al study how to determine the effect of improved design to significant effect on flow analysis and simulation study [19]. They confirmed that the by adding control mechanism to the newly design component delivery and waste valve have enhance about 20% more efficiency than current design. Moreover, they study on waste and delivery valve design modification to the pump performance [20]. The results of the investigation show that the projection of velocity vector for every details case and pressure contour.

In addition, Mondol built a design, manufacture and test a hydraulic ram [21]. Then, he observed that if the water in the tank decreased the time taken between strokes increased and the pump ceased to work. Shortening the stroke increased the frequency of strokes. Thus, a certain amount of water has to be maintained in order to pump water via the ram pump satisfactorily. Furthermore, Guo et al. develop a method for the optimal design and performance analysis of hydraulic ram pump with numerical

IOP Conf. Series: Materials Science and Engineering **539** (2019) 012008 doi:10.1088/1757-899X/539/1/012008

simulation and model experiment [22]. They argue that when the delivery head is less than 50 m, the efficiency of the new product ranges from 50% to 70% while the delivery flow is the largest.

Detailed mechanism of hydraulic ram pump operations are not well understood yet. Therefore, Taye [8] conducted a laboratory test of a hydraulic ram pump to determine some parameters associated to hydraulic ram pump operation. A draft waste valve designed such that the weight and its stroke can be varied depending on the head of driving water supply. Furthermore, Filipan et al. [23] create a mathematical model of hydraulic ram pump system and explain the simplified working cycle of hydraulic ram pump. Because of the flow of water in the hydraulic ram pump system is unsteady, then the equation of unsteady flow in a pipe and method of characteristic (MOC) is used.

In the latest of this decade, CFD (Computational Fluid Dynamics) simulation methods tends to be used, as was done by Maw and Hte [24]. They created hydraulic ram pump design then simulated the phenomenon of water flow through a waste valve with commercial CFD software that is Solid Work software. However, this paper only show the pressure distribution and velocity on the waste valve, as shown in Figure 1, and did not identify any waste valve design influence on the hydraulic ram pump performance. Furthermore, analysis of the hydraulic ram pump performance using CFD was conducted by Shende and Choudary [25]. Analysis of the hydraulic ram pump performance was done by using one of the commercial CFD software ie. Fluent. The simulation results show that the mass flow rate of water through the new design of the waste valve is smaller than the old model, and provide a better performance of hydraulic ram pump system. The analytical calculations can be made further simple by writing a computer programme which will help to determine the optimum solution. Modelling and analysis can be carried out using Ansys Fluent software which determines the various design parameters [26].

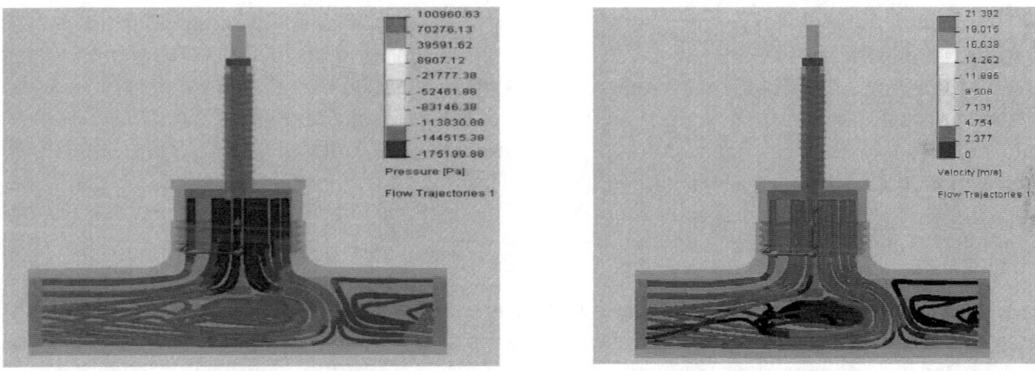

(a) (b)

Figure 2. Pressure (a) and velocity (b) variation in waste valve.

In the latest of this decade, CFD (Computational Fluid Dynamics) simulation methods tends to be used, as was done by Maw and Hte. They created hydraulic ram pump design then simulated the phenomenon of water flow through a waste valve with commercial CFD software that is Solid Work software. However, this paper only show the pressure distribution and velocity on the waste valve, as shown in Figure 1, and did not identify any waste valve design influence on the hydraulic ram pump performance. Furthermore, analysis of the hydraulic ram pump performance using CFD was conducted by Shende and Choudary. Analysis of the hydraulic ram pump performance was done by using one of the commercial CFD software ie. Fluent. The simulation results show that the mass flow rate of water through the new design of the waste valve is smaller than the old model, and provide a better performance of hydraulic ram pump system. The analytical calculations can be made further simple by writing a computer programme which will help to determine the optimum solution. Modelling and analysis can be carried out using Ansys Fluent software which determines the various design parameters [26].

IOP Conf. Series: Materials Science and Engineering **539** (2019) 012008 doi:10.1088/1757-899X/539/1/012008

There is limited information in the literature about experimental flow characterization studies of hydraulic ram pump system. Flow visualization methods are very useful for identifying these deviations from the ideal conditions. Particle Image Velocimetry (PIV), Particle Tracking Velocimetry (PTV), Laser Doppler Anemometry (LDA) and smoke or dye injection are experimental flow visualization methods that have been used for this purpose [27]. One of the most important contributions was the use of aniline dye to produce coloured water in the Osborne Reynolds experiment conducted in 1883.

In a very recent study, Goshayeshi and Chaer [28] experimentally investigated the flow visualization in a closed-loop OHP which is constructed using a Pyrex glass with kerosene as the working fluid. The phase change phenomena and bubble generation in the heating section were recorded using a high-speed video camera (Panasonic VX87) with frame rate of 100 frames per second. The internal flow pattern in pipe or tube could be studied using flow visualization.

The waste valve mainly affects the magnitude of the impulse force generated. In the implementation of the adjustment of this valve waste takes a long time, because there is no reference or sufficient reference in making the design of the valve waste. The characteristics of the waste valve are affected by parameters such as the orifice diameter of the valve, the diameter of the valve disk, the step length, and the valve weight and the discharge and water head of the hydraulic ram pump drive. Besides all these parameters affect each other. Therefore, in this experiment, experimental testing was conducted in the laboratory on hydraulic ram pump waste valve model to investigate the effect of waste valve orifice and disk diameter on hydraulic ram pump.

2. Research Methods

2.1. *Experimental Setup*

In this study, an experimental work was carried out on the installation of hydraulic ram pump system as shown in Figure 3. All hydraulic ram pump components are created using the clear acrylic material. Flow visualization is carried out in water flow throughout the hydraulic ram pump model. The water is injected with dyes of three different color liquids through small orifices in the drive pipe of the hydraulic ram model so as to act as streamline tracers. The water flow phenomena in the hydraulic ram pump was recorded using a high-speed video camera (Sony RX100-IV) with frame rate 960 frames per second. Body and air vessel of the hydraulic ram pump have 55 mm of diameter, and its drive pipe diameter is 25 mm with length of 2270 mm.

Model desain waste valve adalah seperti pada Figure 4. Variasi diameter orifice waste valve adalah 25 mm, 30 mm, 35 mm dan 40 mm pada diameter disk of the waste valve 45 mm. Sedangkan variasi diameter disk of the waste valve adalah 40 mm, 41 mm, 42 mm dan 43 mm pada diameter orifice of the waste valve 32 mm.

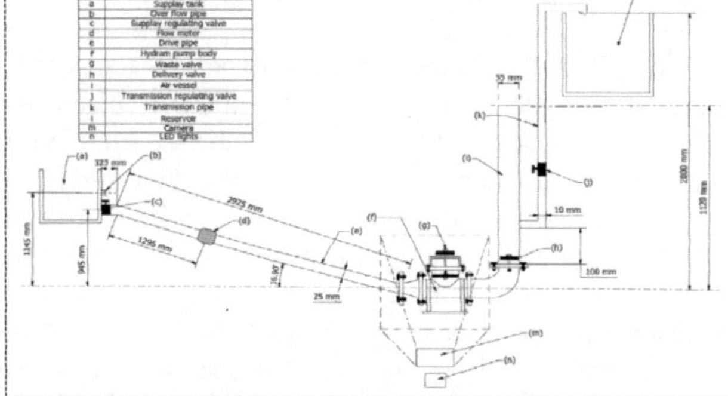

Figure 3. The experimental setup of hydraulic ram pump.

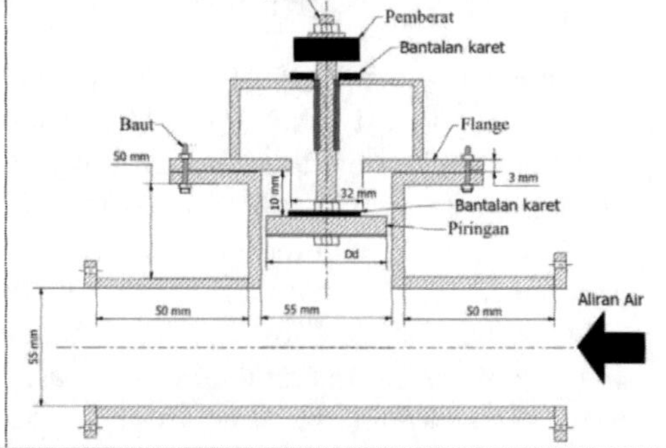

Figure 4. The waste valve model

2.2. Experimental Procedure

The initial observation of the video, which has been taken using a Sony Digital camera RX100-IV. The video is first processed or converted into an image. So the technique used in this research is image processing techniques because the data is processed data in the form of images. The video that has become the image will be processed to create flow patterns and frequency of hydraulic ram pump valve movement. Measurement of flow velocity is done by measuring the length of trace of the implanted particles using ImageJ software, and then the track length of the particle movement divided by the speed of the camera taking the video (shutter speed), in this study using shutter speed 0.001 seconds.

3. Results and Discussions

3.1. Flow pattern

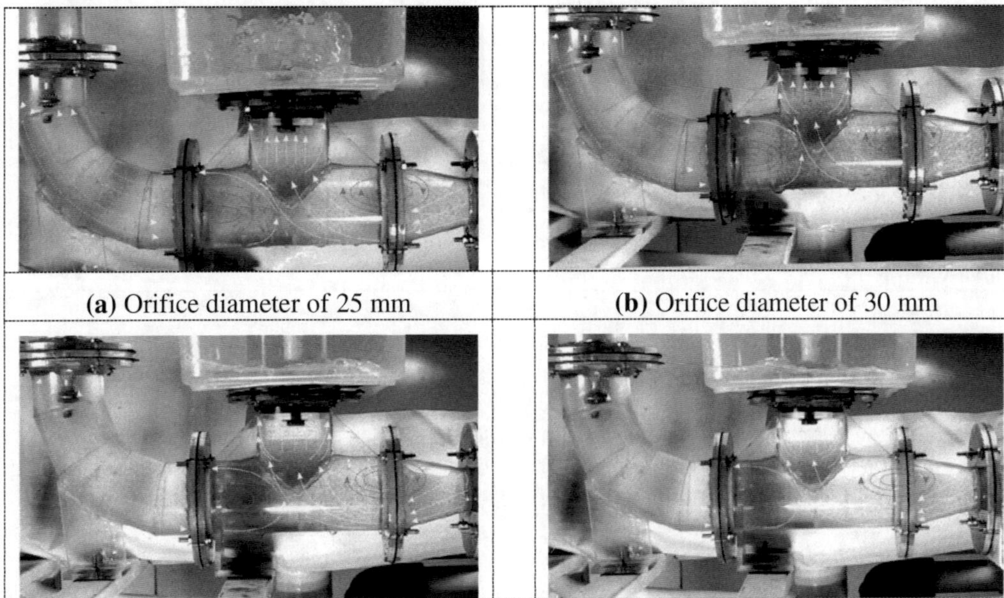

(a) Orifice diameter of 25 mm

(b) Orifice diameter of 30 mm

(c) Orifice diameter of 35 mm

(d) Orifice diameter of 40 mm

Figure 5. Water flow pattern in hydraulic ram body at acceleration stage on the orifice waste valve variation.

Movement of paint particles on the hydra pump is illustrated in each cycle of the hydra pump, so that the 2.07-second video must be converted to split into 2,070 images, and then selected four images representing each cycle of the hydra pump at each variation of the diameter of the valve sewage hole. Figure 5 show the water flow pattern in hydraulic ram body at acceleration stage where the waste valve at opening condition at the orifice of waste valve variation. Water coming out of the moving pipe is illustrated with a dotted line of yellow. The rotating stream (vortex) occurs in two parts illustrated by a dotted line of red. Water begins to enter the pumping body, partially discharged through the sewage valve, as the tap valve is still covered with water again turning towards the waste valve. The flow pattern is almost the same, but the obvious difference occurs in the flow of water that suppresses the waste valve disk, the amount of flow decreases as the diameter of the valve sewer diameter is enlarged, represented by the straight line of yellow dots.

(a) Disk diameter of 40 mm

(b) Disk diameter of 41 mm

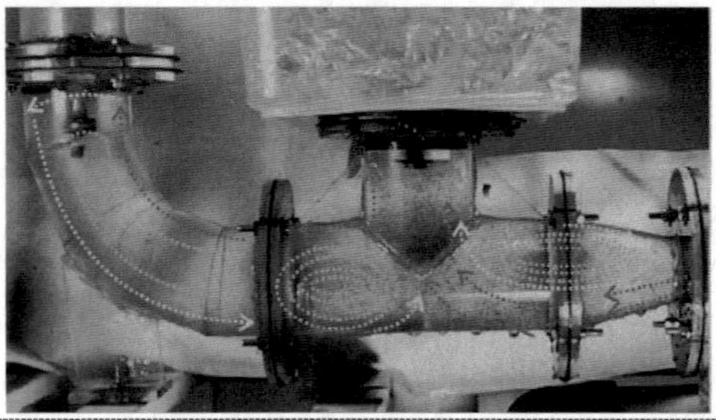

(c) Disk diameter of 42 mm

(d) Disk diameter of 43 mm

Figure 6. Water flow pattern in hydraulic ram body at acceleration stage on the disk waste valve variation.

Figure 6 show the water flow pattern in hydraulic ram body at acceleration stage where the waste valve at opening condition at the disk of waste valve variation. Incoming water will start to get out of the waste valve. The red dotted line indicates the direction of movement of the incoming water from the moving pipe, the water entering outward from the sewage valve and partially to the press valve and turning again to the waste valve. The circular green dotted lines indicate the existence of a rotating stream (vortex), and the yellow color indicates the direction of movement of the water being turned.

3.2. Flow velocity in hydraulic ram pump body

The more rotation of the flow of water that has a vortex of course the distance of the particle closer and closer, because the distance has to do with the speed it is necessary to measure the movement distance of paint particles (glitter) and then later used to determine the speed of water flow that occurs inside the hydra pump body. Speed measurements are made at four points as in Figure 7.

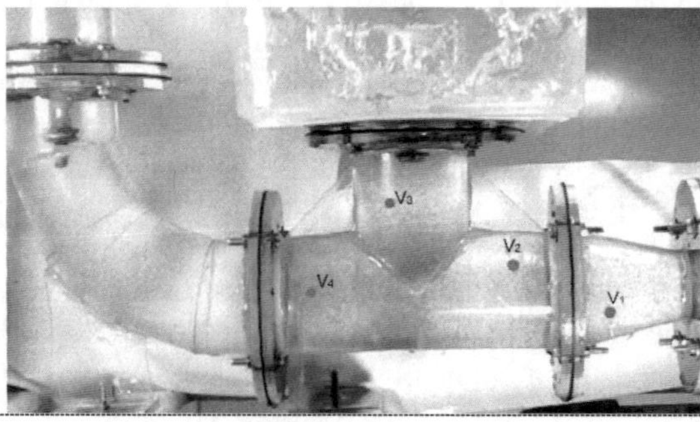

(a) Velocity measuring point in hydraulic ram body

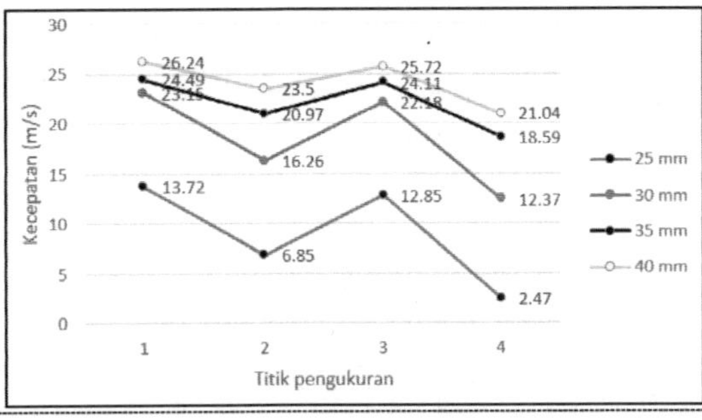

(b) Flow velocity at orifice diameter variation

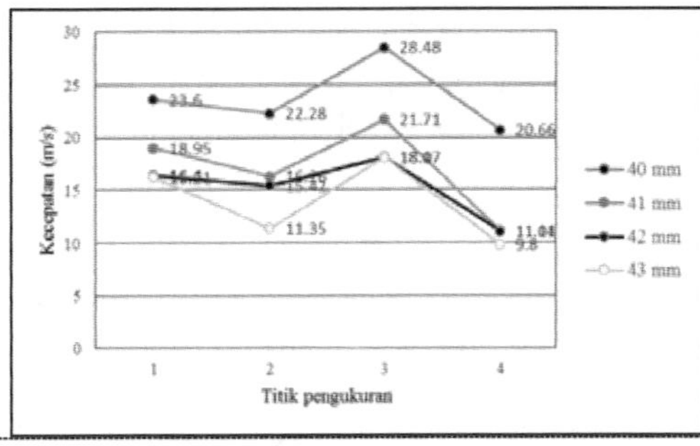

(c) Flow velocity at disk diameter variation

Figure 7. Water flow velocity in hydraulic ram body at acceleration stage on the orifice of waste valve variation.

Point 1 (at the reducer) is the highest speed, due to the water from the reservoir directly to the pump, then at point 2 (before the waste valve) the speed begins to decrease due to the vortex. At point 3 (in the waste valve) the water velocity increases again due to the open sewage valve, the decline occurs at point 4 (after the waste valve), in addition to the occurrence of the whirlpool as well as the closed pressure valve. The enlargement of the waste valve diameter hole resulted in an increase in speed at the 4 points. The distance of the furthest speed increase occurs between holes diameter 25 mm to 30 mm.

In addition, the increase in velocity occurs at point 3 which is directly below the waste valve disk, as it is currently only open sewage valve so that water is subjected to pressure from the reducer pipe or point 1 and from the T pipe after reducer or point 2 and also driven from the flow behind the press valve that is at point 4 located on the T tube after the waste valve. The larger the diameter of the waste valve disk, the water velocity will decrease as the gap of water discharge decreases.

3.3. Duration of the waste valve movement

Frequency of the movement of the waste valve is obtained from the video that has been decoded into the image, the video recorded for 2.07 seconds is decomposed into 2070 images then the image is identified according to the characteristics of each cycle of the hydra pump work. Four cycle of hydram pump work is acceleration, compression, delivery, and recoil consist of 1385 picture (frame). The movement of the next valve is made curve, as for the curve as in Figure 8.

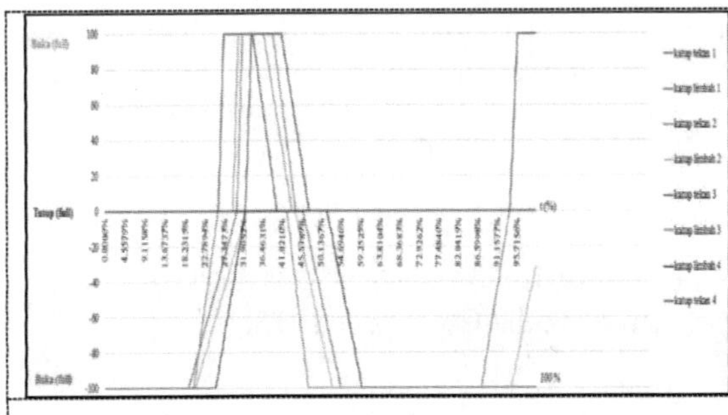

(a) Cycle time at the orifice diameter variation

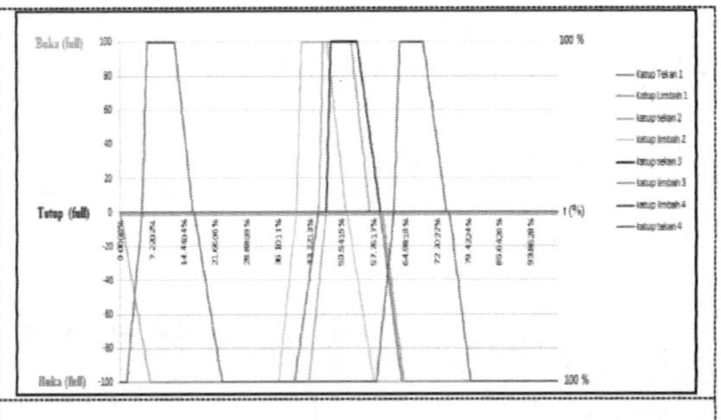

(b) Cycle time at the disk diameter variation

Figure 8. Duration of time each stage of the waste valve movement.

Table 1. Length of time each stage of the waste valve movement at the orifice diameter variation.

DL$_{KL}$ (mm)	Waktu (s)									
	Katup Limbah					Katup Tekan				
	Buka Penuh	Proses Menutup	Tutup Penuh	Proses Terbuka	Jumlah	Tutup Penuh	Proses Terbuka	Buka Penuh	Proses Menutup	Jumlah
25	0.438	0.055	0.167	0.051	0.711	0.583	0.015	0.078	0.066	0.742
30	0.442	0.062	0.179	0.055	0.738	0.595	0.015	0.075	0.063	0.748
35	0.45	0.132	0.126	0.095	0.803	0.65	0.015	0.067	0.072	0.804
40	0.416	0.124	0.172	0.096	0.808	0.654	0.015	0.125	0.058	0.852

Table 2. Length of time each stage of the waste valve movement at the orifice diameter variation.

DL$_{KL}$ (mm)	Waktu (s)									
	Katup Limbah					Katup Tekan				
	Buka Penuh	Proses Menutup	Tutup Penuh	Proses Terbuka	Jumlah	Tutup Penuh	Proses Terbuka	Buka Penuh	Proses Menutup	Jumlah
25	0.438	0.055	0.167	0.051	0.711	0.583	0.015	0.078	0.066	0.742
30	0.442	0.062	0.179	0.055	0.738	0.595	0.015	0.075	0.063	0.748
35	0.45	0.132	0.126	0.095	0.803	0.65	0.015	0.067	0.072	0.804
40	0.416	0.124	0.172	0.096	0.808	0.654	0.015	0.125	0.058	0.852

Table 1 indicated that there is an increase in time due to the magnification of the diameter of the valve waste hole, this is because the larger hole allows more water to go directly to the sewer valve hole rather than pressing the waste valve disk first so that the movement of the valve in one cycle of the hydraulic ram pump becomes longer. In addition, Table 2 presents the greater the diameter of the waste valve disk, the faster the frequency of the hydraulic ram pump valve, when the time is faster the hiccups also increase.

4. Conclusion

The diameter of orifice and disk of the waste valve have significantly influencing the flow pattern in the hydraulic ram pump body and its time movement each stage of the cycle. The greater the diameter of the orifice and the disk of the waste valve the velocity of water that has vortex more and more, so the speed of water flow in the body of the pump also decreases. Furthermore, the larger the diameter of the waste valve disk the shorter the interval of the hydraulic ram pump valve movements, so the frequency of the hydraulic ram pump valve movement increases. The pumping discharges and efficiency are also influenced by the diameter of the hydraulic ram pump waste valve disk, the larger the diameter of the hydraulic ram pump waste valve disk, the pumping discharge and the efficiency of the hydraulic ram pump are also increasing.

Acknowledgment

The authors would like to express their gratitude to the Faculty of Engineering of Udayana University for the fund support on the Unggulan Udayana research scheme for the fiscal year 2018.

References

[1] Glover, P.B.M., 1994. *Computer Simulation and Analysis Method in the Development of the Hydraulic Ram Pump*. PhD Thesis, University of Warwick Department of Engineering.

[2] Jennings, G.D., 1996. *Hydraulic Ram Pumps*, North Carolina Cooperative Extension Service, North Carolina.

[3] P. Diwan, A. Patel and L. Sahu, "Design and Fabrication of Hydraulic Ram with Methods of Improving Efficiency", *in International Journal of Current Engineering and Scientific Research (IJCESR)*, Vol. 3, No. 4, (Technical Research Organization India, 2016), pp. 5-13.

[4] A. Pathak, A. Deo, S. Khune, S. Mehroliya and M.M. Pawar, 2016. "Design of Hydraulic Ram Pump", *in International Journal for Innovative Research in Science & Technology*, Vol 2, No. 10, (IJIRST, 2016), pp. 290-293.

[5] A. Deo, A. Pathak, S. Khune and M.M. Pawar, "Design Methodology for Hydraulic Ram Pump", *in International Journal for Innovative Research in Science & Technology*, Vol 5, No. 4, (IJIRST, 2016), pp. 4737- 4745.

[6] P. Nambiar, A. Shetty, A. Thatte, S. Lonkar and V. Jokhi, 2015. "Hydraulic Ram Pump Maximizing efficiency", *in International Conference on Technologies for Sustainable Development (ICTSD)*, 4-6 Feb. 2015, Mumbai – India, (IEEE Explore Digital Library, 2015).

[7] A.A. Tessema, "Hydraulic Ram Pump System Design and Application", *in ESME 5th Conference on Manufacturing and Process Industry.* September 2000, (ESME, Addis Ababa, Ethiopia, 2000).J. van der Geer, J.A.J. Hanraads, R.A. Lupton, The art of writing a scientific article, J. Sci. Commun. 163 (2000) 51-59.

[8] Taye, T., 1998. Hydraulic Ram Pump, *Journal of the ESME*, Volume II, July 1998, Addis Ababa, Ethiopia.

[9] Young, B., 1997. Design of Homologous Ram Pump, Journal of Fluids Engineering, Transaction of the ASME, Volume 119, June 1997, pp. 360-365.

[10] M. Suarda, dan I G.K. Sukadana, "Perancangan dan Pengujian Unjuk Kerja Katup Tekan Pompa Hydraulic ram Model Plat, Membran, Bola dan Setengah-Bola", *Prosiding: Seminar Nasional Tahunan Teknik Mesin (SNTTM – XII).* Hal. 387 - 394. Bandar Lampung, 22-23 Oktober 2013.

[11] Szaltys, P., Chrusta, M., Przadka, A., Goujon-Durand, S., Tuckerman, L. S., Wesfreid, J. E., 2011. Nonlinear Evolution of Instabilities behind Spheres and Disks. *Journal of Fluids and Structures*, April 11, 2011.

[12] Natarajan, R., Acrivos, A., 1993. The Instability of the Steady Fflow Past Spheres and Disks. *Journal of Fluid Mechanics*, Volume 254, pp. 323–344.

[13] Johnson, T.A., Patel, V.C., 1999. Flow Past a Sphere up to a Reynolds Number of 300. *Journal of Fluid Mechanics*, Volume 378, pp. 19–70.

[14] Tomboulides, A.G., Orszag, S.A., 2000. Numerical Investigation of Transitional and Weak Turbulent Flow Past a Sphere. *Journal of Fluid Mechanics*, Volume 416, pp. 45–73.

[15] Achenbach, E., 1972. Experiments on the Flow Past Spheres at Very High Reynolds Numbers. *Journal of Fluid Mechanics*, Volume 54, pp. 565-575.

[16] Constantinescu, G. and Squires, K., 2004. Numerical Investigation of Flow over a Sphere in the Subcritical and Supercritical Regimes. *Phys. Fluids*, Volume 16, pp. 1449-1466.

[17] Poon, E.K.W., Iaccarino, G., Ooi, A.S.H., Giacobello, M., 2009. Numerical Studies of High Reynolds Number Flow Past a Stationary and Rotating Sphere. *Seventh International Conference on CFD in the Minerals and Process Industries CSIRO*, Melbourne, Australia. 9-11 December 2009.

[18] D. Grygo, "Effect of the Height of the Delivery Water on Performance of Water Ram", *Technical Sciences*, 2016, 19(2), pp. 139–149.

[19] M. N. Harith, R. A. Bakar, D. Ramasamy and M. Quanjin, "A Significant Effect on Flow Analysis & Simulation Study of Improve Design Hydraulic Pump", *in 4th International Conference on Mechanical Engineering Research (ICMER2017)*, IOP Conf. Series: Materials Science and Engineering 257 (2017) 012076., IOP Publishing.

[20] M. N. Harith, R. A. Bakar, D. Ramasamy, K. Kardigama and M. Quanjin, "A Study of Waste and Delivery Valve Design Modification to the Pump Performance", *in iCITES 2018*, IOP Conf. Series: Materials Science and Engineering 342 (2018) 012090, IOP Publishing.

[21] S.S. Mondol, "Design, manufacture and test a hydraulic ram", *ResearchGate Artikel*, May 2017.

[22] X. Guo, J. Li, K. Yang, H. Fu, T. Wang, Y. Guo, Q. Xia and Wei Huang, "Optimal Design and Performance Analysis of Hydraulic Ram Pump System", *Proceeding of Institution of Mechanical Engineering* 2018, Part A: J Power and Energy 0(0), pp. 1-15.

[23] Filipan, V., Vireg, Z., dan Bergant, A. 2003. Mathematical Modelling of a Hydraulic Ram Pump System. *Journal of Mechanical Engineering*. In Strojniski Vestnik. Vol 49 No. 3, pp 137-149.

[24] Maw, Y.Y. dan Hte, Z.M. 2014. Design of 15 meter Head Hydraulic Ram Pump. *International Journal of Scientific Engineering and Technology Research*. Vol 03 No. 10, pp 2177-2181.

[25] Shende, P.B., Choudhary, S.K. dan Ninawe, A.P. 2015. Analysis and Enhancement of Hydraulic Ram Pump Using Computational Fluid Dynamics (CFD). *International Journal for Innovative Research in Science & Technology*. Vol 2 No. 3, pp 109-133

[26] Sampath, S.S., et al. 2015. Estimation of Power and Efficiency of Hydraulic Ram Pump with Re-circulation System. *International Journal of Computer-aided Mechanical Design and Implementation*. Vol 1 No. 1, pp 7-18.

[27] Karadeniz, H.K., Kumlutas, D. and Ozer, O. 2013. Experimental Visualization of the Fow Characteristics of the Outflow of a Split Air Conditioner Indoor Unit by Meshed Infrared Thermography and Stereo Particle Image Velocimetry. *Experimental Thermal and Fluid Science*, Elsevier Inc. Vol. 44, pp. 334–344.

[28] Goshayeshi, H.R. and Chaer, I. 2013. Experimental Study and Fow Visualization of Fe2O3/Kerosene in Glass Oscillating Heat Pipes. *Applied Thermal Engineering*, Elsevier Inc. Vol. 103, pp. 1213–1218.

[29] M. Suarda, A. Ghurri, M. Sucipta and I G.B.W. Kusuma, 2018, "Investigation on characterization of waste valve to optimize the hydraulic ram pump performance", *AIP Conference Proceedings* 1984, 020023 (2018); doi: 10.1063/1.5046607

IOP Conf. Series: Materials Science and Engineering **539** (2019) 012009 doi:10.1088/1757-899X/539/1/012009

The effect of using poly (methyl methacrylate) (pmma) in the preparation of free standing 3d graphene

P Pransisco[1]
[1]Ph.D. student at Universiti Teknologi PETRONAS

Email: prengkipransisco@gmail.com

Abstract. In order to get a free standing 3D graphene during the synthesis process, poly (methyl methacrylate) (PMMA) is often used to dissolve Ni agent. However, the use of this polymer is giving defect on structural and chemical composition of 3D graphene. The aim of this article is to study the effect of using PMMA in removing Ni agent on 3D graphene synthesis process. The 3D graphene was successfully synthesized through low pressure chemical vapor deposition (LPCVD) system. The structural defect and chemical composition study was examined using field emission scanning electron microscopy (FESEM), X-ray diffraction (XRD) and Raman spectroscopy. Based on the findings, the use of PMMA and hot acetone was not really recommended method. The result was confirm that Ni element and oxide are still presence after the treatment process, and these structural and chemical defect might be effect to the properties of the free standing 3D graphene.

1. Introduction

3D Graphene is a multilayer of carbon atoms in honeycomb shape. 3D Graphene has attracted great attention due to its unique structure, graphene possess, extraordinary physicochemical properties, such as high carrier mobility, high thermal conductivity, promising elasticity, stiffness and high surface area. However, these promising properties will not completely utilize if the free standing 3D graphene is not properly prepared. Structural and chemical defect on 3D graphene would give very significant effect to its properties. In several nanomaterial researches, the treatment process to the material such as high annealing temperature will result high significant changes on quality of the material such as big grain size [1, 2] . The other case, treatment to the material by applying certain chemical is also affecting the properties of the material [3, 4]. Therefore, the right method is really needed to prepare particular material, in order to have high quality of the material. There are various methods have been developed in preparation of 3D graphene such as: low pressure (LP) and atmospheric pressure (AP) chemical vapor deposition [5]. These 2 methods were developed to increase the quality of 3D graphene and decreasing malformation of layers in 3D graphene [6]. These methods usually using Ni foam as the substrate to grow a free standing 3D graphene [7]. In the process of Ni removal as a substrate, poly (methyl methacrylate) (PMMA) and hot acetone are often used to make the 3D graphene remain together. However, the use of this polymer could affect the properties of 3D graphene.

This article aims to study the effect of PMMA and hot acetone in the process of synthesizing 3D graphene to get high quality of free standing 3D graphene. FESEM, XRD and Raman spectra were used to characterize the 3D graphene. According to our findings, it was confirmed that the Ni element and oxide was presence in the 3D graphene. Through these findings, we can conclude that the use of PMMA and hot acetone in the process of getting 3D graphene process is not really recommended due to its lack on structural and chemical defects.

IOP Conf. Series: Materials Science and Engineering **539** (2019) 012009 doi:10.1088/1757-899X/539/1/012009

2. Methodology

The preparation of 3D graphene was done out by using low pressure chemical vapor deposition (LPCVD) method. The pressure of LPCVD system was maintained by using vacuum pump at 3.3 mTorr and 3 mTorr for mixture of Ar and H_2 at constant flow. Ni foam was used as a substrate in order to have a free standing 3D graphene. The 2 x 4 cm^2 Ni foam was cleaned up before it used as a substrate using deionized (DI) water and ethanol and afterwards it dried at room temperature. The cleaned and dried microporous nickel foam placed in the quartz tube of CVD system and heat up at 1023 °C for 80 minutes. The temperature was maintained at 1023 °C for 20 minutes under Ar gas (100 sccm). H_2 and CH_4 gas flowed into the tubes around 90 minutes to grow 3D graphene on nickel foam.

In the process of removal Ni element, poly methyl methacrylate (PMMA) was used as the support layer to keep the 3D graphene attached. The PMMA was coated on the surface of 3D graphene and heat up at 80 °C for an hour and cooled down at room temperature. The solution of hydro chloride and distilled water was used as etching solution and afterwards nickel foam/3D graphene/PMMA template was immerged into the solution. The template edges were cut to allow etching solution to penetrate and after that the sample was put in the room temperature until it dried. The hot acetone at 55 °C was used for a couple of times in order to complete PMMA removal.

After the free standing 3D graphene was obtained, some characterizations were used to examine this 3D graphene. FESEM was used to investigate the microstructure of surface of the material and its element through EDX analysis. The phase of 3D graphene was determined by using XRD and the last of characterization was using Raman spectroscopy analysis to study its chemical analysis.

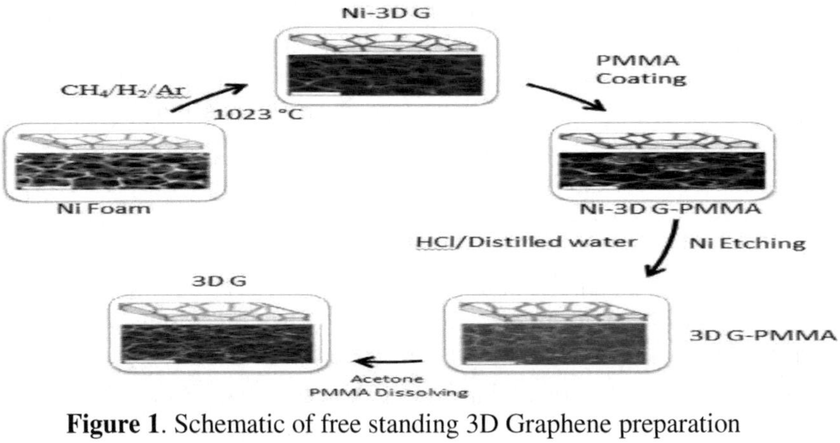

Figure 1. Schematic of free standing 3D Graphene preparation

3. Results and Discussions

3.1. Field Emission Scanning Electron Microscopy (FESEM) Analysis

3.1.1 Microstructure Analysis

IOP Conf. Series: Materials Science and Engineering **539** (2019) 012009 doi:10.1088/1757-899X/539/1/012009

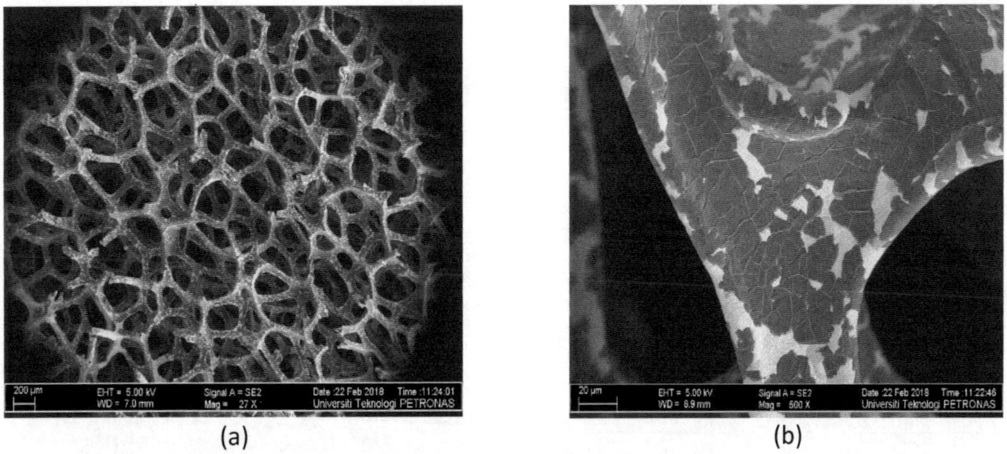

(a) (b)

Figure 2. FESEM images of 3D graphene growth on Ni foam (a) Magnification 200 μm and (b) 20 μm

The honeycomb sheet of 3D graphene on the top of Ni foam is shown in Figure 2. At the magnification 20 micro meter (b), we can see that 3D graphene (black color) was scattered on the top of the Ni foam (white color). Before the treatment using PMMA, the Ni foam seen very thick and very sharp. The 3D graphene looks having very high in porosity and very thin which consist of layers.

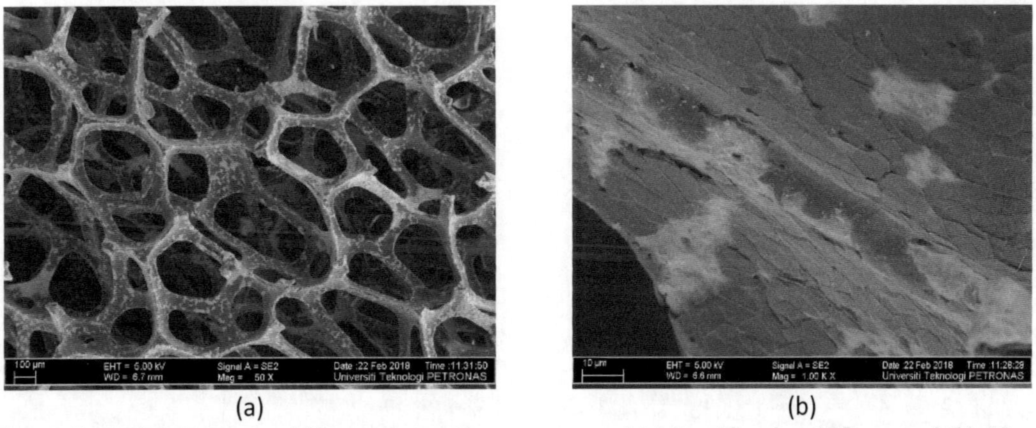

(a) (b)

Figure 3. FESEM images of 3D graphene after treatment, (a) Magnification 100 μm and (b) 10 μm

As compared to microstructure of 3D graphene before treatment, after the PMMA was applied in order to support the free standing 3D graphene, the 3D graphene seems to be very thin as can see in Figure 3b. The thin layer of 3D graphene sheet was obtained after the Ni element was dissolve in the hydro chloride + deionize water solution and further treatment with hot acetone to remove the PMMA. However, it seems like the layer of Ni foam is actually not completely removed. The 3D graphene is not really free standing, it was still sticking on the Ni foam event though the amount of it was not much as compared before it treated.

3.1.2 Element Analysis

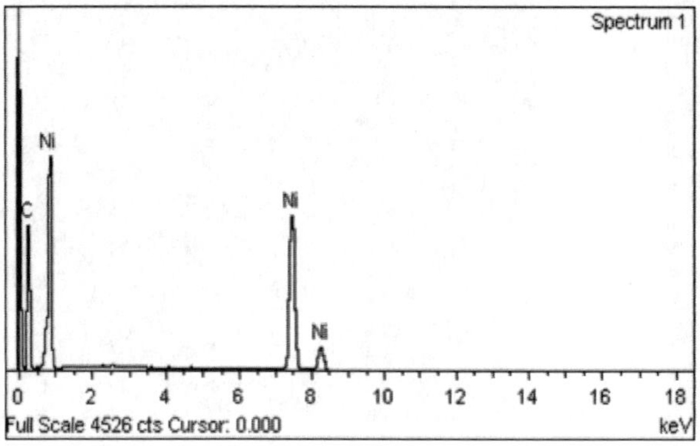

Element	Weight (%)	Atomic (%)
C	69.86	91.89
Ni	30.14	8.11
Totals	100%	

Figure 4. EDX spectrum of 3D graphene growth on Ni foam

The highest atomic percentage of the elements is carbon, this is because the 3D graphene is actually consisting of multi-layer carbon atom and it is confirmed in the Figure 4. Another element is Ni with amount 30.14%, this high percentage value also confirmed through FESEM result, where the 3D graphene was seems having very thick substrate which is the Ni foam.

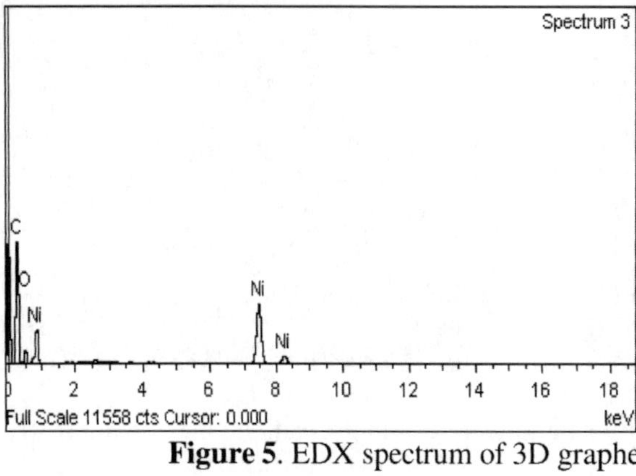

Element	Weight (%)	Atomic (%)
C	58.54	81.95
Ni	33.39	9.56
O	8.07	8.49
Totals	100%	

Figure 5. EDX spectrum of 3D graphene after treatment

After the treatment on 3D graphene with Ni foam by applying PMMA and hot acetone, it is confirm through EDX that the Ni element in Figure 5 is not completely removed and the percentage also still high. It is even produce a new element which is oxide, the presence of this element might be due to the sample of 3D graphene was not keep in vacuum environment.

3.2 X-Ray Diffraction (XRD) Analysis

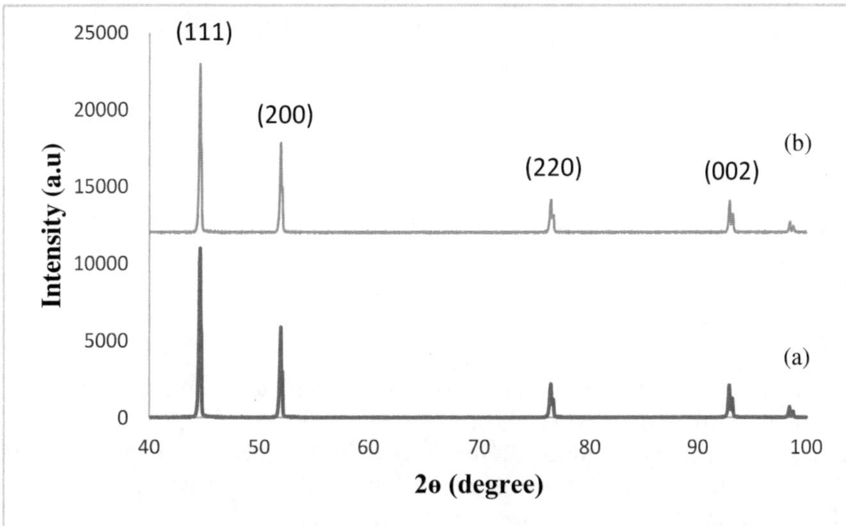

Figure 6. XRD patterns of 3D graphene (a) With Ni foam, (b) After treatment

The single phase of 3D graphene was obtained and it is confirm through XRD patterns as shown in Figure 6. As shown on the XRD patterns, the intensity each of major peaks were decreasing after the 3D graphene was treated with PMMA and hot acetone. According to [8], the highest intensity is belongs to Ni element. Through XRD and EDX we can conclude that the element of Ni is not completely removed even after treated with PMMA and hot acetone.

3.3 Raman Spectroscopy Analysis

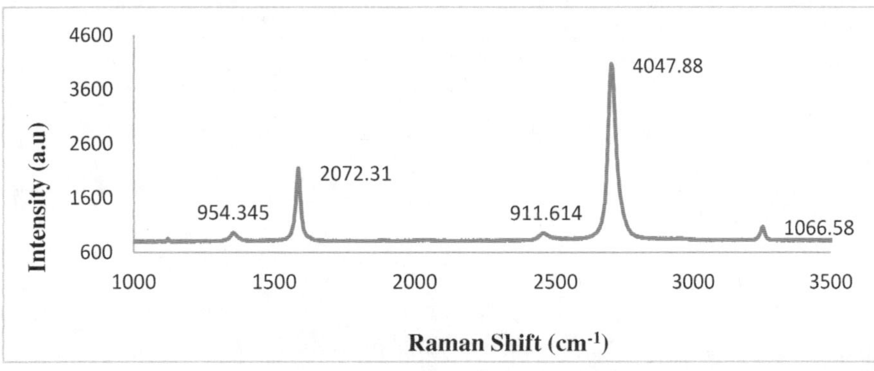

Figure 7. Raman spectrum of 3D graphene

The Raman spectrum of 3D graphene as shown in Figure 7, the peaks of Ni atom is clearly observed at 4047.88 cm^{-1} and 2072.31 cm^{-1}. The peaks at 1066.58 and 911.614 cm^{-1} are belongs to atom carbon and the rest which is band 954.345 is for oxide element.

4. Conclusion

In summary, we have demonstrated that the 3D graphene was successfully made through LPCVD. However, in order to get the free standing 3D graphene by applying PMMA and hot acetone to the material was not really right method. Based on our findings through FESEM, XRD and Raman spectra, it was confirm that Ni element and oxide element was presence in the material. For further study, it is really important to observe a new method in preparation of free standing 3D graphene.

Acknowledgements

The authors wish to thank Universiti Teknologi PETRONAS (UTP) and MOHE for funding the research through FRGS grant with grant number 015 3AB L-20.

References

[1] Prengki Pransisco et al, Effect of Calcination Temperature on Microstructure and Magnetic Properties of $Ni_{0.5}Zn_{0.25}Cu_{0.25}Fe_2O_4$ Nanoparticles Synthesized by Sol-Gel Method. American Institute of physics, 1621, 619(2014).

[2] Prengki Pransisco et al, Investigation of Nanostructure and Magnetic Properties of Nanocrystalline NiZnCu Ferrite synthesized by Sol-Gel Method, Journal of Applied Mechanic and material, 754-755 pp 1169-1174 (2015).

[3] Xiaogung Li et al, 3D graphene/ZnO nanorods composite networks as supercapacitor electrodes, Journal of Alloys and Compounds, 620, 31-37 (2015).

[4] S. Khamlich et al, Rapid microwave-assisted growth of silver nanoparticles on 3D graphene networks for supercapacitor application, 493, 130-137 (2017).

[5] Bei bei zan et al, Free- standing electrochemical electrode based on $Ni(OH)_2$/3D graphene foam for nonenzymatic glucose detection. Nanoscale royal society of chemistry, (2014).

[6] Peng Wu et al, Novel 3D porous graphene /Ni_3S_2 nanostructures for high performance supercapacitor Electrodes, Journal of Alloys and Compounds, 731 (2018)

[7] Beibei zan et al, Free- standing electrochemical electrode basd on $Ni(OH)_2$/3D graphene foam for nonenzymaticglucose detection, Nanoscale, (2014)

[8] Delong li et al, Preparation of sandwich-like $NiCo_2O_4$/rGO/NiO heterostructure on Ni foam for high performance supercapacitor electrodes, Nano-Micro letter springer, 9-16 (2017).

IOP Conf. Series: Materials Science and Engineering **539** (2019) 012010 doi:10.1088/1757-899X/539/1/012010

Proximate and morphology properties of swat bamboo activated carbon carburized under different carbonization temperature

I M Astika[1], D N K P Negara[1], C I P K Kencanawati[1], T G T Nindhia[1], F Hidajat[2]
[1] Mechanical Engineering Department of Udayana University, Street of Bukit Jimbaran, Badung, Bali, Indonesia
[2] PPPTMGB "Lemigas", Jl. Ciledug Raya Kav. 109, Jakarta Selatan, 12230

E-mail: devputranegara@gmail.com

Abstract. Activated carbon (AC) has an important role in many life fields. It has high porosity and very useful for gas mixtures separation and purification of air and water. Every application requires specific properties of AC. Characteristics of AC are strongly influenced by raw material and parameters of manufacturing process. This paper is focused to characterize of activated carbons (ACs) derived from swat bamboo (*Gigantochloa verticillata*) which is manufactured under different carbonization temperatures. Prepared samples of swat bamboo were carbonized by heating variations of 550, 650 and 750^{0}C and held at such temperature for 1 hour. Char yielded is powdered and meshed to maximum grain size of 250 *u*m and then physically activated by heating up to 800^{0}C, held for 1 hour by flowing of 150 mL / min nitrogen. The results show that there was an increase in fix carbon and carbon contents from raw material to char and from char to ACs yielded; there were a little bit increase of fix carbon and carbon contents proportional to increasing of carbonization temperature. The maximum fix carbon (74.73%) and C (75.52) were obtained at carbonization temperature of 750^{0}C. SEM observation showed that there has been a change in the morphology microstructure from raw material to activated carbons, wherein the ACs the pores structures clearly can be observed.

1. Introduction

Activated carbon (AC) is material having high adsorption capacity [1] due to its high porosity [2] and very large surface area [3]. Due to its specific characteristics, Activated Carbons (ACs) are widely used to storage of gas [4-8], to purify the various types of pollutants from industrial waste [5] and waste water [9], to separate of heavy metals [10], to be used as supercapacitor cathode material [11] and others. Coal is raw material generally used to produce ACs, however due to it is a non-renewable material make it becoming has relatively expensive price [12]. Today, it has been widely developed ACs from other resources for AC raw material alternative, such as biomass. Some biomass sources have been used as AC such as palm [13], coconut shell [14-15], and bamboo [5, 13, 16-20].

Bamboo is widely studied as a raw material source for production of activated carbon. Bamboo has some advantages as AC source such as very fast growth and matures in three years [18]; it is very easy to be found in a huge volume; and its chemical composition is very suit for AC. Bamboo has lignin, cellulose and hemicellulose contents more than 90% of its total mass. Generally, bamboo contains of 40-50% alpha cellulose and lignin content in the range of 20% to 26% [19]. Elemental analysis show that bamboo has carbon content around 48.64 % and nitrogen, hydrogen and sulphur contents are low 0.14 %, 6.75% and sulphur 0.11 %, respectively [20]. Bamboo can be converted into activated carbon through carbonization and activation processes. The aim of carbonization is to increase carbon content of material by reducing non-carbon elements by use of thermal decomposition [21] and to produce initial porosity in the char [22]. In order to develop advanced and high porosity of activated carbon, the char yielded at the carbonization stage is activated through physical or chemical activation [7]. Parameters of carbonization are one of factor can affect the characteristics of ACs. In this paper, properties of activated carbon prepared from swat bamboo with different carbonization temperatures were investigated. The properties of activated carbon to be observed were the content of ash, volatile, fix carbon, moisture, carbon (C), hydrogen (H), nitrogen (N) and morphology microstructure.

2. Method

The swat bamboo was cut into approximately small pieces of 1 x 1 x 1 cm and dried under the sunshine for 10 days. The samples were then dehydrated by heating in the electric furnace for 1 hour at a temperature of 110^0 C. Carbonization was carried out by heating up to temperatures of 550, 650 and 750^0C and soaking during 1 hour. The char yielded was powdered with maximum grain size 250 *u*m and physically activated by heating up to 800^0C, held for 1 hour by flowing of 150 mL/min nitrogen. The ACs produced were signed as 550F1, 650F1 and 750F1 for carbonization temperatures of 550, 650 and 750^0C respectively. Proximate test (TGA 701), elemental analysis (CHN628S), Van Soest Analysis and SEM observation were conducted in order to find out of parameters of proximate test (ash, volatile, fixed carbon, moisture and carbon), elemental contents (C, H, N), chemicals composition of swat bamboo (lignin, hemicellulose, cellulose) and morphology microstructure, respectively.

3. Results and discussion

3.1. Chemical compositions of swat bamboo

Figure 1 presents the chemical composition of swat bamboo. As shown in figure 1, cellulose (44.22 %), lignin (22.99 %) and hemicellulose (14.97 %) are the main chemical contents of swat bamboo. This cellulose content is closed to cellulose content of Japanese bamboo (43.30 %) as reported by Scurlock et al [23] and closed to cellulose content of 4 types of *Guadua* bamboo (37 – 44 %) [24]. Cellulose and lignin have significant contribution to final carbon content of AC and porosity formed. Cellulose content is associated with producing of AC micropores structure and lignin content is related with formation of AC macropores structure [25].

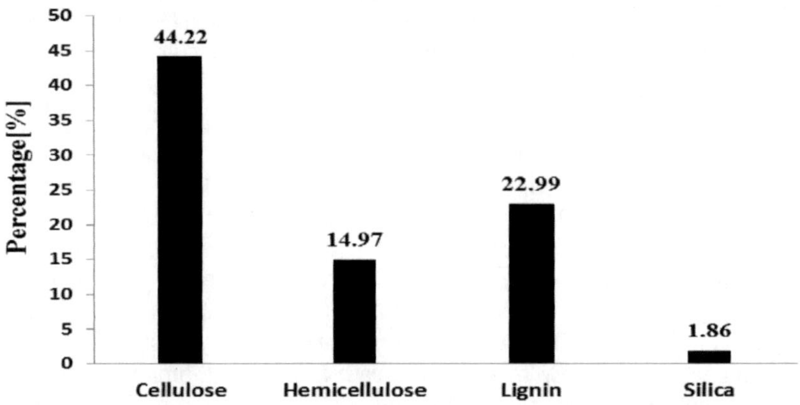

Figure 1. Chemical compositions of swat bamboo

3.2. Proximate and elemental analysis

Table 1 and 2 exhibits the proximate and ultimate analysis of sawt bamboo, char and activated carbos produced, respectively. Fix carbon of raw material (1.99 %), as shown in table 1, has improved significantly after carbonization and activation processes. There are an increase of 97.10 % and 97.33 % in average of fix carbon content from swat bamboo to all chars and ACs produced, respectively. The increase of fix carbon is due to conversion process of volatile become fix carbon during carbonization and activation processes 26. However, the ash content yielded for all ACs produced are relatively high so having potency will produce low porosity and surface area. The highest fix carbon (74.49 %) is obtained when raw material is carbonized at 750^0C. Elemental analysis also shows that C content of swat bamboo increases significantly after carbonization and activation, as shown in table 2. This is due to separation of elements such as H, N and O from fix carbon during elemental analysis. The highest C content (75.52%) is also gained at carbonization temperature of 750^0C.

Figure 2 illustrates the fix carbon and carbon contents of swat bamboo, chars and ACs produced at variation of carbonization temperature. It can be seen that fix carbon and carbon contents increase proportional to increasing of carbonization temperature although the increase is very small. This is probably due to that pyrolysis process of swat bamboo has completed at 500^{0}C, meanwhile the lowest carbonization temperature applied in this carbonization is 550^{0}C. First decomposition occurs at hemicellulose (200-260^{0}C), followed by cellulose at 240-350^{0}C and next lignin at 280-500^{0}C 27. However, the uniform of lignin decomposed can be slowly from ambient temperature to 900^{0}C. [16]. This gives probability that fix carbon and carbon contents will improve significantly when carbonization temperature used is upper then 750^{0}C. Table 1 also show that increasing of temperature carbonization also affects improving of moisture and ash of char, but decreasing of volatile content. For ACs produced, the higher carbonization temperature the higher fix carbon and carbon contents obtained, the lower moisture yielded and no trend to be found at volatile and ash.

Table 1. Proximate analysis of swat bamboo, chars, and ACs

Samples	Proximate Analysis Contents [%]			
	Moisture	Volatile	Ash	Fix carbon
Swat bamboo	7.86	88.32	1.85	1.99
Char 550	5.40	14.92	12.51	67.17
Char 650	6.03	10.65	13.56	69.76
Char 750	7.86	8.58	14.33	69.23
550 F1	2.86	8.36	14.33	74.45
650 F1	1.96	8.6	14.95	74.49
750 F1	1.61	7.75	15.92	74.73

Table 2. Elemental analysis of swat bamboo, chars, and ACs

Samples	Elemental contents [%]		
	C	H	N
Char 550	71.48	2.51	1.19
Char 650	72.32	1.88	0.92
Char 750	71.21	1.27	0.77
550 F1	74.70	0.56	1.09
650 F1	75.31	0.34	1.08
750 F1	75.52	0.37	1.01

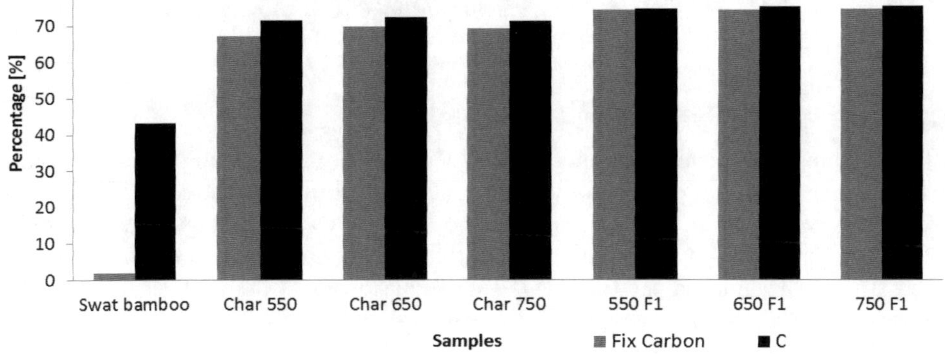

Figure 2. Comparison of the fixed carbon and C contents of swat bamboo, chars, and ACs

3.3. Morphology microstructure

Morphology microstructure of swat bamboo and ACs produced are illustrated in Figures 3 to Figure 6. Figure 3 shows the surface structure of raw material (swat bamboo) that is no structural porosity can be observed. This is due to the swat bamboo is still containing of original chemical composition that trough pyrolysis such chemical composition will be decomposed so that the porosity start to be formed and will be advanced and expanded during activation process. This is shown in Figure 4, 5 and 6 for 550F1, 650F1 and 750F1 respectively, where all of ACs manufactured have possessed porosity. However, it is not clear can be distinguished the porosity of ACs activated with different carbonization temperature. This due to SEM observation shows the qualitative data. The advance analysis (surface area analyser) is needed to find out the quantitative data such as pore volume, surface area and pore diameter so the difference properties of ACs due to applied of different carbonization temperature can be accurately compared.

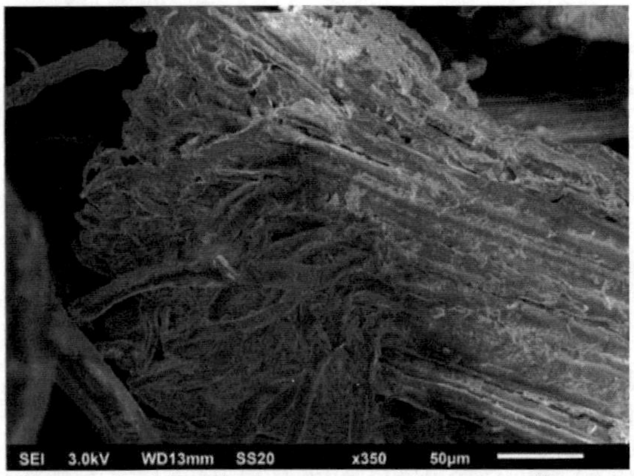

Figure 3. Morphology microstructure of swat bamboo

Figure 4. Morphology microstructure AC carbonized at 550^{0}C (550F1)

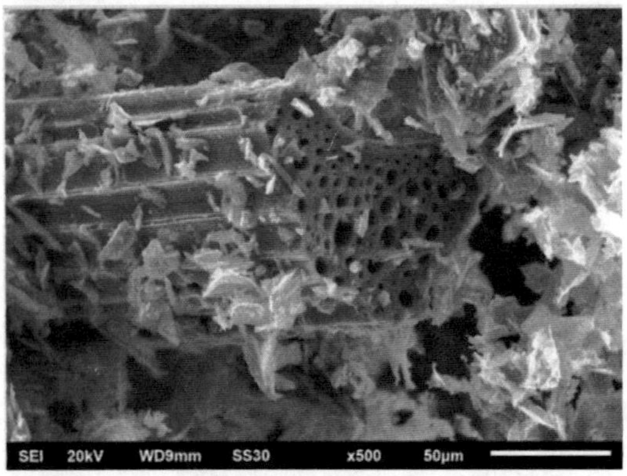

Figure 5. Morphology microstructure AC carbonized at 650^{0}C (650F1)

Figure 6. Morphology microstructure AC carbonized at 750^{0}C (750F1)

4. Conclusion

Carbonization temperatures give different properties of ACs produced. Increasing of carbonization temperature creates higher fix carbon and carbon contents, although for temperatures of 550, 650 and 750^{0}C their increase is very small. On the other hand, in this case, the higher carbonization temperature the lower of moisture obtained. AC requires the high fix carbon and carbon contents so that AC produced at carbonization temperature of 750^{0}C is considered as the best result due to the highest fix carbon (74.73%) and carbon (75.52%) are gained. Observation of morphology microstructure show that clearly can be seen that structural porosity has been formed after carbonization and activation processes. This porosity is specific property of AC making it having ability to adsorb or absorb a substance.

Acknowledgment
Deep gratitude is addressed to Directorate of Research and Community Service of Directorate General of Research and Development of Kemenristekdikti through LPPM Unud which has funded this activity through Contract Implementation of Research through PDUPT scheme of Fiscal Year 2018 No: 171.34 / UN 14.4A / LT / 2018.

References

[1] Hu Z and Srinivasan M P 2001 Mesoporous high-surface-area activated carbon *Micro. Meso. Mater.* **43** 267-75.
[2] Jin X J, Yu Z M and Wu Y 2010 Preparation of activated carbon from lignin obtained by straw pulping by KOH and K_2CO_3 chemical activation *Cellul. Chem. Technol.* **46** 79-85.
[3] Idris S, Iyaka Y A, Dauda B E N, Ndamitso M M and Umar M T 2012 Kinetic study of utilizing ground nut shell as an adsorbent in removing chromium and nickel from dye effluent *Am. Chem. Sci. J.* **2** 12–24.
[4] Lozano-Castello, D, D. Cazorla-Amoro´Sa, A and Quinn, D F 2002 Activated carbon monoliths for Methane Storage: Influence of Binder. *Carbon*. **20** 2817–25.
[5] Cheung W H, Lau S S Y, Leung S Y, Ip A W M and McKay G 2012 Characteristics of chemical modified activated carbons from bamboo scaffolding *Chinese Journal of Chemical Engineering* **20** 515-35.
[6] Isabel A A C E, Marta S S L, Pedro M C N and Jos´e P B M 2008 Adsorption of natural gas and biogas components on activated carbon *Separation and Purification Technology* **62** 281–96.
[7] Inomata K, Kanazawa K, Urabe Y, Hosono H and Araki T 2002 Natural gas storage in activated carbon pellets without a binder *Carbon* **40** 87-93.
[8] Diana C S A, J Cassia S A, Moisés B N, A Eurico B T, Emerson F J and Celio L C 2007 Microporous activated carbon prepared from coconut shells using chemical activation with zinc chloride *Microporous and Mesoporous Materials* **100** 361-64.
[9] Koo W K, Gani NA, Shamsuddin M S, Subki N S, and Sulaiman M A 2015 Comparison of wastewater treatment using activated carbon from bamboo and oil palm: an overview *Journal of Tropical and Resource Sustainable Science* **3** 54-60.
[10] Sheng F L, Song Y W, Ming J T and Lang D L 2012 Adsorption capacity and removal efficiency of heavy metal ions by moso and ma bamboo activated carbons *Chemical Engineering Research and Design* **90** 1397–406.
[11] Tianfu H, Zehai Q, Dewu W and Zhibiao H 2015 Bamboo-based activated carbon @ MnO2 nanocomposites for flexible high-performance supercapacitor electrode materials *Int. J. Electrochem. Sci.* **10** 6312 - 23.
[12] Patil B S and Kulkarni K S 2012 Development of high surface area activated carbon from waste material *International Journal of Advanced Engineering and Studies (IJAERS)* **1** 109-13
[13] Koo W K, Gani NA, Shamsuddin M S, Subki N S, and Sulaiman M A 2015 Comparison of wastewater treatment using activated carbon from bamboo and oil palm: an overview *Journal of Tropical and Resource Sustainable Science* **3** 54-60.
[14] Dipa D, Debi P S and Meikap B C 2015 Preparation of activated carbon from green coconut shell and its characterization *J. Chem. Eng. Process Technol.* **6** 248 doi:10.4172/2157-7048.1000248
[15] Ahmadpour A, Okhovat A, and M J Darabi M 2013 Pore size distribution analysis of activated carbons prepared from coconut shell using methane adsorption data *Journal of Physics and Chemistry of Solids* **74** 886-91.
[16] Mahanim S M A, Asma I W, Rafidah J, Puad E and Shaharuddin H 2011 Production of activated carbon from industrial bamboo waste *Journal of Tropical Forest Science* **23** 417-24.
[17] Ademiluyi F T and Braide O 2012 Effectiveness of Nigerian bamboo activated with different activating agents on the adsorption of *BTX J. Appl. Sci. Environ. Manage.* **16** 267-73.

[18] Itoh T and Shimaji K 1981 Lignification of bamboo culm (phyllostachys pubescens) during its growth and maturation *Bamboo Production and Utilization Proc. XVII IUFRO Congress Group 5.3. Ed. T. Higuchi Kyoto Japan* **10** 101-4

[19] Li X B, Shupe F T, Peter G F, Hse C Y and Eberhardt T L 2007 Chemical changes with maturation of the bamboo species phyllostachys pubescens *Journal of Tropical Forest Science* **19** 6-12.

[20] Edward L K M, Cheung W H, Vinci K C L and Gordon M 2008 Kinetic study on bamboo pyrolysis *International Engineering Chemical Resources* **47** 5710-22.

[21] Norhusna M N, Lau L C, Lee K T and Abdul R M 2013 Synthesis of activated carbon from lignocellulosic biomass and its applications in air pollution control: a review *Journal of Environmental Chemical Engineering* **1** 658–66.

[22] Daud WMAW, Ali WSW and Sulaiman MZ 2000 The effects of carbonization temperature on pore development in palm-shell-based activated carbon *Carbon* **38** 1925-32

[23] Scurlock J M O, Dayton D C and Hames B 2000 Bamboo: an overlooked biomass resource *Biomass & Bioenergy* **19** 229-44.

[24] Sánchez-Echeverri L A, Aita G, Robert D and Garcia M E R 2014 Correlation between chemical compounds and mechanical response in culms of two different ages of guadua angustifolia kunth *Verano* **20** 87-94.

[25] Ozcimen D, Ersoy-Mericboyu A 2010 Adsorption of copper (II) ions onto hazelnut shell and apricot stone activated carbons *Adsorp Sci Technol* **28** (4) 601-604

[26] Mohd Adib Yahya, Z. Al-Qodah, C.W. 2015 Zanariah Ngah. Agriculture bio-waste materials as potential sustainable precursors used for activated carbon production: A review *Renewable and Sustainable Energy Reviews* **46** 218-35.

[27] Krzesinska M, Zachariasz J, Muszynski J and Czajkowska S 2008 The thermal decomposition studies of solid iron bamboo (*Dendrocalamus stricus*)-potential precursor for eco-materials *Bioresources Technology* **99** 5110-114.

IOP Conf. Series: Materials Science and Engineering **539** (2019) 012011 doi:10.1088/1757-899X/539/1/012011

Activated carbon characteristics of tabah bamboo that physically activated under different activation time

D N K P Negara[1,2], T G T Nindhia[2], I W Surata[2], M Sucipta[2] and F Hidajat[3]

[1] Doctoral Study Program of Engineering Science, Faculty of Engineering, Udayana University, Kampus Sudirman, Denpasar-Bali, Indonesia

[2] Mechanical Engineering Department of Udayana University, Street of Bukit Jimbaran, Badung, Bali, Indonesia

[3] PPPTMGB "Lemigas", Jl. Ciledug Raya Kav. 109, Jakarta Selatan, 12230

Email: devputranegara@gmail.com

Abstract. Activated carbon is a versatile porous material that widely applied to purify, deodorize, decolorize, remove or reduce some gases harmful constituents and storage of gas. For different application different characteristics of activated carbon is required. This paper concerns to find out the characteristics of activated carbons derived from tabah bamboo (*Gigantochloc nigrociliata*) that physically activated with different activation holding time. Prepared tabah bamboo was carbonized by heating up to the temperature of 800 ^{O}C for two hours. The char produced was powdered and then activated physically by heating up to 800 ^{O}C, soaked at that temperature during each 1, 1.5 and 2 hours under 100 mL/min N_2 flow. Activated carbons manufactured were characterized by proximate and ultimate analyses and SEM observation. The results show that the highest fix carbon (82.52%) and lowest ash (5.91 %) was obtained at 1 hour soaking time with moisture and volatile contents were 3.56% and 8.00 % respectively; the highest carbon (C) content of 84.16 % was yielded at soaked time of 1.5 hours; SEM observation shows that there were significant differences of morphology structure between raw material and activated carbons for all of the soaking time variations. The pores structures have been formed in the activated carbons produced.

1. Introduction

Activated carbon is a multifunction and versatile material. It is a very unique material due to having porosity in the size of molecules [1]. The pores of activated carbon have intense van der Waals forces and associated with its adsorption ability. Because of its high porosity [2] and very huge surface area [3], activated carbon has high adsorption capacity [4]. Activated carbons are widely used to purify water from industrial waste [5, 6], to storage of gas [7, 8], for water treatment [9], as a catalyst [10], as a material of double layer capacitor [11] and many others. Due to the wider range of its use, the need for activated carbon will continually increase.

Generally, commercial activated carbon is manufactured from coal, a non-renewable source, having a limitation of availability. To overcome this limitation, biomass has become one of alternative raw material for the production of AC. Various of biomass have been researched and produced as activated carbons, including palm [6], coconut shell [12, 13], bamboo [5, 6, 14-16], coffee endocarp [17] and fine cone [18]. Bamboo is a biomass resource naturally growth in Indonesia and traditionally used as a structural material at traditional building and as traditional tools. Especially in Bali, bamboo has been used to make handicraft and it is an important material in ceremony religion. Commonly, bamboo is very fast growth [19] and available in a large volume. Another aspect causing it suitable for a resource of activated carbon is its chemical composition. Commonly, the contents of lignin and alpha cellulose of bamboo are around 20 – 26% and 40-50%, respectively [20]. These cellulose and lignin contribute to the micropores and macropores formed of activated carbon, respectively.

In order to convert bamboo to activated carbon is undertaken through carbonization and activation processes. Carbonization is a process to convert of raw material become charcoal [21] and to create initial porosity [22]. This process eliminates non-carbon elements through thermal decomposition creating the higher material carbon content [23]. Activation is indeed to develop advanced porosity [24] by opening inaccessible initial pores, development of new pores and widening of existing pores [25]. Carbonization temperature is generally carried out around 400 to 850 ^{O}C and activation temperature is usually in the range of 600 – 900 ^{O}C [26]. Characteristic of activated carbon is influenced by the chemical composition of raw material and parameters of the production process. In this paper, the characteristic of activated carbon prepared from tabah bamboo produced with different activation holding time was investigated. Tabah bamboo is original species of Balinese bamboo (Indonesia) having height, diameters and thickness are around 10 m, 5 cm and 6 mm, respectively. They can easily be found in almost all regency in Bali [27].

2. Method

Tabah bamboos (Gigantochloc Nigrociliata) are used as the precursor. They are cut into small pieces, dehydrated by heating under the sunshine during 8 days and then heating for 1 hour in the electric furnace at a temperature of 110 ^{O}C. Carbonization was carried out by heating the samples until the temperature of 800 ^{O}C during 2 hours. Chars produced were powdered and continued to activation process which was undertaken by heating the samples up to a temperature of 800 ^{O}C, holding for each 1, 1.5 and 2 hours under 100 mL/min nitrogen flow. Activated carbons produced were signed as FA1, FA1.5, and FA2 for holding activation of 1, 1.5 and 2 hours respectively. They are then characterized by proximate test (TGA 701, 0.02 % RSD Precision), ultimate test (elemental determination machine CHN628S) and SEM observation. The chemical composition of tabah bamboo was determined by Van Soest analysis.

3. Results and discussion

3.1. Characteristics of tabah bamboo

Characteristics of tabah bamboo have been evaluated at previous work [28] and illustrated in Figure 1. With its chemical contents (44.94 % cellulose, 22.92 % lignin), C content of 42.47% and low ash content (2.92%), tabah bamboo has great potential as a precursor in the production of high quality of activated carbon. Formation of micropores and macropores of activated carbon are strongly influenced by cellulose and lignin content of raw material [29]. The relatively high content of C has also high potency to yield activated carbon with high C content.

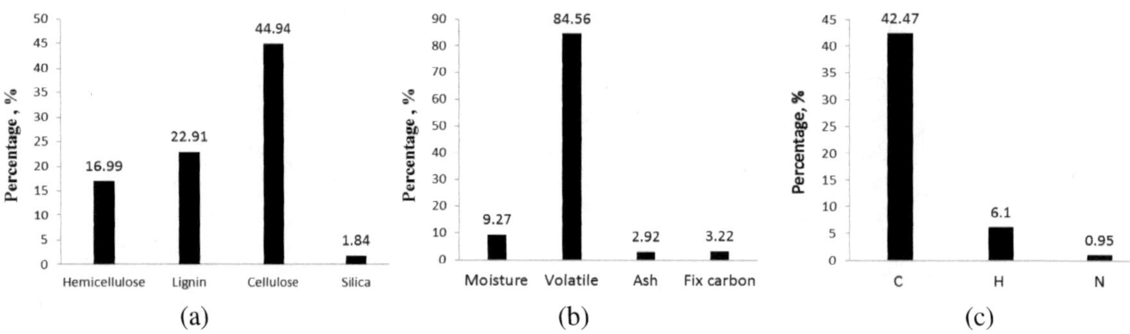

Figure 1. Characteristics of tabah bamboo, (a) Chemical composition (b) Proximate analysis (c) Ultimate analysis.

3.2. Characteristics of activated carbons produced

3.2.1. Proximate and ultimate analysis

Proximate and ultimate analysis of char and activated carbons produced (FA1, FA1.5, and FA2) are respectively shown in tables 1 and 2. From Figure 1 (b) and table 1 can be observed that fix carbon of swat bamboo increases around 96 % from 3.22 % to 80.55 % and ash content also increase by

IOP Conf. Series: Materials Science and Engineering **539** (2019) 012011 doi:10.1088/1757-899X/539/1/012011

approximately 53.35 %. After activation, fix carbon content increases 2.26 % on average from fix carbon of char. The fixed carbon of activated carbons produced has fulfilled the fixed carbon minimal requirement of Indonesian National Standard for activated carbon (SNI 06-3730-1995). Based on SNI 06-3730-1995, the content of fixed carbon is minimal of 65% and the maximal contents of ash, moisture and volatile is 10 %, 15 %, and 25% respectively. All of the activated carbons yielded qualified for this entire standard requirement.

Table 1. Ultimate analysis of char and activated carbons produced

Samples	Proximate Analysis Contents (%)			
	Moisture	Volatile	Ash	Fix Carbon
Char	5.58	7.61	6.26	80.55
FA1	3.56	8.00	5.91	82.52
FA1.5	3.31	7.59	6.81	82.28
FA2	3.58	7.48	6.50	82.44

Table 2. Proximate analysis of char and activated carbons produced

Samples	Elemental Contents (%)		
	C	H	N
Char	82.85	1.99	0.41
FA1	84.08	1.46	0.70
AF1.5	84.16	1.30	0.57
FA2	83.39	1.40	0.49

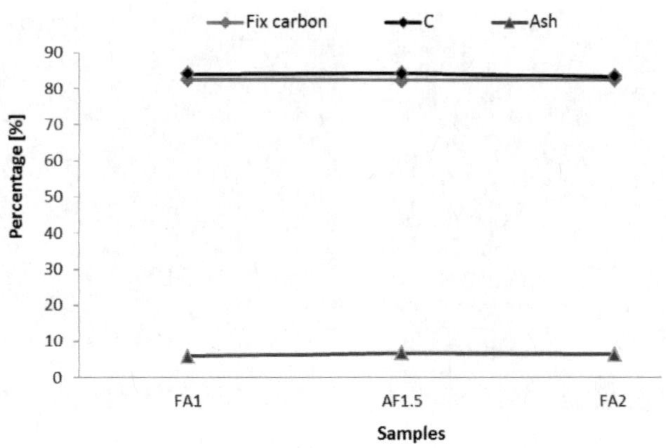

Figure 2. The fixed carbon, carbon and ash contents of activated carbons produced

Ultimate analysis (Table 2) shows that C content of activated carbons also increases compared to the C content of char. However, there is no trend can be observed associated relation between activation holding time and fix carbon and carbon contents of activated carbons yielded. The highest fix carbon is obtained when activation holding time is 1 hour (sample FA1), meanwhile, the highest C content is reached at activation holding time of 2 hours. From figure 2 can be seen that there is no significant difference in fix carbon and carbon contents of activated carbons under different activation soaking time (1, 1.5 and 2 hours).

IOP Conf. Series: Materials Science and Engineering **539** (2019) 012011 doi:10.1088/1757-899X/539/1/012011

3.2.2. Morphology microstructure

SEM observations on the char and activated carbons result in the morphology microstructure as shown in Figure 3, 4, 5 and 6 for char, FA1, FA1.5, and FA2 respectively. Morphology microstructure of tabah bamboo, as shown in Figure 3, is smooth and no porosity can be observed. After activation, the morphology microstructure changed significantly, as shown in Figures 4, 5, and 6. Porosities appear as a result of carbonization and activation. The porosities formed due to different activation soaking time are also difficult to be distinguished. These morphology microstructure images are qualitative data that is difficult to make a comparison accurately. Advanced study is needed in order to find out quantitative data such as pore diameter, pore volume and surface area of activated carbons. However, clearly can be said that in the activated carbons produced have formed porosities, the place where the adsorption process occurred.

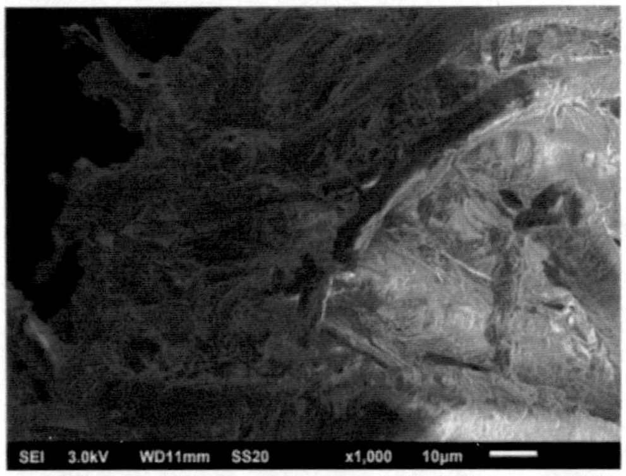

Figure 3. Morphology microstructure of tabah bamboo

Figure 4. Morphology microstructure of FA1

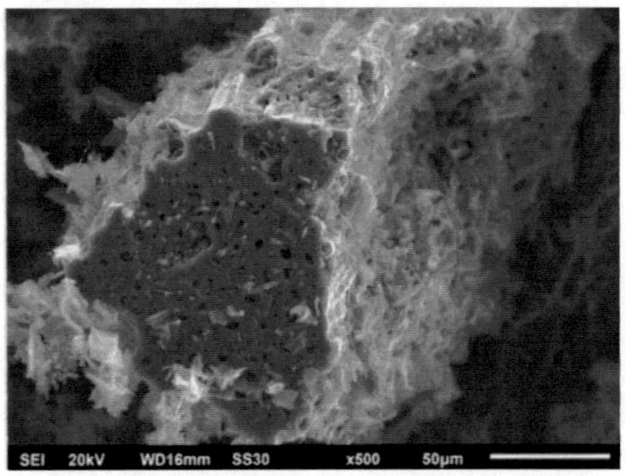

Figure 5. Morphology microstructure of FA1.5

Figure 6. Morphology microstructure of FA2

4. Conclusions

Activated carbons produced have qualified to fulfill the Indonesian National Standard of activated carbons (SNI 06-3730-1995) such as the contents of ash, moisture, fix carbon, volatile and carbon. The effect of activation soaking time on the proximate and elemental components of activated carbon yielded is no obviously can be observed. Based on SEM observation, it clearly can be said that in the activated carbons yielded have formed of pores wherein these pores the adsorption process take place.

References

[1] Jin X J, Yu Z M and Wu Y 2010 Preparation of activated carbon from lignin obtained by straw pulping by KOH and K_2CO_3 chemical activation *Cellul. Chem. Technol.* **46** 79-85.

[2] Hu Z and Srinivasan M P 2001 Mesoporous high-surface-area activated carbon *Micro. Meso. Mater.* **43** 267-75.

[3] Sahu J N, Acharya J and Meikap B C 2010 Optimization of production conditions for activated carbons from tamarind wood by zinc chloride using response surface methodology *Bioresource Technology* **101** 1974-82.

[4] Idris S, Iyaka Y A, Dauda B E N, Ndamitso M M and Umar M T 2012 Kinetic study of utilizing ground nut shell as an adsorbent in removing chromium and nickel from dye effluent *Am. Chem. Sci. J.* **2** 12–24.

[5] Cheung W H, Lau S S Y, Leung S Y, Ip A W M and McKay G 2012 Characteristics of chemical modified activated carbons from bamboo scaffolding *Chinese Journal of Chemical Engineering* **20** 515-35.

[6] Isabel A A C E, Marta S S L, Pedro M C N and Jos´e P B M 2008 Adsorption of natural gas and biogas components on activated carbon *Separation and Purification Technology* **62** 281–96.

[7] Inomata K, Kanazawa K, Urabe Y, Hosono H and Araki T 2002 Natural gas storage in activated carbon pellets without a binder *Carbon* **40** 87-93.

[8] Diana C S A, J Cassia S A, Moisés B N, A Eurico B T, Emerson F J and Celio L C 2007 Microporous activated carbon prepared from coconut shells using chemical activation with zinc chloride *Microporous and Mesoporous Materials* **100** 361-64.

[9] Zhang T, Walawender W P and and Fan L T 2010 Grain-based activated carbons for natural gas storage *Bioresource Technology* **101** 1983-91.

[10] Joanna S N, Weronika K, Beata M and Zvi C K 2013 Production, characterization and methane storage potential of KOH-activated carbon from sugarcane molasses *Industrial Crops and Products* **47** 153-59.

[11] Patil B S and Kulkarni K S 2012 Development of high surface area activated carbon from waste material *International Journal of Advanced Engineering and Studies (IJAERS)* **1** 109-13

[12] Dipa D, Debi P S and and Meikap B C 2015 Preparation of activated carbon from green coconut shell and its characterization *J. Chem. Eng. Process Technol.* **6** 248 doi:10.4172/2157-7048.1000248

[13] Ahmadpour A, Okhovat A, and M J Darabi M 2013 Pore size distribution analysis of activated carbons prepared from coconut shell using methane adsorption data *Journal of Physics and Chemistry of Solids* **74** 886-91.

[14] Koo W K, Gani NA, Shamsuddin M S, Subki N S, and Sulaiman M A 2015 Comparison of wastewater treatment using activated carbon from bamboo and oil palm: an overview *Journal of Tropical and Resource Sustainable Science* **3** 54-60.

[15] Mahanim S M A, Asma I W, Rafidah J, Puad E and Shaharuddin H 2011 Production of activated carbon from industrial bamboo waste *Journal of Tropical Forest Science* **23** 417-24.

[16] Ademiluyi F T and Braide O 2012 Effectiveness of Nigerian bamboo activated with different activating agents on the adsorption of *BTX J. Appl. Sci. Environ. Manage.* **16** 267-73.

[17] Tianfu H, Zehai Q, Dewu W and Zhibiao H 2015 Bamboo-based activated carbon @ MnO2 nanocomposites for flexible high-performance supercapacitor electrode materials *Int. J. Electrochem. Sci.* **10** 6312 - 23.

[18] Itoh T and Shimaji K 1981 Lignification of bamboo culm (phyllostachys pubescens) during its growth and maturation *Bamboo Production and Utilization Proc. XVII IUFRO Congress Group 5.3. Ed. T. Higuchi Kyoto Japan* **10** 101-4

[19] Lus A S C, Giovanna A, Diana R and Mario E R G 2014 Correlation between chemical compounds and mechanical response in culms of two different ages of guadua angustifolia kunth, *Madera Bosques* **20** 87-94.

IOP Conf. Series: Materials Science and Engineering **539** (2019) 012011 doi:10.1088/1757-899X/539/1/012011

[20] Keith K H C, John P B and Gordon M 2005 Production of activated carbon from bamboo scaffolding waste-process design, evaluation and sensitivity analysis *Chemical Engineering Journal* **109** 147–65.

[21] Norhusna M N, Lau L C, Lee K T and Abdul R M 2013 Synthesis of activated carbon from lignocellulosic biomass and its applications in air pollution control: a review *Journal of Environmental Chemical Engineering* **1** 658–66.

[22] Sheng F L, Song Y W, Ming J T and Lang D L 2012 Adsorption capacity and removal efficiency of heavy metal ions by moso and ma bamboo activated carbons *Chemical Engineering Research and Design* **90** 1397–406.

[23] Lozano-Castello D, Alcan˜iz-Monge J, M. A. de la Casa-Lillo, Cazorla-Amoro´s D and Linares-Solano A 2002 Advances in the study of methane storage in porous carbonaceous materials *Fuel* **81** 1777-803.

[24] Edward L K M, Cheung W H, Vinci K C L and Gordon M 2008 Kinetic study on bamboo pyrolysis *International Engineering Chemical Resources* **47** 5710-22.

[25] Li X B, Shupe F T, Peter G F, Hse C Y and Eberhardt T L 2007 Chemical changes with maturation of the bamboo species phyllostachys pubescens *Journal of Tropical Forest Science* **19** 6-12.

[26] Zhao R S, Yuan J P, Jiang T, Shi J B, and Cheng C C 2008 Application of bamboo charcoal as solid-phase extraction adsorbent for the determination of atrazine and simazine in enviromental water samples by high-performance liquid chromatography-ultraviolet detector *Talanta* **76** 956-59.

[27] Scurlock J M O, Dayton D C and Hames B 2000 Bamboo: an overlooked biomass resource *Biomass & Bioenergy* **19** 229-44.

[28] Putra Negara, D N K, Tirta Nindhia, T G, Surata, I W and Sucipta, M. Chemical, Strength and Microstructure Characterization of Balinese Bamboos as Activated Carbon Source for Adsorbed Natural Gas Application. 7th International Conference on Key Engineering Materials (ICKEM). IOP Conf. Series: Materials Science and Engineering 201 012033, 2017.

[29] Sánchez-Echeverri L A, Aita G, Robert D and Garcia M E R 2014 Correlation between chemical compounds and mechanical response in culms of two different ages of guadua angustifolia kunth *Verano* **20** 87-94.

IOP Conf. Series: Materials Science and Engineering **539** (2019) 012012 doi:10.1088/1757-899X/539/1/012012

Hardness and wear rate of basalt/alumina/ shellfish powder reinforced phenolic resin matrix hybrid composite brake lining pads

I K A Atmika[1], I D G A Subagia[1], I W Surata[1] and I N Sutantra[2]
[1]Mechanical Engineering Departement, Udayana University, Bali-Indonesia 80362
[2]Mechanical Engineering Departement, Institut of Sepuluh Nopember, Surabaya-Indonesia 60117

Email: tutadi@unud.ac.id

Abstract. This paper explain the behavior of hybrid composites phenolic resin with basalt/alumina/ shellfish powder reinforced on brake lining pad. Hybrid composite brake has been manufactured use compaction and sintering process through any steps, that an emphasis of 2,000 kg for 30 minutes at a constant temperature of 150° C. The research aims to investigate hardness characteristic and wear rate of hybrid composite that test using the Vickers according to standard ASTM E-384. The reinforced materials and matrix polymer composition is 60%: 40%. The results show for the average hardness VHN to 24.18, 25.11, 26.34, 27.21 and 28.83. The average hardness hybrid composite shows the hardness harder than asbestos materials. The value of asbestos lining pads wear is 0.00011 g/m, while the wear value of composite hybrid brake lining pads is 0.00009 g/ m to 0.000071 g/m.

1. Introduction

One component of the vehicle that has a very important role for the comfort and safety of the rider is the braking system. This system serves to slow or stop the vehicle by providing a braking force that is forwarded to the friction plate and brake pad. Friction plates and brake lining are an important part of the braking system in motor vehicles so deserve more attention to protect consumers and reduce the percentage of causes of road accidents. Brake lining is a component that directly rubs against the rotating parts in the braking system according to the type and brake used. In general, the brake lining material is made of asbestos because it is superior in temperature resistance up to 800^{0}C [1] and is also good in sound absorption and low in water absorption [2]. However, asbestos material has been closed because it has carcinogenic properties that have a negative impact on the environment and human health [3]. Then, the asbestos fiber material is replaced with aramid fibers, Kevlar, Twaron, rock wool, fiberglass, potassium titanate, carbon fiber, graphite, cellulose, vermiculite, steel fiber, BaSO4, resin, nitrile butadiene rubber.

Utilization of hybrid composite materials on motor vehicle components has become popular because it effectively increases the stress and strain due to the imposition of impact as well as reduce the cost of reinforcing material [4]. Hibridisasi is the process of forming a composite material with new traits which made merging two or more materials that have characteristics which differ in a matrix binder (thermoplastic or thermosetting) or otherwise of the reinforcing fibers bonded with two or more types / phase matrix of different [5]. Nowadays a lot of research on hybrid composite material with epoxy basalt reinforced that has been done. Czigany [6], has conducted research on the characteristics of epoxy composite with basalt fiber reinforcement, and the following year a hybrid composites reinforced basalt fiber, polypropylene matrix has been produced, in which the strength properties of the hybrid composites due to increased surface treatment after mechanical tests and microscopic analysis. Besides it also studied the characteristics of basalt fiber hybrid composite laminates due to the imposition of tensile and flexural [7]. Liu, et al [8] and Atmika, et al [9] conduct research of mechanical properties of basalt fiber composite with the transport system where the basalt is very possible material applied to the transportation environment.

In connection with the development of new materials for brake lining, basalt was introduced as one of the hybrid composite formers. Basalt is a natural fiber produced by volcanic eruptions that have heat resistance up to 150^{0}C [10], resistant to corrosion, low water absorption and resistant to chemical treatments and non-toxic [11]. In the last five years, research related to basalt characteristics has been carried out with the conclusion that basalt material has excellent physical and mechanical properties, high ductility, and high wear resistance [12], and can replace glass fibers [13]. Then the most important basal properties are having low thermal conductivity. This property is used to overcome potential brake lining problems lately.

In the context of the development of brake lining materials, the basalt powder hybridization method is used. Basalt powder is combined with other natural ingredients, namely shellfish powder and alumina which are bonded with epoxy as a composite material. Composite hybrids have been manufactured using five types of weight fraction ratio. This study aims to determine the characteristics of hardness and wear rate of the developed hybrid composite.

2. Materials and Method

Basalt as described above, comes from hardening of the lava mountain (figure 1). In this study, basalt powder, alumina powder and shellfish powder with a size of 0.0074 millimeters were used. Mechanical properties of basalt are shown in Table 1. Shellfish shell powder material consisted of 66.70% CaO, 22.28% MgO, 7.88% SiO_2, 1.25% Al_2O_3 and 0.03% Fe_2O_3. The phenolic resin material used in this study coded PR-51 510, which is one type of resin in powder form.

In this study brake pad material is made by mixing, compacting and sintering processes. The shape and size of the test specimen are as shown in Figure 2. Composite composition shown in Table 2 is based on ASTM D standard 3171-09. In the process of making brake lining material it is pressed to load 2000 N for approximately 30 minutes, and the sintering process at a temperature of 150°C.

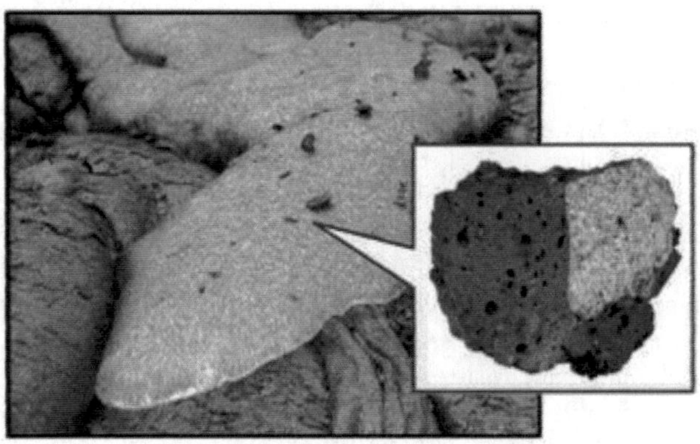

Figure 1. Basalt

Table 1. Mechanical properties of basalt material

Properties of Basalt	Value (unity)
Density	2600-2630 (kg/m^3)
Tensile strength	500k-550k (psi)
Sintering Temperatur	1050 (°C)
Operation Temperatur	-265-+700 (°C)
Modulus of elastisitas	9100-1100 (kg/mm^3)
Mohs Hardness @20°C	5-9
Melting point	1170(°C)
Heat resistance	700-1000 (deg.C)
Elongation at break	3.15(%)

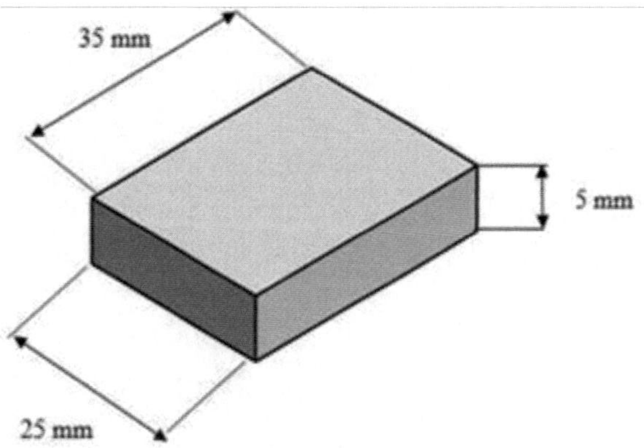

Figure 2. Specimen geometry

Table 2. The weight fraction ratio of hybrid composite

Composition	Basalt powder	Clamshell powder	Alumina powder	Phenolic Resin
I	45	5	10	40
II	40	10	10	40
III	35	15	10	40
IV	30	20	10	40
V	25	25	10	40

Hardness test use Vickers method according to ASTM E384-99 (Standard test method for micro-indentation hardness of materials). The wear rate is tested by the pin on disc method. Each specimen was tested with a 36 N load in 83.3 minutes or 200 meters with a 120 rpm engine speed.

3. Result and Discussion

The average result of VHN (Vickers Hardness Number) of each composition and comparison with the hardness of the brake lining pads that asbestos material (X) are shown in figure 3, while the wear rate of brake pads is shown in Figure 4

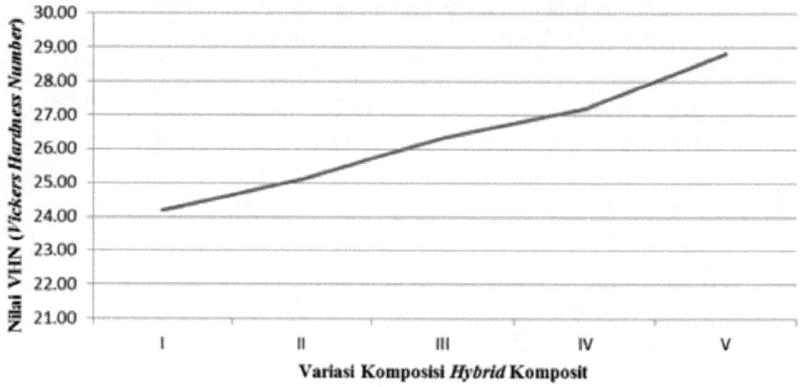

(a)

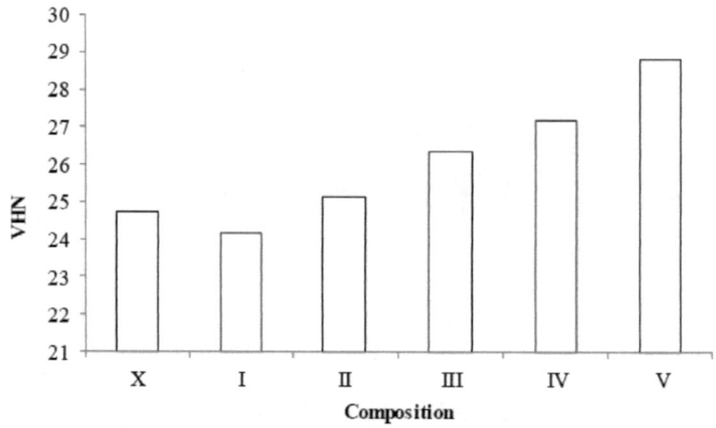

(b)

Figure 3. Vicker Hardness Number of the hybrid composite and asbestos

Figure 3a shows a hardness diagram of hybrid composite brake pads and brake pads from asbestos material. From the diagram the hardness value of composition I is 24.72, the hardness value of composition II is 25.40, the hardness of composition III is 25.67, the hardness of composition IV is 25.74, and the hardness of composition V is 26.55. Thus the amount of basalt powder has a significant influence on the level of hardness of the brake lining material made. Then from Figure 3b shows that the asbestos brake pad has a hardness value between composition I and composition II, so that basalt powder and shellfish powder are natural alternative materials that can be used as an alternative brake lining material for asbestos brake pads.

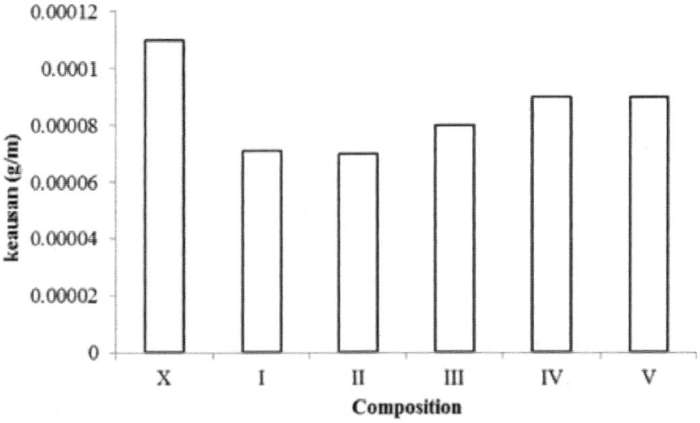

Figure 4. Wear rate of the hybrid composite and asbestos

Figure 4 shows a graph of wear value for the five ingredients of hybrid composite brake pads and asbestos brake pads. From the picture shows that the decreasing amount of basalt content in the specimen the greater the wear value. The value of asbestos lining wear is 0.00011 g / m, while the wear value of composite hybrid brake pads is 0.00009 g / m to 0.000071 g / m. Thus the composite hybrid brake pads made have a better wear value compared to the wear of asbestos brake pads.

4. Conclusion

The reduction in the percentage of basalt powder and the addition of the percentage of shellfish powder, and the percentage of constant alumina results in a higher hardness value. The hardness value of composite hybrid brake pads is not much different from the value of asbestos brake pads. The reduction in basalt percentage in each composition results in increased wear value but not so significant. The wear value of hybrid composite brake pads made is still smaller than the wear value of asbestos brake pads, therefore hybrid composite brake pads have the potential as a substitute for asbestos materials for environmentally friendly brake pads.

References

[1] Sivarao, M. Amarnath, M.S.Rizal, and Kamely, A. 2009. An investigation toward development of economical brake lining wear alert system. International Journal of Engineering & Technology IJET-IJENS vol. 09, No. 09, pp. 63-68
[2] Aigbodion, V. S. , Akadike, Hassan, U.S.B., Asuke, F., and Agunsoye, J.O.. 2010. *Development of asbestos - free brake pad using bagasse*, Tribology in industry, vol. 32, No.1, pp. 12-18.
[3] Louis St, 2004. U.S. Survey shows imports of asbestos brake materials increasing. St. Louis.
[4] Kedar S Pandya, Ch. Veerraju, and N. K. Naik. 2011. Hybrid composites made of carbon and glass woven fabrics under quasi-static loading. Materials and Design, vol. 32 pp. 4094–4099.
[5] Ashby, M. F. and Y. J. M. Brechet,. 2003. Designing hybrid materials. Acta Materialia, vol. 51, pp. 5801-5821.
[6] Czigany, T. 2005. Basalt fiber reinforced hybrid polymer composites. Materials Science, Testing and Informatics Ii, vol. 473-474, pp. 59-66.
[7] Ary Subagia, I.D.G., and Y. Kim. 2013. A study on flexural properties of carbon-basalt/epoxy hybrid composites. Journal of Mechanical Science and Technology, vol. 27 (4), p. 987~992.
[8] Qiang Liu, Montgomery T. Shaw, and Parnas, R. S. 2006. Investigation of basalt fiber composite mechanical properties for applications in transportation. POLYMER COMPOSITES, vol. 27, pp. 41-48.

[9] Atmika I.K.A, Subagia I.D.G.A, Surata I.W, Sutantra I.N. 2017. Study of the mechanical properties of hybrid composite basalt/alumina/shells for brake lining pads. IOP Conf. Series: Materials Science and Engineering 201(2017) 012009.
[10] Singha, K. 2012. A short review on basalt fiber. International Journal of Textile Science, vol. 1 (4), pp. 19-28.
[11] Fiore, V. G. Di Bella, and Valenza, A. 2011. Glass–basalt/epoxy hybrid composites for marine applications, Materials and Design, vol. 32, pp. 2091–2099.
[12] Dalinkevich A.A., Gumargalieva K.Z, Marakhovsky S.S, and Soukhanov A.V. 2009. Modern basalt fibrous materials and basalt fiber-based polymeric composites. Journal of Natural Fibers, vol. 6, pp. 248-271.
[13] Lopresto, V.C. Leone, and I. De Iorio,. 2011. Mechanical characterisation of basalt fibre reinforced plastic. Composites Part B: Engineering, vol. 42, pp. 717–723.
[14] Hyndman R.D. and Drury M. (2013). Physical Properties Of Basalts, Gabros, and ultramafic Rocks From DSDP Leg 37. 395-401.

IOP Conf. Series: Materials Science and Engineering **539** (2019) 012013 doi:10.1088/1757-899X/539/1/012013

Experimental study on combustion characteristics of an indirect injection diesel engine fuelled by biodiesel blends

A Ghurri[1] and S K Keun[2]
[1] Department of Mechanical Engineering, Udayana University, Bali-Indonesia
[2] Chonbuk National University, RCIT, 657 Baekje-daero Deokjin-Gu Jeonju-si Jeollabuk-do 561-756, Republic of Korea

Email: a_ghurri@unud.ac.id

Abstract. An experimental investigation was conducted to evaluate the performance of an indirect injection (IDI) diesel engine using diesel (D100) and diesel-biodiesel blends (BD25, BD45, BD65) separately. The engine was run in various engine loads at constant engine speed ranging from 1000 to 2400 rpm with an interval 200 rpm. The results showed that the maximum pressure inside a cylinder of D100 is slightly higher than that of BD65 and other biodiesel blends. The biodiesel blends exhibited higher pressure change rate at a moment before peak pressure inside the combustion chamber. It was also clearly observed that the heat release rate of D100 is higher than those of biodiesel blends. This result is consistent with the higher engine power delivered by D100 compared to the biodiesel blends. In related to the combustion period, it was recorded that BD65 started the combustion slightly earlier than that of D100. This resulted in the shorter ignition delay of BD65 compared to D100. On the other hand, from the burnt mass data it was found that the BD65 ended the combustion later than the D100. This means the combustion duration of biodiesel blends are longer than that of diesel fuel which is worse combustion condition.

1. Introduction

The prospect of the use of biodiesel to replace fossil fuels has received much attention since the past decades. Beside its ability as an alternative for diesel fuel in engine performance side, the environmental-friendly properties of biodiesel also encouraged the use of biodiesel. The term biodiesel commonly refers to fatty acid methyl or ethyl esters made from vegetable oils or animal fats, whose properties are good enough to be used in diesel engines [1]. There have been many reports on the effect of biodiesel on diesel engine performance. The results of these works are sometimes quite diverse and even contradictory. For example, when the most of reports showed that the use of diesel and biodiesel blend decreased the engine power, Altipamark et.al [2] reported the increases in power and torque when using a blend with 70% biodiesel. Lapuerta et.al [3] conducted a massive review on the effect of biodiesel on diesel engine emissions with an enclosed brief review on the engine performance and showed varied results on the engine performance and emissions. The lower LHV of biodiesel compared to diesel fuel got much attention in the engine performance analysis. Meanwhile, the role of the cetane number and diesel engine type had less attention. In more specific biodiesel type, Enweremadu & Rutto [4] reviewed the effect of used cooking oil biodiesel on combustion, emission, and engine performance characteristics. In spite of that, relatively high disparity results were still found especially for emission characteristics. More recently Xue et.al [5] conducted a similar review and still reported inconsistent trends for biodiesel engine performances and its emissions due to the different tested engines, the different operating conditions or driving cycles, the differently used biodiesel or reference diesel, the different measurement techniques or instruments, etc. The comparison with other reports in such review paper [3-5] gave valuable insight into the role of fuel properties to the engine performance and emission characteristics.

In relation to the following review to other reports, we gave deeper attention to the diesel engine type (i.e. DI or IDI), and three fuel properties i.e. fuel density/viscosity, fuel Cetane number, and fuel heating value to get a more comprehensive explanation on the contribution of these parameters to engine performance characteristics. The higher density and viscosity of biodiesel will have an effect to fuel compression process in the injection pump i.e. causes faster increase to reach the nozzle opening pressure. As a result, this can cause an advance injection and/or ignition timing. This also causes poor atomization compared to lower density fuel. The higher density and viscosity will also result in longer spray tip penetration and lower spray cone angle that is important in DI than IDI diesel engine. Cetane number is a measure of ignition quality. Cetane number is actually a measure of ignition delay. Higher cetane number will have a shorter ignition delay that will improve the combustion process. Higher cetane numbers indicate shorter times between timing of injection and ignition of the fuel [6]. The different type of biodiesel may have higher or lower cetane number compared to diesel fuel. Biodiesel generally has oxygen content around 10% to 12% compared to zero oxygen content in diesel fuel. The oxygen content in the biodiesel has two impacts; *firstly*, lowered the fuel heating value; and *secondly*, improved combustion by providing more complete combustion. The lower heating value of the fuel is a very important property that is related to the available energy in the fuel. Biodiesel generally has lower LHV, higher density, and viscosity, but may have higher or lower cetane number compared to diesel fuel. These fuel properties may come contrarily into the biodiesel, so it is important to give more attention to assess the overall effect of these properties to the combustion process.

As for engine type, it is important to distinguish DI and IDI diesel engine because of the difference in nature of the air-fuel mixing in the combustion chamber. In an IDI diesel engine, air is pushed into the swirl pre-chamber on the cylinder head by the piston and starts swirling rapidly, which promotes a good mixing when the fuel is sprayed. This makes a better mixture of the air and fuel, which improves combustion. Preliminary combustion of the mixture starts and heat rises, forcing the remaining unburned fuel into the chamber at high velocity, where it mixes well with the air and continues the complete combustion [7, 8]. In the indirect injection combustion process, air-fuel mixing is more important. Meanwhile, in a direct injection system, fuel is introduced directly into the combustion chamber, and as a result, the fuel spray characteristics have a bigger contribution to improve the combustion.

The decrease of torque was reported by most of the authors [9-14] for IDI and [15-22] for DI diesel engines with all mentioned the lower LHV of biodiesel as the main cause. Laforgia and Ardito [9] used IDI diesel engine and neat rapeseed biodiesel with 5.1% higher density and 12.9% lower LHV compared to those of the diesel fuels but with a higher cetane number of the biodiesel in their experiment; and reported the decrease of torque at level 2%. Ryu and Oh [10-11] showed similar results when used biodiesel with 5.1% higher density, 14.6% lower LHV, and higher of cetane number of biodiesel in an IDI diesel engine; and the results showed a slight decrease of torque with BD100. A relatively bigger decrease of torque occurred for biodiesel with significant lower both cetane number and LHV compared to that of diesel fuel are presented by Cetin & Yuksel [12] that used hazelnut oil with 5.8% higher density, 27.2% lower LHV, and lower cetane number of biodiesel in an IDI diesel engine and reported around 7% decrease of the torque with BD100. Similarly, Buyukkaya [21] reported a 4.9% decrease of torque when using BD100 in DI diesel engine with 9.9% higher density, 14.4% lower LHV, and lower cetane number, respectively. The lower cetane number of biodiesel might result in the much worse combustion as shown in the higher decrease of torque. It is difficult to find the difference of the torque resulted by DI and IDI diesel engine since both showed a different level of the decrease due to the difference in LHV and cetane number.

The reduced of power was reported by the most of authors [9-12, 23] for IDI and [15-22, 24-25] for DI diesel engines in various level of decrease with all mentioned the lower LHV of biodiesel as the possible cause. There is an exception result from Altiparmak et.al work [2] that reported the increase of power when using very high density of biodiesel (922 kg/m^3; compared to 835 kg/m3 of the diesel fuel) and higher cetane number of biodiesel (54; compared to 43.76 of the diesel fuel). Similar to the torque, the bigger power decrease is found in the present study and Cetin and Yuksel [12] and Cetinkaya et.al [16] when the lower LHV and lower cetane number of biodiesel to be present together in the biodiesel. Cetin & Yuksel [12] reported the power decrease up to 15% while Cetinkaya et.al

[16] reported the decrease up to 9% when using hazelnut oil and used cooking methyl esters, respectively. In the reports mentioned above the level of power, decrease varied strongly from almost no difference to less than 8% compared to the diesel fuel.

In related to brake-thermal efficiency, some authors noted the increase when using biodiesel blends in IDI engine [9, 13, 26] and DI engine [21, 27]. The possible reasons for the increase of brake thermal efficiency are better combustion due to the additional oxygen content [13, 26] and better-improved lubricity by the biodiesel content [21, 28, 29]. Rakopoulos et.al [27] reported the increase of brake thermal efficiency for medium load and inconsistent trend for high load, and this was down to the small uncertainty in the measurements of the fuel heating value and consumption rates. On the other hand, Ozsezen et.al [18] and Behcet [22] noted the decrease of brake thermal efficiency when using biodiesel blends. Ozsezen et.al [18] simply mentioned the higher brake specific fuel consumption and lower energy content; while Behcet [22] mentioned the longer ignition delay due to high viscosity and density of biodiesel that resulting in poor atomization and incomplete combustion as the possible reasons. In respect to the diesel engine type, generally, the DI diesel engine [18, 21] has higher thermal efficiency than the IDI diesel engine [9, 13, 26]. This is understandable since the IDI diesel engine has additional pre-chamber that enhanced the surface area and the heat loss.

The increase of the brake specific fuel consumption and the decrease of the brake specific energy consumption were reported by all authors of reviewed paper in the present study both in IDI [9-14, 26, 30] and DI [2, 17-19, 21-22, 24-25, 27, 31] diesel engine regardless of the cetane number of biodiesel lower or higher than the diesel fuel.

Several authors conveyed that the peak cylinder pressure by diesel fuel is higher compared to the biodiesel and its blends regardless of the cetane number and diesel engine type [12-13, 21, 30-32] and explained the higher energy supplied by diesel fuel as the reason. Some other authors reported that biodiesel and its blends resulted in higher peak pressure than the diesel fuel in spite of the power decreased with the biodiesel blends [9-11, 18]. Ryu and Oh [10-11] mentioned the contribution of oxygen, while Laforgia [9] stated as the worse combustion due to the longer ignition delay and combustion duration. Ozsezen et.al [18] described the higher BSFC amount, higher cetane number, oxygen content and advanced of the start of injection timing as the reasons. As a note, in all those reports [9-11, 18] the cetane number of biodiesel is higher than that of diesel fuel. Sahoo and Das [33] reported in their work that biodiesel and its blends from Jatropha, Karanja, and Polanga resulted in higher peak cylinder pressure without information on the cetane number and engine power.

Several authors proved that biodiesel resulted in an earlier start of combustion compared to diesel fuel regardless of the value of cetane number [9-11, 18, 21, 31]. It seemed caused by higher density and viscosity that affected faster pressure increase in fuel delivery and faster injector opening time.

In related to the emission, by Altiparmak et.al [2] reported that NOx increased up to 30% when using biodiesel blends. However, several reports showed a lower increase at around 6%-10% [21, 22, 25]. In related to smoke opacity, another main pollutant from biodiesel, some reports showed the small decrease of smoke opacity at around 10%-25% [13, 22, 29], while some authors reported the decrease of smoke opacity with the biodiesel contents reached 30%-46% [2, 11, 21].

In the present work, the blend of biodiesel derived from soybean oil would be tested in an IDI diesel engine. The additional raw material like waste edible oil or animal fats might be mixed to meet the biodiesel standard in Korea. The objective of this work is to investigate the combustion characteristics compared to standard diesel fuel. The data resulted from this work would be discussed and compared to other reports with giving more attention to the density, LHV, and cetane number of the biodiesel and diesel engine type and its role to the engine performance.

2. Experimental setup

The experiments were conducted on three cylinders, four-stroke, natural aspirated indirect injection diesel engine. The schematic of the experimental facility is shown in Fig. 1. The engine specifications are shown in Table 1. The engine was connected to an engine dynamometer providing maximum engine power of 74 kW. The engine was tested at various engine load, fuelled with diesel and diesel-biodiesel blends at constant engine speed ranging from 1000 to 2400 rpm with interval 200 rpm. At each speed of testing, the maximum torque of each fuel was recorded. The engine speed, torque, crankshaft position, and the cylinder pressure were recorded simultaneously into a computer connected to the engine dynamometer. The fuel consumption was measured by a fuel meter by recording the fuel delivered to the engine for a given time the engine operated. The exhaust gases produced from the diesel engine are then directed to a sampling pipe connected to the emission monitoring device MRU MGA5 during 5 - 10 minutes until the relatively constant value of NOx emission is reached. Another sampling pipe is used to be connected to opacimeter (OP-160) that monitors the smoke opacity of the exhaust gas.

The biodiesel used in this experiment was derived from soybean and waste edible oil produced by local producer. The diesel and biodiesel properties are shown in Table 2. At the engine running, the power, torque and engine speed were monitored and then recorded from 1000 rpm to 2400 rpm. The order of testing was firstly diesel fuel, then the blends of diesel and biodiesel with 65% (BD65), 25% (BD25) and 45% (BD45) in volumetric weight of biodiesel.

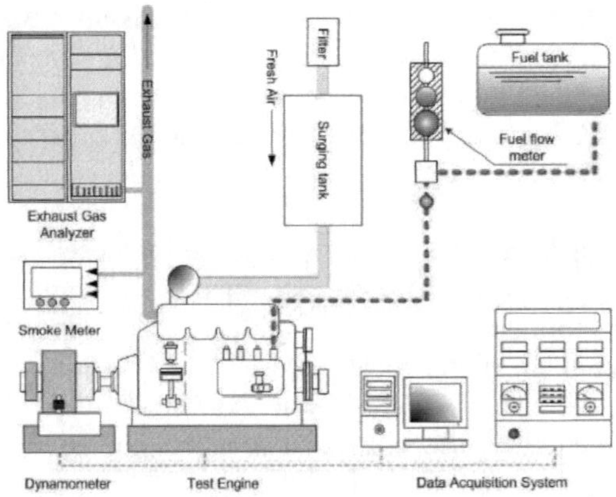

Figure 1. The experimental facilities

Table 1. Engine specifications

Item	Specification
Engine type	In-line, vertical, 4-stroke
Cylinder number	3
Bore, stroke, capacity	75 mm x70 mm, 927 cm^3
Combustion system	In-direct, vortex chamber
Compression ratio	22: 1
Injection timing (BTDC)	24°
Injection pump	Bosch K type mini pump
Injection pressure	13.7 MPa
Injection nozzle	Throttle type

Table 2. Diesel and biodiesel properties

Fuel properties	Diesel	Biodiesel
Density (kg/m^3)	850	882
Viscosity (mm^2/s)	3.25	4.3
Flash point (^oC)	68	177.9
Cetane number	54.6	54.2 ~ 58
LHV (MJ/kg)	43.15	39

3. Results and Discussion

3.1 Heat release rate

Figure 2 shows the variations of cylinder pressure with a crank angle for diesel fuel and BD65 at 2400 rpm. The pressure change rate in bar/°CA is also presented in the same graph. From this graph, it is shown that the maximum pressure inside the cylinder of D100 is slightly higher than that of BD65 and other biodiesel blends. This could be down to the higher energy supplied by diesel fuel. The cylinder pressure is deliberately shown only for D100 and BD65. The cylinder pressure for the BD25 and BD45 are between D100 and BD65. On the other hand, the biodiesel blends exhibited higher pressure change rate at a moment before peak pressure inside the combustion chamber. This could be considered as the role of the higher oxygen content in the biodiesel that improved combustion at the main combustion period.

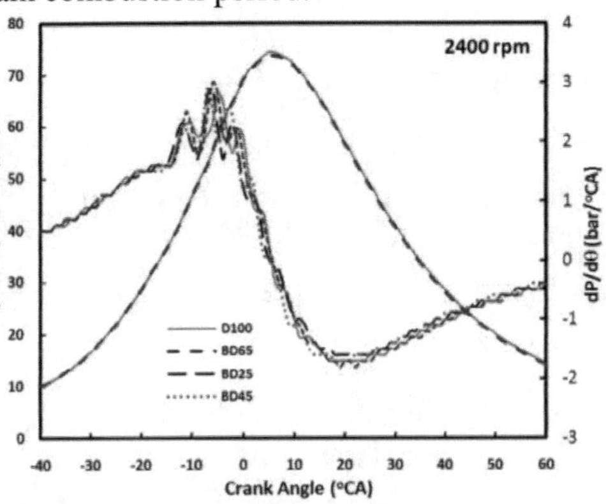

Figure 2. In-cylinder pressure & rate of pressure change of D100 & BD65 at 2400 rpm

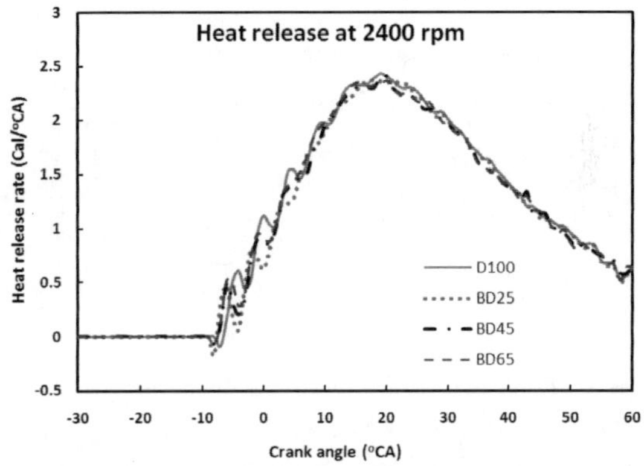

Figure 3. The heat release rate with respect to crank angle and fuel type at 2400 rpm

The heat release rates of all tested fuels are presented in Figure 3. It is shown that biodiesel blends started the combustion earlier than that of D100. From Figure 3 it is also observed that the heat release rate of D100 is slightly higher than those of biodiesel blends. This result is consistent with the higher engine power delivered by D100 compared to the biodiesel blends.

Table 3. Combustion parameters at 2400 rpm and full-load engine operation

Fuel type	Max pressure (bar)	Start of injection (°BTDC)	Start of combustion (°BTDC)	Ignition delay (°)	End of combustion (°ATDC)	Combustion duration	Max heat release rate (cal/°CA)
Diesel	74.4	24	7	17	91	98	2.43
BD25	74.2	24	8	16	97	105	2.37
BD45	74.1	24	8	16	98	106	2.40
BD65	73.8	24	8	16	98	104	2.37

The related combustion parameters to the in-cylinder pressure and the heat release as shown in Figure 2 and 3 is shown in Table 3. The start of injection is assumed same with the engine's manufacturer specification, while the start and end of combustion are decided from heat release and burnt mass data. The maximum pressure resulted by D100 is 74.4 bar, slightly higher than 73.8 bar of BD65. However, the maximum heat release rate of D100 is higher than that of BD65. In related to the combustion period, it was recorded that BD65 started the combustion slightly earlier than that of D100. This resulted in the shorter ignition delay of BD65 compared to D100. On the other hand, from the burnt mass data we found that the BD65 ended the combustion later than the D100. This also occurs with other biodiesel blends. This means the combustion duration of biodiesel blends are longer than that of diesel fuel which is worse combustion condition. It suggested that the higher density and viscosity of biodiesel played its role in faster-increasing pressure in the fuel line caused the earlier start of combustion, and on the other side resulted in poor atomization and slower fuel-air mixing.

3.2 Ignition delay and combustion duration

Figure 4 and 5 show the ignition delay of the tested fuels at various engine speed and various engine load (at constant 1600 rpm), respectively. It is clearly observed that the biodiesel blends have shorter ignition delay compared to those of diesel fuel, with no difference between biodiesel blends. The ignition delay increased slightly with the increase of the engine speed as a consequence of more fuel injected that resulted in a longer time to break into droplets and its chemical readiness to be burned. From Figure 5 it can be said that the ignition delay doesn't change due to the engine load.

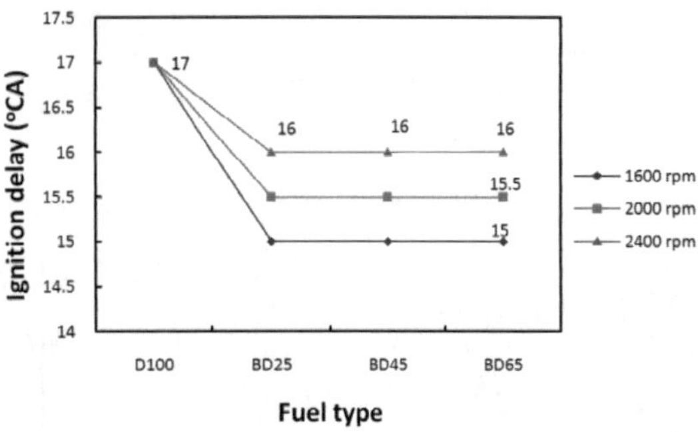

Figure 4. The ignition delay at various engine speeds with the fuel type

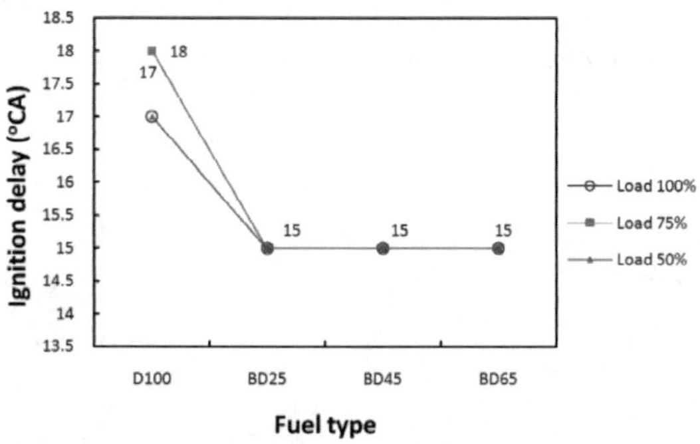

Figure 5. The ignition delay at various engine loads with the fuel type

The combustion duration of the tested fuels at various engine speed and various engine load (at constant 1600 rpm) show similar phenomenon, as shown in Figure 6 and 7, respectively. The combustion duration increased with both the engine speed and load. As mentioned before, this might be caused by more fuel injected at higher engine speed. It is also clearly observed that the combustion duration of biodiesel blends is always longer than those of diesel fuel. This proved again the contribution of the higher density of the biodiesel fuel compared to the diesel fuel.

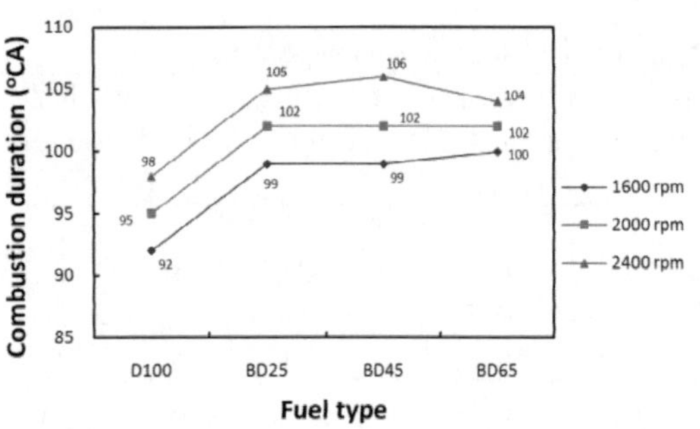

Figure 6. The combustion duration at various engine speeds with the fuel type

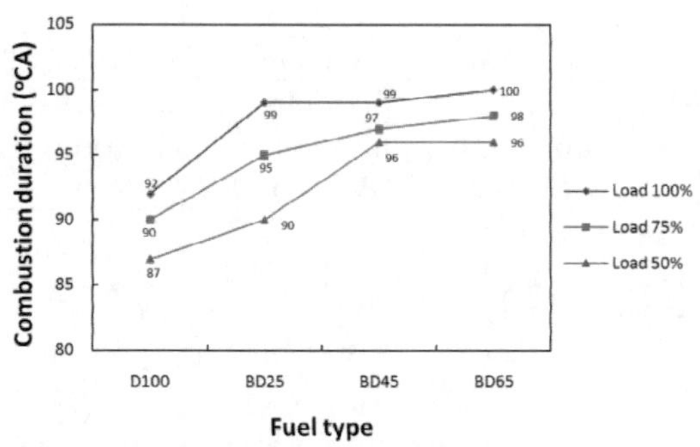

Figure 7. The combustion duration at various engine loads with the fuel type

4. Conclusions

The combustion characteristics of an IDI diesel engine fuelled with diesel and diesel-biodiesel blends were experimentally conducted. The results can be concluded as follows:

a. The maximum pressure inside the cylinder of D100 is slightly higher than that of BD65 and other biodiesel blends. On the other hand, the biodiesel blends exhibited higher pressure change rate at a moment before peak pressure inside the combustion chamber. This could be considered as the role of the higher oxygen content in the biodiesel that improved combustion at a moment before the main combustion period.
b. From the heat release analysis, it is found that the heat release rate of D100 is higher than those of biodiesel blends. The biodiesel blends started earlier and ended the combustion later than the D100. This means the combustion duration of biodiesel blends are longer than that of diesel fuel which is worse combustion condition.
c. The biodiesel blends have shorter ignition delay compared to those of diesel fuel, with almost no difference between biodiesel blends. The ignition delay increased slightly with the increase of the engine speed as a consequence of more fuel injected that resulted in a longer time to break into droplets and its chemical readiness to be burned. The combustion duration increased with both the engine speed and load. The combustion duration of biodiesel blends is always longer than those of diesel fuel. This proved again the contribution of the higher density of the biodiesel fuel compared to the diesel fuel.

References

[1] Ayhan Demirbas 2008 *Biodiesel - a realistic fuel alternative for diesel engines* (Springer)

[2] Duran Altiparmak, Ali Keskin, Atilla Koca, Metin Guru, Alternative fuel properties of tall oil fatty acid methyl ester-diesel fuel blends 2007 *Bioresource Technology* **98** 241-246

[3] Magin Lapuerta, Octavio Armas, Jose Rodriguez-Fernandez, Effect of biodiesel on diesel engine emissions 2008 *Progress in Energy and Combustion Science* **34** 198-223

[4] C.C. Enweremadu, H.L. Rutto, Combustion, emission and engine performance characteristics used cooking oil biodiesel – A review 2010 *Renewable and Sustainable Energy Reviews* **14** 2863-2873

[5] Jinlin Xue, Tony E. Grift, Alan C. Hansen, Effect of biodiesel on engine performances and emissions 2011 *Renewable and Sustainable Energy Reviews* **15** 1098-1116

[6] National fuel quality standards 2004 *Measuring cetane number: options for diesel and alternative diesel fuel* (The Australian department of the environment and heritage)

[7] John B. Heywood 1988 *Internal Combustion Engine Fundamentals* (McGraw-Hill, Inc.)

[8] V. Ganesan 1994 *Internal Combustion Engine* (McGraw-Hill, Inc.)

[9] D. Laforgia, V. Ardito, Biodiesel fueled engines: performances, emissions and heat release investigation 1995 *Biosource Technology* **51** 53-59

[10] Kyunghyun Ryu, Youngtaig Oh, A study on the usability of biodiesel fuel derived from rice bran oil as an alternative fuel for IDI diesel engine 2003 *KSME International Journal*, **Vol. 17 No. 2** 310-317

[11] Kyunghyun Ryu, Youngtaig Oh, Combustion characteristics of an agricultural diesel engine using biodiesel fuel 2004 *KSME International Journal* **Vol. 18 No. 4** 709-717

[12] Murat Cetin, Fikret Yuksel, The use of hazelnut oil as a fuel in pre-chamber diesel engine 2007 *Applied Thermal Engineering* **27** 63-67

[13] Mustafa Canakci, Ahmet Necati Ozsezen, Ali Turkcan, Combustion analysis of preheated crude sunflower oil in an IDI diesel engine 2009 *Biomass and Bioenergy* **33** 760-767

[14] Ahmet Necati Ozsezen, Mustafa Canakci, The emission analysis of an IDI diesel engine fueled with methyl ester of waste frying palm oil and its blends 2010 *Biomass and Bioenergy* **34** 1870-1878

[15] A. Senatore, M. Cardone, V. Rocco, M.V. Prati, A comparative analysis of combustion process in DI diesel engine fueled with biodiesel and diesel fuel 2000 *SAE Technical Paper Series* **2000-01-0691** 1-10

[16] Merve Cetinkaya, Yahya Ulusoy, Yucel Tekin, Filiz Karaosmanoglu, Engine and winter road test performances of used cooking oil originated biodiesel 2005 *Energy Conversion and Management* **46** 1279-1291

[17] Sehmus Altun, Husamettin Bulut, Cengiz Oner, The comparison of engine performance and exhaust emission characteristics of sesame oil-diesel fuel mixture with diesel fuel in a direct injection diesel engine 2008 *Renewable Energy* **33** 1791-1795

[18] Ahmet Necati Ozsezen, Mustafa Canakci, Ali Turkan, Cenk Sayin, Performance and combustion characteristics of a DI diesel engine fueled with waste palm oil and canola oil methyl esters 2009 *Fuel* **88** 629-636

[19] Sungyong Park, Hwanam Kim, Byungchul Choi, Emission characteristics of exhaust gases and nano particles from a diesel engine with biodiesel-diesel blended fuel (BD20) 2009 *Journal of Mechanical Science and Technology* **23** 2555-2564

[20] Oguntola J. Alamu, Ezra A. Adeleke, Nurudeen O. Adekunle, Salam O. Ismaila, Power and torque characteristics of diesel engine fuelled by palm-kernel oil biodiesel 2009 *Leonardo Journal of Sciences* **14** 66-73

[21] Ekrem Buyukkaya, Effects of biodiesel on a DI diesel engine performance, emission and combustion characteristics 2010 *Fuel* **89** 3099-3105

[22] Rasim Behcet, Performance and emission study of waste anchovy fish biodiesel in a diesel engine 2011 *Fuel Processing Technology* **92** 1187-1194

[23] M.A. Kalam, H.H. Masjuki, M.H. Jayed, A.M. Liaquat, Emission and performance characteristics of an indirect ignition diesel engine fuelled with waste cooking oil 2011 *Energy* **36** 397-402

[24] Yuan-Chung Lin, Wen-Jyh Lee, Tser-Son Wu, Chih-Ta Wang, Comparison of PAH and regulated harmful matter emissions from biodiesel blends and paraffinic fuel blends on engine accumulated mileage test 2006 *Fuel* **85** 2516-2523

[25] M. Gumus, S. Kasifoglu, Performance and emission evaluation of a compression ignition engine using a biodiesel (apricot seed kernel oil methyl ester) and its blends with diesel fuel 2010 *Biomass and Bioenergy* **34** 134-139

[26] H. Raheman, S.V. Ghadge, Performance of diesel engine with biodiesel at varying compression ratio and ignition timing 2008 *Fuel* **87** 2659-2666

[27] C.D. Rakopoulos, K.A. Antonopoulos, D.C. Rakopoulos, D.T. Hountalas, E.G. Giakoumis, Comparative performance and emissions study of a direct injection diesel engine using blends of diesel fuel with vegetable oils or biodiesel of various origins 2006 *Energy Conversion and Management* **47** 3272-3287

[28] Labeckas G, Slavinskas S., The effect of rapeseed oil methyl ester on direct injection diesel engine performance and exhaust emissions 2006 *Energy Conversion Management* **47** 1954-1967

[29] Ramadhas AS, Jayaraj S., Muraleedharan C., Performance and emission evaluation of a diesel engine fueled with methyl esters of rubber seed oil 2005 *Renewable Energy* **30** 1789-1800

[30] C.D. Rakopoulos, Comparative performance and emission studies when using olive oil as a fuel supplement in DI and IDI diesel engines 1992 *Renewable Energy* **Vol. 2 No. 3** 327-331

[31] M.G. Bannikov 2011 *Combustion and emissions characteristics of mustard biodiesel*: *Proc. 6th International Advanced Technologies Symposium* (Elazig, Turkey)

[32] M. Gumus, A comprehensive experimental investigation of combustion and heat release characteristics of a biodiesel (hazelnut kernel oil methyl ester) fueled direct injection compression ignition engine 2010 *Fuel* **89** 2802-2814

[33] P.K. Sahoo, L.M. Das, Combustion analysis of Jatropha, Karanja and Polanga based biodiesel as fuel in a diesel engine 2009 *Fuel* **88** 994-999

[34] Gerhard Knothe and Kevin R. Steidley, Lubricity of components of biodiesel and petrodiesel - The origin of biodiesel lubricity 2005 *Energy & Fuels* **19** 1192-1200

[35] Jianbo Hu, Zexue Du, Changxiu Li, Enze Min, Study on the lubrication properties of biodiesel as fuel lubricity enhancers 2005 *Fuel* **84** Issues 12-13 1601-1606

Wear of nickel-tungsten layer on carbon steel with varying contact surface temperature

I M Widiyarta[1], I M Parwata[1] and I P Lokantara[1]
[1]Study Program of Mechanical Engineering, Faculty of Engineering, Udayana University

Email: m.widiyarta@unud.ac.id

Abstract. The aim of Nickel-Tungsten coatings on the material of carbon steel is to improve the wear resistance of the material. However, with the elevated contact temperature due to friction force in the surface area of sliding may alter the wear resistance of Nickel-Tungsten layers. If the elevated temperature is quite high to reduce the hardness performance of Nickel-Tungsten layers, the layers may be degraded more quickly. In this work, Nickel-Tungsten was deposited on the surface of medium carbon steel in the process of flame powder spray coating with layer thickness of about 0.2 mm. The wear test was conducted using linear reciprocating ball-on-plat with varying surface temperature of about 30ºC up to about 250ºC. The wear rate and wear mechanism of Nickel-Tungsten layer is investigated.

1. Introduction

Wear is failure of material at the surface as material removed from the surface due to the mechanical action of a contacting bodies [1, 2]. This failure of surfaces can change the profile of the component and cause its working performance to decrease. In the sliding contacting bodies, the two contacting bodies will develop a friction load and generate surface temperature. When the friction load increase, the elevated of surface temperature will increase and this may cause the surface of the material to fail more quickly.

The wear is not an essential property of the material and occurs due to interaction of two contacting working bodies, therefore the wear of materials is possible to be controlled by changing surface properties [3]. Altering surface properties of the material by increasing the hardness of the material surfaces can improve the wear resistant of the material. This can becomes an alternative solution to increase the working life of the components by improving part of the surface components/materials that come into direct contact. One method that can be used is by coating the surface of material with a material that has a better hardness. For this purpose, thermal coating can be selected.

The thermal coating is a coating method by spraying the melted deposited material to the surface of the substrate. The deposited coating material is melted by the combustion of gasses, an electric arc or a plasma [2]. The quality of the result of a coating layers depends on the variable coating process which includes temperature, spraying distance (distance of tip of the nozzle to the surface of the substrate) [4], the speed of the gun over the surface of the substrate, and the speed of the spray coating particles (flow rate of particle deposited) out of the nozzle spray-torch [5]. The quality of the coating layer can be identified from the physical coating characteristics, such as number of porosity, unmelted particles and oxidation state [4, 7].

In this work, powder flame spray coating is selected and performed with Ni-W alloy powder as a deposit material and commercial carbon steel ST 70 is selected as a substrate. Wear test using linear reciprocating ball-on-plat with varying surface temperature of about 30ºC up to about 250ºC was conducted to determine the wear rate of the Ni-W coating and the mechanism of wear is investigated.

IOP Conf. Series: Materials Science and Engineering **539** (2019) 012014 doi:10.1088/1757-899X/539/1/012014

2. Method

In this work, commercial medium carbon steel ST 70 with the carbon content is about 0.5% [8] and hardness of about 273 HVn was selected as the substrate and Nickel Tungsten (Ni-W) with 53.61% Ni, 19.0% W (Table 1) was used for the coating material.

Table 1. Chemical composition of the Ni-W powder

Compound	SI	P	Ca	Cr	Mn	Fe	Ni	W	others
Cons. unit %	2.20	0.47	0.23	13.70	0.31	3.50	53.61	19.00	6.98

The test specimen (medium carbon steel) was formed in the size of 20x15x8 mm (Figure 1). Some preparations of the surface of the specimen (substrate) was performed before the coating process, such as cleaning the surface from any contaminations, roughening the surface and pre-heating of the substrate. The surface of the substrate is roughen by the process of sand-blasting with particles blasting of Aluminum Oxide (Al_2O_3), with air pressure of 8 bar, spraying distance of 40 mm and within about 60 seconds. Pre-heating to the surface of the substrate with temperature of about 250°C was done just before performing the coating process.

Figure 1. The form of test specimens

The thermal coating is conducted by using the powder flame spray coating. The melted Ni-W powder was sprayed into the surface of the substrate through the powder-spray-torch with spraying distance of 40 mm and with the powder-spray-nozzle torch is perpendicular to the surface of substrate. The parameter of coating process are organize with powder-feeder of about 300 g/minute, Oxygen pressure of about $4x10^5$ Pa and Acetylene pressure of about $0.9x10^5$ Pa. Before the wear test was performed, the surface of Ni-W coating layer was grinded and polished for creating a roughness surfaces about 0.1 – 0.3 μm.

Wear test was conducted using linear reciprocating ball-on-plat with varying contact surface temperature, i.e., 30°C, 50°C, 100°C, 150°C, 200°C and 250°C as shown in the schematic drawing of wear test (Figure 2). Load W is given through ball indenter with a diameter of 8 mm, with a load of 25 N, length 10 mm step of loading. The test was done up to 100 m which is equal to 5000 cycles loading.

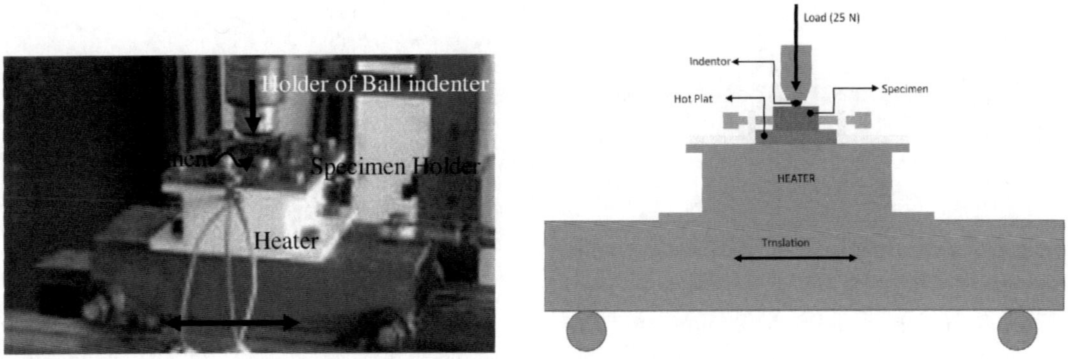

Figure 2. The diagram of the wear test.

3. Results and Discussion

Figure 3 shows the hardness and wear rate of the Ni-W coating layer with varying temperature of the contact surfaces. One graph with dot marks represents the hardness of the Ni-W layer with varying surfaces temperature and the second graph with triangle marks represents the wear rate of the Ni-W layer with varying surfaces temperature. In general, the hardness of the Ni-W coating decreases with increasing the surface temperature. For the surface temperature of about 30°C, the hardness of the Ni-W coating layer is about 857 HVn. With increasing the surface temperature up to 100°C, the hardness of the Ni-W coating layer decreases to 715 HVn and the hardness of the Ni-W layer decreases to 567 HVn with increasing surface temperature up to 200°C. However, with increasing the surface temperature from 100 to 150°C, the surface hardness of the Ni-W coating relatively increases from 715 up to about 736 HVn. The similar hardness performance was also found in the increase of surface temperature from 200 up to 250°C, which is the surface hardness increases from 567 up to 656 HVn with increasing surface temperature from 200 to 250°C. This may be caused by the form of contaminant layer such as oxide layer on the surface of Ni-W due to dry oxidation which is relatively strong.

From the second graph (triangle marks), the increase of the surface temperature from 30 up to about 100°C cause the wear rate of Ni-W layer to decrease from 0.0040×10^{-3} for surface temperature of 30°C to 0.0017×10^{-3} gr/m for surface temperature of 100°C. Then, with increasing the surface temperature up to 250°C, the wear rate of the Ni-W gradually increase, i.e., the wear rate of about 0.0027×10^{-3} for surface temperature of 150°C, the wear rate of about 0.0050×10^{-3} and 0.0055×10^{-3} gr/m for the surface temperature of 200°C and 250°C respectively. The increase of wear rate with increasing surface temperature is caused by the reduction of the hardness of the material. At this stage, the surface temperature is relatively high enough to reduce the hardness of the material.

The decrease of hardness with increasing surface temperature that occurred in the range of surface temperature 30 to 100°C was not followed by the reduction of the wear rate. This condition may be caused by existing the contaminant (i.e., oxide layer) at the surface due to a surface temperature which is reducing the friction, resulting wear rate to decrease. At this range of surface temperature (i.e., from 30 to 100°C), the oxide layer maybe gives a greater effect to the wear rate so that the wear rate decrease although the hardness of the Ni-W decreases.

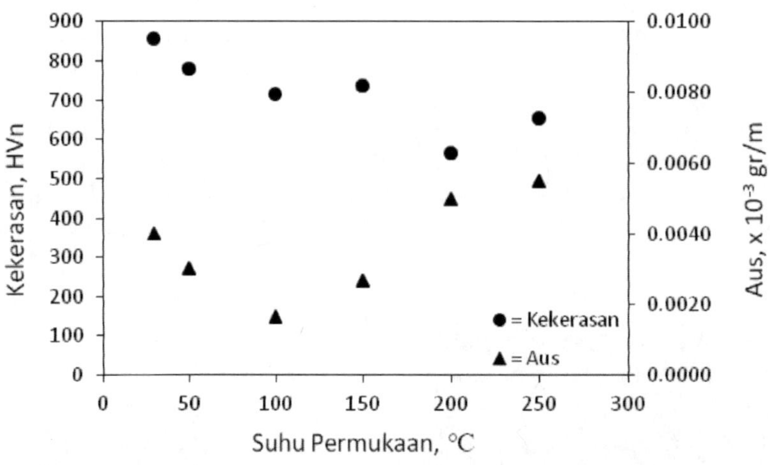

Figure 3. Hardness and wear rate of NiW coating layer with varying temperature of contact surfaces

Figure 4 to 9 show the SEM photograph of the part of the worn surfaces of Ni-W coating with a varying surface temperature in the range of 30 to 250°C. The worn surfaces look like many little fracture particles that released off the surface and no visible long scratches or long and wide debris, this shows that the Ni-W coating is a brittle material and well resistance to wear failure. This mean that the surface temperature in the range of 30 to 250°C is not high enough to soften the Ni-W material become a ductile material.

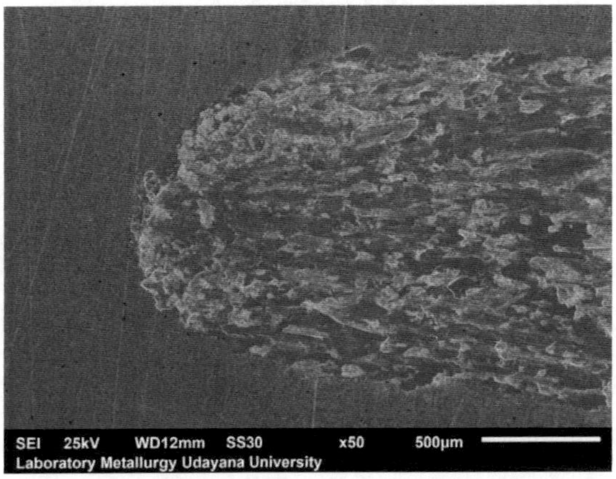

Figure 4. SEM photograph of wear surface of NiW layers with surface temperature of about 30℃.

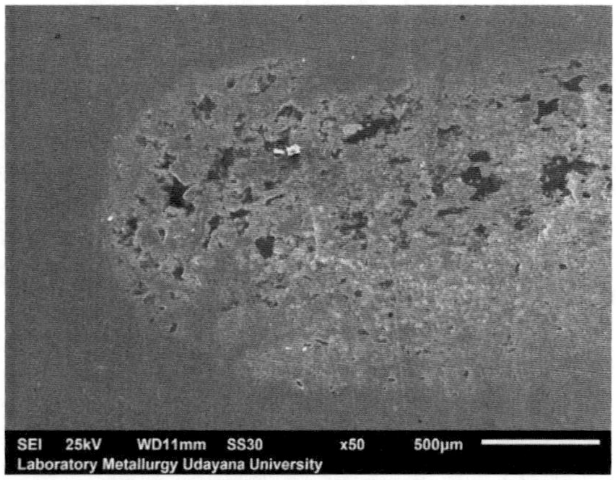

Figure 5. SEM photograph of wear surface of NiW layers with surface temperature of about 50℃.

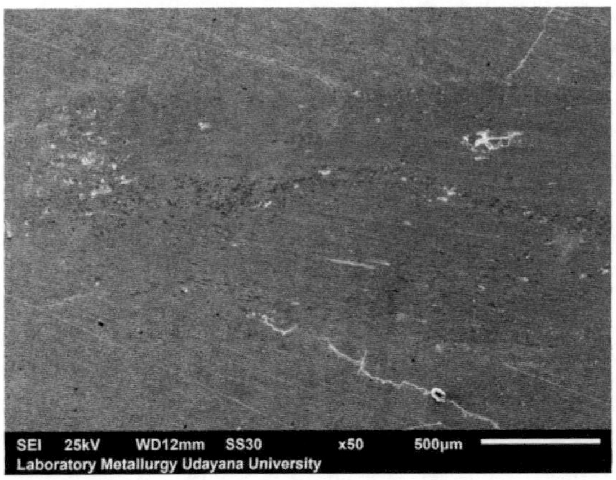

Figure 6. SEM photograph of wear surface of NiW layers with surface temperature of about 100℃.

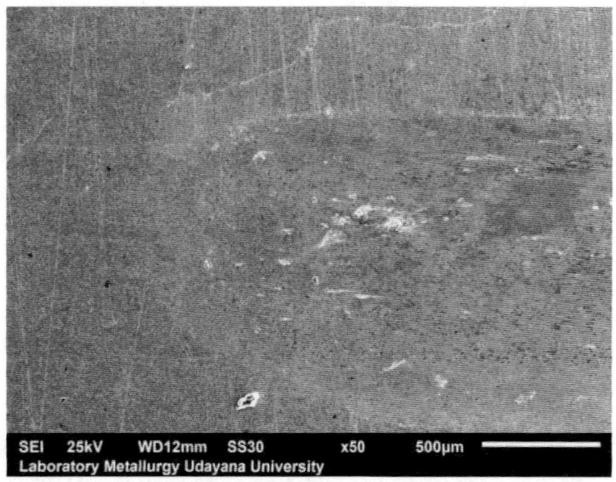

Figure 7. SEM photograph of wear surface of NiW layers with surface temperature of about 150℃.

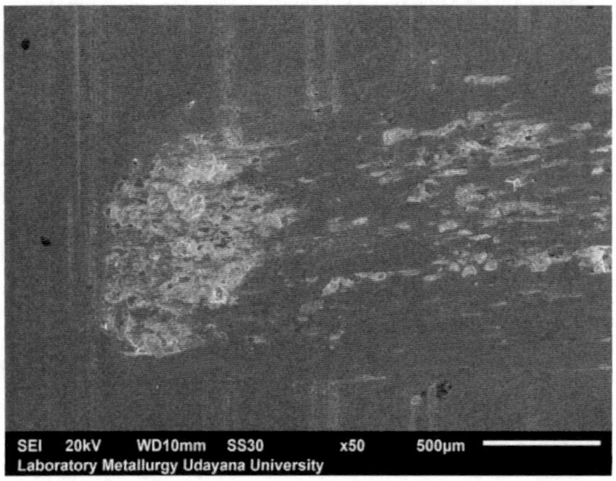

Figure 8. SEM photograph of wear surface of NiW layers with surface temperature of about 200℃

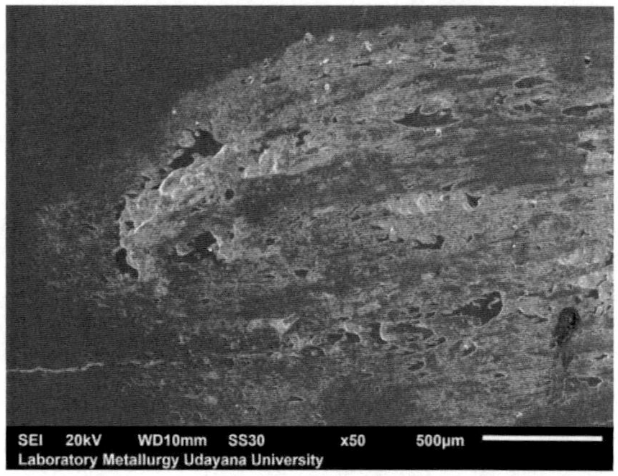

Figure 9 SEM photograph of wear surface of NiW layers with surface temperature of about 250℃

4. Conclusions

The increase of surface temperature in the range of 30 to 250°C cause the hardness of the Ni-W layer to slightly decrease. Even though the hardness of the Ni-W layer decreases with increasing surface temperature from 30 to 100°C, the wear rate decrease. This is because the existing of oxide layer due to dry oxidation at those surface temperature maybe cause the contact friction to decrease and resulting the wear rate to decrease.

The mechanism of wear of Ni-W coating with a varying surface temperature in the range of 30 to 250°C was identified as abrasive wear with very small wear debris that removed from the surface without introduction any plastic deformations.

Acknowledgment
The authors would like to thank the Ministry of Research, Technology and Higher Education for the support of the research grants through the National Competitive Research Grant 2018.

References

[1] Williams, J. A., "Engineering tribology", 1994, New York: Oxford University Press Inc.
[2] R.S.C. Paredes, S.C. Amico, A.S.C.M. d'Oliveira,"The effect of roughness and pre-heating of the substrate on the morphology of aluminium coatings deposited by thermal spraying" Surface & Coatings Technology 200 (2006) 3049 – 3055.
[3] Ramnarayan Chattopadhyay-Advanced Thermally Assisted Surface Engineering Processes-Springer (2004)
[4] Widiyarta I M., Parwata, I M., Lokantara, I P. and Davin P. A., Karakteristik lapisan NiCr pada baja karbon sedang dengan metode pelapisan *flame spray coating* dengan variasi jarak semprotan, Jurnal METTEK Volume 2, No. 1, 2016.
[5] R.C. Tucker, Jr., Thermal Spray Technology, ASM Handbook, 2013, Volume 5A
[6] Sarjito, Analisa kekuatan puntir, lentur putar dan kekerasan baja ST60 untuk poros propeller setelah diquencing, ROTASI, Volume 11 Nomor 2 April 2009.
[7] I M. Widiyarta, I M. Parwata, I P. Lokantara, D. Perangin-Angin and N. A. Suryawiranata, Wear of NiCr Coating Deposited on Medium Carbon Steel with Powder Flame Spray Coating Method, Journal of Materials Engineering and Processing Technology, Vol. 1, No. 2, September 2017, ISSN 2549-6891.

[8] I M. Widiyarta, I M. Parwata, I P. Lokantara dan D. Ranto, "Kekerasan dan Aus Lapisan Nikel-Tungsten pada Baja Karbon Sedang Akibat Beban Gelinding-Luncur" Proceeding Seminar Nasional Tahunan Teknik Mesin XVI (SNTTM XVI), Oktober 2017

IOP Conf. Series: Materials Science and Engineering **539** (2019) 012015 doi:10.1088/1757-899X/539/1/012015

Impact Strength and fluid absorption at the brake lining of hybrid composite

I G A K Suriadi[1] and I K A Atmika[1]
[1]Mechanical Engineering Departement, Udayana University, Bali-Indonesia 80362

Email: tutadi2001@yahoo.com

Abstract. Brake lining is one of the active safety components in the vehicle that serves to reduce and stop the vehicle speed. Brake lining is arranged in brake system made of asbestos material and other materials such as metal oxide, sulphate, Mn or Co and silikan. In research to realize the environmentally friendly brake pads have developed hybrid composite materials with reinforced basalt, clamshell and alumina as the material of brake pad. The purpose of this study was to examine sea water absorbtion and impact strength in brake pad and can look for alternative brake pads that are environmentally friendly. In this test obtained the value of impact strength of asbestos 0,000374867 J/mm². From the calculation of the hybrid composite test with variations in composition, the highest impact strength value is obtained on hybrid composite HC2 with a value of 0.000339547 J/mm^2. This is due to the hybrid composite with a variation of 40% basalt, 10% clamshells, 10% alumina and 40% resin, which has a stronger and more perfect bond between metric and basalt compared to other hybrid composite. Sea water absorption by composite materials and asbestos brek lining materials, showing that the trend after the 24th day immersion starts to be constant for each specimen. From these conditions, the sea water absorption capacity of specimen I was higher than specimen II of 0.51% and Specimen III was 1.01%, but lower than the asbestos materials at 1.14%.

1. Introduction

One of the factors of failure in braking is the loss of the ability of the brake lining to stop the vehicle. Therefore, along with the development of motor vehicle speed, it must be balanced with a higher brake system and quality. The quality of brake lining is influenced by several factors, namely the composition of the material, the type of material and the hardness of the brake lining material. The ability of brake pads has an important role is to absorb the amount of kinetic energy during braking. The problem that often occurs in brake lining is wear due to shock loads by sudden braking [1].

In general, brake lining is made of asbestos, because it has several advantages such as resistance to high temperatures reaching 800° Celsius [2], while the weaknesses in wet conditions will experience slippery effects. Asbestos brake pads are made of 60% asbestos or fiber material as the main fiber, and other additional elements such as Silicon Carbide (SiC), Manganese (Mn), Cobalt (Co), Resin, Additive Friction, Filler [3]. Brake pads from asbestos which are components that cause carcinogenic properties and are also not environmentally friendly [4].

The development of composite technology has now begun to lead to the concept of returning to nature. The composite technology has begun to shift by using natural materials such as natural fibers, natural wood, coconut shell powder, basalt fiber and waste fly ash from the combustion of coal [5], to apply powder coconut shell as friction material in which the hardness and tensile strength of the brake increases with the optimum level of 14.82% for grade resin 15%, holding time of 6 hours and the temperature of the oven 160 degree Celsius and get a hardness of 60-70 HRB. Czigany [6], has been conducting research on the characteristics of epoxy composite with basalt fiber reinforcement, in which the strength properties of the hybrid composites due to the increase of surface treatment after mechanical tests and microscopic analysis. Then the laminate characteristics of basalt fiber composite laminates have been studied because of the imposition of tensile and flexural [7]. Liu, et al. [8] conducted a study on the mechanical properties of basalt fiber composites in vehicle components.

Based on the results of research that has been done, in this study using volcanic sediment material called Basalt. Basalt which has corrosion resistance, low water absorption, and resistance to chemical treatment and non-toxic [9]. Based on laboratory analysis, Basalt content consists of chemical elements such as: SiO, Al_2O_3, Fe_2O_3, MgO, CaO, Na_2O, K_2O, TiO_2, P_2O_5, MnO, and Cr_2O_3 [10]. Then the clamshell powder is a powder produced from the burning of crushed clamshells and based on laboratory analysis shows the chemical content of: CaO, MgO, SiO_2, Al_2O_3 and Fe_2O [11]. While alumina is a lightweight metal that has good corrosion resistance and is easily formed [12]. Composite formation with basalt powder, clamshell and alumina reinforcement is carried out by hybridization method on single binder, epoxy. In the formation of the test material carried out by sintering process with temperature control 150°C which will produce strength and hardness as well as increasing frictional force.

This study used basalt powder, alumina and clamshell powder and epoxy as hybrid composite composers. The specimens tested with Charpy impact refer to ASTM D6110-04 standard, to determine the characteristics and toughness of hybrid composite materials that have been made. Analysis of surface fault characteristics using SEM test. While sea water absorption testing refers to ASTM D 570 - 98

2. Materials and Method

This research was conducted by hybridizing 3 types of material as reinforcement and one material as a matrix. Strengthening materials include basalt, clamshells and alumina. While as a composite matrix material used phenolic resin. All reinforcing materials are particles. In this study, basalt powder, alumina powder and clamshell powder with a size of 0.0074 millimeters were used. Mechanical properties of basalt are shown in Table 1. Clamshell powder material consisted of 66.70% CaO, 22.28% MgO, 7.88% SiO_2, 1.25% Al_2O_3 and 0.03% Fe_2O_3. The phenolic resin material used in this study coded PR-51 510.

This brake pad research is made by mixing, compacting and sintering. The shape and size of the specimen as shown in Figure 1. Composite composition shown in table 2 and table 3 is based on ASTM D 3171-09 standard. In the process of making brake lining material this is done with a compressive load of 2000 N for approximately 30 minutes, and a sintering process at 150°C.

Table 1. Mechanical properties of basalt material

Properties of Basalt	Value (unity)
Density	2600-2630 (kg/m^3)
Tensile strength	500k-550k (psi)
Sintering Temperatur	1050 (°C)
Operation Temperatur	-265-+700 (°C)
Modulus of elastisitas	9100-1100 (kg/mm^3)
Mohs Hardness @20°C	5-9
Melting point	1170(°C)
Heat resistance	700-1000 (deg.C)
Elongation at break	3.15(%)

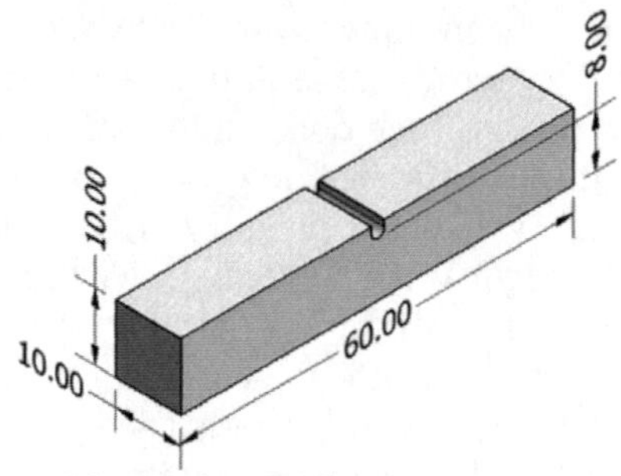

Figure 1. Specimen geometry

Table 2. Hybrid composite weight fraction ratio for impact test

Composition	Basalt powder	Clamshell powder	Alumina powder	Phenolic Resin
HC1	45	5	10	40
HC2	40	10	10	40
HC3	35	15	10	40
HC4	30	20	10	40
HC5	25	25	10	40

Table 3. Hybrid composite weight fraction ratio for fluid absorption test

Composition	Basalt powder	Clamshell powder	Alumina powder	Phenolic Resin
I	30	50	10	20
II	40	40	10	20
III	50	30	10	20

3. Result and Discussion

The average result of the impact of each composition and comparison with the impact strength of the brake lining pads that asbestos material (C) are shown in figure 2 and figure 3, while the sea water absorption of brake pads is shown in Figure 4.

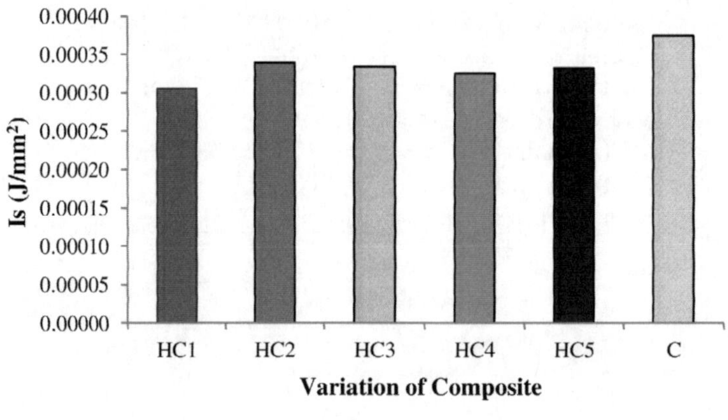

Figure 2. Impact strength of the hybrid composite and asbestos

Figure 2 shows the relationship of variation in impact strength specimens that experience an increase and decrease in composition variation. In this test obtained the value of impact strength of asbestos materials(C) 0,000374867 J/mm². From the calculation of the hybrid composite test data with variations in composition, the highest impact strength value is obtained on composite hybrid HC2 with a value of 0.000339547 J/mm^2. This is due to the hybrid composite with a variation of 40% basalt, 10% clamshells, 10% alumina and 40% resin, which has a stronger and more perfect bond between metric and basalt compared to other hybrid composites.

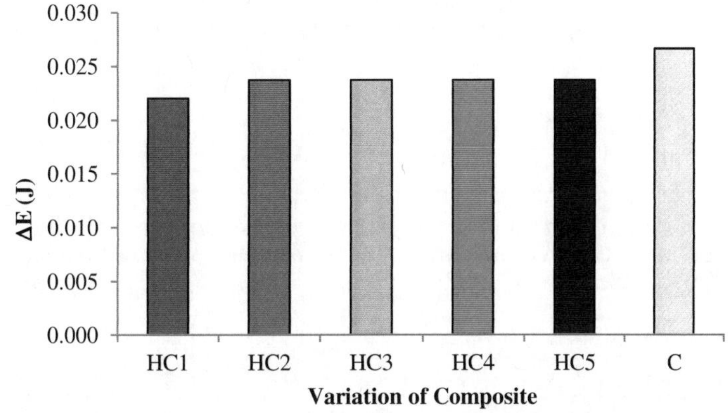

Figure 3. Value of energy absorbed that occurs in impact testing

From Figure 3, the comparison of energy values absorbed by specimens in impact testing occurs. It can be seen from the graph of the absorbed energy value that the HC1 absorbed energy value is 0.022012589 J, for HC2 to HC5 which is 0.023739901 J, while the energy in asbestos material (C) is 0.026656442 J. This means that the energy absorbed by the asbestos material (C) is greater than hybrid composites brake lining pads.

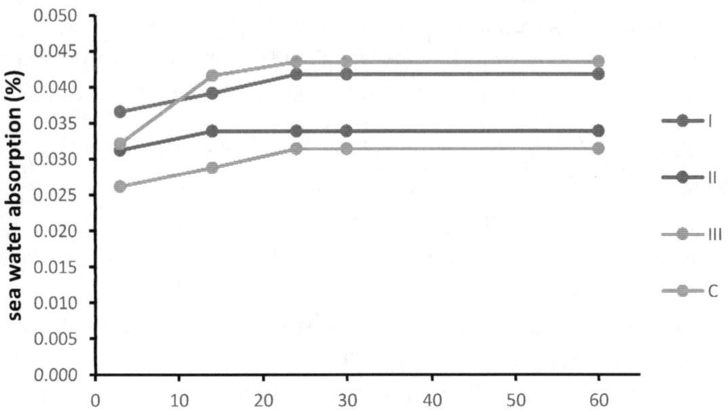

Figure 4. Sea water absorption of the hybrid composite and asbestos

Figure 4 shows a graph of seawater absorption by hybrid composite materials and asbestos material, showing that the trend after the 24th day immersion starts to be constant for each specimen. From these conditions, the absorption capacity of specimen I was higher than specimen II of 0.51% and specimen III was 1.01%, but lower than the asbestos material at 1.14%. This shows that specimen I has a higher absorptive capacity than specimen II and specimen III but has a lower absorption capacity than asbestos material specimens, so that specimen I reaches a faster saturation point on day 14 compared to specimen II and specimen III which reached saturation on 24th day.

4. Conclusion

The effect of the weight fraction on each hybrid composite variation on the impact strength is a reduction in the percentage of basalt and the addition of percentage of clamshell powder, and a constant percentage of alumina which in each composition produces varying strength values. Seawater absorption in the specimens brake lining pads are strongly influenced by the basalt content, where the greater the basalt content in the specimen, the sea water absorption capacity of the specimens will be smaller.

Acknowledgment

This research is part of the PTUPT scheme research funded by the Directorate of Higher Education in 2018 with an Assignment Agreement Letter in the Framework of National Competitive Research Implementation. For this reason the author wishes to express his gratitude for the support of this research fund. Thanks are also conveyed to the Udayana University LPPM which facilitated the research activities.

References

[1] Aigbodion, V. S. , Akadike, Hassan, U.S.B., Asuke, F., and Agunsoye, J.O.. 2010. *Development of asbestos - free brake pad using bagasse*, Tribology in industry, vol. 32, No.1, pp. 12-18.
[2] Louis St, 2004. U.S. Survey shows imports of asbestos brake materials increasing. St. Louis.
[3] Kedar S Pandya, Ch. Veerraju, and N. K. Naik. 2011. Hybrid composites made of carbon and glass woven fabrics under quasi-static loading. Materials and Design, vol. 32 pp. 4094–4099.
[4] Ashby, M. F. and Y. J. M. Brechet,. 2003. Designing hybrid materials. Acta Materialia, vol. 51, pp. 5801-5821.
[5] Czigany, T. 2005. Basalt fiber reinforced hybrid polymer composites. Materials Science, Testing and Informatics Ii, vol. 473-474, pp. 59-66.
[6] Ary Subagia, I.D.G., and Y. Kim. 2013. A study on flexural properties of carbon-basalt/epoxy hybrid composites. Journal of Mechanical Science and Technology, vol. 27 (4), p. 987~992.
[7] Qiang Liu, Montgomery T. Shaw, and Parnas, R. S. 2006. Investigation of basalt fiber composite mechanical properties for applications in transportation. POLYMER COMPOSITES, vol. 27, pp. 41-48.
[8] Atmika I.K.A, Subagia I.D.G.A, Surata I.W, Sutantra I.N. 2017. Study of the mechanical properties of hybrid composite basalt/alumina/shells for brake lining pads. IOP Conf. Series: Materials Science and Engineering 201(2017) 012009.
[9] Singha, K. 2012. A short review on basalt fiber. International Journal of Textile Science, vol. 1 (4), pp. 19-28.
[10] Fiore, V. G. Di Bella, and Valenza, A. 2011. Glass–basalt/epoxy hybrid composites for marine applications, Materials and Design, vol. 32, pp. 2091–2099.
[11] Dalinkevich A.A., Gumargalieva K.Z, Marakhovsky S.S, and Soukhanov A.V. 2009. Modern basalt fibrous materials and basalt fiber-based polymeric composites. Journal of Natural Fibers, vol. 6, pp. 248-271.
[12] Lopresto, V.C. Leone, and I. De Iorio, 2011. Mechanical characterization of basalt fiber reinforced plastic. Composites Part B: Engineering, vol. 42, pp. 717–723.

IOP Conf. Series: Materials Science and Engineering **539** (2019) 012016 doi:10.1088/1757-899X/539/1/012016

Properties of thermal conductivity, density and hardness of aluminium matrices with reinforcement of sicw / al2o3 hybrid after sintering process

K Suarsana[1], I M Astika[1] and P W Sunu[2]
[1]Mechanical Engineering Department, Engineering Faculty, Udayana University, 80361 Denpasar, Bali, Indonesia.
[2]Mechanical Engineering Department Bali, State Polytechnic Kampus Bukit Jimbaran, Badung Bali, Indonesia.

Email: suarsana@unud.ac.id

Abstract. The material of composite hybrid Al / (SiCw + Al_2O_3) is a mixture of fine aluminium powder as a matrix while Silicon Carbide whiskers (SiCw) and Alumina (Al_2O_3) as reinforcement. Powder metallurgy method is used to manufacture composites according to the form of test specimens. Testing without treatment and after sintering treatment with temperature parameters and sintering time of 500 and 600°C at holding time 6 hours. This research was conducted to know the properties of thermal conductivity, hardness, density and SEM analysis after sintering process. The results show that the sintering process that has been carried out can provide a percentage increase and affect the composite properties. Improved thermal, hardness and density occurs due to stronger or denser interface bonds between the matrix and reinforcement influenced by increasing sintering temperatures, so sintering processes can affect composites on physical and mechanical properties.

1. Introduction

The availability of conventional metal-based materials with limited quantity and quality raises the thought of developing new materials consisting of a combination of two or more different materials, such as metals with ceramics, these materials are called composites [1, 2]. In general, research and development of composite technology aims to improve the characteristics of material properties that are significant, such as for the application of lightweight but very strong materials. Research on hybrid composites showed that the highest density was obtained when lamination of 6 hours holding time and the lowest porosity was obtained when the volume fraction was 40% and the lowest density was achieved at a holding time of 6 hours at a volume fraction of 30% [3].

The strength of the composite surface bond is determined by the processing temperature and the formation of the interface compound. Hard values increase in the matrix region due to the presence of Si and C elements. With the increase in temperature, the effect of unidirectional strengthened SiC-oriented fibre increases, but the effect of grain size, wear rate and linear friction coefficient decreases [4, 5]. The mechanical and thermal properties of the aluminium alloy 2024 (Al_2O_4) composite matrix are reinforced with multi-walled carbon nanotubes or multi-layer graphene at temperature range 250-430°C, using powder metallurgy process, showing a high voltage yield of about 110 Mpa, due to uniform dispersion from the strong nano-scale interface, it inhibits the dislocation movement and the coarse and softening eutectic phase occurs in the matrix [6]. The mixing casting method used to construct aluminium metal matrix composites reinforced with different weight fractions of 2.5, 5 and 7.5% SiC and Fly Ash. In this case it was found that there was a good improvement in terms of the most recent tensile strength, hardness and wear properties with increasing the weight fraction of SiC and Fly Ash in a hybrid composite aluminium matrix. The thermal property is also increased, useful for improving the appropriate design parameters for developing the product [7, 8, 9].

The behaviour of heat deformation and the composite aluminium hybrid matrix mechanism reinforced with micro-SiC and nano-TiB_2 show that true stress increased rapidly with an increase in true strain and then stabilizes or decreases slightly after the true stress reaches peak value. The effect of creep thermal residual stress on the microstructure and tensile properties of SiC whiskers reinforced by the combined aluminium matrix indicates that the dislocation state in the matrix is greatly altered, especially in the region adjacent to the interface between the SiC whiskers and the matrix caused by stress thermal residues. Thermal residual stress creeping treatment is a new method to effectively reduce stress thermal residues, and increase strength [11]. The multiwall carbon nanotubes reinforced composites are harder than composites reinforced by vapour-grown carbon fibres and are shown higher final tensile strength. Research on the utilization of silicon carbon fibre material and alumina particles on Aluminium Matrix Composite can improve the mechanical properties of density, hardness and otherwise decrease porosity percentage.

The effect of silicon on the in-situ Al_2O_3-SiCw composite ceramic properties synthesized by the carbothermal reduction of the kaolin aluminum coal series for solar thermal storage shows that Si additives are superior in terms of improving SiCw yields and composite properties [14].

Composite manufacturing uses powder metallurgy process which includes several stages such as mixing which uses wet mixing method using ethanol, compacting to compress powder into raw material form and finally sintering process is a heating process, with or without compressive application so the particles will be interrelated chemically becomes a wireless structure [3,15]. The sintering process will improve mechanical properties, this is due to the union of the particles so as to increase the density of the product. To be able to increase the value of the hardness properties of metal materials, the addition of reinforcing material to the metal is needed by adding hard materials, such as: TiC Al_2O_3, SiCw, and ZrO_2. Materials. The SiCw material can be readily bonded to the mixing process and does not cause oxidation to aluminium [2]. Aluminium is used very widely in various life as well as technical material where the raw material can be obtained in the electrolysis process based on the stoichiometric balance. Other users are for the manufacture of chemicals, abrasives and ceramic fibres. Alumina (Al_2O_3) is stored in aluminium oxide compounds obtained through the process of purifying bauxite (Al_2O_3 x H_2O). This process is referred to as the Buyer Process [15].

2. Experimental

2.1 Material

Aluminium fine powder p.a Merck with purity above 90% (≥90%) is used as a composite matrix. SiC whisker commercial diameter (d ≈ 0.5 μ m), length (l ≈ 40 μ m) and powder $A_{12}O_3$ p are used as composite reinforcement. Ethanol 96% solution (CH_3COOH) was used as a mixing medium. The mixing is wet and stirred with a magnetic stirrer at 50°C. The materials of Al_2O_3p to be used are: 3, 6 and 9% wt weight, then inserted in measuring cup. At the time of mixing used ethanol solution as solvent of aluminium fine powder matrix and SiCw / Al_2O_3p amplifier with ratio between composite volume to ethanol solution is 1: 2 then stirred for 30 minutes. After mixing, inserted into the oven for 30 minutes at a temperature of 200°C until dry perfectly, then inserted into a silentric mould and pressurized using a 2.5 ton hydraulic press for 15 minutes to obtain specimens with uniform density and pressure. The sample size is 0.8 cm in diameter and 1 cm high with volume ± 0.5 cm^3.

2.2 Sintering treatment

Sintering process is a process of heating with pressure or without pressure so that the particles will be chemically bonded to each other into a homogenous or coherent structure [17]. The sintering process will usually be followed by an increase in mechanical properties when compared to compaction material that has not gone through the sintering process. This is due to the joining of bonding particles and amplifiers so as to increase the density of the structure of the product or so-called compaction process [18]. The sintering process is carried out in a Fluidized Bed kitchen without pressing on the Argon gas conduction. Sintering temperature varies from 500°C and 600°C with a holding time of 6 hours. The specimen is inserted into the fluidized bed kitchen according to the temperature and holding time. Temperature setting and holding time for each treatment on the specimen.

2.3 Thermal Conductivity
The heat conductivity coefficient k is a constant connecting heat flux Q with a temperature gradient of $\Delta T / \Delta x$. Thus the heat conductivity unit (k, thermal conductivity), k is J / (mm^2.s) / (°C / mm) or (W / mm^2) / (°C / mm) [19]. The thermal conductivity value of a material indicates the rate of heat transfer flowing in a material.

$$q = -kA \frac{dT}{dx} \quad (1)$$

Which:

q = rate of heat transfer (Btu/h or W)
A = area where the heat moves (ft^2 / m^2)
$\frac{dT}{dx}$ = gradient or slope temperature (Gradient temperature) in the direction of heat transfer (ft /m)
k = thermal conductivity (Btu/h.ft. or W/m.)

2.4 Determination of density
This solid or bulk density measurement uses Archimedes method.

$$\rho = \frac{m_s}{(m_b - m_g)} x\rho H_2 0 \quad (2)$$

Which:

ρ = material density (gr/cm^3)
m_s = mass of the dry specimen (gr)
m_g = mass of the specimen suspended in water (gr)
m_b = mass of the specimen after soaking water (gr)
ρH2O = water mass (1 gr/cm^3)

2.5 Hardness Testing
Hardness test using Vickers with pyramid indenter, the large angle between the pyramid facing surfaces is 136°. The ASTM used is ASTM E384.

$$VHN = = \frac{1{,}854 \times P}{d^2} \quad (3)$$

Which:

VHN = Vickers Hardness Number
P = load (kgf)
D = average of diagonal length

3. Results and Discussion

3.1. Percentage improvement of thermal conductivity after sintering

Figure 1 shows the composite thermal conductivity relationship with Al_2O_3 composite composition with weight variation: 0, 3, 6 and 9%, sintering temperature 500 and 600°C at holding time 6 hours as well as thermal conductivity results without sintering process. Increased temperature sintering causes increased thermal conductivity of composite, it can be seen that the increase of sintering temperature in Aluminum matrix composite is influenced by the addition of Al_2O_3p composition. Composite density occurs in addition to SiCw composition. Increased density is related to the thermal conductivity in which the pores formed in the composite are decreasing, where the holding time and sintering temperature increase in each percentage of the Al_2O_3p weight. The effect of sintering temperature on thermal conductivity at temperatures of 500°C and 600°C, where with increasing temperature in the reinforcing composite (SiCw+Al_2O_3p) the thermal conductivity increases. Figure 1 (a) shows the result of the research before the sintering process is shown the thermal conductivity respectively is composition I, composition II, composition III and composition IV is data before sintering. In Figure 1 (b) and (c) shows that each composite composition with a temperature treatment of 500°C and 600°C with a holding time of 6 h, there was an increase in thermal conductivity. In this is caused by increasing temperature so that the composition density, which allows better thermal conductivity. Thermal conductivity increased from before sintering with after sintering process where in composition I, composition II, composition III and composition IV was 40% (0.137434544 Kcal / mh°C), 33% (0.113368327 Kcal / mh°C), 23% (0.081727405 Kcal / mh°C) and 4% (0.0147033 Kcal / mh°C), at a temperature of 500°C, also at a temperature of 600°C. In this case increases the largest percentage of heat conductivity in composition I and the smallest of the composition IV. Thus the composition of Al_2O_3p and SiCw influences heat conduction on the composite due to a good heat conduction SiCw.

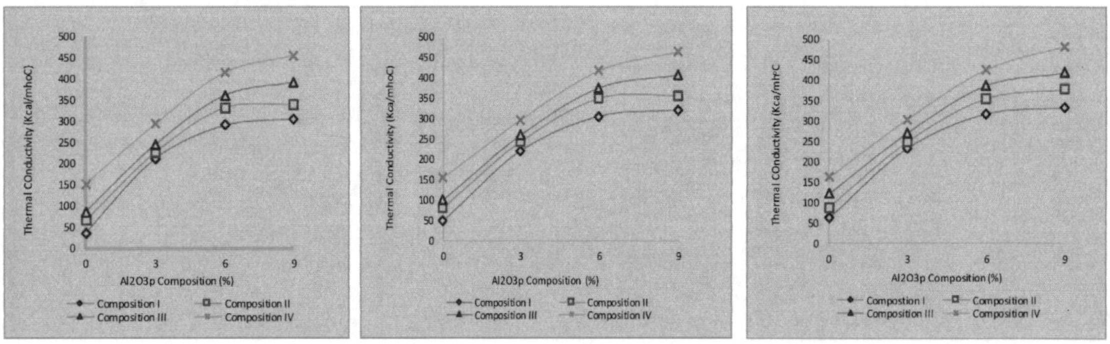

(a) (b) (c)

Figure 1. (a) Before sintering. (b) After sintering 500°C, 6 hours. (c) After sintering 600°C, 6 hours

Next seen in Figure 2 is the percentage increase in thermal conductivity from before sintering to after the sintering process.

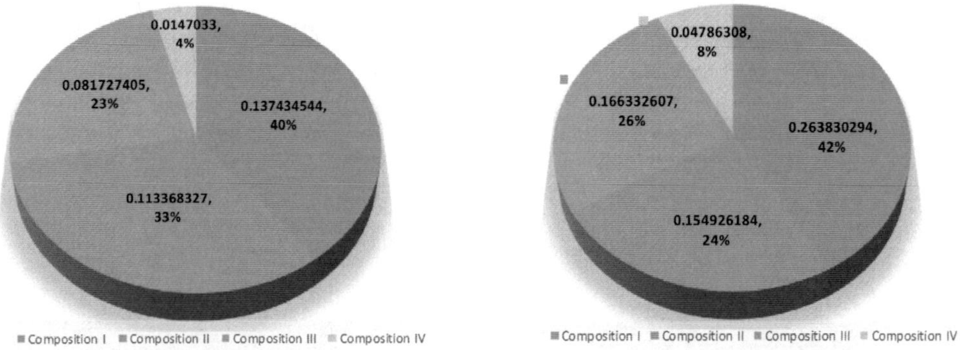

(a) (b)

Figure 2. (a) Percentage of increase in thermal conductivity after sintering 500°C, 6 hours
(b) Percentage of increase in thermal conductivity after sintering 600°C, 6 hours

3.2. Percentage Improvement of Density properties after sintering process

The results of density testing that has been carried out is known that the changes that occur are the increasing percentage of alumina material (Al_2O_3) in each variation of the composition of the specimen affecting the increase in density value. This is because Al_2O_3 has a high density between two other specimens. In Figure 3 shows the percentage increase in density seen from before flashlight treatment until after sintering process temperature of 500°C and also on the 600°C temperature each with a holding time of 6 hours. The temperature increases the composition density which allows a reduction in the composite pores formed. The density properties increased from before sintering and after the sintering process wherein composition I: 26% (0.091336076 gr / cm^3), composition II: 26% (0.0932675 gr / cm^3), composition III: 27% (0.096692 gr / cm^3) and composition IV : 21% (0.074839 gr / cm^3) at a temperature of 500°C, also 600°C. In this case, each composition increased the percentage before and after sintering, where the increase in the percentage of small density in each composition 1, II, III and IV.

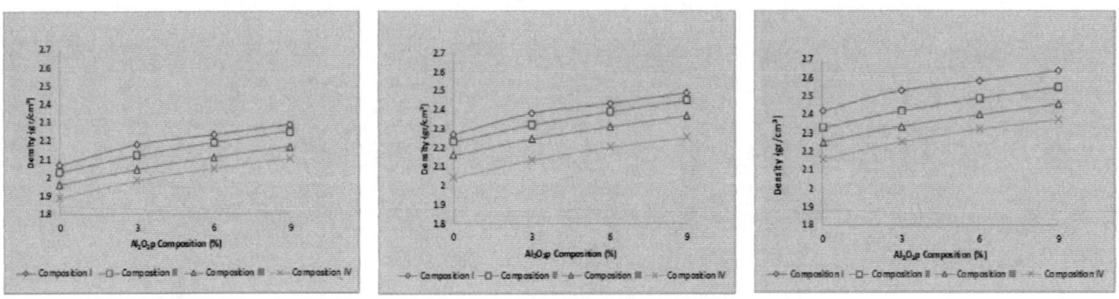

(a) (b) (c)

Figure 3. (a) Before sintering. (b) After sintering 500°C, 6 hours. (c) After sintering 600°C, 6 hours

Next seen in Figure 4 is the percentage increase in density before sintering and after the sintering process.

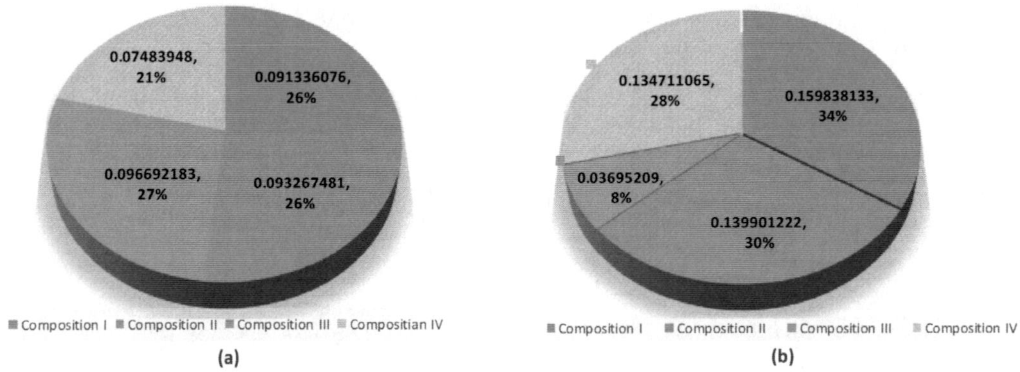

Figure 4. (a) Percentage of increase in sintering density 500°C, 6 hours
(b) Percentage of increase in sintering density 600°C, 6 hours

3.3. Percentage Improvement of Hardness properties after sintering process

Figure 5 shows the hardness relationship with Al_2O_3 composition with weight variation: 0, 3, 6 and 9%, sintering temperature 500 and 600°C at holding time 6 hours. In general, there is an increase in hardness from Al and SiC whisker compositions without the addition of Alumina or with the addition of Alumina. The percentage increase in hardness was seen from before the flashlight treatment until after the sintering process at temperatures of 500°C and 600°C, each with a holding time of 6 hours. This is due to the increasing temperature of the composition density which allows increased hardness in the hard-formed composite with the bonds between the particles of the composite constituent particles. Along with the increase in density a hard composite is formed due to an increase in temperature in the sintering process. The percentage increase in hardness was seen without flash lighting until after sintering process at 500°C and also at 600°C with a holding time of 6 hours. This is due to the increasing temperature of the composition density which allows the occurrence of high hardness. Hardness increased from before sintering and after sintering process where in composition I, composition II, composition III and composition IV is 14% (0.143399936 VHN), 24% (0.238526311 VHN), 26% (0.263107767 VHN) and 36% (0.367843991 VHN) at 500°C, also at 600°C.

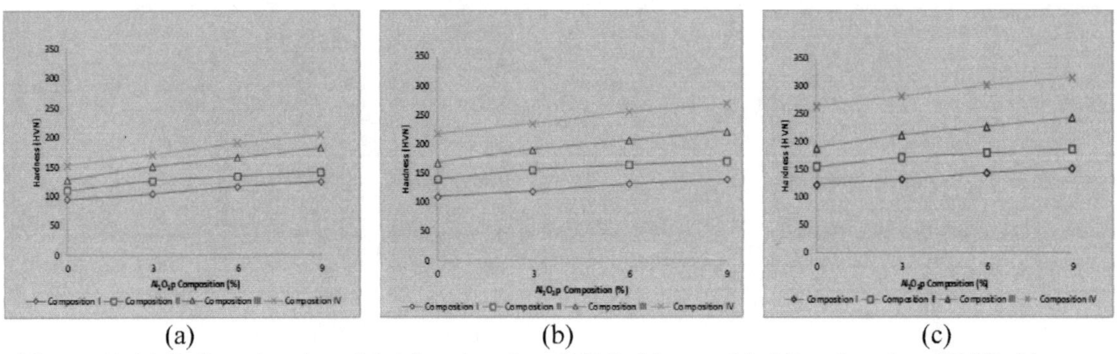

Figure 5. (a) Before sintering. (b) After sintering 500°C, 6 hours. (c) After sintering 600°C, 6 hours

Next seen in Figure 6 is the percentage increase in hardness before sintering and after the sintering process.

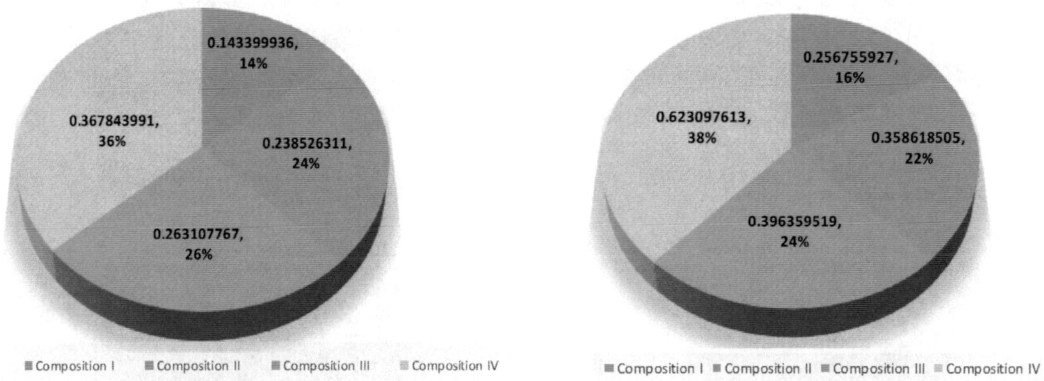

(a) (b)

Figure 6. (a) The percentage of increase in sintering hardness 500°C, 6 hours
(b) The percentage of increase in sintering hardness 600°C, 6 hours

3.4. Observation Analysis Scanning Electron Microscope (SEM)

Composites made from powder metallurgy, in general, tend to have high porosity compared to cast metallurgy. This is because all stages of the powder metallurgy process allow air or lubricants to be trapped between powder particles such as when weighing, mixing and compacting. In Figure 7, SEM is seen from the composition I, II, III, and IV at a temperature of 500°C and also at 600°C each with a holding time of 6 hours.

Based on observations using SEM, in compositions I, II, III, and IV with a mean temperature of 500°C, the SEM results of the composite have bigger pore apertures compared to SEM results at 600°C. In this case that the sintering regulator affects the shrinkage of the formation of pores / cavities in the composite. It can also prove that the hardness and density also increase seen in the results of SEM with the density formed.

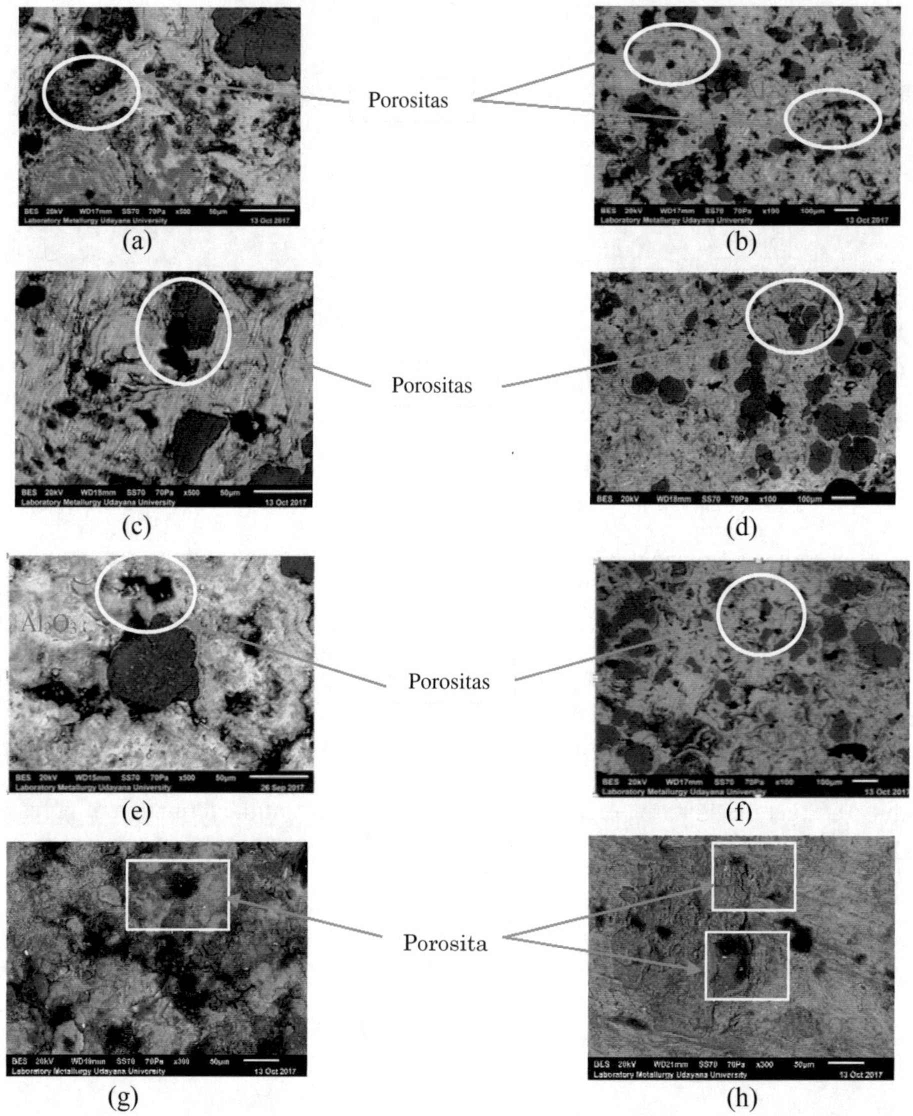

Figure 7. (a). SEM composition I, 500°C, 6 hours, (b). SEM composition I, 600°C, 6 hours, (c). SEM composition II, 500°C, 6 hours, (d). SEM composition II, 600°C, 6 hours, (e). SEM composition III, 500°C, 6 hours, (f). SEM composition III, 600°C, 6 hours, (g). SEM composition V, 500°C, 6 hours, (h). SEM composition IV, 600°C, 6 hours.

4. Conclusion

The properties of heat conductivity in the composite are affected by the temperature sintering treatment. The increase in thermal percentage occurs after compositions I, II, II and IV are subjected to a sintering process at both sintering temperatures of 500 and 600°C at a holding time of 6 hours. Hardness and density properties of composites are influenced by sintering treatment both at sintering temperatures of 500 and 600°C at a holding time of 6 hours. Increasing the percentage of hardness and density occurs in each composite composition after sintering. Specimens observed using SEM, found that specimens that have a high percentage of alumina have high hardness and density values as well as increased heat conductivity. This was corroborated by SEM photos on the sintering temperature treatment of 500 and 600°C at a 6 hours holding time in each composite composition.

Acknowledgment
Thanks to the Head of Metallurgical Laboratory of Udayana University, Head of Metallurgy Laboratory of Institut Teknologi (ITN) Malang, Head of Metallurgy Laboratory of Mechanical Engineering Department of Brawijaya University of Malang and Institute of Research and Service of Udayana University who has assisted in this research.

References

[1] Froyen, L. B. Verlinden, *Aluminum Matrix Composites Materials*. Talat 1402. Belgium. European Aluminum Associations (EAA), (1994).

[2] Garnier, V., Fantozzi, G., Nguyen, D., Dubois, J., & Thollet, G. *Influence of SiC whisker morphology and nature of SiC / Al_2O_3 interface on thermo mechanical properties of SiC reinforced Al_2O_3 composites*. Journal of the European Ceramic Society, 25, 3485-3493. doi : 10.1016/ j. jeurceramsoc. (2004).09.026.

[3] Widyastuti, Dedi Priadi., *Struktur mikro daerah laminasi Komposit Lamina Isotropik Al/SiC-Al/Al_2O_3, Dengan Variasi Waktu Tahan Sinter.* Prosiding Seminar Nasional Sains dan Teknologi-II. (2008). Universitas Lampung in Indonesia

[4] Lim, D., Park, D., Han, B., Kan, T., & Jang, H. *Temperature effects on the tribological behavior of alumina reinforced with unidirectionally oriented SiC whiskers.* Science, (2001). 251, 1452-1458.

[5] Sozhamannan, G. G., & Prabu, S. B. *Influence of interface compounds on interface bonding characteristics of aluminium and silicon carbide*. Materials Characterization, *60*(9), 986-990. Elsevier Inc. doi: 10.1016/j.matchar. (2009).04.005.

[6] Shin, S. E., Ko, Y. J., & Bae, D. H. (2016). Mechanical and thermal properties of nanocarbon-reinforced aluminum matrix composites at elevated temperatures. *Composites Part B: Engineering*, *106*, 66–73. https://doi.org/10.1016/j.compositesb.2016.09.017

[7] Rajendran, M., & Suresh, A. R. (2018). Characterisation of aluminum metal matrix composites and evaluation of thermal properties. *Materials Today: Proceedings*, *5*(2), 8314–8320. https://doi.org/10.1016/j.matpr.2017.11.523

[8] Reddy, B. R., & Srinivas, C. (2018). Fabrication and Characterization of Silicon Carbide and Fly Ash Reinforced Aluminium Metal Matrix Hybrid Composites. *Materials Today: Proceedings*, *5*(2), 8374–8381. https://doi.org/10.1016/j.matpr.2017.11.531

[9] Homeny, J., Vaughn, W. L., & Ferber, M. K. *Silicon Carbide Whisker/Alumina Matrix Composites: Effect of Whisker Surface Treatment on Fracture Toughness*. Journal of the American Ceramic Society, 73(2), 394-402. doi: 10.1111/j.1151-2916.(1990).tb06524.x.

[10] Chen, X., Fu, D., Teng, J., & Zhang, H. (2018). Hot deformation behavior and mechanism of hybrid aluminum-matrix composites reinforced with micro-SiC and nano-TiB2. *Journal of Alloys and Compounds*, *753*, 566–575. https://doi.org/10.1016/j.jallcom.2018.04.223

[11] Fei, W. D., Hu, M., & Yao, C. K. (2003). Effects of thermal residual stress creeping on microstructure and tensile properties of SiC whisker reinforced aluminum matrix composite.

Materials Science and Engineering A, *356*(1–2), 17–22. https://doi.org/10.1016/S0921-5093(02)00827-4

[12] Ogawa, F., Yamamoto, S., & Masuda, C. (2018). Strong, ductile, and thermally conductive carbon nanotube-reinforced aluminum matrix composites fabricated by ball-milling and hot extrusion of powders encapsulated in aluminum containers. *Materials Science and Engineering A*, *711*(September 2017), 460–469. https://doi.org/10.1016/j.msea.2017.11.077

[13] Suarsana K, Soenoko R. (2015). Hardness, Density and Porosity of Al/(SiCw+Al2O3p) Composite by Powder Metallurgy Process without and with Sintering. Trans Tech Publications, Switzerland Accepted: 2015-04-10 doi:10.4028/www.scientific.net/AMM.776.246.

[14] Lao, X., Xu, X., Wu, J., Xu, X., Zhang, Y., & Li, K. (2017). Effect of silicon on properties of Al2O3-SiCwcomposite ceramics in-situ synthesized by aluminium-assisted carbothermal reduction of coal series kaolin for solar thermal storage. *Journal of Alloys and Compounds*, *692*, 825–832. https://doi.org/10.1016/j.jallcom.2016.09.107

[15] Kalkanli, a, & Yilmaz, S. *Synthesis and characterization of aluminum alloy 7075 reinforced with silicon carbide particulates.* Materials & Design, *29*(4), 775-780. doi: 10.1016/j.matdes. (2007). 01.007.

[16] Srinivasa, C. (1996). 2014 and 6061 Aluminium powder metallurgy composites containing silicon carbide particles / fibres. *Materials & Design*, *3069*(96), 359-366.

[17] Randall M., 1984, **Powder Metallurgy Science** USA: Metal Powder Industries Federation. German.

[18] Fritz V. Lenel, 1980, **Powder Metallurgy, Principles and Aplication**. New Jersey: Princeton.

[19] Falcon, Rafael., 2008, ***Analisis Karakteristik Termal dan Resistansi Konduktor Pada Kabel Inti Ganda NYM 2x1.5 mm^2***, Depok, Universitas Indonesia.

IOP Conf. Series: Materials Science and Engineering **539** (2019) 012017 doi:10.1088/1757-899X/539/1/012017

Pyrolysis study of coconut leaf's biomass using thermogravimetric analysis

I M Rajendra[1,5], I N S Winaya[2], A Ghurri[3], and I K G Wirawan[4]
[1]Study Program of Doctoral Engineering Sciences, Faculty of Engineering, University of Udayana
[2,3,4] Study Program of Mechanical Engineering, Faculty of Engineering, University of Udayana
[5]Mechanical Engineering Department, Bali State Polytechnic

Email: made_rajendra2508@yahoo.co.id

Abstract. Coconut leaf is one of the most potential biomass to be converted into bio-oil through pyrolysis process and the availability in Indonesia is abundance. The mechanism of decomposition of coconut leaf into bio-oil productions requires further research because of the complexity of pyrolysis and differences in biomass composition. Therefore, the design, optimization and modeling of pyrolysis processes is strongly influenced by biomass characteristics. The purpose of this study was to find the characteristic differences in pyrolysis behavior of the three main parts of coconut leaf based on its constituent parts; leaflets, midrib and whole leaf. Moisture is removed by drying the sample in an electric oven at 110oC for 24 hours. Characteristics were tested using Cellulose Analysis, Ultimate Analysis, and Heat Value, whereas pyrolysis behavior used Thermogravimetric Analyzer (TGA). The results show that leaflets, midrib and whole leaf exhibit different pyrolysis behavior. In terms of considering flow process of separation, the whole leaf becomes an option as a fuel for further pyrolysis processes. The maximum temperature that as a reference in the pyrolysis process of coconut leaves is 500oC with temperature rate of 20oC/min and the process lasts as long for 130 minutes.

1. Introduction

The future of energy is recently an emerging issue. Anxiety arises as fossil fuel reserves keep running low, urging the human to find other new energy resources (Mohan et al., 2006). One of potential renewable energy resources is biomass since the availability in Indonesia is abundance. Some biomass are commonly used such as the sugarcane and corn grain, lignocellulosic biomass, algae and oil seeds (Zaimes et al., 2015). The main materials of lignocellulosic biomass include rice husks, wood waste such as pine wood sawdust, potential energy corps such as Mahua seeds, castor seeds, and industrial and municipal solid waste (Abu Bakar dan Titiloye, 2013), (Miandad et al., 2016), (Dhanavath et al., 2017), (Aguiar et al., 2008). Biomass fulfills 9% of market demand upon energy in the world and this percentage is expected to increase as concerns on global warming and sustainable energy keep emerging (Gadonneix et al., 2010). The attempt has been implemented the last decades, in which the spotlight is on the importance of converting biomass to biofuel through thermochemical conversion processes such as pyrolysis, gasification, and combustion (Patel et al., 2016). Pyrolysis conversion process appears to be a promising technology as it is able to convert biomass to syngas, bio-oil and bio-char. Syngas and bio-oil have ideal heating value to be used as a potential alternative energy. Bio-oil can also be upgraded to become renewable diesel fuel (Quan et al., 2016).

Pyrolysis is a thermochemical decomposition that produces several products which process absence of oxygen. During the pyrolysis process, the large and complex hydrocarbon bonds break down into smaller molecules in the forms of gases, liquids, and charcoal. Pyrolysis shares similarities with cracking, devolatilization, carbonization, torrefaction, dry distillation, destructive distillation, and thermolysis processes. However, pyrolysis is not similar to gasification process that involves external agents in its chemical reaction which is known as medium gasification. Pyrolysis is usually administered at a temperature that ranges from 300-650 °C, while gasification occurs at between 800-1000 °C, and between 200-300 °C for torrefaction (Basu, 2013).

The characteristics of pyrolysis products are strongly influenced by the type of biomass, pre-treatment of the materials, type of reactor and its parameters including temperature, heating rate, pressure, residence time, catalyst, and air in the environment. Among those factors, temperature is the key factor in pyrolysis process (Auta et al., 2014). Biomass pyrolysis is a complex process that involves a number of chemical reactions. Biomass mainly contains cellulose, hemicellulose, and lignin, as well as other organic extracts and inorganic minerals (Mohan et al., 2006). The main contents of biomass vary depending on the type of the biomass. Commonly, cellulose, hemicellulose, and lignin share 40-60%, 20-40% and 10-25% respectively in the total weight of dry biomass(Burhenne et al., 2013). Cellulose, hemicellulose, and lignin are also known to have different thermal behaviors. As the complexity of biomass conversion determines the variety of products, it is necessary to exclusively investigate the conversion of each composer. Adequate knowledge about the pyrolysis characteristics of those three main components is the key to better understanding of chemical conversion of thermal biomass conversion, which eventually makes the selection of the most appropriate biomass sources and pyrolysis technique to produce certain intended products easier.

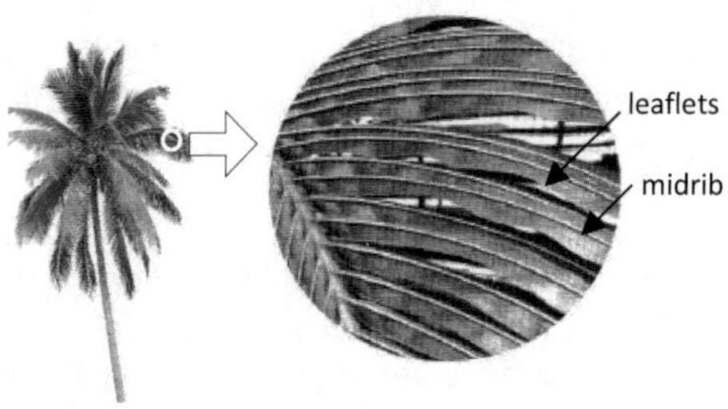

Figure 1. The three main parts of coconut leaves; leaflets,midrib and whole leaf

Biomass from coconut leaves is a potential biomass that can be used as an alternative energy source. Coconut palm (Cocos Nucifera L.) is a commodity that is widely available in tropical areas. Coconut palm can be found in more than 80 tropical countries. Indonesia is an agricultural country that ranks third in after the Philippines and India in coconut production. According to the statistical data on Indonesian agriculture released by the Directorate General of Farming, the Ministry of Agriculture in 2015 - 2017, the coconut plantation area in Indonesia in 2017 reached 3.544.393 Ha. It implies that Indonesia has a massive coconut biomass potential. Coconut leaf consists of two parts; leaflets and midrib that contain different amount of cellulose, hemicellulose and lignin. Biomass from materials that have high amount of cellulose and hemicellulose can be decomposed faster, resulting in more fractions of liquid and gas compared to biomass that contain greater amount of lignin (Quan et al., 2016).

This study was conducted to investigate the different characteristics of pyrolysis behaviors of the three main parts of coconut leaves; leaflets, midrib and whole leaf as illustrated in Figure 1. Due to the abundant presence and complex nature of coconut leaves make it very interesting to studyand as a recommendation of the next energy conversion process where this is still little found in the open literature.

IOP Conf. Series: Materials Science and Engineering **539** (2019) 012017 doi:10.1088/1757-899X/539/1/012017

2. Materials and methods

2.1 Materials

In this study, the materials used were 3 samples in form of powder made by parts of coconut leaves. The leaves were obtained from coconut tree (Cocos nucifera) ecologically in Bali-Indonesia. Coconut leaves were separated between the leaflets and midrib in order to obtained three variations. They are the samples of midrib, leaflets, and whole leaf. Each sample was chopped and blended, and then they were sifted with 80 Mesh. The sample powder then was dried in an electric oven in 110°C holding temperature for 24 hours to get rid the humidity.

2.2. Experimental methods

Cellulose composition was determined through extraction method in accordance to ASTM D1107-96, in which cellulose, lignin and ash in accordance to ASTM D 1103-60, ASTM D1106-56, and ASTM D1102-84 respectively. To get a good validity value, the extraction containing cellulose, hemicellulose, lignin, and ash was tested three times in a kind of sample, and then the result was averaged. The data from the testing result are presented in Table 1.

The chemical elements of C, H, N, and O were tested with LECO elementer CHN add-on O type 628. The testing results are presented in Table 2 as an ultimate analysis. A testing on moisture, volatiles, fixed carbon and ash contents was measured using TGA 701. The analysis method used was in accordance to ASTM D7582 MVA biomass. The data of the testing result are presented in Table 3 as a proximate analysis.

The implementation of thermogravimetric analysis (TGA) to investigate the nature of thermal was very useful in determining thermal behavior experienced by biomass in various temperatures. The understanding of the thermal behavior of biomass was important because the process parameter had a direct impact on chemical composition (quality and quantity) of the product produced. A fast, simple, and reliable analysis method used to investigate thermal decomposition was TGA. It has been widely used by researchers to investigate generally and specifically thermal characteristic of organic materials (Burhenne et al., 2013)(Quan et al., 2016), (Gunasee et al., 2016), (Lazdovica et al., 2017). A test using Thermogravimetric Analysis (TGA) was conducted differentially using TGA 701. The accuracy for temperature measurement was ± 2°C. In each sample, 5 mg dry biomass was heated from 50°C to 900°C in dynamic condition with 20°C/min heating rate. TGA Test was administered to atmosphere pressure and 20 ml/min nitrogen flow (N_2).

Table 1. Composition of dry biomass by wt.%compared to Ref.

	Cellulose	Hemicellulose	Lignin	Ref.
leaflets	45.58	10.35	39.64	
midrib	60.75	11.42	22.10	
whole leaf	56.71	10.66	28.44	
Coconut leaf	59.39	8.45	27.97	(Poddar et al., 2016)

Table 2. Ultimate analysis of the biomass by wt.%compared to Ref.

	C	H	N	O	Ref.
leaflets	47.78	6.13	1.16	28.21	
midrib	47.49	5.95	0.29	39.94	
whole leaf	47.63	6.10	1.07	38.87	
Coconut leaf	47.89	6.19	1.66	37.93	(Phichai et al., 2013)

Table 3. Proximate analysis by wt.% and HHV-LHV of the biomass samples compared to Ref.

	Moisture	Volatiles	Fixed carbon	Ash	HHV (MJ/kg)	LHV (MJ/kg)	Ref.
leaflets	2.74	82.10	7.69	7.48	18.631	18.029	
midrib	2.24	93.69	2.06	2.02	17.848	17.181	
whole leaf	2.55	83.32	7.16	6.97	18.369	17.696	
Coconut leaf	4.77	87.75	12.92	6.33	20.828	-	(Phichai et al., 2013)

IOP Conf. Series: Materials Science and Engineering **539** (2019) 012017 doi:10.1088/1757-899X/539/1/012017

3. Result and Discussion

3.1 Characterization of Biomass

The result of biomass biochemical analysis presented in Table 1 explained that the content of cellulose, hemicelluloses, and lignin were around 93-96%. The other contents such as extractive substance and ash which was dissolved in water or the organic solvent which usually got up to 10% of the total biomass, can be ignored its role in characterization. This extractive substance was phenolic, alkaloid, non-proteinogenic amino acids, acid, terpene, and fatty acid compounds (Ranzi et al., 2008).

Leaflets and midrib had cellulose content as much as 24.58% and 60.75%, respectively. Each of them also had lignin content as much as 39.64% and 22.10%respectively. From its cellulose, midrib had higher cellulose content compared to leaflets. This had made cellulose to form stronger and more rigid tissue. Meanwhile, leaflets had higher lignin content which made the decomposition process became longer. Hemicellulose evaporates easily which cause on the number of volatile products produced. In this study, midrib had higher hemicelluloses than leaflets.

The composition of cellulose, hemicellulose, and lignin in biomass was formed the atom balance of its component. Cellulose and hemicellulose substances formula for all biomass were assumed as $C_6H_{10}O_5$ and $C_5H_{10}O_5$, respectively. Lignin was polymerase product of three kinds of monolignols such as paracoumaryl, coniferyl, and synapyl alcohols, which were put on each lignin in form of p-hydroxyphenyl (H lignin), guaiacyl (G lignin), and syringyl (S lignin). Each formula of H, G, and S lignin was $C_9H_{10}O_2$, $C_{10}H_{12}O_3$ and $C_{11}H_{14}O_4$ (Burhenne et al., 2013).

Referring to the test results of C, H, N, O and S components presented in Table 2, it can be inferred that the percentage of C and H were not comparable with cellulose or hemicellulose percentages in wither leaflets or midrib. C and H in leaflets refer to the lignin formation while in midrib, they refer to cellulose formation.

3.2 Pyrolysis kinetics in TGA

In the experiment of thermogravimetric analysis (TGA), the average mass changes in all samples were declared as temperature function between 50°C and 750°C. The target temperature in TGA was set up to 900°C. However, because there were no mass changes any longer start from 650°C temperature, automatically, the analysis process was stopped on 750°C. The results were presented in Figure 2.During 50°C until 110°C temperature, there was evaporation process. Leaflets had lost its mass bigger than midrib due to its moisture substance which was higher in leaflets compared to in midrib and whole leave.

Pyrolysis decomposition of all biomass of coconut leaves ideally occurred during 230°C until 650°C temperature. Referring to DTGA Curve in Figure 3, decomposition process dominantly occurred during 230°C until 500°C temperature. Above those, the process tended to be constant. This data could be a future consideration in pyrolysis coconut leaves into bio-oil. The maximum temperature which could be referenced was 500°C. It can be a reference in the planning of pyrolysis reactor design which saves energy during the process.

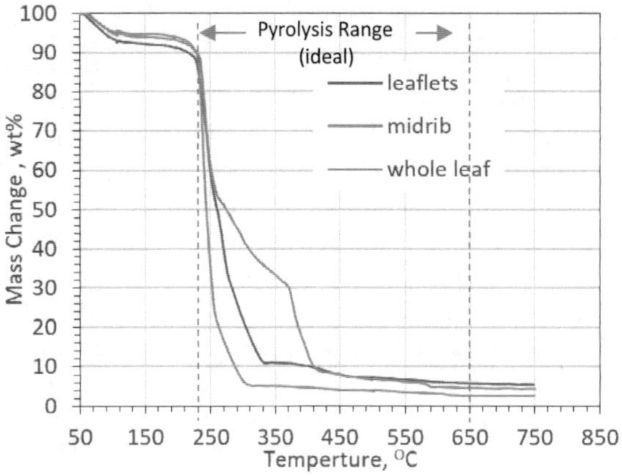

Figure 2. TGA curvesof three coconut leaves with a heating rate of 20°C / min.

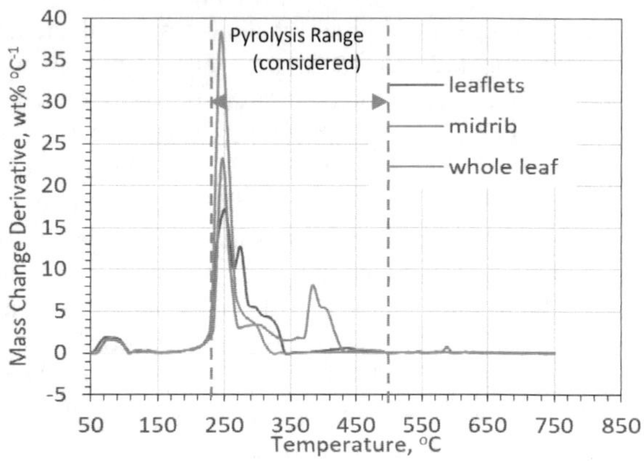

Figure 3. DTGA curvesof three coconut leaves with a heating rate of 20°C / min.

Midrib experience decomposition at a lower temperature compared to leaflets. It explained that biomass with high cellulose content could be easily elaborate. Meanwhile, biomass with dominant lignin substance was elaborated longer as visualized in Figure 4. The process time needed to reach 500°C temperature with 20°C/minute temperature rate was 130 minutes.

Thermogravimetry for leaflets curve shifted for around 30°C compared to midrib curve. This shows that the bigger energy is needed to disentangle leaflets since it has bigger and stable lignin substance compared to its cellulose and hemicellulose.

The behavior in this situation was not systematically. Especially in whole leave, there was a significant deviation. Seen from the contents of cellulose, hemicellulose, and lignin, whole leave was in the position between leaflets and midrib. However, its decomposition curve in TGA shifted for around 110°C compared to midrib.

The composition of the product produced by coconut leaves pyrolysis was moisture, volatile, fixed carbon and ash, visualized in Figure 5. Those three kinds of biomass produced a high volatile matter content. This product was in form of steam which could be condensed into a liquid form known as bio-oil. This bio-oil was future energy potential. Midrib produced higher volatile than leaflets. It can be seen in DGTA curve where mass changes were bigger in a narrow temperature range, 230°C until 250°C. The three kinds of the biomass were suitable to proceed into bio-oil. However, from their separation process between leaflets and midrib, whole leave was more likely to be considered.

One of the products of pyrolysis which needed to get more attention was ash. Ash which experienced further heating would form hard lumps attached to the bottom of the reactor. Ash factor would be a hindrance of the pyrolysis process in the reactor.

Each ash produced by leaflets and midrib was 7.48% and 2.02%respectively. This difference is significant, almost 5%. If midrib was chosen as pyrolysis material, there would be fewer problems caused by ash. However, if leaflets or whole leave was chosen instead, a special mechanism needs to be applied in the pyrolysis reactor in order to run well.

3.3 HHV of Biomass

The heat value in bio-oil was one of the important characteristics. The standard measurement of energy from heat could be reported as higher heat value (HHV) and lower heat value (LHV). The difference between HHV and LHV was the same as the heat of water evaporation formed by burning fuel.

The results of HHV testing showed high heat value of leaflets, midrib, and the whole leaf were 18.631MJ/kg, 17.848MJ/kg and 18.369 MJ/kg, respectively. Those three values were considered almost the same. However, there was significantly different compared to the HHV reported by (Phichai et al., 2013) as much as 20.828 MJ/kg. This difference possibly influenced by the type of biomass which was from different varieties of coconut leaves.

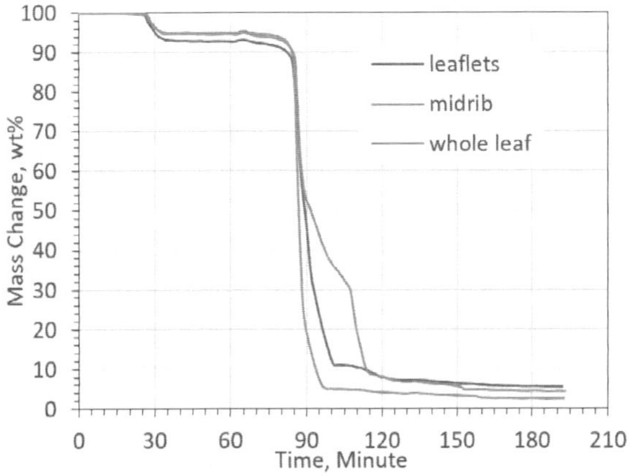

Fig.4. TGA curves of three coconut leaves of time heating

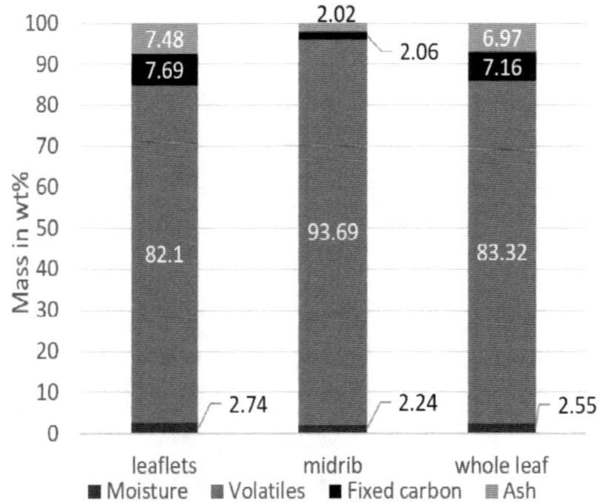

Fig.5. Pyrolysis product composition by wt.%. in TGA

In determining the appropriate kind of biomass based on HHV value, whole leave was worth to be considered. Besides HHV value which was similar to leaflets, the shorter process in separating leaflet and midrib also became the consideration as previously discussed.

4. Conclusion

A study on pyrolysis of coconut leaves using Thermogravimetric Analysis (TGA) had been conducted with a conclusion that leaflets, midrib, and whole leave showed the different testing result as well as different pyrolysis behavior. Considering the separation process, thus, leaflets were chosen as the fuel for the next pyrolysis process. The maximum value which can be used as a reference in the pyrolysis process of coconut leaves was 500°C and 20°C/minute temperature rate. Thus, the process will be 130 minutes. Further research which is needed to support this study is ash characterization produced by the above parameter.

Acknowledgements

The authors acknowledge gratefully the very valuable support of I Made Agus Putrawan in the Materials Analysis Laboratory of Udayana University, Bali-Indonesia, is acknowledged for placing the TGA equipment at our disposal.This research was supported the financial by Bali State Polytechnic.

References

[1] Abu Bakar, M. S. & Titiloye, J. O. 2013. Catalytic pyrolysis of rice husk for bio-oil production. *Journal of Analytical and Applied Pyrolysis,* 103, 362-368.
[2] Aguiar, L., Ma´Rquez-Montesinos, F., Gonzalo, A., Sa´Nchez, J. L. & Arauzo, J. 2008. Influence of temperature and particle size on the fixed bed pyrolysis of orange peel residues. *Journal of Analytical and Applied Pyrolysis,* 83 124–130.
[3] Auta, M., Ern, L. M. & Hameed, B. H. 2014. Fixed-bed catalytic and non-catalytic empty fruit bunch biomass pyrolysis. *Journal of Analytical and Applied Pyrolysis,* 107, 67-72.

[4] Basu, P. 2013. *Biomass Gasification, Pyrolysis, and Torrefaction, Practical Design and Theory,* 32 Jamestown Road, London NW1 7BY, UK, 525 B Street, Suite 1800, San Diego, CA 92101-4495, USA, Published by Elsevier Inc.

[5] Burhenne, L., Messmer, J., Aicher, T. & Laborie, M.-P. 2013. The effect of the biomass components lignin, cellulose and hemicellulose on TGA and fixed bed pyrolysis. *Journal of Analytical and Applied Pyrolysis,* 101**,** 177-184.

[6] Dhanavath, K. N., Shah, K., Bankupalli, S., Bhargava, S. K. & Parthasarathy, R. 2017. Derivation of optimum operating conditions for the slow pyrolysis of Mahua press seed cake in a fixed bed batch reactor for bio–oil production. *Journal of Environmental Chemical Engineering,* 5**,** 4051-4063.

[7] Gadonneix, P., Castro, F. B. D., Medeiros, N. F. D., Drouin, R., Jain, C. P., Kim, Y. D. & Frei, C. 2010. Survey of Energy Resources. Regency House 1-4 Warwick Street London W1B 5LT United Kingdom: World Energy Council.

[8] Gunasee, S. D., Carrier, M., Gorgens, J. F. & Mohee, R. 2016. Pyrolysis and combustion of municipal solid wastes: Evaluation of synergistic effects using TGA-MS. *Journal of Analytical and Applied Pyrolysis,* 121**,** 50-61.

[9] Lazdovica, K., Kampars, V., Liepina, L. & Vilka, M. 2017. Comparative study on thermal pyrolysis of buckwheat and wheat straws by using TGA-FTIR and Py-GC/MS methods. *Journal of Analytical and Applied Pyrolysis,* 124**,** 1-15.

[10] Miandad, R., Nizami, A. S., Rehan, M., Barakat, M. A., Khan, M. I., Mustafa, A., Ismail, I. M. I. &Murphy, J. D. 2016. Influence of temperature and reaction time on the conversion of polystyrene waste to pyrolysis liquid oil. *Waste Management,* 58**,** 250-259.

[11] Mohan, D., Pittman, C. U. & Steele, P. H. 2006. Pyrolysis of Wood/Biomass for Bio-oil: A Critical Review. *Energy & Fuels* 20, 848-889.

[12] Patel, M., Zhang, X. & Kumar, A. 2016. Techno-economic and life cycle assessment on lignocellulosic biomass thermochemical conversion technologies: A review. *Renewable and Sustainable Energy Reviews,* 53**,** 1486-1499.

[13] Phichai, K., Pragrobpondee, P., Khumpart, T. & Hirunpraditkoon, S. 2013. Prediction Heating Values of Lignocellulosics from Biomass Characteristics. *Journal of Chemical and Molecular Engineering,* 7**,** 532-535.

[14] Poddar, P., Asad, M. A., Islam, M. S., Sultana, S., Nur, H. P. & Chowdhury, A. M. S. 2016. Mechanical and Morphological Study of Arecanut Leaf Sheath (ALS), Coconut Leaf Sheath (CLS) and Coconut Stem Fiber (CSF). *Advanced Materials Science,* 1**,** 1-4.

[15] Quan, C., Gao, N. & Song, Q. 2016. Pyrolysis of biomass components in a TGA and a fixed-bed reactor: Thermochemical behaviors, kinetics, and product characterization. *Journal of Analytical and Applied Pyrolysis,* 121**,** 84-92.

[16] Ranzi, E., Cuoci, A., Faravelli, T., Frassoldati, A., Migliavacca, G., Pierucci, S. & Sommariva, S. 2008. Chemical Kinetics of Biomass Pyrolysis. *Energy & Fuels* 22, 4292–4300.

[17] Zaimes, G. G., Vora, N., Chopra, S. S., Landis, A. E. & Khanna, V. 2015. Design of Sustainable Biofuel Processes and Supply Chains: Challenges and Opportunities. *Processes,* 3**,** 634-663.

IOP Conf. Series: Materials Science and Engineering **539** (2019) 012018 doi:10.1088/1757-899X/539/1/012018

The metal gasket sealing performance of bolted flanged with fem analysis

I M G Karohika[1] and I N G Antara[1]

[1]Mechanical Engineering Department University of Udayana

Email: gatot.karohika@unud.ac.id

Abstract. There is a wide variation of flange style and configuration, but only four types of facing are widely used, which are raised face, flat face, ring joint facing and lap joint flange. Corrugated metal gaskets 400MPa-mode has better performance than gaskets 0Mpa-mode for flat face flange type. In this study we investigated the performance of corrugated metal gasket 400MPa-mode for three types of flange facing which are raised face, flat face and combination raised-flat face. The finite element method employed to analysis the effect of different types of flange on contact width and contact stress in relation with axial force of corrugated metal gasket. The result shows corrugated metal gasket 400MPa-mode give good performance for all type of flange face.

1. Introduction

In the piping system there are several types of pipe connection methods. One of them is a mechanical joining system. In mechanical systems there are several methods used and this system continues to evolve as technology advances. A flange joining is one of the mechanical connection systems. Flange is a non-permanent tightening mechanism, it can be dismantled and installed using bolts as a fastener. Pipes that use this method are usually because they often have to be dismantled when repairs. The flange that is used to connect also will differ in type. In general there are two types of flange, namely flat face type and Rise face type.

However, due to flange surface imperfections, there is still a leak in the pipe connection. So we need components that can cover the imperfections of the flange surface to prevent leakage. Gaskets are components that are inserted between flanges, which when in contact with the surface of the flange are deformed plastic and fill the gap of the flange surface which is not perfect so as to prevent leakage. With advances in technology research on gasket design and material is growing rapidly too. Research on corrugated metal gaskets producing super seal gaskets has been carried out to obtain an optimal design [1]. The research then continued by considering the influence of the forming process [2]. The research about the corrugated metal gasket by utilizing simulation method [3, 4] which is validated also by experiment [5]. The surface roughness of the flange also affects the performance of the gasket

[6].

Previous research is a study to produce new design products from corrugated metal gaskets based on full face flange. The forming process use open dies type. Referring to the design concept from previous research that the contact width and contact stress are the important sealing performance parameter. In this research the aim to determine the contact stress and contact width by considering forming by closed die in contacted with full face and rise face flange. The analysis using computer simulations design and manufacturing processes so that the cost and time for trial-error processes will be reduced.

2. Material and Method

The gasket used in this study is a corrugated metal gasket 400 MPa mode mounted between two flanges on the upper and lower sides, as illustrated in Figure 1. The flanges type used in this paper are flat face and rise face flange. When the gasket material is tightened between the two flanges, then the two bead on the gasket will come into contact with the flange and creating contact width lines that have a high contact stress to prevent leakage (Figure 2). Although the contact width that occurs in this corrugated type gasket is small, it will produce contact stress which is high so that the tightening load smaller is needed.

This research was carried out using computer simulation. The material used is SUS 304 steel material because it is resistant to high temperature and corrosion resistance. This material using JIS standards based on strength tests drag JISZ 2241, as shown in table 1.

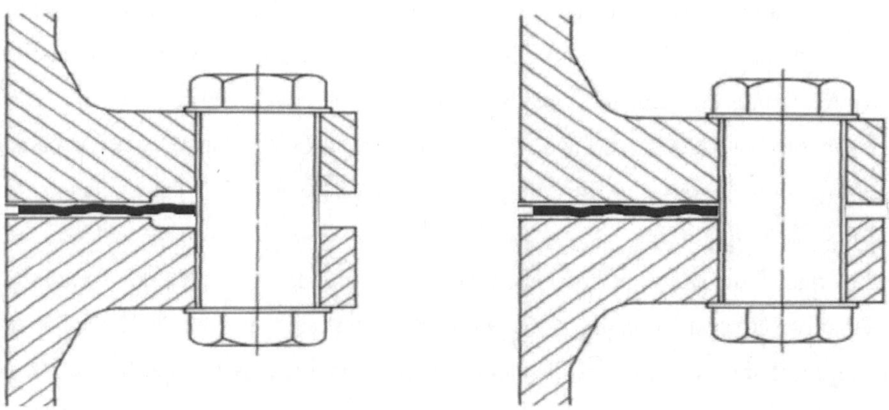

a) Rise Face type b) Flat face type

Figure 1. Flange type

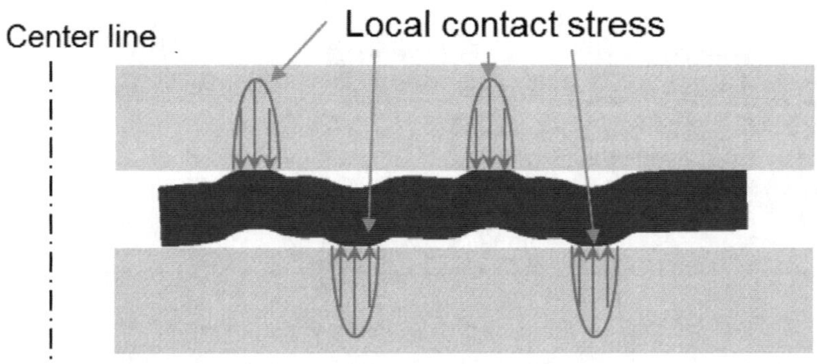

Figure 2. High local contact stress on small contact width

Table 1. Mechanical properties of SUS304

Properties	Value
Yield stress (Mpa)	398,83
Modulus tangen (Mpa)	1900,53
Modulus elasticity (E) GPa	210
Poisson ratio (v)	0.3

Material is modeled isotropic bilinear hardening because it is used to analyze plasticity experienced by workpieces as a result high strain. Stress-strain curve bilinear was obtained by including the yield strength and tangent values of the modulus. Segment slope first on a curve equivalent to young modulus while the slope of the second segment is equivalent to tangent modulus. Figure 3 shows material model relationship from Stainless Steel SUS304.

The corrugated gaskets model analyze uses the finite element method based simulation. Both flanges are assumed to be rigid body, while gaskets are assumed to be deformable. Modeling is made with 2 dimensions assuming an axe-symmetrical model. Displacement in the axial direction is used as a model tightening the upper flange and lower flange against the gasket (Figure 4 and Figure 5).

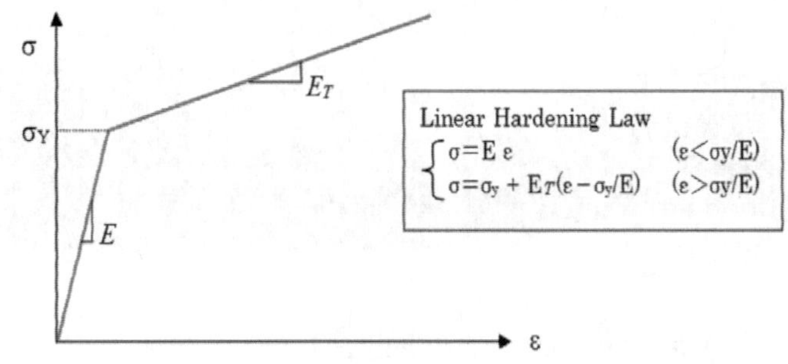

Figure 3. Linear strain hardening model for SUS304

IOP Conf. Series: Materials Science and Engineering **539** (2019) 012018 doi:10.1088/1757-899X/539/1/012018

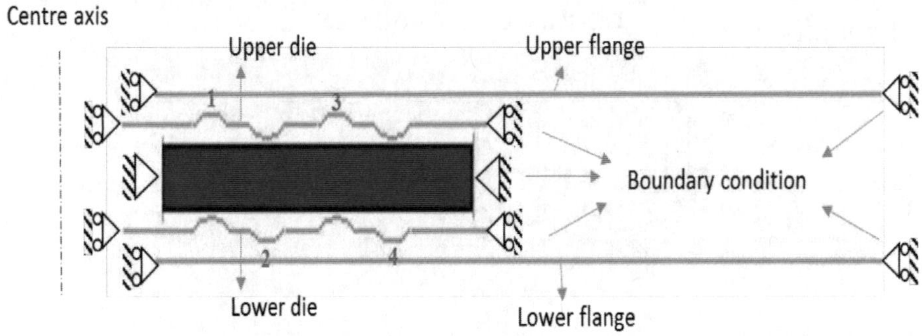

Figure 4. Simulation for flat face flange axisymmetric gasket model

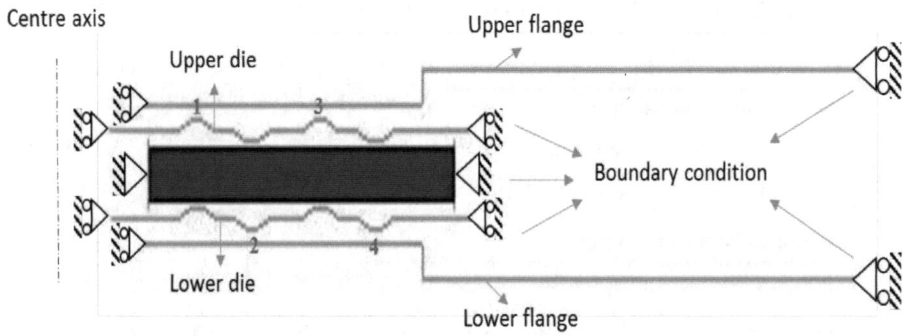

Figure 5. Simulation for rise face flange axisymmetric gasket model

In previous studies, it was known that contact width affects the resistance level gasket to prevent leakage with consider contact stress that occurs [2]. The contact width is measured by the number of gasket elements that have contact with the flange. Determination of contact width is done by considering contact stress by removing contact stress values below the yield stress of 398.83 MPa. Sealing lines will form on contact widths that have plastic contact stress conditions (Figure 6).

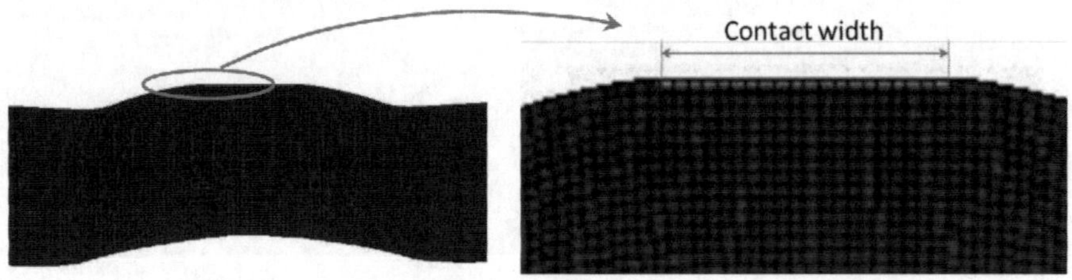

Figure 6. Contact width on the convex gasket

3. Result and Discussion

The result of computer simulation, there is a distribution of contact stress and contact width that occurs. The magnitude of the flange tightening load on the gasket is accompanied by the stress distribution patterns and gasket deformation that occur. Based on previous research that contact width that have high plastic contact stress condition occur most on convex 2 and convex 3. Therefore our focus analysis on convex 2 and convex 3 because influence gasket sealing performance to reduce leakage.

Firstly we plot graphics with horizontal axis gives the axial force for tightening flange while contact width or average contact stress is given along the vertical axis. Figure 7 shows the axial force relation to contact width for convex portion 2. It shows for axial force under 80 KN the contact width between three types of flange almost the same value. For the same closed dies type the contact width value seems the same for flat face and rise face flange for each axial force. For the same flat face type the contact width value for open dies is higher than closed dies for each axial force. Overall the flat face flange type with open dies has highest contact width value.

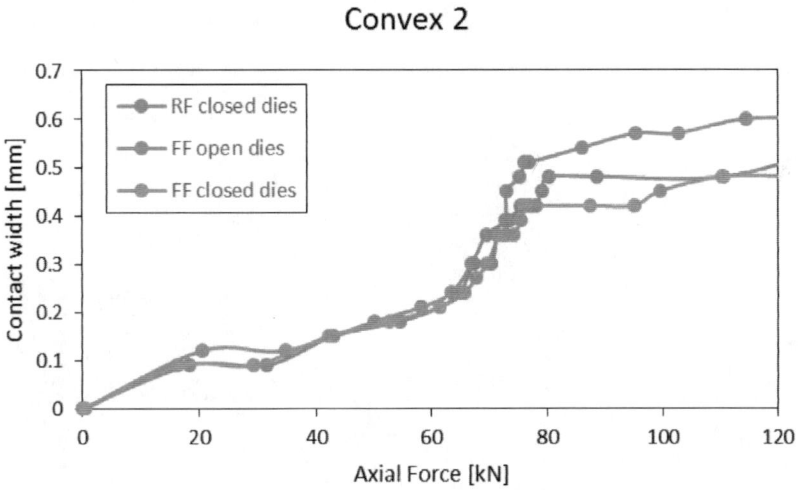

Figure 7. Contact width versus axial force for convex 2

Figure 8 shows the axial force relation to contact width for convex portion 3. It shows for axial force under 80 KN the contact width between three types of flange almost the same value. For the same closed dies type the contact width value seems the same for flat face and rise face flange for each axial force. For the same flat face type the contact width value for open dies is higher than closed dies for each axial force. Overall the flat face flange type with open dies has highest contact width value.

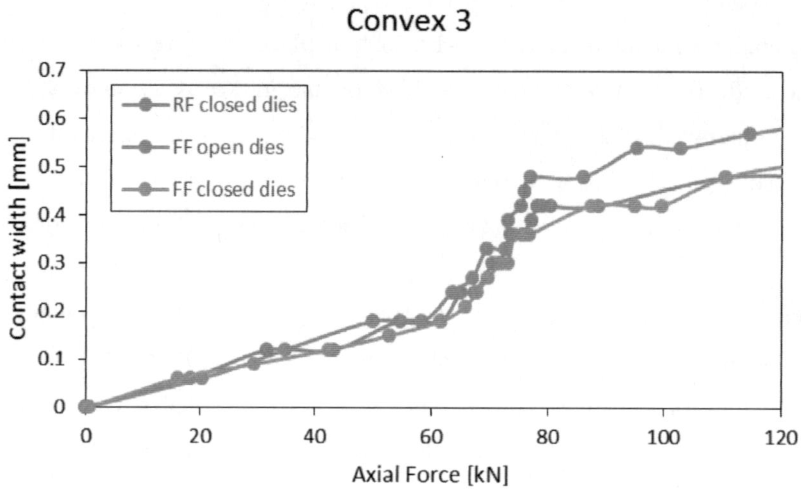

Figure 8. Contact width versus axial force for convex 3

Figure 9 shows the relation between axial force and average contact stress for convex portion 2. For the same closed dies type the average contact stress value seems the same distribution for flat face and rise face flange for each axial force. For the same flat face type the average contact stress value for closed dies is higher than open dies for each axial force. Overall the closed die type has highest average contact stress value.

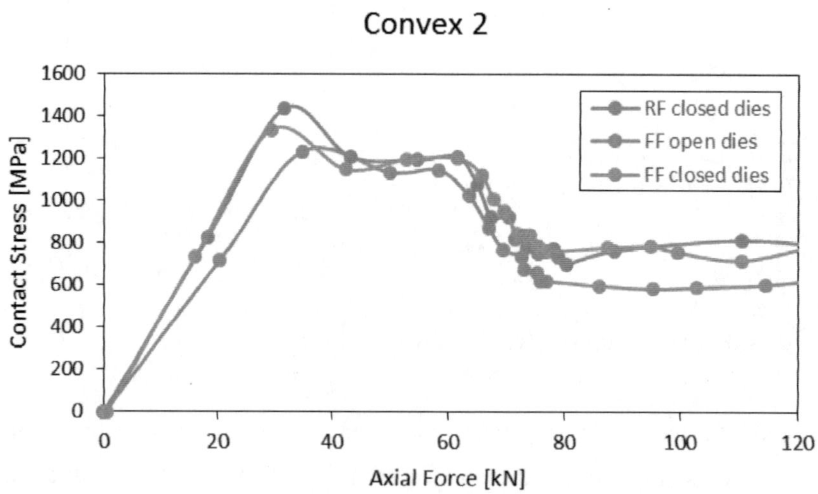

Figure 9. Average contact stress versus axial force for convex 2

Figure 10 shows the relation between axial force and average contact stress for convex portion 3. For the same closed dies type the average contact stress value seems the same distribution for flat face and

IOP Conf. Series: Materials Science and Engineering **539** (2019) 012018 doi:10.1088/1757-899X/539/1/012018

rise face flange for each axial force. For the same flat face type the average contact stress value for closed dies is higher than open dies for each axial force. Overall the closed die type has highest average contact stress value.

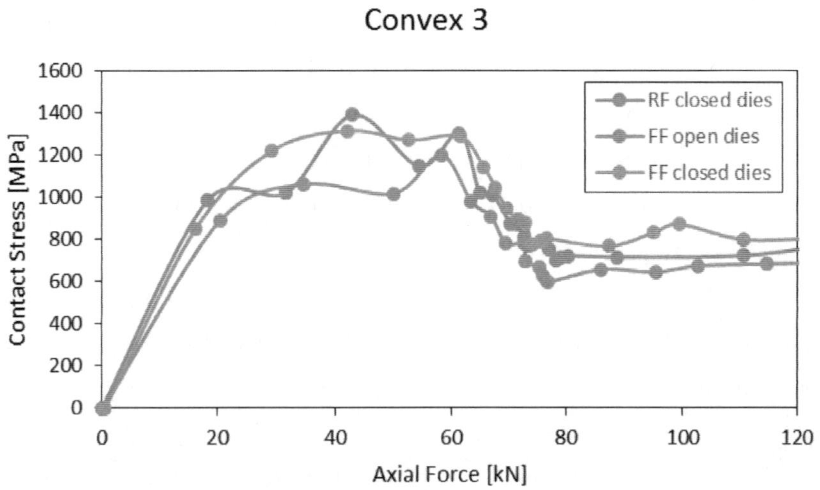

Figure 10. Axial force versus Average contact stress for convex 3

Metal gasket made by closed die has higher average contact stress than open dies because when forming process gasket made by open dies has lack of die fill, especially on the convex portion.

4. Conclusion

Based on simulation result, corrugated metal gasket 400 MPa has the good sealing performance for flat face and rise flat face type of flange. Gasket made by closed dies will give better sealing performance because has higher average contact stress than gasket made by open dies.

Acknowledgment

The research/ article's publication is supported by LPPM University of Udayana through Udayana Invention Research and partially funded by the United States Agency for International Development (USAID) through the Sustainable Higher Education Research Alliance (SHERA) Program for Universitas Indonesia's Scientific Modelling, Application, Research and Training for City-centered Innovation and Technology (SMARTCITY) Project, Grant#AID-497-A-1600004, Sub Grant#IIE-00000078-UI-1

References

[1] Saeed, H.A., Izumi, S., Sakai, S., Haruyama, S., Nagawa, M., Noda, H., 2008, ***Development of New Metallic Gasket and its Optimum Design for Leakage Performance***, Journal of Solid Mechanics and Material Engineering, Vol. 2, No. 1, , pp. 105-114.

[2] Haruyama, S., Choiron, M.A., Kaminishi K., 2009, ***A Study of Design Standard and Performance Evaluation on New Metallic Gasket,*** Proceeding of the 2nd International Symposium on Digital Manufacturing, Wuhan China, , pp. 107-113.

[3] Choiron, M.A., Haruyama, S., Kaminishi, K., 2011, ***Simulation and Experimentation on the Contact Width of New Metal Gasket for Asbestos Substitution***, International Journal of Aerospace and Mechanical Engineering, Vol. 5, No. 4, , pp. 283-287.

[4] Nurhadiyanto, D., Choiron, M.A., Haruyama, S., Kaminishi, K., 2012, ***Optimization of New 25A-size Metal Gasket Design Based on Contact Width Considering Forming and Contact Stress Effect***, International Journal of Mechanical and Aerospace Engineering, Vol. 6, , pp. 343-347.

[5] Haruyama, S., Nurhadiyanto, D., Choiron, M.A., Kaminishi, K., 2013, ***Influence of surface roughness on leakage of new Metal Gasket,*** International Journal of Pressure Vessels and Piping, 111-112, pp. 146-154.

[6] Haruyama, S., Nurhadiyanto, D., Kaminishi, K., 2014, ***Contact Width Analysis of Corrugated Metal Gasket based on Surface Roughness***, Advanced Materials Research, Vol. 856, pp. 92-97.

[7] Nurhadiyanto, D., Haruyama, S., Kaminishi, K., Karohika I M. G., Mujiono, 2015, ***Contact Stress and Contact Width Analysis of Corrugated Metal Gasket***, Applied Mechanics and Materials, Vols. 799-800, pp. 765-769.

[8] Haruyama, S., Karohika, I M. G., Sato, A., Nurhadiyanto, D., Kaminishi, K., 2016, ***Development of 25A-Size Three-Layer Metal Gasket by Using FEM Simulation***, International Journal of Mechanical, Aerospace, Industrial, Mechatronic and Manufacturing Engineering Vol:10, No:3, pp.555-561.

[9] MSC Marc 2010, User Manual.

IOP Conf. Series: Materials Science and Engineering **539** (2019) 012019 doi:10.1088/1757-899X/539/1/012019

Influence of dies type for gasket forming shape

I M G Karohika[1], I N G Antara[1] and I M D Budiana[1]

[1]Mechanical Engineering Department University of Udayana

Email: gatot.karohika@unud.ac.id

Abstract. Performance of metal gasket influenced by the final design of the forming result. Good shape of the gasket will give optimal leakage performance. It is difficult to produce good shape because of the spring back effect. In this study, we investigated the forming process of metal gasket 400MPa-mode for two types of dies which are open dies and closed dies. The finite element method employed to analyze the effect of different types of dies on the final design of the corrugated metal gasket. The simulation shows corrugated metal gasket 400MPa-mode with closed dies result in better final shape than open dies.

1. Introduction

Leaks in piping connections due to flange surface imperfections must be avoided. For this reason, a gasket is needed as a component inserted between two flange surfaces, in which there is a pressurized fluid, to prevent leakage. When tightening the fastening bolt between two flanges, the gasket will react by changing shape according to the level of elasticity. This change in shape will fill the space produced by the two flanges due to uneven workmanship. This change in shape causes the connection between the two flanges to become tight and allows to prevent leakage. Research on gaskets evolves over time. The development is in terms of material and system performance of the gasket. In material terms, metal, non-metallic and semi-metal materials are known. One of the gaskets made from metal is a corrugated metal gasket [1 – 8]. This gasket has a corrugated surface that produces high contact stress that can prevent leakage. While the flat part serves as a spring so that it can reduce the occurrence of loose bolts.

In the previous study the process of making corrugated metal gaskets using press mold forming method. The dies type used is open dies. There are still found some defects in the press results, especially at the peak of the convex portion that still does not fill the entire surface of dies [5]. This can affect the performance of the gasket when in contact with the flange surface, where the gasket does not work optimally to prevent leakage.

At present, the development of software technology is advancing so that it is possible to predict the results of forming processes using simulation methods. With the right modeling, the number and time of the experiment can be limited so that the costs incurred because the experiment process can be reduced.

2. Material and Methods

In this paper, the simulation method is used to analyze the effect of the type of dies that is open dies and closed dies on the results of the forming process. The material used as a gasket material is SUS304 with mechanical properties indicated in table 1. The type of dies selected dies 400 MPa with a size as in table 2. The geometry of dies is shown in figure 1.

Table 1. Mechanical properties of SUS304

Properties	Value
Yield stress (Mpa)	398,83
Modulus tangen (Mpa)	1900,53
Modulus elasticity (E) GPa	210
Poisson ratio (v)	0.3

Table 2. Dies design gasket model of 400-MPa

Factor	dimension (mm)
Overhang (OH)	3
Pitch 1(p_1)	3.5
Pitch 2 (p_2)	4.5
Pitch 3 (p_3)	3.5
Thickness (Tg)	1.5
Radius (R)	2.5
Radius (R1)	2.3
Height (h)	0.3
Height (h1)	0.33

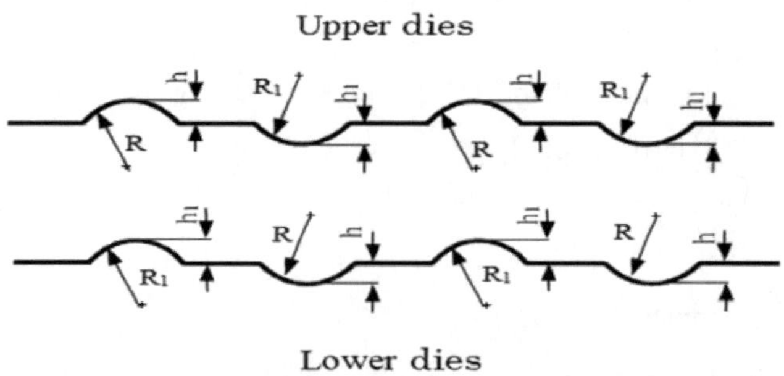

Figure 1. Dies geometry

The simulation divided into 2 stages, the first stage of the plate is made using CAD software. The second stage of the model that has been made is processed in FEM software which consists of process meshing and forming processes. In the meshing process for plate gasket with element size 0.03 mm quadrilateral element type. Next is the forming process assuming two dimensions of the axis-symmetric model in the axial direction between the upper and lower dies as shown in figure 2. Upper and lower dies are assumed to be rigid bodies while deformable bodies gaskets. The dies used are types of open dies and closed dies. The flowchart of the gasket simulation steps is shown in figure 3. In this simulation, the gasket material is Linear hardening law.

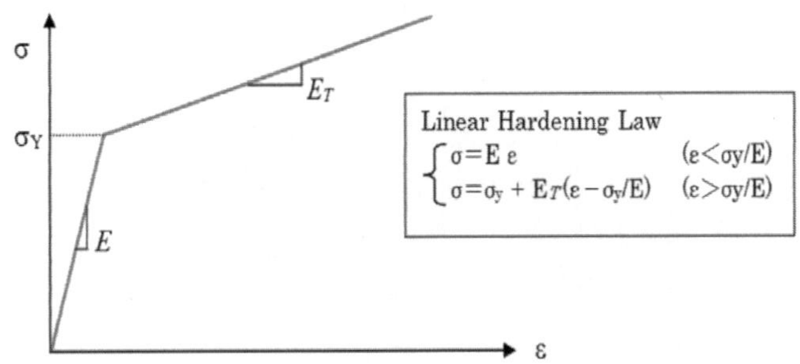

Figure 2. Linear strain hardening model for SUS304

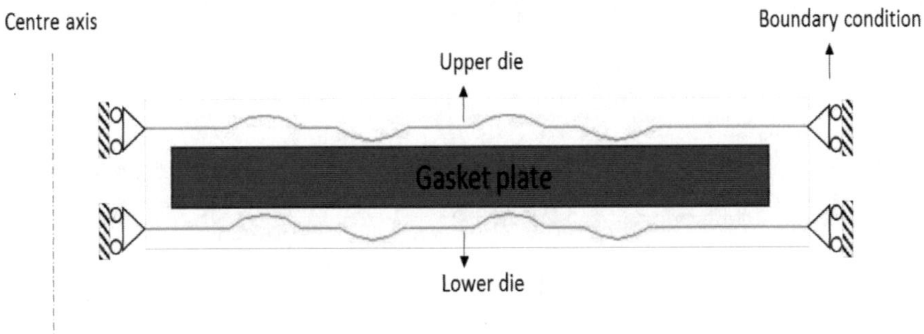

Figure 3. Forming process simulation for open dies axisymmetric gasket model

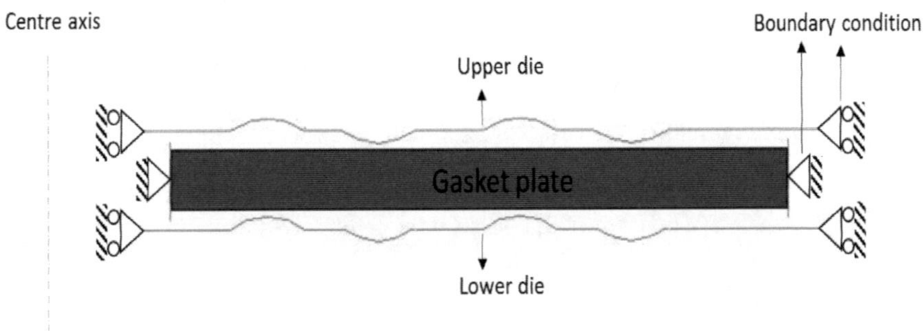

Figure 4. Forming process simulation for closed dies axisymmetric gasket model

IOP Conf. Series: Materials Science and Engineering **539** (2019) 012019 doi:10.1088/1757-899X/539/1/012019

3. Result and Discussion

To analyze the forming results that occur, the focus is on 4 parts of the peak of convex of the gasket because the performance of the gasket is affected by that part [5]. The forming result is said to be good if the gasket formed fill the entire parts of the dies. If it does not fill the contents of the dies it is said the product defects and the defect amount is calculated based on the number of elements that are not in contact with the dies field.

Die fill defect = element size x number of element (mm)

= 0.03 x number of element (mm)

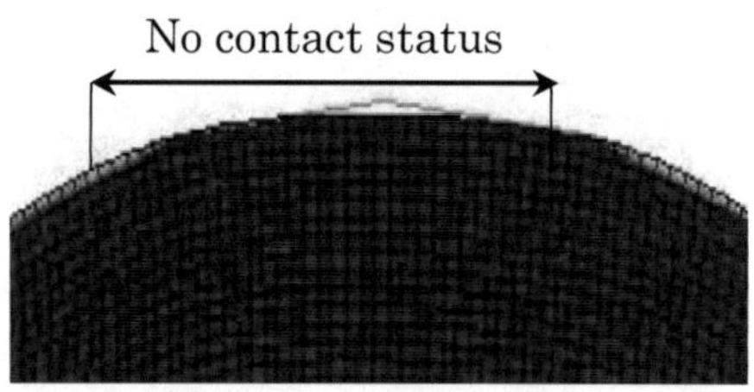

Figure 5. Contact status between the gasket and die

Figure 6. Forming result for open die type

Figure 7. Forming result for closed die type

IOP Conf. Series: Materials Science and Engineering **539** (2019) 012019 doi:10.1088/1757-899X/539/1/012019

Table 3. Die fill defect open dies

	Convex 1	Convex 2	Convex 3	Convex 4
figure	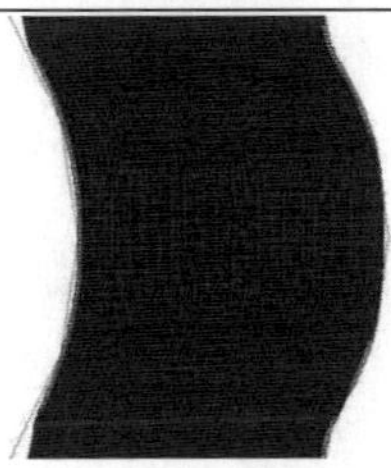			
Die fill defect	0.75	0.57	0.63	0.81

Table 4. Die fill defect closed dies

	Convex 1	Convex 2	Convex 3	Convex 4
figure	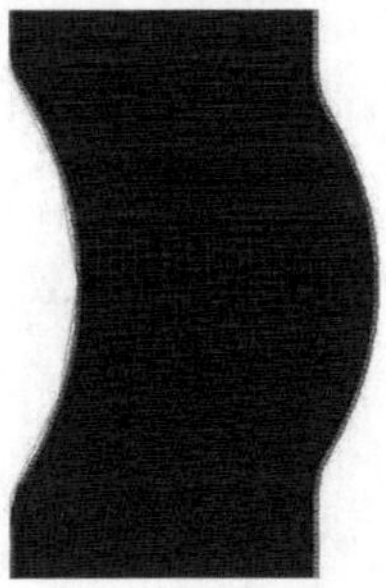			
Die fill defect	0.33	0.39	0.36	0.3

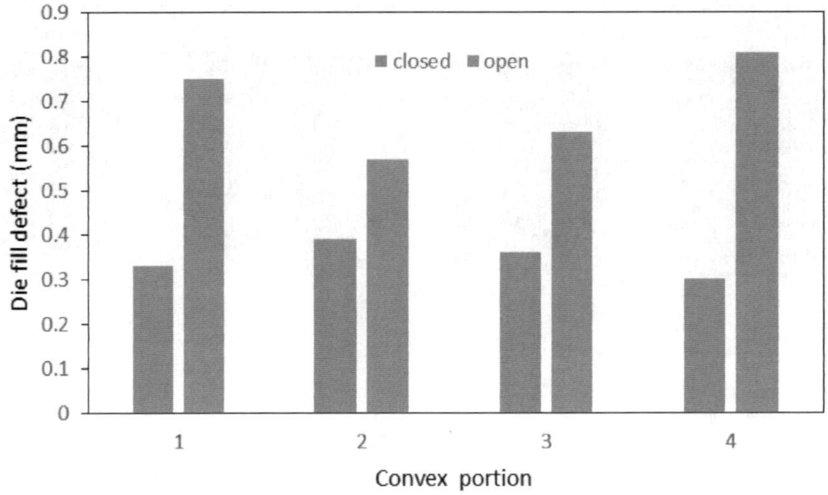

Figure 8. Graphic Comparison die fill defect

Based on table 3 and 4, is showed when forming process, gasket material has plastic deformation. For open dies type the deformation occurs in the vertical and horizontal directions. In the vertical direction the plate changes shape to be corrugated, in the horizontal direction the gasket plate also becomes longer, therefore, the gasket plate does not fill the mold, especially at the peak of the corrugated shape.

For closed dies type, gasket plates only deform in a vertical direction because the horizontal direction is limited by dies. The shape of the gasket becomes more perfect. However, there is still a small element in the peak of corrugated gaskets that do not fill the dies. Figure 8 shows the comparative lack of die fill defect for closed die and open die type. The optimization design for designing closed type dies needs to be done in the future.

4. Conclusion

Based on dies type in the press mold process, by using the simulation method the resulting design was found. The result shows that gasket by closed dies type gives a better result than open dies type. Next, optimization design research is necessary for better gasket shape result.

Acknowledgment

The research/ article's publication is supported by LPPM University of Udayana through Udayana Invention Research and partially funded by the United States Agency for International Development (USAID) through the Sustainable Higher Education Research Alliance (SHERA) Program for Universitas Indonesia's Scientific Modelling, Application, Research and Training for City-centered Innovation and Technology (SMARTCITY) Project, Grant#AID-497-A-1600004, Sub

Grant#IIE-00000078-UI-1

References

[1] Saeed, H.A., Izumi, S., Sakai, S., Haruyama, S., Nagawa, M., Noda, H., 2008, ***Development of New Metallic Gasket and its Optimum Design for Leakage Performance***, Journal of Solid Mechanics and Material Engineering, Vol. 2, No. 1, pp. 105-114.

[2] Haruyama, S., Choiron, M.A., Kaminishi K., 2009, ***A Study of Design Standard and Performance Evaluation on New Metallic Gasket,*** Proceeding of the 2nd International Symposium on Digital Manufacturing, Wuhan China, pp. 107-113.

[3] Choiron, M.A., Haruyama, S., Kaminishi, K., 2011, ***Simulation and Experimentation on the Contact Width of New Metal Gasket for Asbestos Substitution***, International Journal of Aerospace and Mechanical Engineering, Vol. 5, No. 4pp. 283-287.

[4] Nurhadiyanto, D., Choiron, M.A., Haruyama, S., Kaminishi, K., 2012, ***Optimization of New 25A-size Metal Gasket Design Based on Contact Width Considering Forming and Contact Stress Effect***, International Journal of Mechanical and Aerospace Engineering, Vol. 6, pp. 343-347.

[5] Haruyama, S., Nurhadiyanto, D., Choiron, M.A., Kaminishi, K., 2013, ***Influence of surface roughness on leakage of new Metal Gasket,*** International Journal of Pressure Vessels and Piping, 111-112, pp. 146-154.

[6] Haruyama, S., Nurhadiyanto, D., Kaminishi, K., 2014, ***Contact Width Analysis of Corrugated Metal Gasket based on Surface Roughness***, Advanced Materials Research, Vol. 856, pp. 92-97.

[7] Nurhadiyanto, D., Haruyama, S., Kaminishi, K., Karohika I M. G., Mujiono, 2015, ***Contact Stress and Contact Width Analysis of Corrugated Metal Gasket***, Applied Mechanics and Materials, Vols. 799-800, pp. 765-769.

[8] Haruyama, S., Karohika, I M. G., Sato, A., Nurhadiyanto, D., Kaminishi, K., 2016, ***Development of 25A-Size Three-Layer Metal Gasket by Using FEM Simulation***, International Journal of Mechanical, Aerospace, Industrial, Mechatronic, and Manufacturing Engineering Vol:10, No:3, pp.555-561.

[9] MSC Marc 2010, User Manual.

IOP Conf. Series: Materials Science and Engineering **539** (2019) 012020 doi:10.1088/1757-899X/539/1/012020

Sea water absorption on corrosion and hardness properties of lining brake with hybrid particulates composites

I D G A Subagia[1] and N P G Suardana[1]
[1]Department of Mechanical Engineering Udayana University, Kampus Bukit Jimbaran Badung-Bali, Indonesia

Email: arsubmt@gmail.com

Abstract. This work is focused on study water absorption behavior on hybrid particle reinforcement for phenolic resin composites. This experiment works subjected to investigates corrosion and hardness properties effect of seawater absorption. Hybrid composites were constructed by three types of particles such as basalt, seashell, and alumina were mixed for reinforcement, and the phenolic resin was employed as a matrix. Specimens were manufactured by using the hot-press process on temperatures and pressure of 150°C and 2000N. Water absorption test was carried out by immersing specimens in seawater at room temperature for different time durations (3, 14, 24, 30, and 60 days) according to ASTM D 570 standard. The morphology of hybrid particles composites was investigated by SEM. The result showed that the increasing weight fraction of basalt powder has increased the hardness of specimens. Water absorption was shown monotonous trend on all hybrid composites configuration. In addition, hybrid composite immersion in seawater made hybrid particle composites corrosive.

1. Introduction

Particulate composites are the type of composite materials that consisted of a particulate phase as filler in single polymers. Particulate reinforced polymer composites have widely implemented as engineering product in especially on brake shoes of the automobile. Generally, particulate reinforcement has extensively used to expand the properties of matrix material such as to modify the thermal and electrical conductivities. Also, it can improve work temperature, wear, and abrasion resistant, improve machinability, surface hardness, and reduce shrinkage and friction. However, particulate reinforce polymer composite was not effective in improving fracture toughness [1]. All properties of composite particulate reinforced caused by particle size, contour, shapes, and porosity [2].

Particulate reinforced polymer composites (PRPC) have widely involved in engineering product according to the characteristics and the required mechanical properties. Brake shoes on the brake system are one type of PRPC application, which become a major part of active safety for each vehicle [3, 4]. In general, a combination between the asbestos and metallic particulates have been applied as reinforcement and thermosetting polymers have used for matrix such as phenolic resin. The characteristics of asbestos are low in water absorption, excellent mechanical properties, stable at a temperature elevated, and high friction properties. But, Asbestos material has a very negative impact on human health. The asbestosis, mesothelioma, and other cancers are kinds of diseases that infected by asbestos [5-7].

Since a few last decades, environmental issues have become the main reason to develop brake shoes using natural material. The selection of natural material is very important for reflected the fundamental relationships among material properties and be used to find out a range of materials suitable for an engineering application [8-10]. In the present, natural material such as jute, goony, banana fibers, pineapple, coconut fiber [11], and other materials due to their positive properties is possible used as fillers of brake shoes. Several experimental works were performed to investigate the possible uses of natural material as brake shoes. In recent years, the eco-friendly brake friction material has been carried out sporadically [12]. Brake pads using the banana peel for produces eco-friendly brake pads was investigated by Idris, et.al [13]. Ikpambese et.al [14] study for automotive brakes pad free-asbestos by using palm kernel fibers. Ameh et.al [15] study for the effect of date palm seed particle on the mechanical properties of reinforced polyester.

Water absorption and hardness properties is a major problem for natural material in the brake shoe materials. There are not many studies carried out for corrosion behavior currently for seawater absorption impact on the brake shoes using natural reinforcement. In this work, we would like to investigate particulate composites phenolic resin using hybrid reinforcement of basalt, seashell and alumina particles with immersion on sea water. The purpose is to study the hardness, corrosion, and absorption rate of the hybrid particulate polymer composites. In experimental work, sintering molding at 200°C temperatures for produces specimen was carried out. The test was in accordance with ASTM D 570 standard, and hybrid particulate composites morphology investigated using SEM.

2. Material And Method

2.1. Material

In experiment work, hybrid composites had been applied three particulates for reinforcement and thermosetting polymer as a matrix. Basalt, seashell and alumina particles were used due to their positive mechanical properties and also, it's a part of natural material. Whilst phenolic resin was chosen as polymers. Basalt is filler that produced by the melting process of the volcanic rock on temperatures elevated i.e. 1200°C ~ 1700°C [16]. In addition, seashells powder was conducted by crusher process to found 0.25 mesh grains size. Table 1 shows the properties of the reinforcement and polymer.

Table 1. Particulate compound.[17, 18]

Particles Item	Compounds content (%)									
	SiO_2	Al_2O_3	Fe_2O_3	CaO	MgO	Na_2O	TiO_2	K_2O	P_2O_5	MnO
Basalt	55.50	19.50	8.25	11.20	8.50	3.32	1.00	1.50	1.20	0.11
Seashell	7.88	1.25	0.03	66.70	22.28	-	-	-	-	-
Metallic (Alumina)	0.03	98.4	0.03	0.06	-	0.60	0.005	-	-	-

Table 2. Hybrid Particulate phenolic resin composites

Hybrid Composites	Code	Particulate Filler % (wt)			Polymers % (wt)
		Basalt	Seashell	Alumina	Phenolic resin
$B_{30}C_{40}A_{10}P_{20}$	A	30	40	10	20
$B_{40}C_{30}A_{10}P_{20}$	B	40	30	10	20
$B_{50}C_{20}A_{10}P_{20}$	C	50	20	10	20

2.2. Methods

2.2.1. Hybrid Composite Fabrications

In the present work, the composite was fabricated by compression processes in four steps such as preparation polymers and fillers, mold preparation, molding process and testing of the specimen. Filler preparation by mixing three particulates such basalt, calm seashell and alumina. The weight percentage of fillers and polymers was used in the present investigation is according to ratios 80wt% and 20wt% for all variations. The particulate mixtures illustrated in Table 2. In the present study, three hybrid composite variety were produced, and one specimen made from real brake shoe material was used as a control. Mixing particulate was stirred manually until a homogeneous mixture was obtained.

Then, the mixture is kept in the desiccator for removing pollutants content and made homogeneous compound bound between the particles.

Figure 1 shows the molding process, mold temperature curve, and specimen dimension of composites. In this work, the composite was manufactured with hydraulic molding press (hot press). The compression process was set to 60 minutes with a pressure of 2000kgf/cm2 on constant temperature of 150°C. After the mold time has achieved, the hybrid composites cured before removed from the mold.

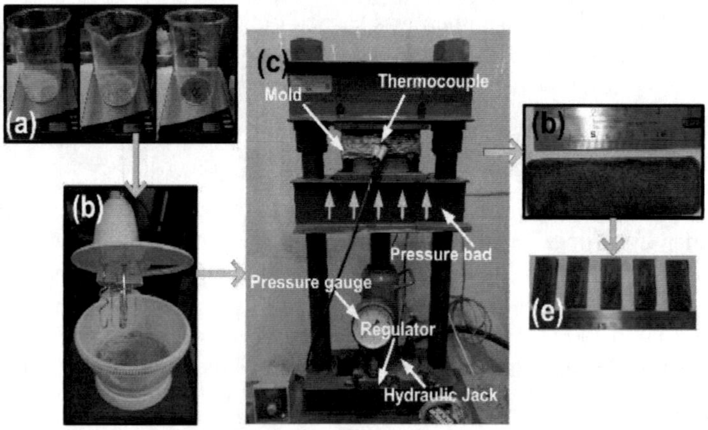

Figure 1. a). Filler particle weighed; b). Particulate mixtures; c). Hot-press machine and molding process; d). Hybrid particulate filler polypropylene polymer composite; e). Specimens test.

2.2.2. Hardness test

Vickers hardness test Zwick was conducted using a square-base diamond pyramid indentation with the angle of the pyramid of 136°. The Vickers microhardness (VHN) was defined as the load divided by the surface area of the indentation. In this experiment, the load indention uses at 2 kg, and it applied for 15 seconds. Hardness test was done according to ASTM E-92 standard.

2.2.3. Water absorption test

In order to obtain the effect of water absorption to each hybrid particulate filler, polymer composites property, the immersion test was carried out. The water absorption test was conducted is immersing specimens in seawater in a beaker at room temperatures as long as 60 days duration. The water absorption on specimens was weighed regularly in each time duration. The moistures absorption was calculated by the weight difference and percentages weight gain. In this test sea water absorption of samples were determined at different day's interval. The measuring of the specimen was started at the 3 days immersion. It process would be repeated at 14, 24, 30 and 60 days according to the ASTM D 570 standard. The water absorption test procedure of the hybrid composites has shown in Figure 2.

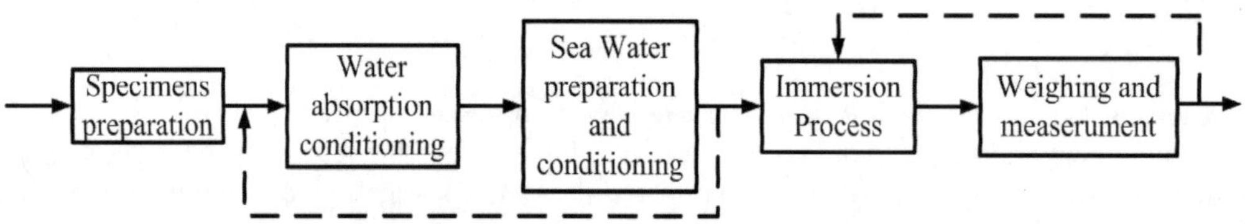

Figure 2. Schematic of the water absorption test

2.2.4. Microstructures test

The microstructure of the hybrid composite was characterized by scanning electron microscopy using JEOL-JSM 6510LA.

3. Result And Discussion

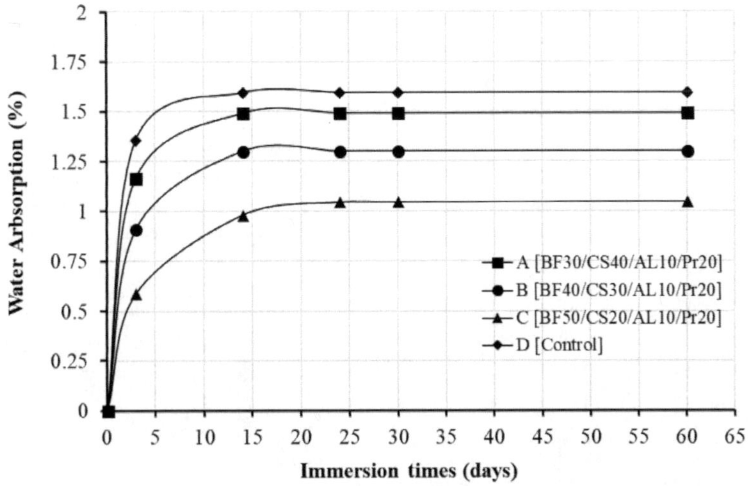

Figure 3. Water absorption of Hybrid particulate reinforced Phenolic resin composites

3.1. Water Absorption properties of hybrid particulate polymer composites

The seawater uptake of hybrid composite phenolic resin with basalt/seashells/aluminum particles reinforced presented in the figure. The water absorption behavior of hybrid composite mainly depends on the voids present in the composites, interfacial adhesion between the reinforcement and matrix, and type of fillers. Figure 3 shows the relationship between immersion time and water absorption of each hybrid particulate reinforced polymer composites which compared with control. It can be seen from the graph that the trend of sea water absorption property shows monotonous. In the first water absorption time, the specimen weight was significantly rising until immersion at fourteen days. Furthermore, immersion duration at 24, 30 and 60 days water absorption were not occurred. It means that there is no add weight on hybrid composites.

On the longer saturated time, the sea water absorption of the hybrid particulate polymer composite is increased for all variation of basalt and seashells particles, respectively. It can be seen that in the 3 and 14 days, the water absorption has increased sharply compared with the next duration days. In other words, the water absorption value increases drastically on the first immersion, due to the specimen in dry condition and the next time the absorption increase slowly, and then become constant. In the research, the high content of seashells particle (specimen A) increases seawater uptake of the specimen and exhibit a high value of weight gain. The hybrid composite ($BF_{30}/CS_{40}/AL_{10}/PR_{20}$) has a value maximum of water absorption is 1.35% (wt) highest than ($BF_{40}/CS_{30}/AL_{10}/PR_{20}$) and ($BF_{50}/CS_{20}/AL_{10}/PR_{20}$). Normally, density and salinity of seawater have influenced the immersion value of hybrid composites, so it can produce less amount of water absorption.

One of the most dominant compositions that increases water absorption on the hybrid composite has caused by CaO and MgO content [19]. Basalt and seashell particles have a relative high content of CaO and MgO can be seen in Table 2, respectively. This characteristic had been phenomena on the natural fillers [20]. However, thermoplastic (phenolic resin) hydrophobic behavior has small value water. According to the result can be concluded that all particles of the filler are bound well by the matrix. Similar observations were conducted by other researchers. [21-25]

3.2. Hardness property of hybrid particulate polymer composites

Figure 4 shows the hardness properties chart of hybrid particulate polymers composite effect of immersion on sea water. Capital letters such A, B, C, and D on the horizontal axis of the graph shows type of specimens in immersion test. Then, the number such as 3, 14, 24, 30, and 60 indicates duration days of immersion test. According to the test result of hardness, it can be seen that the water absorption decreases hardness Vickers number of each hybrid composites. The high value of hybrid composites hardness was exhibited on sample C, which contains basalt particles weight percent highest, and it occurred on each of immersion duration. Vickers microhardness has the highest value on average is 25.5HVN on three days immersion for each hybrid composites configuration. Additionally, the HVN of hybrid composites after immersion twenty days shows decreases for all hybrid composites about 12.5HVN, and it has constant values in the next immersion day's duration, respectively. In contrast, the hybrid composites effect of water absorption shows harder compared to control specimen due to increases weight percent of basalt particles. In present, it can be concluded that the weight percent of basalt particle increases had improved the hardness of hybrid composites.

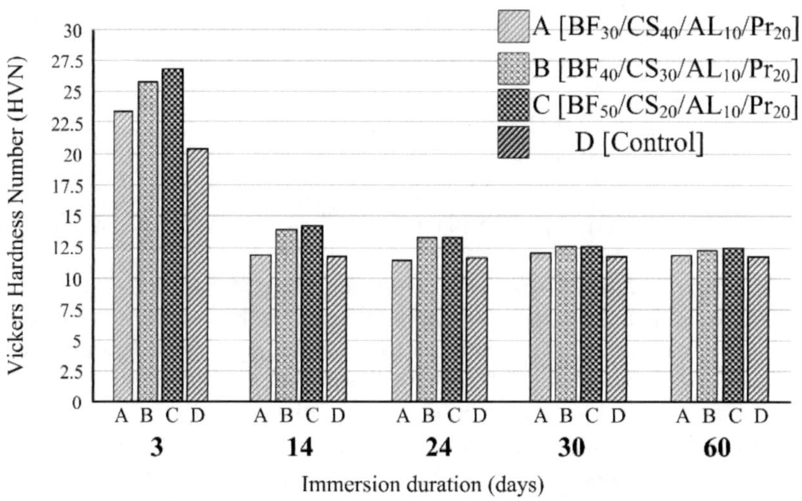

Figure 4. Bar chart of Vickers microhardness hybrid composite after the immersion in seawater

3.3. Corrosion properties of hybrid particulate polymer composites

Figure 5 shows the corrosive damages of a hybrid particulate polymer composite that effect of seawater. It was observed that corrosion occurred after seawater immersion on each hybrid composites variety. Figure 5a, corrosion of hybrid composite with [BF_{40}/CL_{30}/Al_{10}/Pr_{20}] as pointed by yellow color arrows. Then on Figure 5b and 5c were exhibited corrosion on hybrid composite with [BF_{50}/CL_{20}/Al_{10}/Pr_{20}] and [BF_{50}/CL_{20}/Al_{10}/Pr_{20}], respectively. As can be seen in figure delamination, crack, and particles removing of bound are corrosion behavior effect of hybrid composites immersion in seawater.

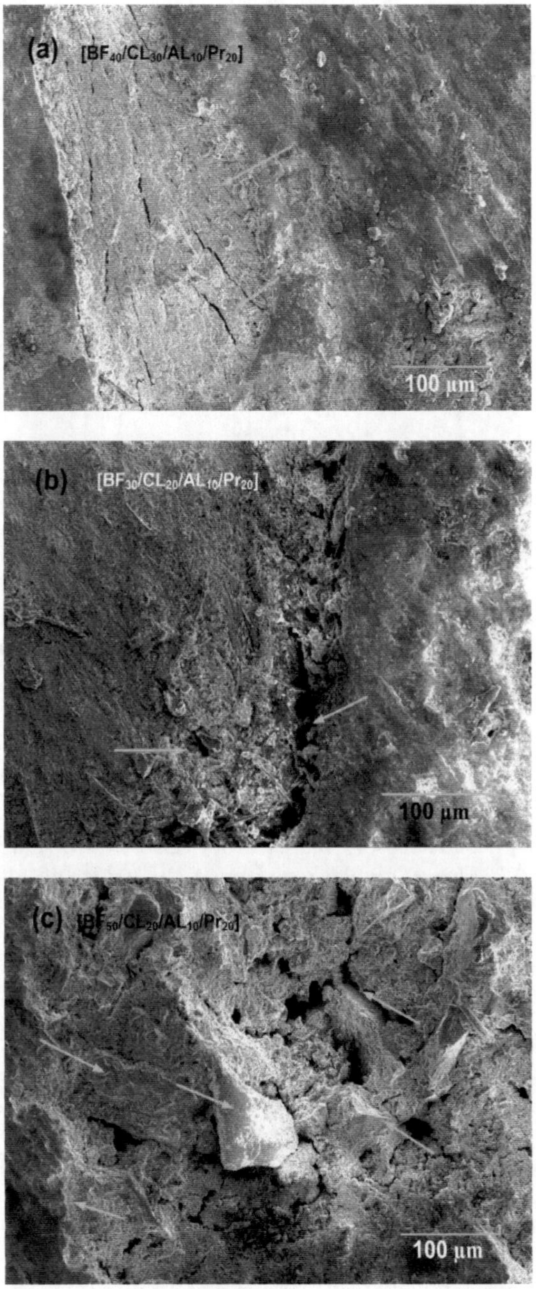

Figure 5. corrosion behavior hybrid composites in seawater at 30 days immersion

4. Conclusion

The corrosion and hardness properties of hybrid composite effect of seawater absorption have been investigated. The hybrid composite with a varied weight ratio of basalt particulates and seashells particulate have used in this test. From the results obtained, the following conclusions are drawn:

1. The amount of water absorbed by hybrid composites are constant after 14 days immersion in seawater at duration days.

2. Increasing basalt particle content has been increased Vickers microhardness for all hybrid composite configuration. However, all composition of the hybrid composite was reduced their hardness after immersion in sea water.
3. The sea water immersion for all configuration hybrid composite was effective made samples corrosion, due to the salination. The corrosion was influenced on immersion at 30 days. The corrosion was shown by many delaminations, crack and particles removed from the bond that occurred on the hybrid composite surface.
4. Hybrid particulate polymer composite property has potential used alternative material of brake shoes of the vehicles.

Acknowledgment
The author would like to express thanks to Udayana University (UNUD) and The Ministry, Research, Technology, and High Education Department, providing a fund to carry out of the research.

References
[1] David Arencón and Velasco JI; Fracture toughness of polypropylene-based particulate composites. Materials, 2009; 2(4): 2046-2094.
[2] Fu S-Y, Feng X-Q, Lauke B, and Mai Y-W; Effects of particle size, particle/matrix interface adhesion and particle loading on mechanical properties of particulate–polymer composites. Composites Part B: Engineering, 2008; 39(6): 933-961.
[3] Aku S, Yawas D, Madakson P, and Amaren S; Characterization of periwinkle shell as asbestos-free brake pad materials. The Pacific Journal of Science and Technology, 2012; 13(2): 57-63.
[4] K. W. Liew and Umar Nirmal; Frictional performance evaluation of newly designed brake pad materials. Materials & Design, 2013; 4825-33.
[5] von Uexküll O, Skerfving S, Doyle R, and Braungart M; Antimony in brake pads-a carcinogenic component? Journal of Cleaner Production, 2005; 13(1): 19-31.
[6] Ramazzini C; The global health dimensions of asbestos and asbestos-related diseases. Journal Occupational Health, 2016; 58220-223.
[7] Ramazzini C; Comments on the 2014 helsinki consensus report on asbestos. Journal of Occupational Health, 2016; 58224-227.
[8] M.A. Maleque, S.Dyuti, and Rahman MM, Material selection method in design of automotive brake disc; in: The World Congress on Engineering 2010 Vol III WCE 2010, London, U.K; 2010.
[9] Lo J; Designing a composite material for use in brake applications. Materials Science Forum, 2005; (475-479)1109-1112.
[10] Telang A K, Rehman A, Dixit G, and S D; Alternate materials in automobile brake disc applications with emphasis on al composites- a technical review. Journal of Engineering Research and Studies (JERS), 2010; I(I): 35-46.
[11] J. Olumuyiwa Agunsoye, Talabi S. Isaac, and Samuel SO; Study of mechanical behaviour of coconut shell reinforced polymer matrix composite. Journal of Minerals and Materials Characterization and Engineering, 2012; 11774-779.
[12] Yun R, Filip P, and Lu Y; Performance and evaluation of eco-friendly brake friction materials. Tribology International, 2010; 432010-2019.
[13] Idris UD, Aigbodion VS, Abubakar IJ, and Nwoye CI; Eco-friendly asbestos free brake-pad: Using banana peels. Journal of King Saud University - Engineering Sciences, 2015; 27185-192.
[14] Ikpambese KK, Gundu DT, and Tuleun LT; Evaluation of palm kernel fibers (pkfs) for production of asbestos-free automotive brake pads. Journal of King Saud University - Engineering Sciences, 2014; xxxxxx-xxx.

[15] Alewo Opuada AMEH MTI, Ibrahim SANUSI Effect of particle size and concentration on the mechanical properties of polyester/date palm seed particulate composites. Leonardo Electronic Journal of Practices and Technologies 2015; (26): 65-78.

[16] Subagia IDGA and Kim Y; A study on flexural properties of carbon-basalt/epoxy hybrid composites. Journal of Mechanical Science and Technology, 2013; Volume 27(4): 987.

[17] Lapena MH and Marinucci G; Mechanical characterization of basalt and glass fiber epoxy composite tube. Materials Research, 2017; 21(1):

[18] Abiodun Ademola Odusanya, Babatunde Bolasodun, and Madueke CI; Property evaluation of sea shell filler reinforced unsaturated polyester composite International Journal of Scientific & Engineering Research, 2014; 5(11): 1343 - 1349.

[19] Genevive C. Onuegbu and Igwe IO; The effects of filler contents and particle sizes on the mechanical and end-use properties of snail shell powder filled polypropylene. Materials Sciences and Applications, 2011; 02(07): 810-816.

[20] Pandian A, Vairavan M, Jebbas Thangaiah WJ, and Uthayakumar M; Effect of moisture absorption behavior on mechanical properties of basalt fiber reinforced polymer matrix composites. Journal of Composites, 2014; 20141-8.

[21] Alaneme KK, Oke SR, and Omotoyinbo JA; Water absorption characteristics of polyester matrix composites reinforced with oil palm ash and oil palm fiber. Usak University Journal of Material Sciences, 2013; 2(2): 109-109.

[22] Bansal G and Vk S; Water absorption and thickness swelling characterization of chicken feather fiber and extracted fish residue powder filled epoxy based hybrid biocomposite. International Journal of Waste Resources, 2016; 6(3):

[23] Correlo VM, Pinho ED, Pashkuleva I, Bhattacharya M, Neves NM, and Reis RL; Water absorption and degradation characteristics of chitosan-based polyesters and hydroxyapatite composites. Macromol Biosci, 2007; 7(3): 354-63.

[24] Mohammed AA and Issa TT; The water absorption effect on the hardness of composites polyester. 2016; 1096020016.

[25] Dhakal H, Zhang Z, and Richardson M; Effect of water absorption on the mechanical properties of hemp fiber reinforced unsaturated polyester composites. Composites Science and Technology, 2007; 67(7-8): 1674-1683.

IOP Conf. Series: Materials Science and Engineering **539** (2019) 012021 doi:10.1088/1757-899X/539/1/012021

Utilization plastic waste using pyrolysis fixed bed

I B A Swamardika[1], I N S Winaya[2] and R S Hartati[3]
[1,3]Study Program of Electrical Engineering, Faculty of Engineering, University of Udayana
[2] Study Program of Mechanical Engineering, Faculty of Engineering, University of Udayana

Abstract. This investigative research aims to convert urban non-organic waste fuels such as plastic waste into energy using fixed bed pyrolysis technology. Pyrolysis fixed bed (PFB) is one of the most efficient technologies for converting various types of undegradable solid waste into thermal energy and liquid fuels with high efficiency. The study was conducted on a lab-scale PFB reactor with dimensions of 15 cm and 55 cm in diameter and bed height respectively. The reactor is equipped with a cooling and decomposition system to accommodate liquid and gas products. Reactor temperature variations using an external heater were tested and quartz sands were selected as a bed material to improve the temperature function of the PFB reactor so as the maximum quality and quantity of liquid product were obtained. The fuel properties of the derived oil were tested including their physical properties, calorific value, elemental measurement (CHNOS) analysis. The results of the study showed that the increase in temperature resulted in an increased pyrolysis performance which was indicated by the increased volume and liquid caloric value.

1. Introduction

Urban wastes such as plastic debris have become a common problem in the big cities over the world due to the massive increase of plastic usage in daily life as it is versatile, lightweight, flexible, moisture resistant, strong and considered cheap (Guern C.L.,2018). Plastic is a polymeric compound with a very large molecular shape, in the chemical sense, includes synthetic or semi-synthetic polymerization products which are formed from organic condensation or the addition of polymers and can also consist of other substances to improve their performance or economic value.

Some plastic wastes cannot be recycled so that the solution comes to a conventional method by transporting them to the landfill disposal which is not suitable to be called a waste problem-solving. In its development, waste management using landfill and open-dump methods is not appropriate because not only requiring a large site area but also raising various adverse impacts on the environment and community (Salem-Al S.M., *et.al.* 2017). In addition, the wastes such as plastic, tire, and other materials cannot be easily decomposed over time (non-biodegradable waste), so they have a strong impact on increasing the volume of landfill area.

One of advanced technology is by using a thermo-chemical process such as pyrolysis technic converting solid plastic waste into liquid fuel in the absence of oxygen or with a limited supply that does not permit reduction process to an appreciable extent (Basu, 2010). Pyrolysis is one of the alternatives for processing solid plastic waste which is considered quite prospective to be developed because it has several advantages in its delivery with a high conversion ratio and energy content. The low temperature of the pyrolysis in the range of 300-650°C makes it flexible to choice, while gasification and combustion occur at between 800-1000°C (Basu, 2013).

Research on the characteristics of the gas fraction resulting from the pyrolysis process has been performed by many investigators (Islam NM. and Beg M.R.A., 2004, Moinodim KAM., *et.al.*, 2012, Papuga *et.al.*, 2016, Miandad et.al., 2016). Therefore, in this study, the decomposition characteristics of plastic at pyrolysis temperature were conducted to obtain liquid yield as an alternative fuel. The study was carried out in a bench scale fixed bed pyrolyzer using quart sand as catalyst bed material in the variation of reactor temperature of 300 to 600°C. The research is carried out using plastic polypropylene

(PP) pellet with the aim to get an optimal content and quantity of liquid products. The liquid yield is measured and analyzed by the calorimetric test, and the reactor PFB efficiency is determined based on fuel conversion attained from the proposed investigations.

2. Materials and Methods

2.1. Materials

The sample materials used in the study of fixed bed pyrolysis are in the form of granular polypropylene plastic (PP) pellet as shown in Figure 1.

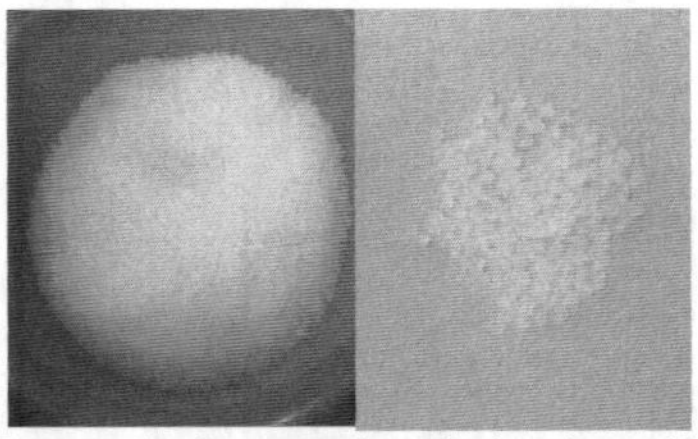

Figure 1. A sample of plastic pellet

In the initial process, the calorific value of plastic PP was determined using the Parr 6200 Calorimeter for fuel based on ASTM D5865, standard test method for the coal and coke. The following are the results of the measurement as seen in Table 1 with two times repetitions.

Table 1. The calorific value of plastic PP (as received)

No.	Sample	Mass(gr)	LHV (cal/gr)	LHVaverage (cal/gr)
1.	Plastic 1	1.01606	10205.242	10184.394
2.	Plastic 2	1.01938	10163.547	

2.2. Experimental setup and methods

Figure 2 shows the detail part of the PFB reactor while Figure 4 shows the schematic diagram of the pilot PFB conversion system of plastic waste into liquid oil and char. The FB reactor is 15 cm, 55 cm, 400 cm^3 of diameter, high and bed volume respectively.

IOP Conf. Series: Materials Science and Engineering **539** (2019) 012021 doi:10.1088/1757-899X/539/1/012021

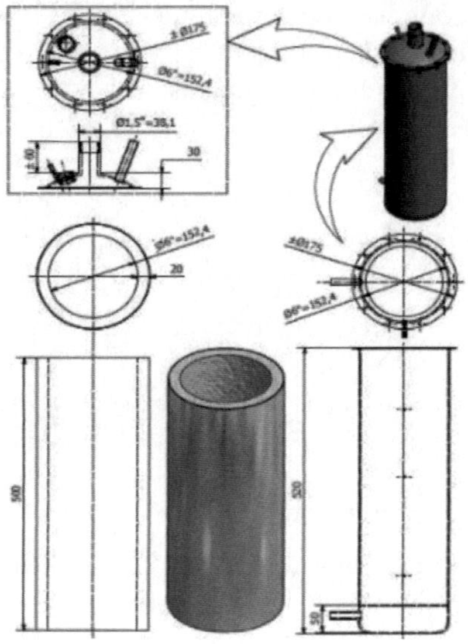

Figure 2. Detail part and dimension of a PFB apparatus

For the heating purpose, an electrical source heater was used. The heater supplied the heat to the pyrolysis reactor and when the desired temperature attained, PP particles were fed into the reactor from the reactor feeder. The reactor temperature was varied using thermostat control from the box panel. The condense vapor product and the gases passed through the condenser and then collected at the oil condensate tank, while the uncondensed products in the form of gases coming from same liquid oil pipe were collected in the gas container or directly exhausted from the reactor. Bed material as a catalyst of quartz sand was employed in this study as it can be stood at high-temperature o 1400°C.

The FB reactor is made of stainless steel and covered with a loop of an electric heater, which allows a maximum temperature of 800°C. The reactor was equipped with a K-type thermocouple inside to monitor the temperature when heated by an external furnace. The experiments in this study were designed at the temperature variation of 300 (I), 400 (II), 500°C (III) and 600°C (IV), while nitrogen gas was inserted to sterilize the reactor from oxygen.

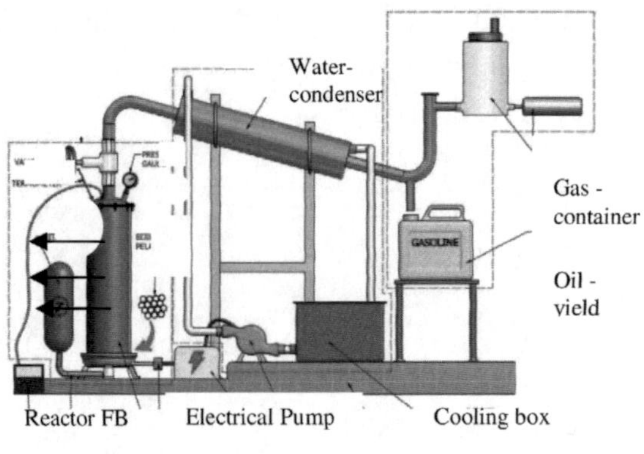

Figure 3. Schematic of fixed bed pyrolysis

In all of the experiments, 1000 gr of PP plastic feedstock was used in the form of pellets with an average diameter of 0.4 cm. The heating chamber of the pyrolysis reactor was heated at a rate of 5°C per min to achieve the set temperature.

3. Result and Discussion

3.1. Effect of temperature on liquid oil yield

Experiments were carried out at three different temperatures (300, 400, 500 and 600°C) to investigate the effect of temperature on the yield and quality of produced liquid oil. After the determination of optimum temperature, experiments were carried out at similar reaction times of 120 min to investigate the effect on the feedstock decomposition solely based on temperature differences. The purpose of using different temperature was to find the optimum temperature at the same reaction time for pyrolysis of PP pellet so that the effect of the reaction time on the composition and quality of liquid oil and the production of char were predicted.

3.2. Temperature PFB profile

Figure 4 shows the characteristic temperature profile along the reactor representing the operation work on the temperature set at 400°C. This study uses 3 thermocouples (T1, T2, and T3) to determine the temperature profile in the reactor. Thermocouples installed on the reactor wall are connected to the data logger which is then recorded in the laptop. The total time needed for the variation II pyrolysis process is 168 mins, based on the details of 48 mins for heating the reactor and 120 mins for the pyrolysis process. Compared to variation I, the time needed for the variation II processes is longer than variation I because the time used to heat the reactor is longer. The position of the T1 thermocouple point is just above the bed material (10 cm), T2 (30 cm) and T3 (45 cm) which are measured from the bottom of the bed. It can be seen from Figure 4 where T1 is the experimental benchmark which when it reaches steady 400°C conditions, PP plastic pellet fuel starts to be employed into the reactor. T2 and T3 temperatures are lower than T1 because heat from the external heater is made more focused near quartz sand bed material.

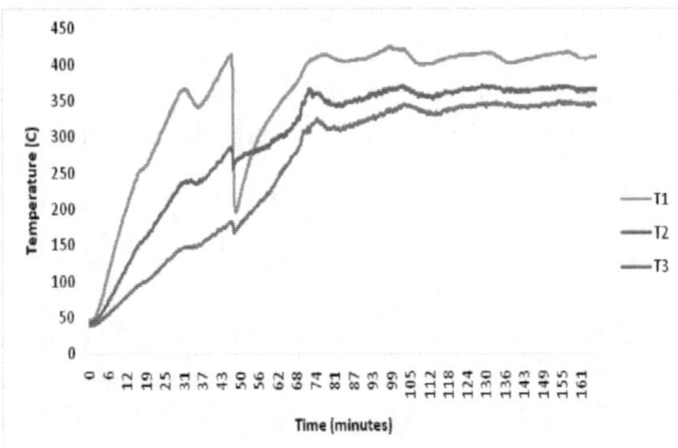

Figure 4. Temperature profile along 3 different positions along vertical bed reactor

3.3. Liquid yield

The gas formed in the PFB process naturally flows upward through a cooling pipe that has been designed to be submerged in a larger pipe containing cold water. The cold water condenser is made in circulating arrangement with the help of a pump located in the bottom of the test rig. After the gas is cooled, the denser of the gas melts and drips on the first tap (oil condensate), while for the lighter gas moves further up through the final tap (gas container/exit). The results of the liquid yield are then measured as can be seen in Figure 4.

From 1000 gr of polypropylene (PP) plastic that was inserted into the fixed bed reactor, the liquid yield was produced as much as 0.9 liters at variation I while 0.99, 1.08 and 1.06 liters for variation II-IV respectively. Figure 5 below shows an increase in liquid yield accompanied by increasing temperature at the reactor. This implies that the higher the reactor working temperature, the higher the number of PP plastic converted into a liquid. However when the operation temperature of 600°C, the liquid yield was found slightly decreased.

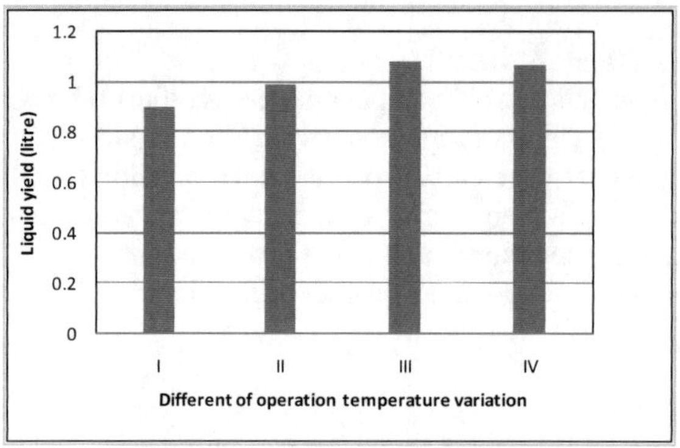

Figure 5. The liquid yield of pyrolysis fixed bed at different temperature operation

According to Basu (2010), in the process of slow pyrolysis tend to form charcoal in large quantities, whereas in the fast pyrolysis process there is a reduction in oxygen resulting in the possibility of forming more liquid hydrocarbons. The series of physical and chemical reactions take place slowly at temperatures below 350°C and runs faster at higher temperatures, this process is heat-absorbing (endothermic). Figure 5 also shows a slight drop of liquid yield when the temperature as high as 600°C. This may indicate more light gas was produced that not condense into the liquid with existing cooling water. Other work from different researchers (Islam et.al., 2010) also showed a decreased oil yield at higher temperature operation.

The purpose of the mass balance calculation is to find out the amount of mass in and out. For the mass out includes the liquid mass and charcoal. Mass balance calculations are separated based on variation I to IV. The following were the measurement of mass equilibrium of 680, 760, 800 and 790 gr for variations I to IV respectively. The remaining mass was considered a loss. Losses rather than the calculation above were gas that comes out of a fixed bed device and was not converted to liquid. For losses in variations, I-IV were as many as 200, 180, 160 and 164 gr respectively.

3.4. Oil yield calorific value

The oil yield produced from pyrolysis fixed bed of PP pellet was then tested to measure their calorific value for a different variation of temperature. The most important factor during pyrolysis is the temperature, with an increase in temperature the polymer is cracked and more light condensed liquid is produced with higher calorific content as shown in Table 2. Taking account in comparison of the initial calorific value of PP pellet from Table 1, the average increased of 2.58% was achieved.

Table 2. Calorific value of oil yield at different variation temperature

Variation	**Oil yield**	**Mass (gr)**	**LHV (cal/gr)**
I	Yield @300°C	0.81027	10205.165
II	Yield @400°C	0.83483	10341.970
III	Yield @500°C	0.88930	10411.661
IV	Yield @600°C	0.88930	10856.661

3.5. Pyrolysis efficiency

The efficiency calculation aims to determine the performance of fixed bed pyrolysis using PP plastic as fuel. The efficiency in this calculation uses the value of Low Heating Value (LHV) since evaporation generated from the heat of condensation of water vapor from the combustion results is not counted.

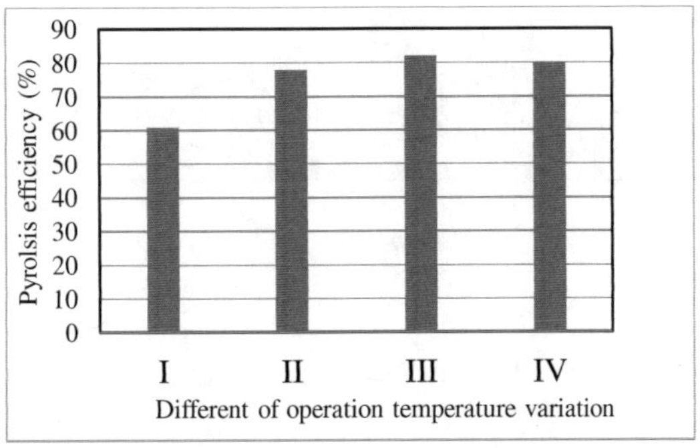

Figure 6. Pyrolysis efficiency of PP fixed bed at different temperature operation

The efficiency at an operating temperature of 500°C results in higher performance than that at operating temperatures of 300°C and 400°C. This shows that the greater the working temperature, the greater the thermal energy to break the PP plastic carbon chain in the reactor so that more carbon molecules (C) and hydrogen (H) are converted into liquid fuel capable of high calorific value. However, at a temperature of 600°C, there is a slight decrease in liquid production, as explained earlier.

3.6. Fuel conversion rate (FCR)

The rate of conversion of plastic PP fuel into liquid and gas in the pyrolysis process can be calculated using the following equation based on variation I to IV:

$$FCR = \frac{\textit{mass of initial PP-mass of char}}{\textit{timeofoperational}} \left({}^{kg}\!/_{s} \right)$$

From the mass measurement and LHV of each PP condition, the maximum of PP plastic conversion was attained at the highest operating temperature of 600°C. Although not all of PP was converted to liquid for the reason of cooling equipment, high temperature allows having high solid cracking of PP into gas/liquid. The complete calculations of FCR for variation I to IV are done as below:

- $FCRI = \frac{1-0,12}{7200} \left({}^{kg}\!/_{s} \right) = 0,000122 \left({}^{kg}\!/_{s} \right)$
- $FCRII = \frac{1-0,06}{7200} \left({}^{kg}\!/_{s} \right) = 0,000130 \left({}^{kg}\!/_{s} \right)$
- $FCRIII = \frac{1-0,04}{7200} \left({}^{kg}\!/_{s} \right) = 0,000133 \left({}^{kg}\!/_{s} \right)$
- $FCRIV = \frac{1-0,03}{7200} \left({}^{kg}\!/_{s} \right) = 0,000146 \left({}^{kg}\!/_{s} \right)$

It is easy to understand as the higher the temperature so the rate of the fuel conversion into liquid/gas is more rapid.

3.7. Yield oil performance

As seen in Figure 7, the liquid yield from the pyrolysis process at 300°C produces a rather dark and cloudy color. Liquid began to trickle in the 70th mins of the total testing time of 120 mins. At temperatures of 400°C, the liquid is rather bright yellowing, not too cloudy. Liquid starts trickling in the

30th min from the total testing time of 120 mins. While at a temperature of 500°C and 600°C liquid it is bright yellow and clear. Liquid began to trickle in the 10th min of the total testing time of 120 mins. The Brightness of liquid results at high temperatures due to the small amount of charcoal (black powder) in the liquid. This also indicates that at the higher temperature, the higher amount of yield oil is achieved.

@300°C dark residue cloudy	@400°C chocolate small residue rather cloudy	@500°C, 600°C yellow no residue bright

Figure 7. Performance of yield liquid at different temperature operation

3.8. Charcoal yield

In this fixed bed pyrolysis research using PP plastic as fuel besides producing liquid also produces charcoal in the reactor as seen in Figure 8. Charcoal is removed from the reactor and weighed in units of mass (gr). The yield of charcoal in the variation I is 120 gram. Charcoal produced in the variation I was hard and dense. This was because plastic PP that was not converted into gas makes it blend with quartz sand.

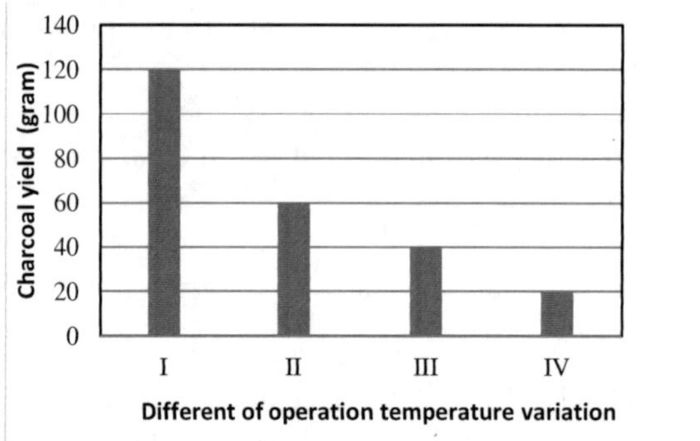

Figure 8. The charcoal yield of PP fixed bed at different temperature operation

In addition, tar was hardened and eventually turned to a solid state when exposed to the room temperature (Sharuddin SDA. *et all*, 2017). In the variation II charcoal was produced as much 60 gram which was different from those of variation I. The charcoal was seen as sand-like. This is indicated by the color of dark quartz sand indicating the presence of charcoal. For variation III and IV, the amount of charcoal produced was 40 and only 20 gram respectively. The observations show that the color of sand/charcoal is brighter in color. This is because the charcoal mass produced is less than the other variations.

4. Conclusion

The study has shown that the fixed bed pyrolysis technology could be used in utilizing the PP plastic pellet into liquid fuel. There is an increased 2.58% of calorific value from PP solid to oil yield using the PFB reactor. Increase in temperature for variation I to IV results in an increase in the FCR and calorific value. The efficiency at an operating temperature of 500°C results in higher performance than that at operating temperatures of 300°C 400°C and 600°C based on the oil yield. In general, the study showed that the increase in temperature resulted in an increased pyrolysis performance which was indicated by the increased volume and liquid caloric value.

Acknowledgment
The authors like to thank the University of Udayana for full funding in this publication through the annual research budget of the Grant Unggulan Udayana (HUU) contract no: 1709/UN14.25.II/LT/2018.

References

[1] Basu, P. 2010. Biomass Gasification and Pyrolisis Practical Design and Theory, USA: Elsevier.p 97-100; 104,110,120,176,195.

[2] Basu, P. 2013. Biomass Gasification, Pyrolysis, and Torrefaction, Practical Design and Theory, 32 Jamestown Road, London NW1 7BY, UK, 525 B Street, Suite 1800, San Diego, CA 92101-4495, USA, Publishedby Elsevier Inc.

[3] Guern. C.L., 2018. The great plastic tide, last updated March 018, plastic-pollution.org, retrieved on September 2018.

[4] Miandad, R., Nizami, A. S., Rehan, M., Barakat, M. A., Khan, M. I., Mustafa, A., Ismail, I. M. I. &Murphy, J. D. 2016. Influence of temperature and reaction time on the conversion of polystyrene waste to pyrolysis liquid oil. *Waste Management,* 58**,** 250-259.

[5] Moinuddin K.A.M., 2012, *Effects of Variation In* Heating *Rate, Sample Mass and Nitrogen Flow on Chemical Kinetics for Pyrolysis*, 18th Australasian Fluid Mechanics Conference.

[6] Papuga, S.V., Gvero, P.M., Vukic L.M., 2016. Temperature and time influence on the waste plastics pyrolysis in the fixed bed. *Thermal Science*, 20:2, 731-741

[7] Islam M.N. and Beg M.R.A., 2004. Fixed Bed Pyrolysis of waste plastic for alternative fuel production (2004). *Journal of Energy and Environment* 3, 69-80.

[8] Salem-Al, S.M., Antelava A., Constantinou, A., Manos, G., Dutta, A., 2017. A review on thermal and catalytic pyrolysis of plastic solid waste (PSW). *Journal of Environmental Management*, 197, 177-198.

Observations of mechanical and manual forging on bronze as a gamelan material

I G N Priambadi[1] and I K G Sugita[1]
[1]Department of Mechanical Engineering Faculty of Engineering Udayana University-Indonesia

Email: priambadi.ngurah@unud.ac.id

Abstract. Gamelan is a traditional musical instrument used by Hindu Balinese in religious and cultural activities. The process of making gamelan begins with the manufacture of copper alloy composition (Cu) and lead (Sn), melting, casting, forging and tone forming by grinding on the surface of the blades. The process of forging on the manufacture of bronze gamelan (CuSn) with an open casting method absolutely must be done because the casting results tend to cause the occurrence of porosity. Porosity that occurs in the bronze as a gamelan material will affect the sound produced. The research conducted is to compare the results of manual forging and mechanical means, where the observations are made on the percentage of deformation that occurs, the density, the hardness and the microstructure of the bronze material. The observations show that mechanical forging provides better mechanical properties, deformation rates, and product microstructure.

1. Introduction

Gamelan is a traditional musical instrument whose existence is used to accompany religious and cultural activities for Hindus in Bali. There are several stages in the making of this gamelan, starting with the manufacture of bronze alloys consisting of copper (Cu) and tin (Sn). The process of melting from the alloy at melting temperature in accordance with its composition, the melting temperature of the bronze is determined based on the composition of the alloy and can be seen in the CuSn phase diagram [1].

The forging process used is forging that still relies on human power. The main disadvantage of this process is that the forging of the gamelan material is uneven. The importance of the compaction process in the series of gamelan production processes is a sufficiently deep consideration to applying mechanical forging machines. Forging machines that are planned to be electrically powered with an ergonomic form of construction. This research is expected to produce the uniform quality of compaction of gamelan material so that the mechanical properties and microstructure of the material are better compared to manually compacted material.

Properties testing for the results of forging both manual and mechanical means include hardness test, density test, microstructure and percentage of deformation level. Hardness testing was carried out by the Vickers method where the test refers to the ASTM E 92-82 standard [2]. The load used in hardness testing is 10 kg, with a 100 x magnification lens. Material hardness is known by the following equation:

$$d = \frac{d_1 + d_2}{2} \qquad (1)$$

where :
d = diagonal on average (mm)
so that

$$VHN = \frac{2 \times P \times \sin(\theta / 2)}{d^2}$$

$$= 1{,}854 \frac{p}{d^2} \ (\mathrm{kg/mm^2}) \qquad (2)$$

Density testing is carried out based on Archimedes's theory. Measurement is done by weighing the weight of the specimen in the air and in pure water on a digital weighing device. The specimen density is determined by the following equation:

$$\rho_b = \frac{w_d}{w_d - w_b} \cdot \rho_w \ (g / cm^3) \qquad (3)$$

where :

ρ_b = Specimen density
ρ_w = liquid density
W_d = the dry weight of the specimen
W_b = Specimen weight in water

This research specifically has a purpose where mechanical forging produces material characteristics (blades) of gamelan namely mechanical properties and microstructure that are better than traditionally produced blades.

Contributions that are expected to be achieved directly or indirectly are perceived in terms of elements of science, society, and the company can be described as follows;

- Gamelan crafters industry; Crafters can apply appropriate technology to increase the quantity and quality of production and produce quality products that are capable of competitiveness.
- Institutions; the results of this study can contribute to the improvement of the gamelan production process, as an effort to assist the craftsmen industry.
- Art and cultural contribution; It is expected that with the improvement of the process, it is expected that the production time can be reduced so that the craftsmen are more productive so that it can lift the crafter's economy which will indirectly continue to develop traditional culture and art.

2. Methodology

Research conducted on the forging process, both manually and mechanically based on the experimental method, where the process of making specimens was carried out in gamelan craftsmen. The specimen is then tested for properties that are owned by the following steps:

1. Conducting forging testing on a designed mechanical forging device. Forging is carried out in hot conditions which are burned in a furnace with a temperature of 300 – 400°C
2. Tests were carried out on bronze material with the same dimensions as the production of gamelan craftsmen in Tihingan village, Banjarangkan district, Klungkung
3. Test specimens in the form of gangse blades made from bronze are then cut according to dimensions adjusted to the test equipment
4. Vickers hardness testing both manual and mechanical forging
5. Conducting density tests to obtain material porosity both manual and mechanical forging results
6. Conduct microstructure testing to get the shape of the structure for both manual forging and mechanical forging
7. Perform an analysis by comparing the results of hardness, porosity and microstructure testing of manual forging material and mechanical forging.

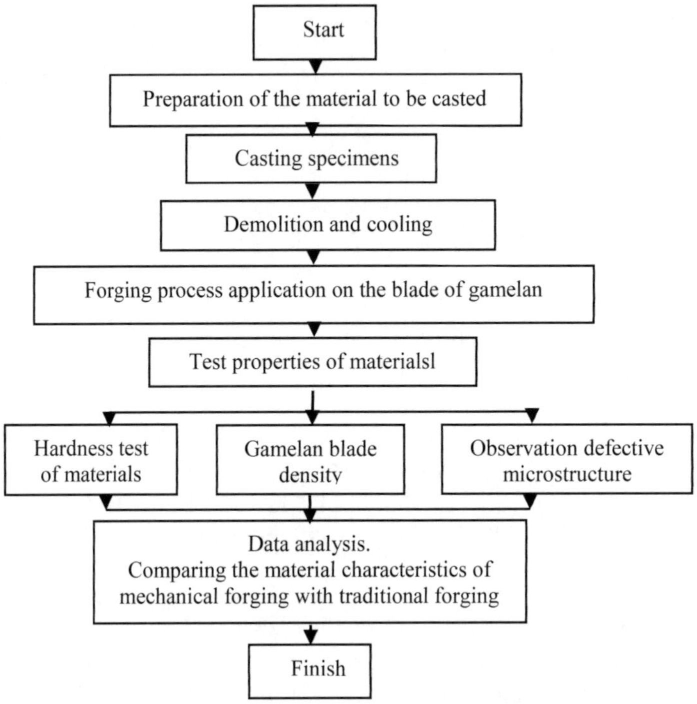

Figure 1. Research step

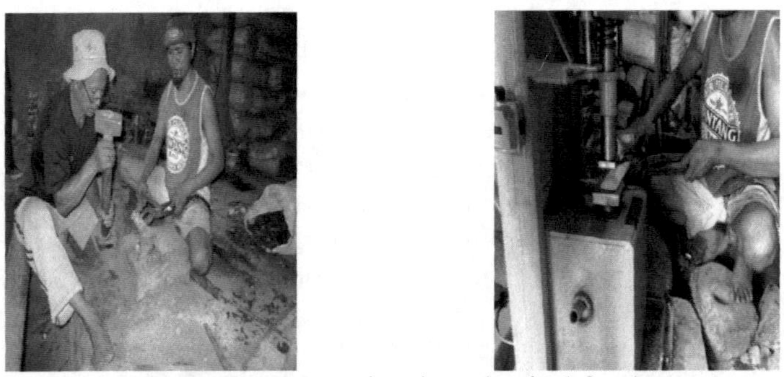

Figure 2. Manual and mechanical forging

Tests for forging bronze material are made with test specimens in the following form :

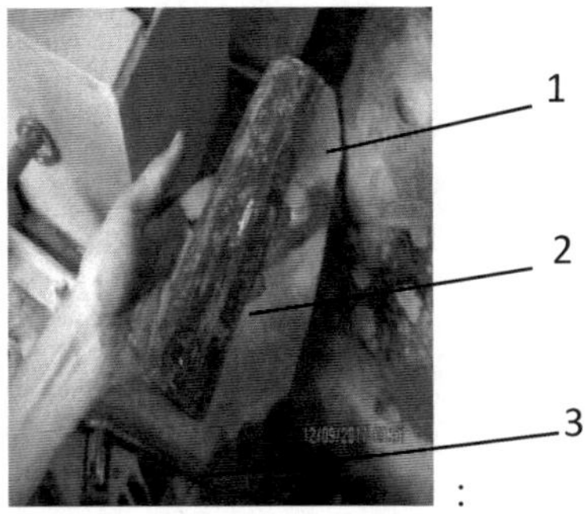

Figure 3. Cutting the specimen on the blade

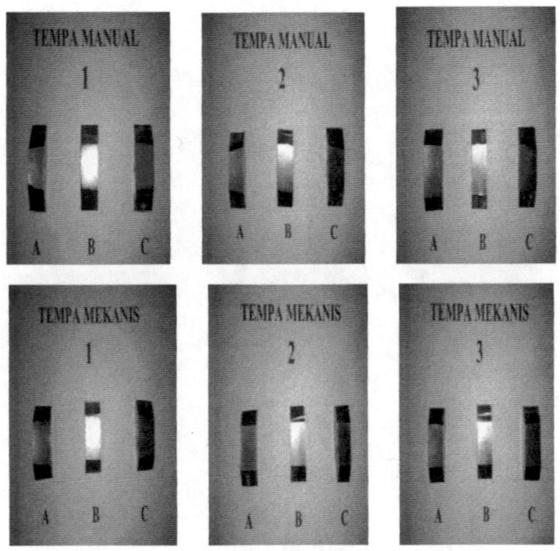

Figure 4. Test Specimens

3. Result and Discussion

3.1. Hardness test results

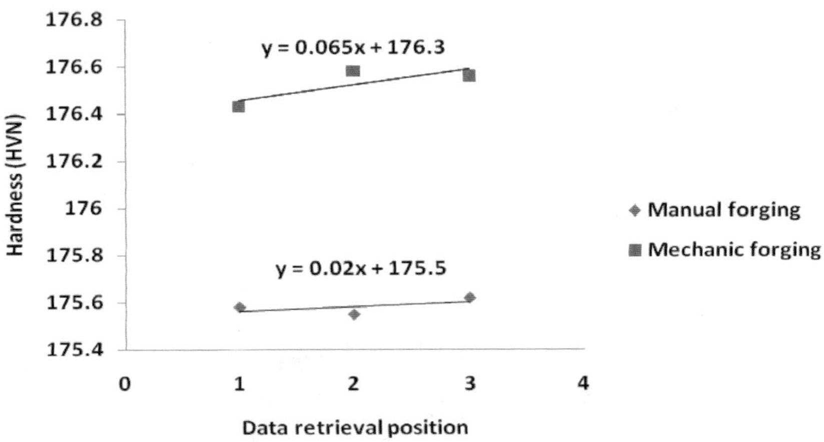

Graph 1. Results of hardness testing

The number of specimens for manual forging as much as 3 and forging with mechanical devices as many as 3 specimens where in each position taken as many as 5 points. The position of taking data as shown in Figure 3. The mean of the hardness test showed that the manual forging was 175.58 VHN while the mechanical forging was 176.52 VHN. The test results show that forging using mechanical devices with a load of 78.5 N gives better hardness results. This statement was supported by other researchers who stated that the forging with the heat process and carried out mechanically gave violence, shape and dimensions and better both [3]. Other studies also state that the forging process mechanically gives a quality product mechanically such as its hardness to be better [4].

3.2. Density Test Results

Test specimens are as shown in Figure 4 of the test results as shown in Graph 2 as follows :

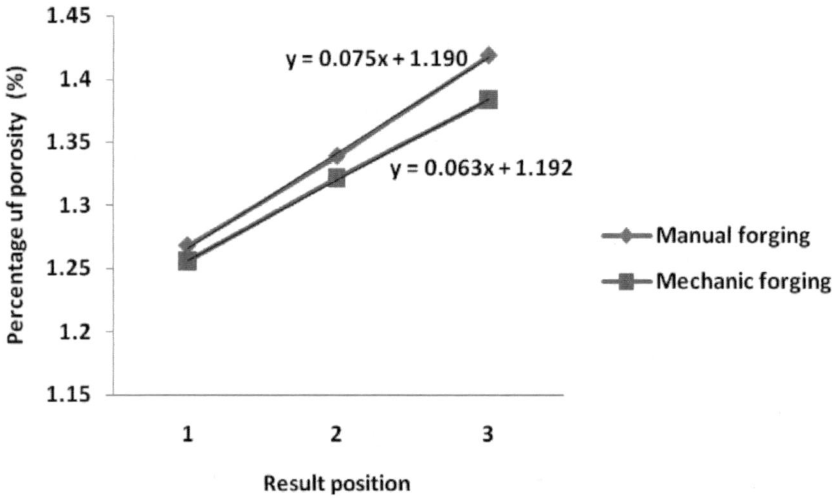

Graph 2. Results of density testing

The results of density testing to determine porosity in the sample where manual forging has a mean percentage of 1.342% while the forging is mechanically averaging 1.320%. This shows that with this level of porosity, it can be said that forged bronze materials using mechanical devices give better results. Such conditions occur because mechanical forging provides consistent loading in all areas forging. The forging process carried out on the material can reduce the porosity that occurs [5]. Mechanical forging that has been carried out also causes the distribution of porosity shrinkage [6].

3.3. Microstructure Testing Results

Testing of microstructure for manual and mechanical forging as shown in Figure 5:

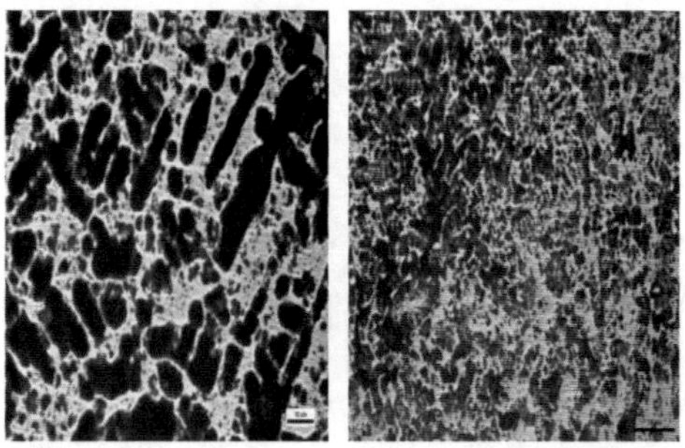

a. manual b. Mechanic

Figure 5. Microstructure size of 10 μm enlargement of 200 x from forging results

Microstructure can be seen in Figure 5, where in part a the result of manual forging shows the long and pointed dendrite structure this happens because the pressure is not constant. Such structures tend to be more ductile. Part b is mechanical forging, it looks like the dendrite shape is shorter and rounder structure like that tends to have hard mechanical properties. The shape of the structure is caused by the pressure applied to the forging to occur more constant and almost evenly distributed in the area of the forged area. The forging process is carried out at a temperature of 300 °C to 400 °C according to the phase diagram of Cu-Sn in Figure 4.8 this condition shows that there is no phase change. The research that has been carried out states that the determination of temperature when conducting hot forging is very influential on the mechanical quality of materials [7].

3.4. Deformation Rate

The deformation level is a change in the dimensions of the material form after the forging process. The forging process is carried out at a temperature of 300 °C to 400 °C with a load of 78.5N where at this temperature phase changes have occurred, making it easy to forging. The deformation rate for forging is shown in the following graph:

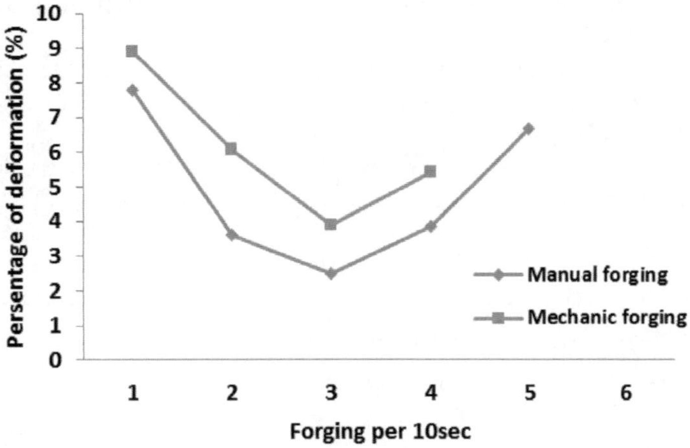

Graph 3. Deformation rates

Graph 3 shows the level of deformation that occurs in the forging process, where the initial dimensions of 9 mm and the end of the forging are expected to be 7 mm. The manual forging process to achieve these dimensions requires 5 times forging with a duration of 10 seconds with an average deformation level of 4.88%. Forging using mechanical equipment with the same conditions requires 4 forging processes with an average deformation rate of 6.07%. Temperature determination according to the phase diagram gives an influence on the degree of deformation as well as the load given. The lower process temperature decreases the material ductility and deformation level [8].

4. Conclusion

The conclusion of the application of ergonomically based mechanical forging equipment that is intended to help gamelan craftsmen in the forging process. Based on the mechanical work of the forging tool it is able to provide results such as hardness, reduction in porosity, microstructure, a better deformation rate than forging done manually.

Acknowledgements

Thank You to the Ministry of Research, Technology and Universities and LPPM-Udayana University, for the Research Costs provided in accordance with the Letter of Agreement on National Decentralization and Competitive Research Implementation Assignments for Fiscal Year 2017, Number: 415.3 / UN 14.4.A / PL / 2017.

IOP Conf. Series: Materials Science and Engineering **539** (2019) 012022 doi:10.1088/1757-899X/539/1/012022

References

[1] Surdia, Tata & Chijiiwa, Kenji. (1982). Teknik Pengecoran Logam. Jakarta: P.T Pradnya Paramita.

[2] ASTM, E 1876-01, 2002, Standard Test Method for Dynamic Young, Shear Modulus, and Poisson's Ratio by Impulse Excitation of vibration, ASTM International.

[3] Vipan Chand.,Dr.SW.S.Sen.,M.S.Sethi .(2014). Taguchi analysis of forging defects for gears. International Journal of Advanced Technology in Engineering and Science www.ijates.com. Volume No.02, Issue No. 08, August 2014 ISSN (online): 2348 – 7550, pp 52-59

[4] Mahendra G. Rathi*, Nilesh A. Jakhade**(2014). An Overview of Forging Processes with Their Defects. International Journal of Scientific and Research Publications, Volume 4, Issue 6, June 2014.ISSN 2250-3153, pp. 1-7

[5] Muhammad Syaiful Nurkholiq*, Helmy Purwanto, Sri M. B. Respati (2013). Analisa Pengaruh variasi tekanan pada pengecoran squeeze terhadap Kekerasan Produk Sepatu Kampas Rem Dengan Bahan Aluminium (Al), Silikon (Si) Daur Ulang. . Momentum**,** Vol. 9, No. 2, Oktober 2013, Hal. 34-37 ISSN 0216-7395.

[6] Jiaqi Wang, Paixian Fu, Hongwei Liu, Dianzhong Li, Yiyi Li. (2012). Shrinkage porosity criteria and optimized design of a 100-ton 30CrMoV forging ingot. Materials and Design. SciVerse Science Direct.35 (2012) 446–456

[7] Yoshimitsu Okazaki. (2012) On the Effects of Hot Forging and Hot Rolling on the Microstructural Development and Mechanical Response of a Biocompatible Ti Alloy. materials ISSN 1996-1944, *5*, pp1439-1461

[8] N.T.Switzne.r,C.J.Van Tyne.,M.C.Mataya (2010). Effect of forging strain rate and deformation temperature on the mechanical properties of warm-worked 304L stainless steel. Journal of Materials Processing Technology Volume 210, Issue 8, , pp 998-1007

Energy analysis of solar farm planning in weh island, aceh, indonesia

S Rahmawati[1], M Iqbal[2] and I D Sara[3]

[1]Megister of Industrial Engineering Program, Syiah Kuala University, Banda Aceh, Indonesia

[2]Department of Mechanical and Industrial Engineering, Syiah Kuala University, Banda Aceh, Indonesia

[3]Department of Electrical Engineering, Syiah Kuala University, Banda Aceh, Indonesia

Email:,mohd.iqbal@unsyiah.ac.id

Abstract. The paper discusses the simulation study and analysis of output energy produced by 5MW solar farm located in Weh Island, Aceh Province, Indonesia. Weh Island is a tourist area. Currently, a diesel power plant supplies all of its electricity requirement. Most of the user were household, hotels and public facilities. Weh Island consists of 8,858 family houses, 124 medium and small hotels. 48 educational facilities (formal schools and traditional Islamic schools) 48 healthcare facilities (consisting of 2 hospitals, 6 public healthcare centre, and 23 public healthcare service units). The daily peak load of electricity usage occurred at two points; 03.00 pm and 08.00 pm, with the peak load reaches 4,667 kW and 5,009 kW respectively. The solar farm was designed based on the peak electrical load (5MWp) using grid-connected system to ensure continuous availability in the area, as solar farms are only able to operate during the day. The system used 300 Wp Monocrystalline technology panels with 16.4% efficiency and inverter with 20Kw capacity and 98.8% efficiency. Each of the inverter was designed to serve 66 to 67 solar panels. Therefore, in order to produce 5MW of electricity, the solar farm was developed with 16,668 solar panels and 250 inverters on 32,309 m2 of land space area. The simulation model of the solar farm was developed using PVsyst simulation software. The simulation study was conducted based on monthly solar radiation, temperature and daily radiation time of the Weh Island during one year. The result of the study showed that the system produced sum annual energy of 7991 MWh/year and the solar panel obtained the average radiation of 5.2 kWh/m2/day. The highest energy output was 792.4 MWh which was produced in January, and the lowest energy output was 603.3 MWh which was produced in September. Furthermore, when the usage time distribution of the electricity was formulated into the model, the simulation result showed that the solar farm system were able to accommodate the electricity requirement in Weh Island.

1. Background

The general energy crisis, including electrical energy, is now a major issue in the world. So far the main source of electrical energy used by the world comes from fossils that are known to be environmentally unfriendly and have limited capacity [1]. In fact, the need for electrical energy consumption continues to increase every year [2]. This limitation makes the number of proposals to process electrical energy from renewable resources such as solar, water, wind, geothermal, wave currents, biomass and hybrid systems as the latest alternative.

Aceh, until 2016 has a peak load requirement of 325 megawatts (MW) derived from fossil, such as Gas Power Plant (PLTG), Steam Power Plant (PLTU) and Diesel power plant (PLTD). The availability of electricity in Aceh should reach 50% of the peak load requirement or should be available around 500 MW, but up to now only 340 MW is available [3], so there is a need for additional power supply, one of them is with a friendly energy proposal environment and renewable (EBT).

One of the easiest EBT to apply and minimal maintenance cost is Solar Farm. The main component used in solar farm system is a solar module which is strung by photovoltaics (PV) cells. Solar modules today can be found easily in the market, although in Indonesia the price of solar panels are relatively expensive.

1.1. Solar Radiation Potential in Research Location

Data from NASA (National Aeronautics and Space Administration - NASA) that has been recorded for 20 years from 1983-2005 explains that the potential of solar radiation in some areas in Aceh is different. The highest solar radiation of 5.1 Kwh / m2 / day occurred in areas such as Weh Island, Nasi Island, Banda Aceh and Aceh Besar. For more detailed explanation of the potential solar radiation in Pulau Weh as the study location are described in Table 1. The highest solar radiation on Weh Island occurs in February-April. Where the sun radiation is in the order of 5.75 - 5.79 - 5.64 kWh / m2 / day. The lowest radiation occurred in October and November, where the radiation is at the number 4.66 - 4.67 kWh / m2 / day.

1.2. Electricity load in Weh Island

The electricity in Pulau Weh is still supplied with diesel power plants (PLTD), and several renewable energy (EBT) are being developed. It is planned to support the existing electricity in Weh Island also in the hope of replacing the working hours of the power plant with all aspects of EBT surplus and also as support to achieve 23% of EBT target in Indonesia [4].

The peak of the electrical load on Weh Island occurs at 15.00 p.m and 20.00 p.m. with average power consumption usage, reaching about 4,667 to 5.009 kW as described in figure 3. The basic production cost (BPP) for this electricity is Rp.2.500, - / kWh and the selling price to the public is Rp.900 / kWh [5].

Table 1. Data of Aceh Potential Solar Radiation (Weh Island) (kWh / m2 / day)

Solar Radiation In Weh Island	Jan	Feb	Mar	Apr	May	Jun	Jul	Aug	Sep	Oct	Nov	Dec	Avarage
Monthly Radiation (kWh/m2/mnth)	167.1	161	179.5	169.2	155.3	149.7	151.9	149.1	142.2	144.8	139.8	151.3	155.1
Daily Radiation (kWh/m2/day)	5.39	5.75	5.79	5.64	5.01	4.99	4.9	4.81	4.74	4.67	4.66	4.88	5.1

Source : Surface Meteorology and Solar Energy (National Aeronautics and Space Administration – NASA

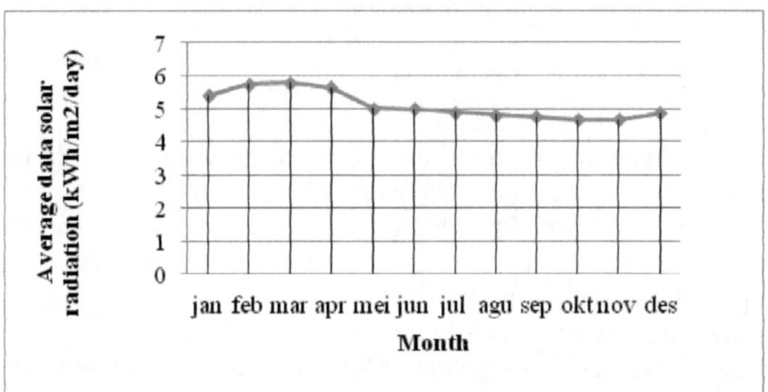

Figure 1. Graph of monthly average data solar radiation in Weh Island (Aceh-Indonesia) in kWh / m2 / day year (NASA 1983-2005)

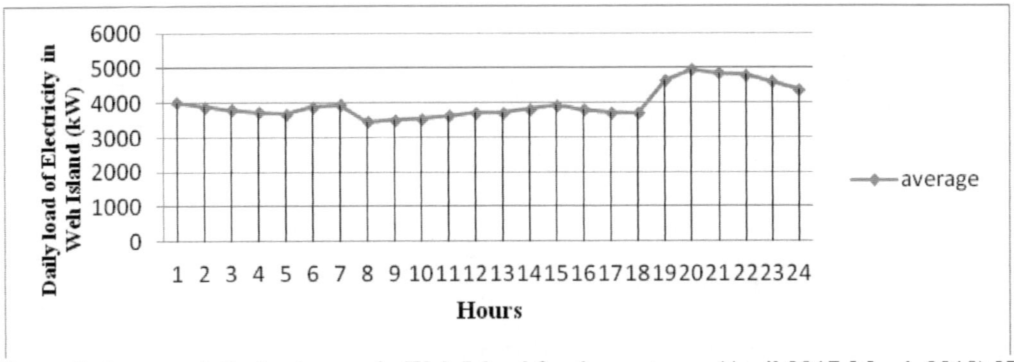

Figure 2. Average daily load curve in Weh Island for the past year (April 2017-March 2018) [5]
Source: PLN Data Centre and PLN Rayon Weh Island- Aceh.

The price of BPP in Weh Island is currently more expensive than the Decree of the Minister of ESDM No.1404 Year 2017 for Weh Island area where Rp 1,733 / kWh [6].One solution offered to lower BPP is to replace the operating hours of the diesel with EBT in the form of Solar Farm. Application of solar farm has started to be done in the region of ASIA although with relatively high investment value.

Research on feasibility studies of solar farm investments conducted in Perlis-Malaysia was investigated by M. Pauzi M. Kassim, et al. (2015) mentions the 5 MW solar farm performance with an investment of RM 41,579,448.10 can produce 1358.11 kWh / kwp of solar energy with a return on investment capital for 11 years. The return on investment capital is closely related to the output of the energy output produced.

The research on energy output performance has been studied by Manoj Kumar et al. (2017) using a 100 Wp Poly-si with a grid-connected system in India. The resultant performance is 161.6 MWh / year, with the specification of energy that can be injected into the grid of 1,616 kWh / kWp / year. The highest energy production is 15, 800 kWh / month and the lowest is 11,060 kWh. The overall performance of the Poly-100 kwp is 80%

With the potential of solar radiation at Pulau Weh averaging at 5.1 kWh / m2 / day, the analysis of energy output from solar farm planning in Pulau Weh is considered important. It aims to see the potential output to be generated as well as the influence of energy output from other environmental variables in Pulau Weh. To meet the needs of the weh island electricity, the solar farm planning on this research is designed based on the daily peak electricity required in Weh Island of 5009 kW, or designed for 5 MWp. The planning in this Solar Farm research, conducted with a grid connected system to ensure continuous availability of electricity in the research area. The solar farm system to be planned is full photovoltaics on-grid system.

1.3. Temperature

Unlike the radiation cases, higher temperatures will actually decrease the power produced by Photovoltaic (PV), this is because the rise in temperature can decrease the voltage on the PV and affect the electrical power generated in PV [7] [8]. Ideally, solar panels work on standard test conditions (STC) with 1000 w / m2 radiation at standard temperature of 25^0C, where for every 1^0C rise in temperature in PV will reduce the working efficiency of solar panels. This decrease depends on the character of each PV, called Temperature coefficient factor (TCF). The decrease in PV efficiency ranges from 0.4 to 0.6 for every 1^0C increase [9]. The average temperature in Indonesia is about 25-35^0C, so it needs to be considered in the construction of solar panels to provide sufficient space under it, so that the air flow can lower the panel temperature [10]. Temperature data obtained from the results of BMKG Weh Island and also available from NASA recorded for 20 years. As shown in table 2.

Table 2. Temperature data in Weh Island from BMKG and NASA

Month	Data BMKG	Data NASA	Month	Data BMKG	Data NASA
Jan	26.75	26.4	July	27.65	27.8
Feb	26.9	26.9	Aug	27.35	27.4
March	27.8	27.5	Sept	27.1	27.1
Apr	27.9	27.9	Oct	26.6	26.8
May	28.3	28.1	Nov	26.4	26.5
Jun	28.1	28.2	Dec	26.4	26.3
Average				27.27	27.24

Source: Secondary data from BMKG Indrapuri , Sabang (Year 2015-2017) and NASA

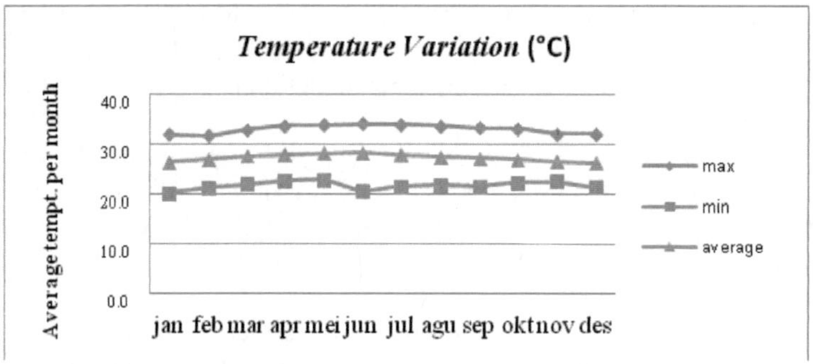

Figure 3. Graph of temperature data in Weh Island
Source: Secondary Data from BMKG Indrapuri and Sabang (Year 2015-2017)

Figure 3. Describes the average temperature as well as the maximum and minimum temperatures that occurred over a year. From the graph it is known that the highest temperature occurred in June at point 34^0C and the lowest temperature at Weh Island was in January at 20.5^0C.

2. Methodology

2.1. General Information About the research location

The location studied is the Islands of Aceh, namely Pulau Weh, where this island is one of the islands that have the potential of solar radiation is quite good. Based on the geographic map of google earth, Weh Island is located at the coordinate point 5.84 ° N 95.28 ° East. Data related to general information about the condition of Weh Island was obtained from the Central Bureau of Statistics (BPS) of Weh-Aceh Island as shown in table 3. From BPS data it is known that, the population of the island is recorded with the number of existing household heads, which is as many as 8,858 heads of households. Since it is a tourist area, the island consists of about 124 hotels. On this island there are 48 educational facilities in the form of schools, foundations and boarding school, 48 units of worship facilities such as mosques and prayer rooms, as well as the number of public health facilities consisting of 2 hospitals, 6 public healthcare centre and 23 public healthcare service unit. All general information shown in table 3.

Table 3. General Information of Weh Island Condition

General Information Weh island-Aceh (2018)	
Research Location (5,84°LU 95,28°BT)	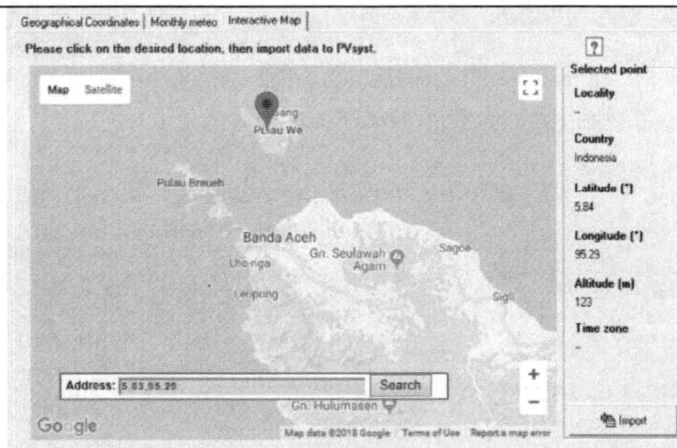
Number of heads of households (2018)	8.858
Number of hotels	124
Number of educational facilities	48
Number of facilities of worship	48
Number of facilities :	
Hospital	2
public healthcare center	6
public healthcare service units	23

Source: Google earth and BPS data in 2018

2.2. *Pvsyst software*

PVsyst (Photovoltaics System) is one of the software used to study the size (sizing) and analyze data of PV system which is complete enough. Available simulation systems consist of grid-connected, stand-alone, pumping and public transportation / DC connected systems [11]. The PVsyst simulation process takes into account the behavior of all system characters and all interruptions for every hour of system operation. The energy generated depends on the meteorological data of site installation by analyzing the data of energy resources, energy production, system configuration and system losses to install the system [12].

Some of the input data on the software, described in figure 4 to 5, consists of meteom data (global radiation, diffuse radiation and temperature), optimal slope of panel construction, the main data product specification, ie solar panel si-monocrystalline 300 Wp and inverter 20 KW.

2.3. *Tilt*

In this study the tilt/slope with optimum loss of 0% is determined by 10^0 west direction. Determination of the slope is based on the latitude of the research location where it is $5{,}84^0$ and the position of the module buffer installation is 10^0. As illustrated in figure 5(b). The slope of the solar panel installation affects the amount of solar radiation received and the temperature. The temperature affected by the slope distance between the panels is ambient temperature. This ambient temperature, a benchmark of the amount of energy generated by the panel based on the influence of temperature in the research location.

2.4. Manufacturer Specification components (Solar panel si- monocrystalline 300 Wp and Inverter 20 kW)

The standard of sunlight that Earth receives on the right conditions above the head is 1000 W / m2. T_{Ref} standard working conditions of solar panels at panel temperature of 25°C. Manual input of product specifications with selected product specifications are described in the table 4 (a) specification of panel module and (b) specification of the inverter.

Table 4. (a) Manufacturer specification of module si-Monocrystalline 300 Wp (b) Manufacturer specification of Inverter

Specification	Product Ability	Symbol	Specification	Product Ability	Symbol
Capacity	300	W	Capacity	20000	W
Max. Power Voltage (Vmp)	37.70	V	MPPT Efficiency	99,9	%
Max. Power Current (Imp)	8.40	A	Maximum Efficiency	> 98,2	%
Open circuit Voltage(Voc)	45.60	V	Input Range	400-850	Vdc
Short Circuit Current (Ics)	8.80	A	Output	230/400 50/60 Hz	VAC

(a) (b)

2.5. PVsyst simulation Software

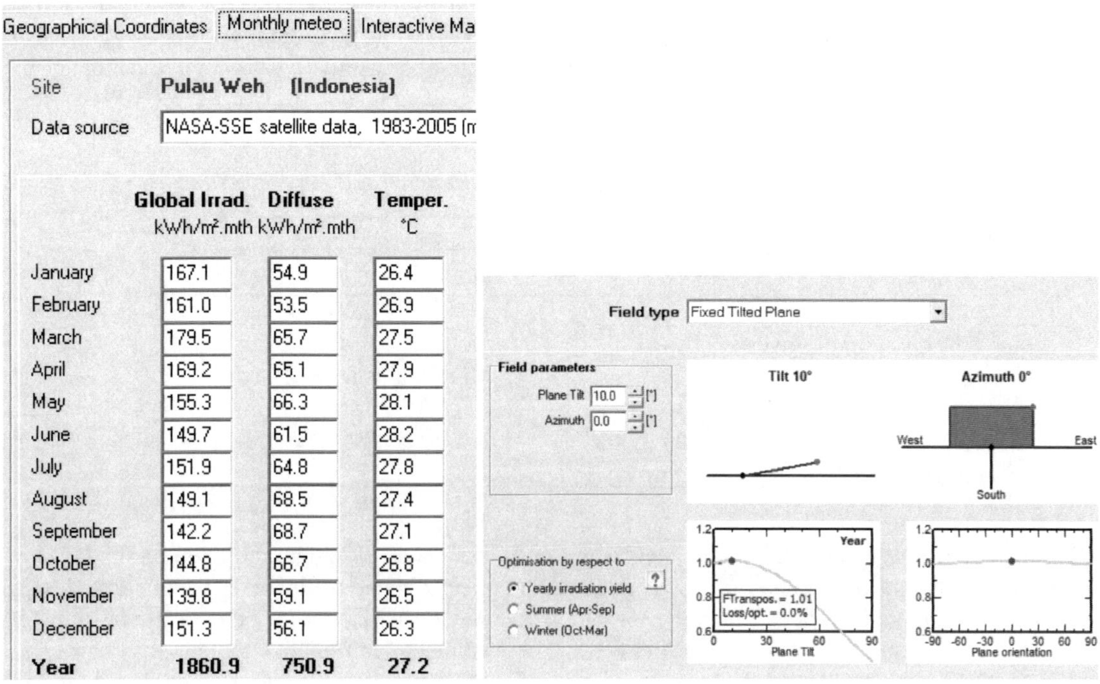

(a) (b)

Figure 4. (a) Input data parameters that affect the output on the solar farm system in PVsyst software (b) Optimization of titl determination at study Location (Weh Island)

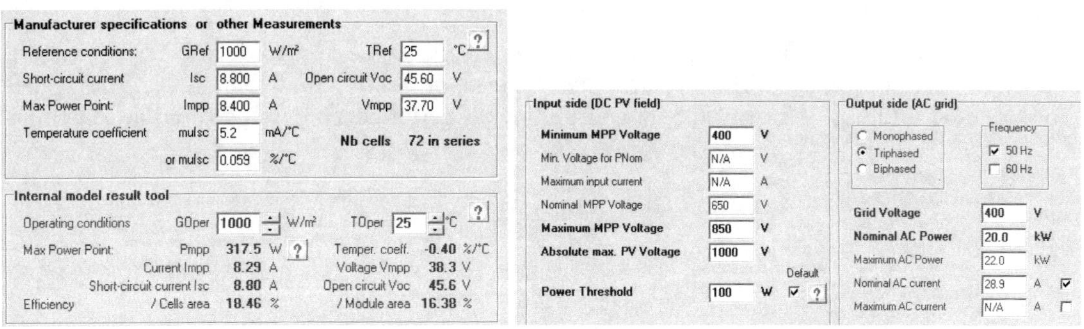

(a) (b)

Figure 5. Input data specification of (a) module and (b) inverter

3. Result and discussion

The results of the 5 MWp solar farm planning on Weh Island using Si-Mono (Crystalline) 300Wp technology with manual data input on PVsyst software shown in table. For the 5 MWp solar farm planning, the system requires as many as 16,667 solar panels with 250 inverters with a capacity of 20 kW. The configuration consists of 18 modules in series with 926 strings, where one inverter will serve 66-67 solar panels.

3.1. Balances and main results of simulation

From the results of the study as shown in table 6, it was known that the average of solar radiation at the horizon line (Horizontal Global Irradiation) was 155.07 kWh / m2 or the sums radiation per year is 1860.9 kWh / m2. The average Ambient temperature of 26.49 ^{0}C, utilizing 300 Wp solar panel by 16,667 panels with an efficiency of 16.4% will produce sums an annual energy of 8173.6 MWh / year (noted as EArray in table 6).

Table 5. Results of Simulation Sizing Solar Farm with PVsyst software

Sizing Simulation	5000 MW Mono (300 Wp ; eff 16,4%)
Number of Modul	16.668
Number of inverter (20kW)	250
Configuration	18 modules in series with 926 string
The area of land required by the panel	32.309 m2

The energy output generated by the panels is not only influenced by global radiation on the horizon surface, but is also influenced by the global incident in collector plane of the 10^{0}C panel slope design effect. Global incident in collector plane generated an average of 5.2 kWh / m^2 / day or by the sums of 1877.8 kW/m^2 (as noted in GlobInc in table 6). The highest solar radiation potential occurred in January in DC from was 792.4 MWh (as noted by E Arry in table 6). This energy then distributed to the inverter to be converted into electrical energy in the form of AC which will occur energy reduction due to the conversion process, so the monthly hourly sums for output generated by inverter in the form of AC energy in January is 775.7 MWh (Noted as E_grid in table 6). The lowest output will occur in September as 603.3 MWh. For more details the data of DC-AC energy conversion can be seen in table 6 (Noted by EArray-E_Grid).

3.2. Performance Ratio

The performance ratio of 5MW solar farm with monocrystalline technology for one year was described as PR in Table 6. The average performance of solar farm was 8.51%. The highest performance occurred in August of 8.55% and the lowest performance ratio occurred in March and May of 8.47%.

3.3. Loss diagram

The various energy losses incurred in the system are described in the diagram above at figure 7. From the PVsyst simulation results, the Global irradiance value on the horizon field occurring in Pulau Weh is 1861 kWh / m2, while the effective irradiance on collector is worth 1817 kWh / m2.The energy conversion that falls on the array with efficiency under standardized test conditions (STC) of 16.38%, yields 9617 MWh of energy in the form of DC. Various losses at this stage occur for -15.4%. The energy generated from the array of virtual energy in MPP is 8174 MWh. The energy then distributed to the Inverter to be converted into AC with a total energy loss of -2.2%. The energy generated by an AC inverter of 7991 MWh / year is the energy that can be channeled to the Grid for distribution to the end-user / load.

Table 6. Balances and Main results of simulation from Pvsyst

	GlobHor	DiffHor	T Amb	GlobInc	GlobEff	EArray	E_Grid	PR
	kWh/m²	kWh/m²	°C	kWh/m²	kWh/m²	MWh	MWh	
January	167.1	54.90	26.19	181.8	176.4	792.4	775.7	0.853
February	161.0	53.50	26.19	169.7	164.9	736.6	720.9	0.849
March	179.5	65.70	26.30	182.3	177.0	789.1	772.1	0.847
April	169.2	65.10	26.58	165.1	159.9	717.1	701.4	0.850
May	155.3	66.30	27.10	146.9	141.6	636.9	622.2	0.847
June	149.7	61.50	27.20	139.6	134.5	608.5	594.3	0.851
July	151.9	64.80	27.03	142.7	137.6	622.2	607.4	0.851
August	149.1	68.50	26.79	143.8	138.9	629.3	614.6	0.855
September	142.2	68.70	26.44	141.6	136.9	617.5	603.3	0.852
October	144.8	66.60	26.13	149.7	144.9	654.4	639.6	0.854
November	139.8	59.10	25.88	149.2	144.6	648.5	634.0	0.850
December	151.3	56.10	26.05	165.3	160.2	721.1	705.5	0.854
Year	1860.9	750.80	26.49	1877.8	1817.4	8173.6	7990.9	0.851

Table 7. The hourly maximum value for E_grid from PVvsyst

	0H	1H	2H	3H	4H	5H	6H	7H	8H	9H	10H	11H	12H	13H	14H	15H	16H	17H	18H	19H	20H	21H	22H	23H
January	0	0	0	0	0	0	0	1	2	3	4	4	4	4	4	3	3	1	0	0	0	0	0	0
February	0	0	0	0	0	0	0	0	2	3	4	4	4	4	4	4	3	1	0	0	0	0	0	0
March	0	0	0	0	0	0	0	1	2	3	4	4	4	4	4	3	3	1	0	0	0	0	0	0
April	0	0	0	0	0	0	0	1	2	3	4	4	4	4	4	3	2	1	0	0	0	0	0	0
May	0	0	0	0	0	0	0	1	2	3	4	4	4	4	3	3	2	1	0	0	0	0	0	0
June	0	0	0	0	0	0	0	1	2	3	3	4	4	4	3	3	2	1	0	0	0	0	0	0
July	0	0	0	0	0	0	0	1	2	3	4	4	4	4	4	3	2	1	0	0	0	0	0	0
August	0	0	0	0	0	0	0	1	2	3	4	4	4	4	4	3	2	1	0	0	0	0	0	0
September	0	0	0	0	0	0	0	1	2	3	4	4	4	4	4	3	2	1	0	0	0	0	0	0
October	0	0	0	0	0	0	0	1	2	4	4	4	4	4	4	3	2	1	0	0	0	0	0	0
November	0	0	0	0	0	0	0	1	2	3	4	5	4	4	4	3	2	1	0	0	0	0	0	0
December	0	0	0	0	0	0	0	1	2	3	4	4	4	4	4	3	2	1	0	0	0	0	0	0
Year	0	0	0	0	0	0	0	1	2	4	4	5	4	4	4	4	3	1	0	0	0	0	0	0

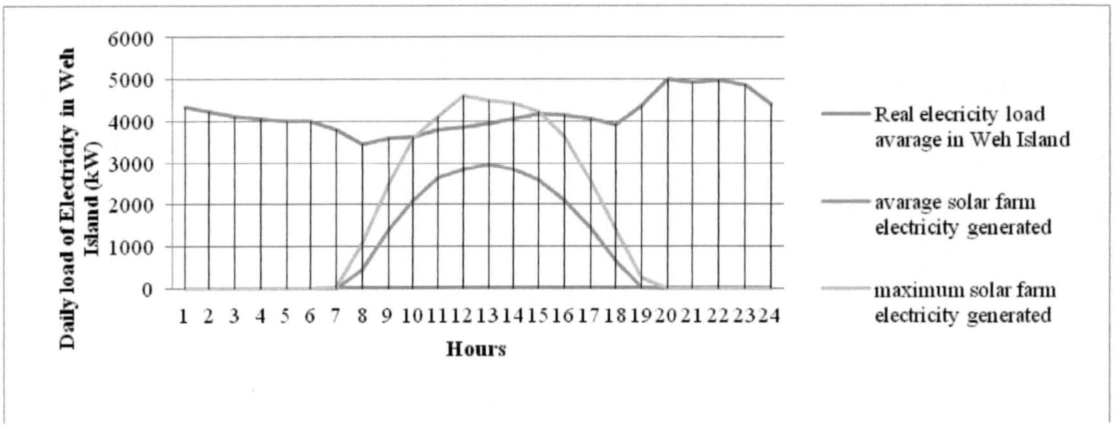

Figure 6. Energy Output Chart to be injected into the Grid per hour in a year

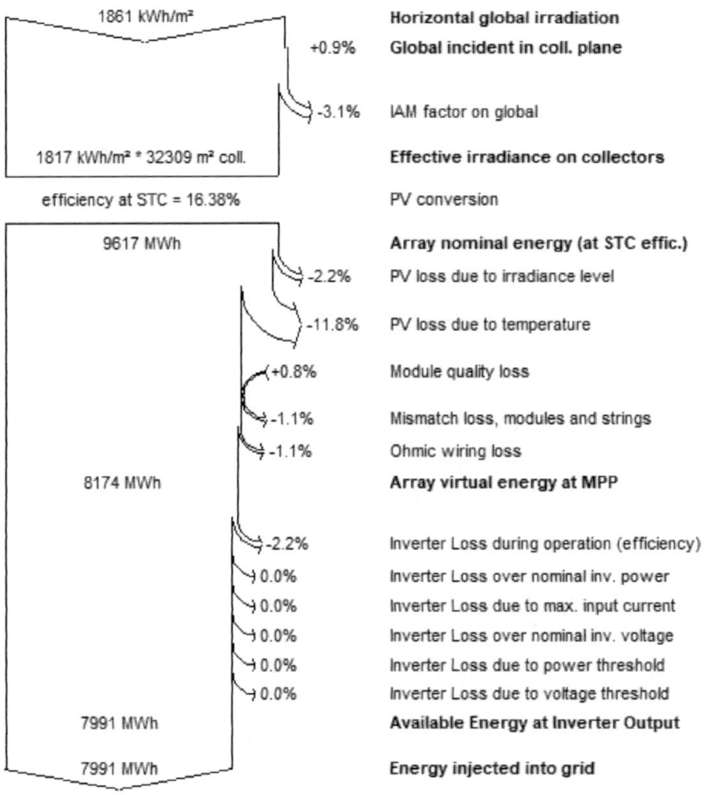

Figure 7. Losses diagram 5000 kWp with Monocrystalline technology

4. Conclusion

From the simulation results using PVsyst software, 5 MWp solar farm planning in Weh Island, Aceh-Indonesia with crystalline technology (monocrystalline) can be concluded the solar-powered Energy will follow the normal solar graph. At the peak of solar radiation, ie during the day, a 5 MWP solar farm is capable of producing an average of hourly electricity of 3-5 MW. In the morning and afternoon, solar farm capable of producing electricity with an average of 1-2 MW. The total conversion in monthly hourly sums (E_Grid) of solar radiation into electrical energy from solar farm can be

IOP Conf. Series: Materials Science and Engineering **539** (2019) 012023 doi:10.1088/1757-899X/539/1/012023

delivered to the grid by 7991 MWh per year. The energy losses occurring in the solar conversion process were 17.6%. Although the highest solar radiation occurred in March but in the conversion process from solar farm affected by other variables showed that the highest production occurred in January at 792,4 MWh / m2. The lowest electricity production occurred in September as 603.3 MWh.

References

[1] M. Pauzi M. Kassim Karam M. Al-Obaidi, M. Arkam C. Munaaim and Abd. Mokhti Salleh feasibility study on solar plant utility grid under Malaysia Feed-in Tariff [Journal]. - malaysia: American Journal Of Engineering and Applied Sciences, 2015.
[2] Badan Pusat Statistik Pulau Weh [Report] 2016-2017, 2017.
[3] Serambi Indonesia [Report], 2016.
[4] (IRENA) International Renewable Energy Agency Renewble Energy Outlook For ASEAN [Report]. - [s.l.] : Asean Centre Of Energy (ACE), 2016.
[5] PT. PLN Pusat Aceh dan PLTD Rayon PulauWeh Sabang 2017. Aceh
[6] The Decree of the Minister of ESDM ESDM No.17 year. 2013 and No. 12 year 2017
[7] Syukri Suriadi dan Mahdi Perencanaan Pembangkit Listrik Tenaga Surya Terpadu Menggunakan Software PVSYST Pada Komplek Perumahan di Banda Aech [Journal]. - Banda Aceh : Jurnal Rekayasa Elektrika, 2010. - Vol. 2.
[8] Sara Ira Devi Analisis Potensi Kondisi Suhu dan Radiasi Sinar Matahari di Kota Banda Aceh untuk Pengembangan Pembangkit Listrik Tenaga Surya [Journal]. - Banda Aceh : Seminar Nasional dan Expo Teknik Elektro , 2014.
[9] Pangestuningtyas D.L Hermawan, and Karnoto Analisis Pengaruh Sudut Kemiringan Panel Surya Terhadap Radiasi Matahari Yang Diterima Oleh Panel Surya Tipe Larik Tetap [Report]. - Semarang : Kampus UNDIP Tembalang, Semarang , 2013.
[10] Akif Karafil Harun Ozbay, Metin Kesler and Huseyin Parmaksiz Calculation of Optimum Fixed Tilt Angle of PV Panels Depending on Solar Angles and Comparison of the Results with Experimental Study Conducted in Summer in Bilecik, Turkey [Journal]. - Turkey : Bilecik Seyh Edebali University.
[11] http://www.pvsyst.com/
[12] Y.M.Irwana A.R.Ameliaa, M.Irwantoa, Fareq.Ma, W.Z.Leowa, N.Gomesha, I.Safwati Stand-Alone Photovoltaic (SAPV) System Assessment using PVSYST Software [Conference] // International Conference on Alternative Energy in Developing Countries and Emerging Economies - Perlish, Malaysia : Energy Procedia , 2015.

IOP Conf. Series: Materials Science and Engineering **539** (2019) 012024 doi:10.1088/1757-899X/539/1/012024

The effect of diameter downcomer in air entrainment process from vertical plunging water jet with downcomer

Warjito[1], Budiarso[1], and R L Jaya[1]
Departemen Teknik Mesin, Fakultas Teknik, Universitas Indonesia, Jalan margonda Raya, Depok, 16425, Indonesia

Email: warjito@eng.ui.ac.id

Abstract. The development of technology on flotation in water-waste treatment is very fast. This environmentally friendly technology using bubbles. The characteristic of the bubble causes the material to separate into the material hydrophilic or hydrophobic. Obviously with the addition of other chemicals. At the time of bubble formation, air entrainment events occur. This study aims to determine the parameters that affect the entrainment of air events that is the relationship of diameter downcomer and jet velocity to air entrainment. Set up tools used are pump, nozzle, downcomer, flow meter, reducer and glass water tank. All that is connected with the pipe, the data obtained in the form of video and photos using the camera. From there obtained quantitative and qualitative data, for quantitative data in the process using image processing program. Results from this research that jet velocity affects air entrainment, whereas the diameter of the downcomer affects the value of the air entrainment.

1. Introduction

The development of technology is growing very fast. This happens also in flotation technology and water-waste treatment, one of microbubble technology. This technology is growing rapidly as human needs on flotation and desalination process. Jameson cell is one of the flotation technology.

Jameson cell was developed by Prof. G. J. Jameson of the University of Newcastle and Mount Isa Mine Ltd. Jameson cell is used in the mining industry as a mineral extractor. The Jameson cell that has downcomers, as the first contact on the bubbles and particles takes place, the high-pressure jet makes the air entering the water and produces bubbles (Harby, Chiva, & Muñoz-Cobo, 2014). These bubbles can be hydrophilic and hydrophobic.

Bubbles are used as a medium for flotation and a good bubble type for flotation is Microbubble. Microbubble is an air bubble with diameter of 10-50μm (Agrawal, Ashutosh, Wun Jern Ng. Yu Liu, 2011). Microbubble can be said to be environmentally friendly technology, because microbubble can produce highly reactive free radicals without using harmful materials.

In every microbubble forming process, there will be an air entrainment phenomenon. Air entrainment is natural occurrence which is often seen, where the air is immersed in water as an external flow of water. This occurs because of the dynamic properties of fluid mechanics. One example of natural occurrences of air entrainment in water is found in rivers, especially in areas where there are waterfalls. air entrainment is responsible for the infusion of oxygen to the water that is intended for the flora and fauna in the river.

This study aimed to document the effects of downcomers diameter on the dynamic properties of fluid mechanics. The downcomers diameter contribute a lot on trapping the air so that the air is not being interfered by its environment, along with variables and constants discussed in the following chapters.

The properties of the water jet are very important as they will be the contact point where the incoming air occurs. These affect the level and momentum of the air below the surface and the bubbles formation. In this air entrainment study on microbubble was done in Fluid Mechanics laboratory of Department of Mechanical Engineering, Universitas Indonesia. This study aimed to determine the characteristics of air entrainment from the perspective of fluid mechanics.

2. Methods

In studies using a model of vertical air entrainment plunging water jets where the liquid will move down towards the surface of the pond. The air entrainment process resulted the air coming through a vertically downcomers pipe. Factors affecting air entrainment are the size of the nozzle and downcomers diameter, the distance between the water surface in the pool against the nozzle tip, the physical properties of the fluid, the jet turbulence, and the jet velocity during the collision. However, the variable which will be used in this study are nozzle diameter, diameter downcomers, jet drop and water discharge.

The tools used are as follows.

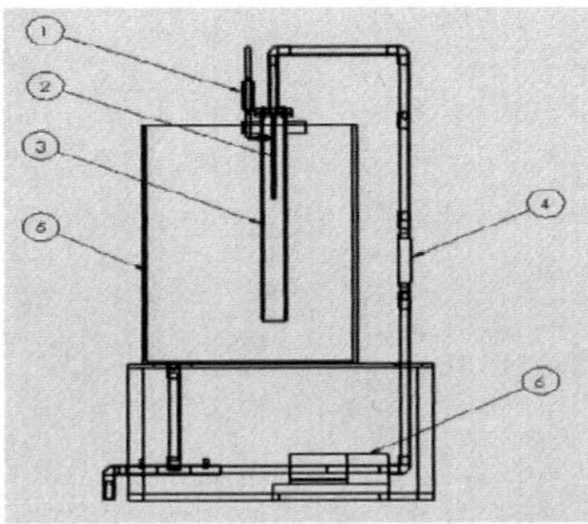

1. Air Flowmeter
2. Nozzle
3. Downcomer
4. Water Flowmeter
5. Water Tank
6. Pump

Figure 1. Schematic Apparatus

The tools used for the study were water tank, pump, water and air flow meter, downcomers, nozzle, pipes, reducer and flanges. Water tank was made by glass, with length of 910 mm, width of 610 mm, and height of 620 mm. This water tank had thickness about 2 mm, and filled by water. The pump had the capacity of 10-18 L / min, with engine pump rotation up to 2900 RPM.

In the piping system, PVC pipes were used with diameter of 1 ". In the suction section, the pipe was connected to the drain of the water tank vertically along the length of 20 cm. This pipe was connected with a 'T' connection, where one side flows to the gate valve which was parallel to the outlet pipe and drainage of the laboratory. On the other side it is connected to horizontal pipe with length of 70 cm, and that also included additional connection with the inlet at 1.5 ".

At the discharge section, the pipe of 1.5 " was reduced its diameter into 1 ". Pipe extended to 70 cm before the gate valve was installed. Gate valve is used to regulate the water discharge coming out of the pump, so that the water discharge could be easily adjusted. After the water had been passed through the flowmeter, the water would be passed through pipe of 1 " and installed the flange at the end. Nozzle connected with pipe of 1 “and nozzle also was covered with downcomer of 700 mm, so that air trapped in downcomer.

In the observation water tank, white LED light was given behind the water tank. At the back of the water tank was covered by polyfoam with the thick of 2 mm. This was useful as the LED light so it could be scattered well. Photography technique used was backlighting.

To get the data accurately as possible, it used applications to process the data (image processing), i.e. 'ImageJ', which is an open source Java image processing software. The software was capable of filtering the image with 2048 x 2048 pixels only in 0.1 seconds, 8-32 bit on RGB color in wide format. 'ImageJ' offered image enhancement as an option, due to the need to take high-speed photo capturing.

3. Results

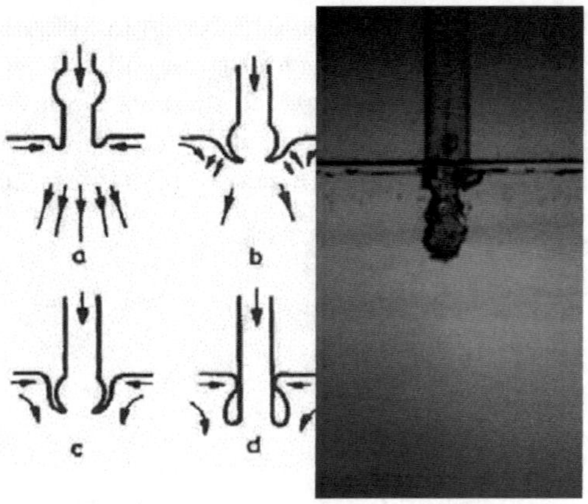

Figure 2. Air Entrainment and Air entrainment mechanism: (a)-(d) show subsequent phases of the phenomenon as a disturbance in the jet moves downwards (Bin, 1993)

Air entrainment visual observation through photography. Air entrainment from vertical plunging water jet is a phenomenon that generates bubbles (Chanson & Manasseh, 2003). When the jet hit the water flat surface, it will form a depression in the meniscus at the water surface (Smith, 2007). As jet velocity increases, depression in the meniscus also increases. Air trapped in the water, and will burst produce bubbles. Air entrainment can be observed alternately.

Figure 3. Air Bag

Air cavity on the vertical jet is caused by energy coming from the jet. This energy comes from jets that hit the water surface (Zhu, Oğuz, & Prosperetti, 2000). Jet which hit make an stress rupture and create a vacuum in the form of water cavity. The breaking of the surface tension causes the meniscus to form a pouch, where it will break apart as air enters into the bag and produces bubbles.

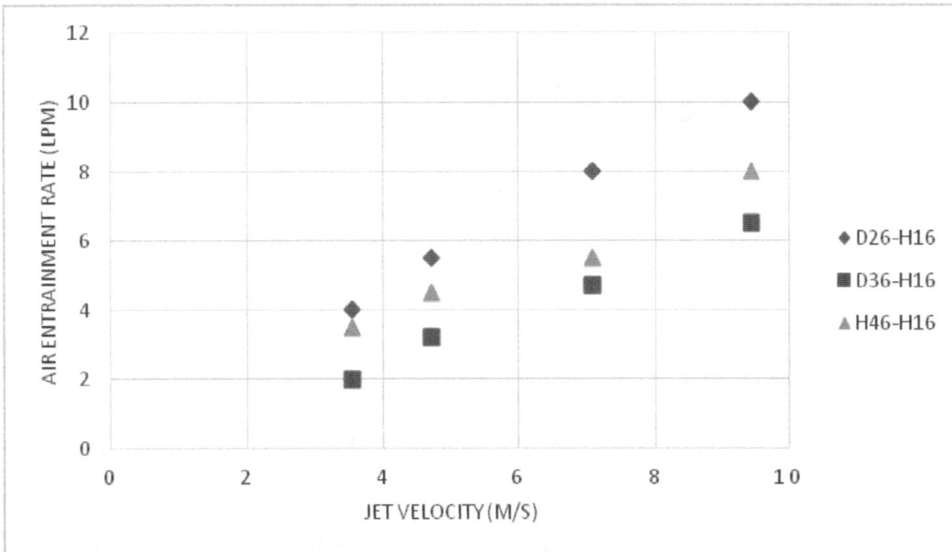

Figure 4. Graph between Jet Velocity and Downcomer Diameter with Air Entrainment

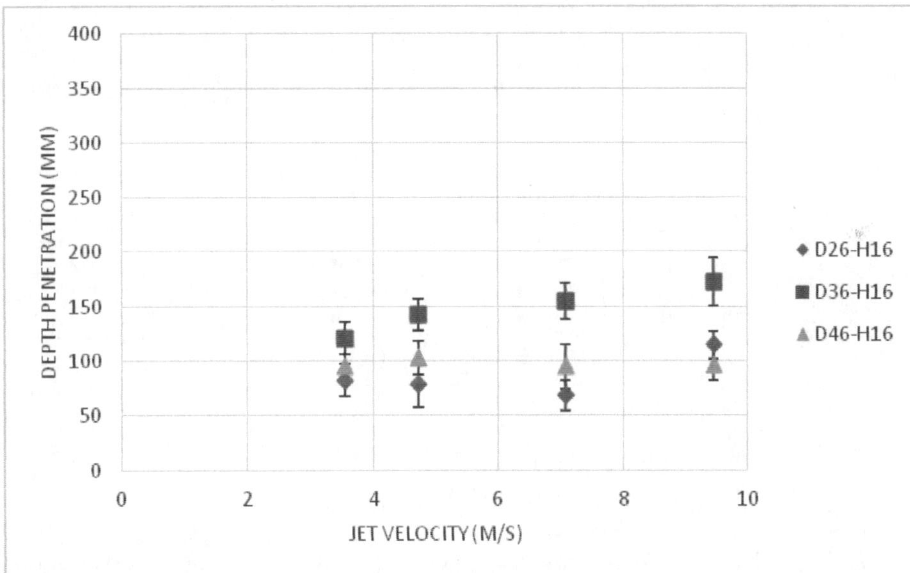

Figure 5. Graph between Jet Velocity and Downcomer Diameter with Depth Penetration

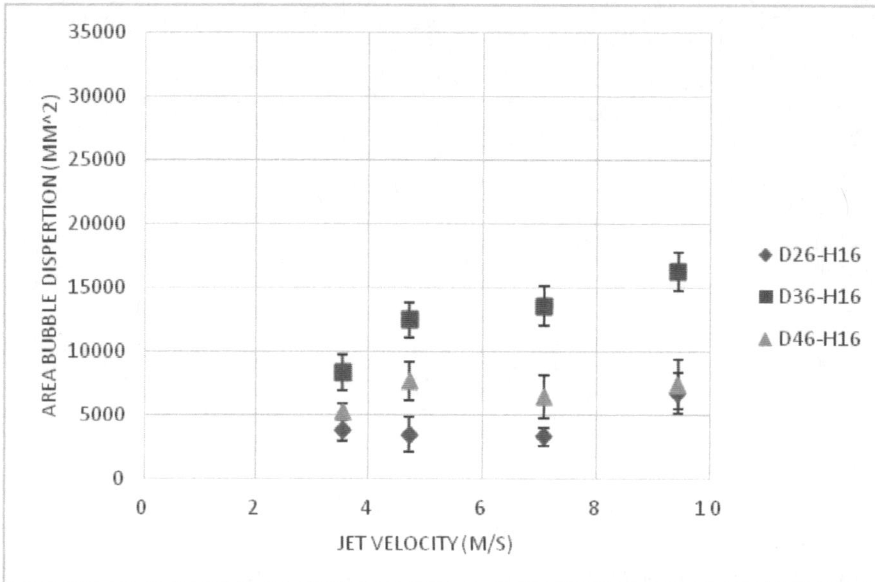

Figure 6. Graph between Jet Velocity and Downcomer Diameter with Bubble Dispersion Area.

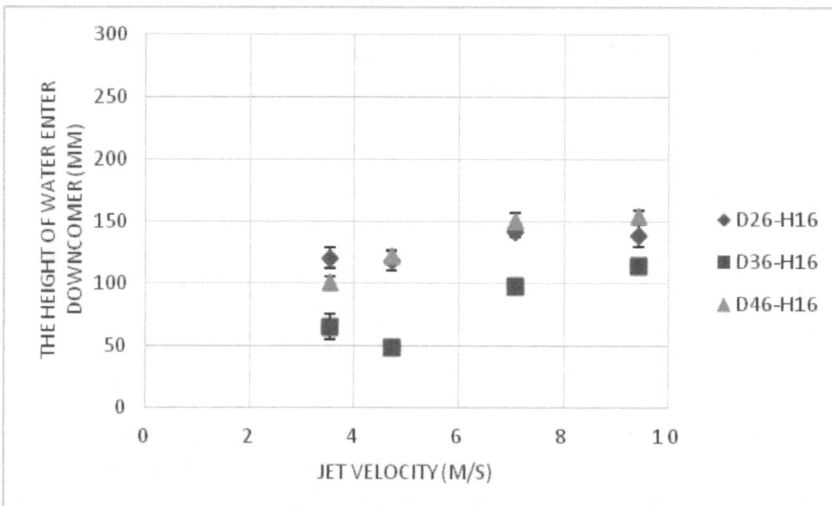

Figure 7. Graph of jet velocity and downcomer diameter to the incoming water level in the downcomer (HDc)

Diameter of downcomer and jet velocity used to determine the value of air entrainment that occurs. Downcomer used measuring 26mm, 36mm, and 46 mm. In the graph with a diameter of 6 mm nozzle, jet velocity of 3.53 m / s up to 9.44 m / s. From graph plotting, gas entrainment tends to rise when the jet velocity is increased, and at 36 mm downcomer diameter, the incoming air is smaller than the 26 mm and 46 mm diameter downcomers. At each different jet height, 36mm diameter downcomers also experience small air entrainment values. For higher jet velocity, air entrainment or air in the vacuum more. For 8 mm and 11 mm nozzle diameters also have the same result with 6 mm diameter nozzle.

Graphs between jet and downcomer velocity to depth penetration, if seen in diameter nozzle 6 decreased to 26 mm and 46 mm downcomer diameter, but on the contrary the 36 mm downcomer has increased. Many anomalies on 36 mm downcomer diameter. Stagnant graph trends are indicated in 8mm and 11mm nozzle diameters

At 6mm nozzle diameter, the bubble spread decreases, due to the water entering the downcomer. For a 46 mm diameter downcomer, the graph declines. The decline occurred in all the conditions of the nozzle, the height of the jet, and the jet velocity. This decrease occurs because, the air is not brought into the water, so the water entrainment is less effective. For graphs with jet velocity and 36 mm downcomer diameter, bubble dispersion area increases in all conditions. With increasing jet velocity, the bubble dispersion area also increases in this condition. And the nozzle diameter values of 6mm, 8mm, and 11mm indicate that jet and downcomer diameter does not significantly affect the dispersion width distribution of bubble.

Based on the graph above shows the experimental data, that the diameter of the downcomer influenced by the volume of air to air entrainment. Jets that hit the surface of the water will form the meniscus, where the meniscus will form an air sac, and burst into bubbles

The effect of the downcomer on bubble production affects the air volume values for air entrainment. Where air entrainment occurs and produces bubbles. In order for bubbles to be produced as desired, it must change the geometry of the experimental apparatus. Changing the nozzle diameter, the height of the jet, the jet velocity, and the diameter of the downcomer.

Graph of the relationship between the water level in the downcomer with the downcomer diameter and jet velocity. The height of the water that goes into the downcomer tends to rise every condition, this is because the difference of downcomers are immersed with the air flow into the downcomer through the feed, consequently water into the downcomer. Another factor is the difference in the diameter of the nozzle, the diameter of the downcomer, and the height of the jet. The result also increases only the difference in height values in each condition. The water that goes into the downcomer is also due to the pressure difference in the downcomer with the pressure in the downcomer environment. The pressure difference gets bigger if the smaller downcomer resulted in much water coming into the downcomer. The incidence of water coming into the downcomer is due to the pressure difference between the downcomer and inside the downcomer, different jet velocity cause different water levels as well. The effect of the large diameter of the downcomer does not cause any change in the number that goes up or down.

4. Conclusions

air entrainment occurs when the jet hit the water level in the pond, the entry of air into hiding due to water jet causes the formation of pouches are then divided into a bubble of air as it goes deeper into the pockets. Jet velocity affects air entrainment, the faster the velocity the more the air gets sucked The diameter of the downcomer can determine the value of a lot of air is sucked, according to (Tasdemir, Tasdemir and Oteyaka, 2011) the diameter of the downcomer does not affect the water entrainment, this is contrary to the results of this study. The velocity of the jet and the downcomer diameter also affect the water that goes into the downcomer. Jet velocity and downcomer diameter do not affect depth penetration. Jet velocity and downcomer diameter do not affect the area of bubble dispersion.

Acknowledgement
The Authors wish to thank Educational Ministry funding us from PITTA and encourage us to publish authors's. Research. Grant number: 2556/UN2.R3.1/HKP.05.00/2018

References

[1] Agrawal, Ashutosh. Wun Jern Ng. Yu Liu.2011. Principle and Application of Microbubble and Nanotechnology for Water Treatment.
[2] Harby, K., Chiva, S., Muñoz-Cobo, J. L. 2014. An Experimental study on bubble entrainment and flow characteristic of vertical plunging water jets.
[3] Chanson, H., Manasseh, R. 2003. "Air Entrainment Proccess in a Circular Plunging Jet: Void-Fraction and Acoustic Measurements". Journal of Fluids Engineering. Vol 125. 910-921.
[4] Zhu,Y., Oğuz, H., and Prosperetti, A. 2000. "On the Mechanism of air entrainment by liquid jets

at free surface". Journal of Fluid Mechanics. 151-177.

[5] Davoust, L., Achard, J., El Hammoumi, M. 2002. "Air Entrainment by Plunging Jet: The Dynamical Roughness Concept and Its Estimation by A Light Absorption Technique". International Journal of Multiphase Flow. Vol 28. 1541-1564.

[6] Amout Smith, A. 2007. "Air entrainment with plunging jets". Thesis, Magister of Science Delft University of Technology.

[7] Biń, A. K. (1993) 'Gas entrainment by plunging liquid jets', Chemical Engineering Science. doi: 10.1016/0009-2509(93)81019-R.

[8] Tasdemir, T., Tasdemir, A. and Oteyaka, B. (2011) 'Gas entrainment rate and flow characterization in downcomer of a Jameson cell', 47, pp. 61–78.

IOP Conf. Series: Materials Science and Engineering **539** (2019) 012025 doi:10.1088/1757-899X/539/1/012025

Two water wheels performance in series for empowerment of irrigation

Asral [1], F Huda [1] and M Akbar [1]
[1] Department of Mechanical Engineering, Universitas Riau Pekanbaru, 28291, Panam, Indonesia

Email: asral@lecturer.unri.ac.id

Abstract. Irrigation system especially in Riau Province are widely available in around of agriculture area. The water flow of irrigation seems to provide continuous energy however many people ignored. The source can be used to drive some water wheels to generate electricity. The head of flow is relatively low with long enough area as the reason many water wheel in series like undershot type was appropriate to apply. This study aims to investigate the possibility of using many water wheels in a canal by installing two units of water wheel together i.e picohydro with series. Then, doubling the amount of energy might be utilized from the resources. For manufacture the corrosion resistant and lightweight material are selected to maintain the wheel from damage. The water wheel was tested from the ground up irrigation keep constant, various level of water to determine the characteristics of water wheel achievement. Power output and efficiency of the water wheel are among other of the results present in this study. Single water wheel produce that the maximum electrical power of about 1300 Watt. However the installation of two water wheel in series caused a decrease the total output power of the system approximately 1065 watt.

1. Introduction

The flow water in small river or irrigation provide the renewable energy but not yet widely used. In last decade, many irrigations system has been developed in Riau province in order to irrigate 250,571 ha of lands. However, the people who has live at around it have a problem with electrical power supply and many cases the irrigation has damaged. The damage lead to water resources has only slightly reach the lands. This condition requires the problem solving because it may affect the people economically and their survival.

The irrigation in good condition are likely to have great potential to be utilized as a power generation. In addition, based on the data from the energy ministry of human resources, throughout of Indonesia's territory available hydropower potential are about 75.5 GW with installed capacity of 4.7 GW [1]. Therefore the potential energy contained in the irrigation is much possible used to drive some small-scale water turbines. While in area with degredaded irrigation might be also develops the spiral water pump. In order to establish both water and electricity issues, the incorporation concept of water turbine with pump are proper devices to developed [9].

The length of irrigation channels were great course also provides a large amount of potential energy as well. Although likely to be able to move small-scale turbines but if made many turbines would generate large amounts of energy. If only every 200 m on irrigation channels built a small-scale plant would have generated thousands of watts of electrical power without disrupting water needs for agriculture. Many studies have been conducted on the utilization of energy as well as setting the energy requirements on small-scale water channel such as efficient water and energy usage for modernization irrigation [2]. Regulation and optimizing the use of water resources for consumption, irrigation and hydroelectric power plant [3]. Then the effort to improve the energy production by water power plant is also a study concern with applied the combined water cycle power generating system [4].

A study that aims to develop a new type of turbine has been delivered by Li *et al* [5]. Turbines are developed according to an open channel area with a small head like a small stream or canal irrigation farming in order to buffer used for hydroelectricity. For the use of water cross flow turbine has undershot type into consideration. As a result, a significant simplification is realized by making the turbine without the casing. However, the flow field of water cross-flow turbine type undershot move somewhat complex with free surface. This means that the depth of the water around the runner turned to variations in rotational speed, and the flow field itself is complex and changes with time. Therefore it is necessary to better understand the flow field around a water turbine with a free surface, in order to improve the performance of a turbine of this type. Furthermore, the performance of the water turbine made determined and the flow field visualized using particle image velocimetry (PIV) techniques. The results showed that, the depth of the water between the outer and inner diameter of the runner decreases when the rotational speed increases. In addition, the speed of fixed-point at a different angle on the inlet and outlet area of the first stage and second extracted.

Pico hydro water turbine is an alternative technology for generating electricity from a hydroelectric plant small size. It is suitable for use in rural areas because of the construction cost and the technology used is conventional. A research report prepared by Keawsuntia [6] presents the design and results of testing of a power plant of water pico crossflow turbine, and to store the energy in batteries 12 V. An examination pico crossflow turbine with turbine wheel diameter of 0.8 m and 20 blade has the shape of a semi-circle diameter of 0.1 m, the length of the blade is 0.8 m, tested in a river. The results show that the system generates a maximum electrical power which is 145.42 watts or 3:49 kW-hours per day, which is enough to supply up to 6 families in a small community in rural areas. While the rate of return for electricity generation is reaching \$ 0023 per kW-hour.

Like a lot of research and studies conducted around the world to explore a variety of alternative and renewable energy sources, this study presents a description of the initial testing performed on the prototype pico hydro generator system for the purpose of investigating the performance [7]. The kinetic energy contained in the flow of water in domestic pipes are known to have the potential to generate electricity for energy storage purposes while performing routine activities such as laundry, cooking and bathing. The water pressure and water flow in the pipe from the main tank utility, which is used for the activities of everyday life, used for small-scale hydro turbines rotate to drive a generator for generating electricity. The test results indicate that the readings are significantly convincing in terms of electrical voltage recorded since it can be calculated to be followed is this system is feasible for electrification aimed at storage of energy and showed good prospects for improvement and further research in the future.

A study was conducted by designing appliances small hydroelectric power plants based on an analysis of the technological characteristics and shortcomings of existing hydropower [8]. As a supplement to conventional power plants, this device can be used in places where the state does not have a dam but it has plenty of water resources. The working principle of the new devices are introduced and the analysis of motion and force are made. Schematic design proves its feasibility through theoretical analysis. Tests on the model showed that the new hydropower has a high efficiency.

In order to explore new energy resources and utilization the irrigation where energy is provided free of charge, the development of small-scale hydro power into a proper object to be investigated. The research will be focused on designing small-sized water wheel rated in accordance with the existing potential based on previous study. Then determine of performance the power generator by testing in a specified place. If possible, some water wheel could be produced when the area is found broad narrow but long enough. Picohydro water wheel can be installed sequentially in the direction of flow is a method that can be done to get the energy from water over the maximum. The method of use of water wheel with more than one in a canal is to be a new thing in this study.

2. Methods

This field study located at Koto Tibun Village. The experiment carried out by various conditions of water wheel placed in the irrigation, namely: the main water wheel, main and second wheel installed in series. The water wheels were fabricated in Mechanics Technology laboratory of Riau University. Second wheel have outside diameter a half than main wheel. Four steps of pully-belt mechanical

IOP Conf. Series: Materials Science and Engineering **539** (2019) 012025 doi:10.1088/1757-899X/539/1/012025

power transmition system are applied. The experimental conditions are described in Figure 1 and Figure 2.

In the main water wheel of the test, parameters measured were rotation, water level, flow, and electrical voltage output. Measurements were conducted for every hour. This works in order to investigate the energy availability of water flow as resources that might affect on water wheel performance.

In the condition of water wheel series, two water wheel are installed sequentially in the direction of flow. One of wheel situated close the entrance side dams, and the second one on behind downstream. Experiment parameters measured equal to the method of one wheel. In throughout study, the data was collected for all conditions respectively.

Figure 1. The arrangement of water wheel apparatus in irrigation

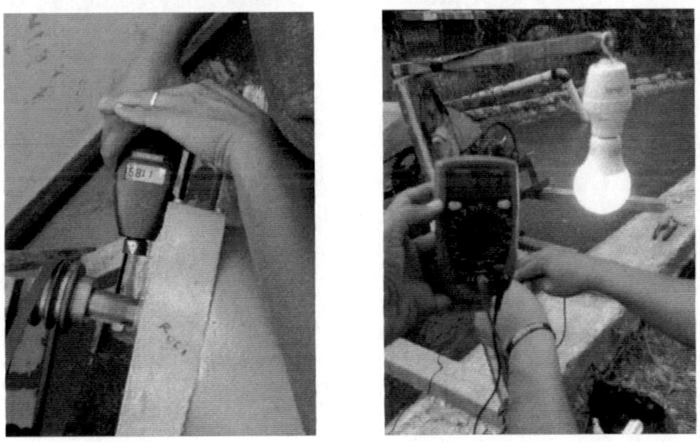

Figure 2. Rotations and Voltages measurement

3. Results and Discussion

In this test, H is the height of the water level was adjusted from the maximum height of water level that can be accommodated in the irrigation channel until the water level could no longer work the water wheel. Meanwhile H is a representation of the state of flow in the irrigation ditch where the height difference (head flow) has remain constant. Then the generator rotation is measure of shaft rotation on the shaft generator. Electricity voltage in volts is the voltage that is measured at the output terminal of the generator. The data collected from the measurements depicted in graphic form as will be described below.

3.1 With Single Water wheel.

The increase in water level resulting in an increases in the rotation of generator, as shown in Figure 3. The image illustrates the relationship between the two parameters of the test for the main water wheel installed single. The increase in the amount of water flowing per second, more and more energy is sourced from water provided for drive water wheel. The overall water will push the wheel blades resulting in a rotation on its shaft is transmitted to the generator. In accordance with the state of irrigation canals on the location of the change in water level can only be achieved as the maximum water level equal to the length of the blade.

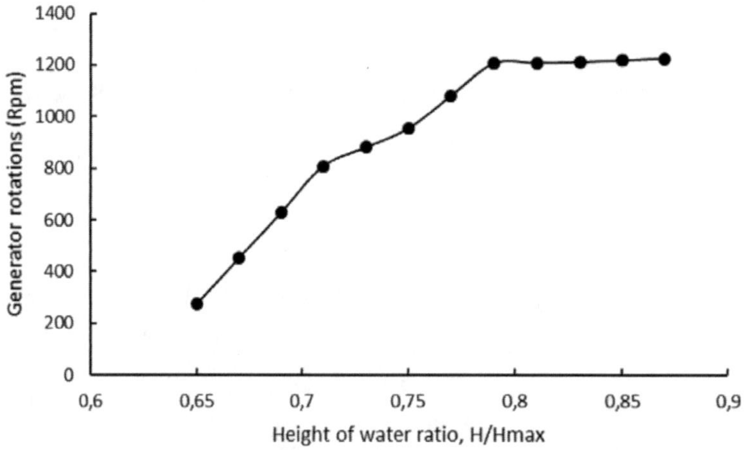

Figure 3. Characteristics of the generator rotation with single wheel to the height of water level

Furthermore, the increase in water level makes the line voltage that is obtained increases, as shown in Figure 4. When the ratio of water height to the maximum channel height, maximum electric generator produces a voltage of about 145 volts. Furthermore, if the ratio of this height is reduced, the electrical energy generated also seem to be significantly until it reaches zero volt. This graph shows that the electrical power can generated by the generator depends solely on the availability of water.

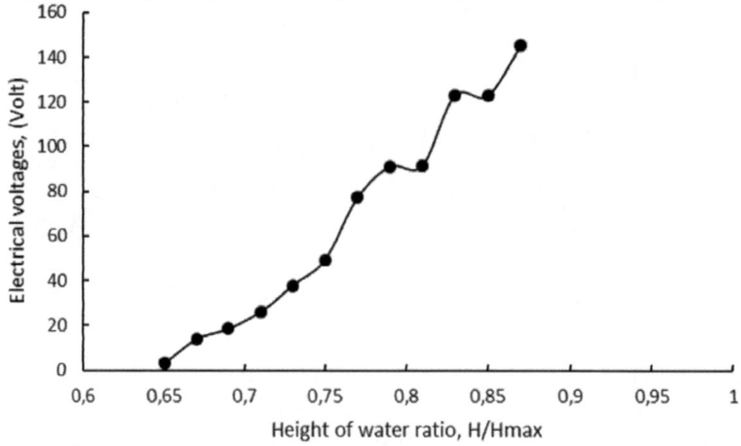

Figure 4. Characteristics of the electrical voltage with single water wheel to the water level

The electricity voltage is a function of the rotation, it is a state that has already prevalent in the science of electricity power. Because the rotation may cause the magnetic field on the plants which were removed in the form of voltage. The data like that depicted in Figure 5, where the relationship between the two in the form of a non-linear function. As apparent from the figure that the maximum voltage is achieved about at 1300 rpm.

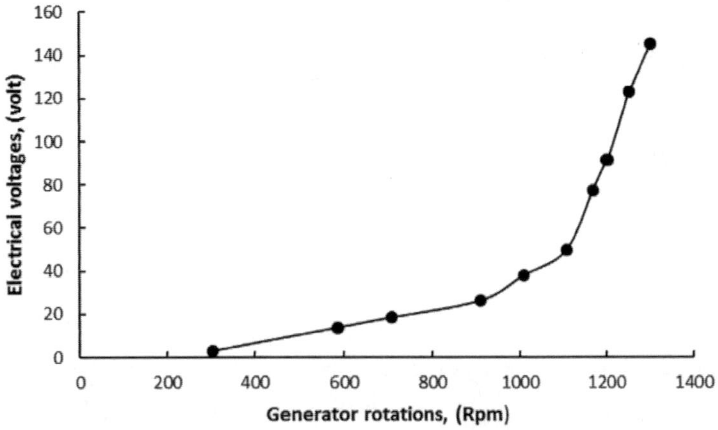

Figure 5. Characteristics of electrical voltage on certain generator rotation with single water wheel

3.2 Two Water wheel in series

Water wheel installed in series its means sequentially towards the flow of running water. As seen in Figure 6, show the characteristics of the main wheel as the height of water level has change as the flow entering the water wheel lap when it installed in series with a second wheel. The trend line shows the same state as water wheel installed itself however seen a decrease in the achievement of the maximum rotation of electricity generator about 1200 rpm. Thisis indicates, the installation of a second wheel though smaller than main water wheel may affect the performance of the main wheel has located at upstream. Water on the downstream side of the main wheel slowdown due to the head flow differences was decreases.

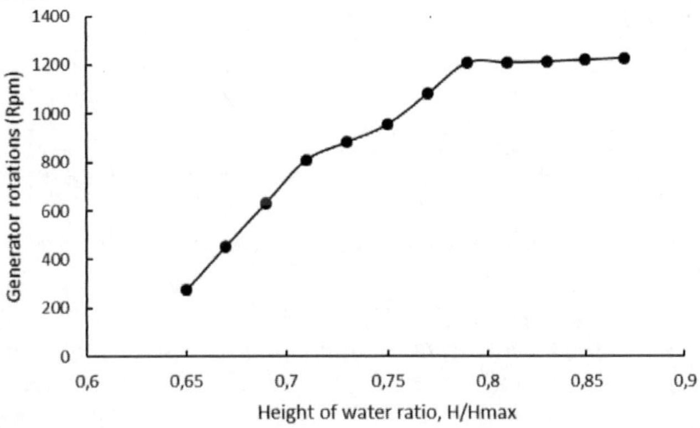

Figure 6. Characteristics of the generator rotation to the height of water level with water wheel in series.

Furthermore, changing the height of the water inlet on the water wheel blades are also cause changes in electrical voltage. However as the main wheel installed in series with the second are shows a decrease in the achievements of its maximum voltage, which in this case is about 115 volts. The data as shown in Figure 7.

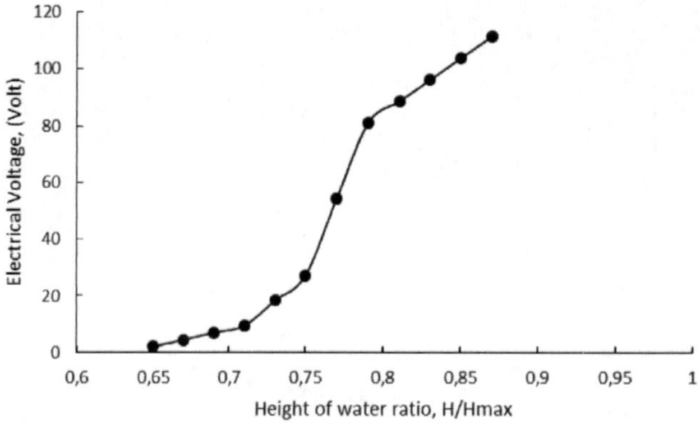

Figure 7. Characteristics of electricity voltage with water wheel in series to the height of water level

The higher the rotation will produce the voltage that is higher as well. As seen in Figure.8, shows the characteristic of rotation against electricity power by the main water wheel on the state installed series with a second wheel. It is evident here that the existence of a second wheel seems to cause a decrease in electricity power output.

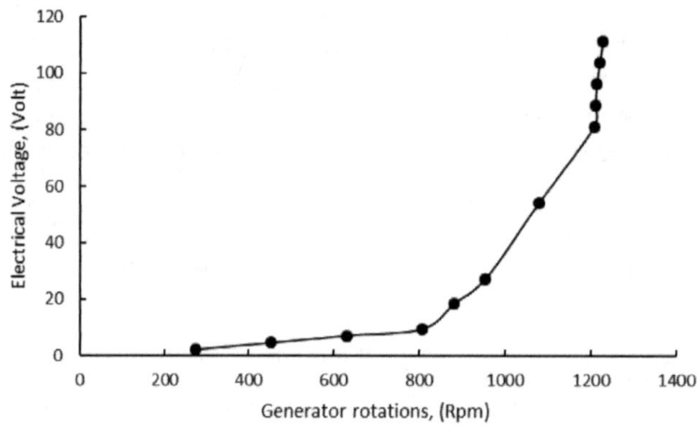

Figure 8. Characteristics of the electricity voltage to the generator rotation with water wheel in series

Meanwhile, when viewed from the side of the second wheel, characteristic changes in the water level of the rotation generator showed the same trend with the main wheel (Figure 9). However in a smaller scale. The rotation decreased significantly when compared with the main wheel. At a height of water ratio of water level achieved maximum at about 0.725, second wheel could cause the generator rotates about 325 rpm. These results indicate that the energy of the water that reaches the wheel was reduced as a part of it has absorbed by the main wheel due to positioned at upstream of flow.

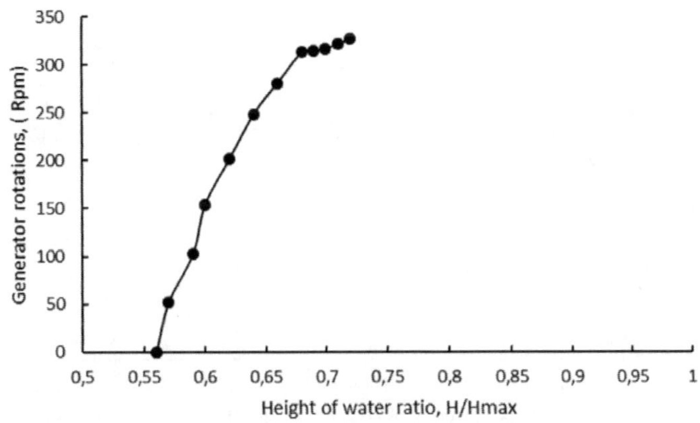

Figure 9. Characteristics of the generator rotation to the height of water level with water wheel in series

Increasing in the height of water ratio causes an increase in the electricity voltage as shown in Figure 10. The ratio of maximum water level can generate the electricity volatage by the second wheel was about 13 volts. This data indicate that as the height of water ratio increase has produce insignificant change on electricity voltage by second water wheel.

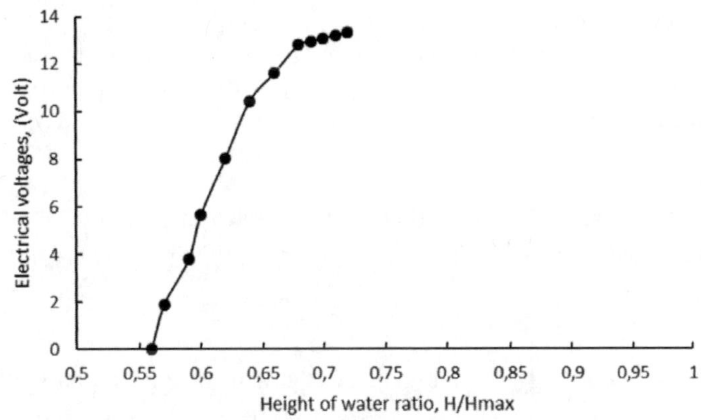

Figure 10. Characteristics of the electricity voltage to the height of water level with water wheel in series

Figure 11 presents the influence of rotation on the electricity voltage output of system with water wheel series. This data showed that the trend of parameters seems to proportional. The data was derived from a rather low rotation so it looks like a linear relationship. As it compared with the result of the main water wheel particularly in condition of low rotation show similarities. Thus it can be predicted here that these symptoms has temporary effect, however as the rotation increased then the trend show as the general relationship between of rotation and voltage.

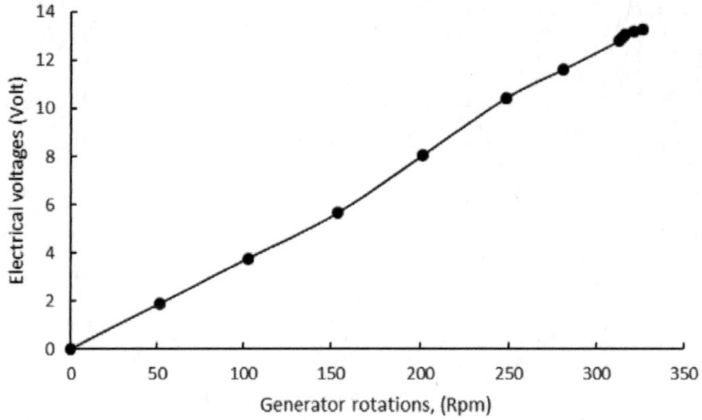

Figure 11. Characteristics of the electricity voltage to the generator rotation with water wheel in series

4. Conclusions

This study reveals the performance of two undershot water wheel installed in series applied particularly on rural irrigation. The conclusion can be drawn are:

1. As electrical generator the water wheel is agree installed in irrigation system more than one sequentially, however take into consideration of spacing between wheel.
2. The maximum power produced with single wheel is about 1300 Watts, however when installed as series are decreases to 1065 Watts.
3. For single and series water wheel conditions, the power are decreases as the water level decreases.
4. The power generation with water wheel of series less than single approximately 23%. Total electricity power produce by water wheel series was 18% lower than single main water wheel.

Acknowledgments

The authors would like to convey deep gratitude to Lembaga Penelitian dan Pengabdian Kepada Masyarakat Universitas Riau (LPPM-UR) which has provide financial support to completion of this research.

References

[1] BPPT, (2012)."Outlook Energy Indonesia". Energy Ministry of Human Resources Republic of Indonesia.

[2] Tarjuelo, José M., Rodriguez-Diaz, Juan A., Abadía, Ricardo, Camacho, Emilio, Rocamora, Carmen and Moreno, Miguel A., (2015). Efficient Water and Energy Use in Irrigation Modernization: Lessons from Spanish Case Studies, Agriculture Water Management, Int. Journal, 162, 67-77

[3] Wu, Yiping, and Chen, Ji , (2013). Estimating irrigation water demand using an improved method and optimizing reservoir operation for water supply and hydropower generation: A case study of the Xinfengjiang reservoir in southern China, Agriculture Water Management, Int. Journal, 116, 110-121

[4] Liu, Yue, and Packey, Daniel J., 2014. "Combined-Cycle Hydropower Systems - The Potential of Applying Hydrokinetic Turbines in The Tailwaters of Existing Conventional Hydropower Stations", Renewable Energy, Int. Journal, 66, 228-231

[5] Li , Yanrong, Nishi ,Yasuyuki, Inagaki, Terumi dan Hatano, Kentarou. 2014 " Study on the Flow Field of an Undershot Cross-flow Water Turbine". Applied Mechanics and Materials Vol. 620, pp 285-291

[6] Keawsuntia, Yuttachai. 2014. "Design and Test of Pico Crossflow Turbine for the Generation of Electricity for Use in the Rural Area". Applied Mechanics and Materials, Vols. 496-500, pp 605-608

[7] Ridzuan, M.J.M., Hafis ,S.M., Azduwin, K., Firdaus ,K.M., dan Zarina, Z. 2014." Development of Pico-hydro Turbine for Domestic Use".Applied Mechanics and Materials, Vol. 695, pp 408-412
[8] Zhang, Lizhen, Cao, Xiaolong, Li, Juntao, and Wu, Shulei. 2012." Design of New Small Hydroelectric Device". Advanced Materials, Vols. 472-475, pp 883-886
[9] Asral, Akbar, Mustafa, and Syafri. 2017. "The Performance of Undershot Water Turbine Combined With Spiral Tube Pump On Empowerment of Energy Resources Local Contiguous Small River" Journal of Ocean, Mechanical and Aerospace-Science and Engineering, Vol. 42, pp 19-23.

IOP Conf. Series: Materials Science and Engineering **539** (2019) 012026 doi:10.1088/1757-899X/539/1/012026

Experimental study of an organic rankine cycle system using r134a as working fluid with helical evaporator and condenser

A Martin[1], Romy[1], D Agustina[1] and A M Ibra[1]

[1]Laboratory of Energy Conversion, Department of Mechanical Engineering, Faculty of Engineering, University of Riau, Indonesia

Email:awaludinmartin01@gmail.com

Abstract. Organic Rankine Cycle (ORC) system is a modification of the conventional Rankine cycle to generate electricity from the utilization of low-grade temperature heat sources. Experiment on ORC systems was conducted to analyse the performance of a small scale ORC system with 1 kW capacity using R134a as working fluid. A helical type of evaporator and condenser were used as heat exchangers and a scroll type car compressor was used to generate power with variation in heat source temperatures of 75°C, 85°C, and 95°C. Turbine power is calculated theoretically based on the experimental data obtained due to the turbine was not able to rotate. This experiment results show the optimum efficiency and theoretical power of ORC is respectively 3.33% and 279.58 Watt with inlet pressure and temperature of turbine are respectively 1.25 MPa and 67.7°C.

1. Introduction

Rapid population growth and industrial development have an impact on increasing world energy demand, especially electrical energy [1]. Fossil energy consumption for steam power plant has been widely used in many countries around the world because of the ease in getting it to suffice human needs in large numbers. About 40% of the electricity generated worldwide comes from power plants with coal fuel [2, 3]. Power plants with larger production capacity require a large supply of coal, while coal supplies nowadays around the world are dwindling due to increasingly higher demand. Coal-burning gas is the largest source of greenhouse gas (GHG) emissions that can lead to climate change problems [3]. Therefore, it becomes the reason for energy sector researchers to be able to lower the dependence on fossil energy and to develop low temperatures heat sources potential that has not been currently practiced optimally [4]. Based on data from the International Energy Agency (IEA) in 2014, the use of coal fossil fuels accounted for 44% of total CO2 emissions worldwide. Coal burned in a steam power plant also produces a number of harmful pollutants such as NOx, SO_2, SO_3, mercury, arsenic, and radioactive particle which are harmful to human health and environment [4, 5]. Air pollution caused by coal combustion leads to an increased risk of lung cancer, stroke, heart disease, premature birth and respiratory diseases [2, 6].

Because of those adverse effects of coal use, the most important step to be taken now is by replacing coal-fired power plants with renewable energy resources such as mini and micro hydro, geothermal, wind energy, solar cell and utilization of industrial waste heat [7]. Moreover, countries around the world are developing the concept of clean energy, especially for power plant, that is Clean Power Plant (CPP) to reduces CO_2 emissions by coal-fired power plants in 2030 by 30 % and encourage more environmentally use of renewable energy [8]. In Indonesia, renewable energy such as geothermal has the greatest potential to be developed. Geothermal energy potential in Indonesia is 40% of the total worldwide but its current use is only 4% [2, 9]. Some causes for the underdevelopment of geothermal power plants are due to high operational costs and the type of most geothermal energy in Indonesia is low-grade temperature. Thus other innovations must be applied to overcome the low temperature problem. Among the solutions is utilization of geothermal energy for power plant with Organic Rankine Cycle System (ORCs).

Organic Rankine Cycle (ORC) system is a modification of the conventional Rankine cycle to generate electricity from the utilization of low-grade temperature heat sources. This technology does not require coal-fire which means no CO_2 emissions are generated. Various studies on the application of the ORC system have been widely studied with varying sources of low temperature heat energy used. Research on possible use of geothermal energy for ORC from multiple hot springs with temperatures of 70-80°C throughout Indonesia using R227ea as working fluid was conducted [10]. Lompio-1 hot springs in Central Sulawesi, which has a temperature of 78°C with the largest mass flow rate of 3000 l / min, is capable of producing a turbine power potential of 130.13 kW. It can be concluded from the research that the mass flow rate from the hot springs has a significant effect on the electrical energy produced. Usman *et al.*and Yuh-Ren *et al.*[4, 11] conducted an experimental design and review for application of ORC system with R245fa as working fluid and utilization of waste heat from steam at temperature of 120°C that capable to generate 1.02 kW of electrical energy and thermal efficiency of 5.64%. The efficiency can be increased by insulating the piping system and increasing the evaporative temperature of the fluid in the evaporator. Yuh-Ren *et al.* [11] conducted experiments to determine the effect of vapor exit of the turbine by using plate-type and shell-and-tube evaporators on performance of ORC system. Results of the experiment showed that ORC system performance is more stable using evaporator with shell-and-tube type than of plate type. Nigusse *et al.* [12] conducted a research to determine the effect of shell-and-tube type condenser on the performance of geothermal power plant with ORC system, which showed that an increase in the flow rate of cooling water in the condenser results in the pressure drop of the working fluid in the condenser. It is thus increasing the incoming work on the pump that leads to decreasing the efficiency of the system. Hung et al. [13] presents a study of the application of a heat source in a ventilation system of solar thermal energy for the development of an ORC system and found that the maximum temperature generated by a solar thermal vent system is 120°C, with an efficiency of 6.2% when utilized for an ORC system. Another study presents the use of biomass heat sources from palm oil mill waste for the development of the ORC system and concludes that it is technically possible to produce higher thermal efficiency than the ORC basic system [14].

The selections of the working fluid employed should also consider the expansion component or turbine type to be used for the ORC system. The consideration must include the capacity, cost and complexity of manufacture and maintenance [15] as shown in Table 1 below.

Table 1. Suitable Expansion Component Type for ORC System [15]

Type	Capacity range (kW)	Rotate speed (rpm)	Cost	Advantages	Disadvantages
Radial-inflow turbine	50–500	8000–80,000	High	Light weight, mature manufacturability and high efficiency	High cost, low efficiency in off-design conditions and cannot bear two-phase
Scroll expander	1–10	<6000	Low	High efficiency, simple manufacture, light weight, low rotate speed and tolerable two-phase	Low Capacity, lubrication and modification requirement
Screw expander	15–200	<6000	Medium	Tolerable two-phase, low rotate speed and high efficiency in off-design conditions	Lubrication requirement, difficult manufacture and seal
Reciprocating piston expander	20–100	–	Medium	High pressure radio, mature manufacturability, adaptable in variable working condition and tolerable two-phase	Many movement parts, heavy weight, have valves and torque impulse
Rotary vane expander	1–10	<6000	Low	Tolerable two-phase, torque stable, simple structure, low cost and noise	Lubrication requirement and low capacity

Oralli *et al.*[16] experimented with using a compressor scroll as an expander on the ORC system to observe the system performance, which concluded that the expander did not work optimally because there was no modification of the original function of the compressor for the turbine. Eicke and Smolen [17] presents a study of the trial use of a commercial scroll expander (Air Squared Inc., E15H22N4.25) for a 1 kW ORC system using the working fluid of R245fa and found that 17 kW of power was generated. The uses of commercial turbine expander with more promising performance are good but generally require larger cost.

From several studies that have been mentioned above, it is known that there has been not sufficient research on the ORC system with a variety of heat sources and types of heat exchangers used. More previous researches applied the use of heat exchangers with shell-and-tube type and plate type. Therefore, the use of heat exchangers with helical type for ORC system needs to be conducted in

IOP Conf. Series: Materials Science and Engineering **539** (2019) 012026 doi:10.1088/1757-899X/539/1/012026

experiments to obtain more information. This study will describe the performance of ORC system that using helical type heat exchangers with a capacity of 1 kW and working fluid of R134a. To function as expansion turbine, a car AC compressor with scroll type that has never been tested for a laboratory scale ORC is used for its low cost. This experiment will also observe the effect of temperature variations of the heat source to the inlet temperature and the inlet pressure of the expander, then calculate its effect on the efficiency of the ORC system.

2. Methodology

This section describes experimental set up, experimental preparation, and data collecting.

2.1. Experimental set up

Power plant with the ORC system has been designed and manufactured at Energy Conversion Laboratory, Department of Mechanical Engineering, University of Riau. ORC system has a generator capacity of 1 kW, the heat source is obtained from heated water by heater in evaporator and air conditioning system to cool water as cooling system on condenser. To keep the temperature constant in the heat exchanger, temperature controller Autonic TC4S was used to regulate heater and AC system. Working fluid used for ORC and AC systems were respectively R134a and R22. Temperature data recording was performed using K type Thermocouple that connected to data acquisition (DAQ) USB TC-08. Pressure measurements were performed manually using refrigerant pressure gauge Clause 1.6 that has measurement accuracy of 0.1 bar, fluid flow rate measurements were performed manually using water flow meter of Wierbock brand that has a measurement accuracy of 0.5 LPM.

Turbine expander used scroll type car compressor Keihin HS-090R to generate power. Heat exchanger helical (condenser and evaporator) and piping system of copper tube with 3/8 inch diameter were manufactured. ORC system used gear pump for ammoniac fluid with capacity of 1.08 m3 / h, maximum pressure of 2.5 MPa, maximum head 25 meter and driven by Modern JY1A-4 1 pasha electric motor with power ½ HP, current 4.24 Ampere, voltage 220 V and maximum rotation of 1400 rotation per minute (RPM). The gear pump serves to drain the working fluid of the ORC system in order to rotate the turbine expander. Schematic and manufactured of the ORC systems can be seen in Figure 1.

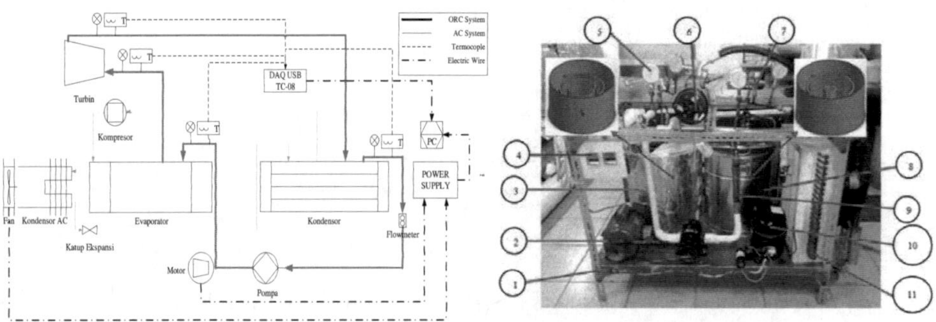

Figure 1. Schematic and installation of ORC system.

Annotation:

1. Gear pump
2. Electric motor
3. Condenser
4. Temperature controller
5. Pressure gauge
6. Thermocouple
7. Expansion turbine
8. Evaporator
9. Flowmeter
10. AC system compressor
11. AC system condenser

2.2. *Experimental preparation*

Prior to the experiment, it is necessary to prepare the ORC system first to obtain accurate data. Water inside the evaporator tank that will be used as a heat source must to be heated by using a heater to reach a temperature in accordance with the experiment to be performed. Meanwhile, the AC system is also turned on to cool the water inside the condenser tank until it reaches a temperature of 10°C. Then turn on the temperature controller to keep the water temperature constant at the heat exchanger. When the desired water temperature of the heat exchanger has been reached, the pressure inside the system is lowered by a vacuum pump to a pressure of -30 bar and after that R134a is inserted into the ORC system slowly until the pressure is 5 bars. The last step is to turn on DAQ TC-08 then make sure the thermocouple can read correctly the temperature at each experiment point and set the time of recording temperature data for 20 minutes with a time interval of 1 second. Data collected can be start when all the experiment preparations have been completed.

2.3. *Data collecting*

Data collecting process can be started by running the gear pump and recording temperature data by DAQ TC-08. Data collecting of pressure is recorded manually by noting the value on the pressure gauge at each experimental point every 30 seconds. The same procedure is done to determine the rate of fluid flow flowing through the flow meter. During the data collection process, keep the heater and AC system on to condition the water temperature in each heat exchanger. The data is collected for 20 minutes until the temperature data recording on DAQ TC-08 ceased and after that the pump is turned off. The resulting temperature data will then be averaged every 30 seconds to adjust to the time of the pressure data and the working fluid flow rate. Data collecting process can be repeated using different temperature variations of different heat sources by resuming from the experiment preparation stage.

3. Results and Discussion

Experimental study was carried on a completed ORC system to determine its performance with variations of heat source temperature of 75°C, 85°C, and 95°C. After the data collecting, the quality of fluid and enthalpy at each experiment point will be checked using REFPROP 9.0.Turbine power is calculated theoretically based on the experimental data obtained because the turbine is not able to rotate. From the experiment, various heat source temperatures resulting in different efficiency of the ORC system. Efficiency of ORC derived from the 75°C heat source temperature can be seen in Figure 2.

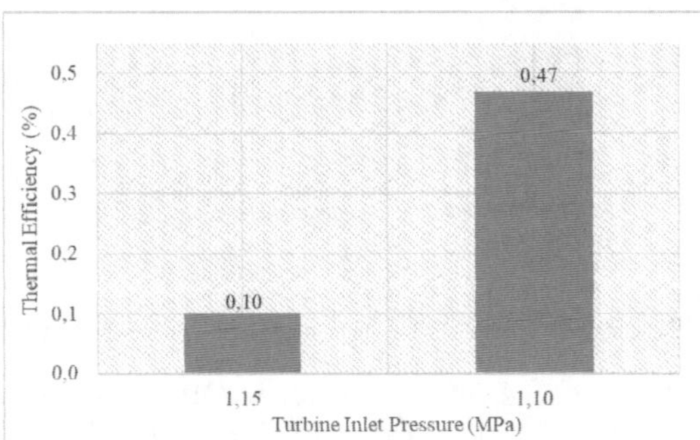

Figure 2. Efficiency of an ORC system at heat source temperature of 75°C

Figure 2 displays the system performance of ORC efficiency at the heat source temperature variation of 75°C. There were only two valid data of the total data obtained during the experiment that can be used in calculation of the efficiency. At heat source temperature variation of 75°C, the highest efficiency that ORC system can produce is 0.47% at turbine pressure and temperature of 1.1 MPa and 57 C; while the increased turbine inlet pressure also increases the efficiency of ORC system. This condition occurred because condenser did not work optimally to reject heat from the working fluid after exiting the turbine thus the fluid quality is still superheated after passing the condenser. This is not desirable because it will cause a decrease in the pump's ability to increase the fluid pressure flowing into the evaporator. Decreasing in evaporative pressure will affected on exit pressure of evaporator to further enter the turbine and decrease the turbine work output and efficiency of the ORC system.

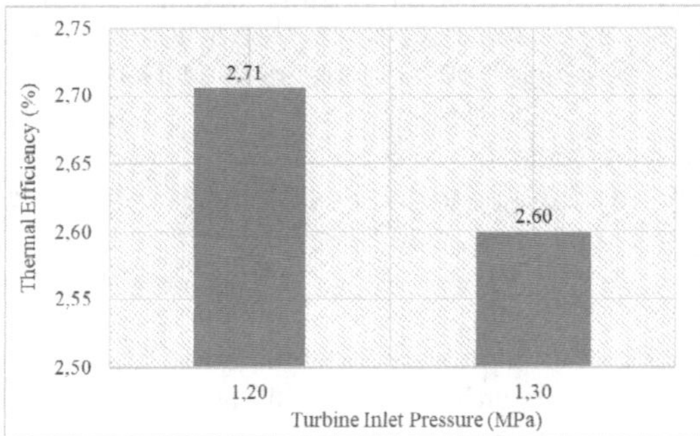

Figure 3. Efficiency of an ORC system at heat source temperature of 85 °C

Figure 3 displays the efficiency of the ORC system at 85°C heat source temperature which obtained the highest efficiency of 2.71% at turbine inlet pressure and temperature of 1.2 MPa 62.33°C, respectively. During the experiment, efficiency of the ORC system as whole was decreased when turbine inlet pressure increased to 1.3 MPa. This phenomenon is caused by temperature of the heater water that did not conditioned which continued to decrease caused by the working fluid that has a cooler temperature than hot water, thus resulted a large temperature difference at the evaporator. It increases the heat input on evaporator. During the same process, the two different fluid temperatures in the evaporator also attempt to reach the equilibrium temperature, in addition to the heat wasted to the environment due to an evaporator that was not well insulated.

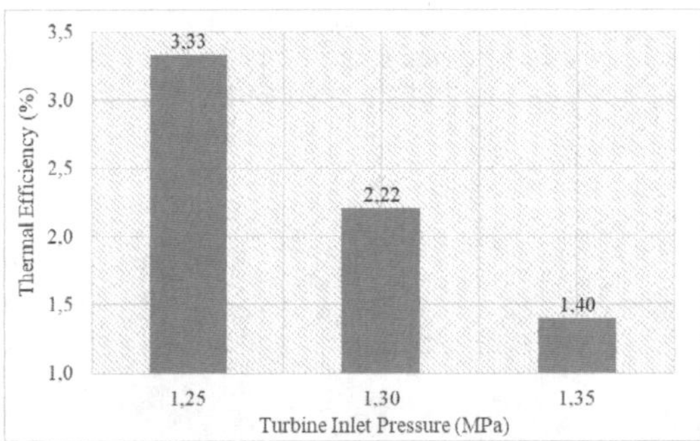

Figure 4. Efficiency of an ORC system at heat source temperature of 95 °C

Figure 4 shows the effect of turbine inlet pressure of ORC system efficiency at heat source temperature of 95°C which is the highest temperature variation used in this experiment. From the graph shown in Fig 4, the efficiency of the ORC system is fluctuating and unstable. At the beginning of the data collection, efficiency increased to an optimum point at 3.3% and turbine inlet pressure of 1.25 MPa. After that, turbine inlet pressure increases up to 1.35 MPa but overall ORC system efficiency has decreased significantly to 0.59%.This is due to increased turbine inlet pressure also resulting in an increase of fluid flow temperature thus causes condenser heat output to increase, decreasing on net heat of the system and also decreasing in the thermal efficiency of ORC system.

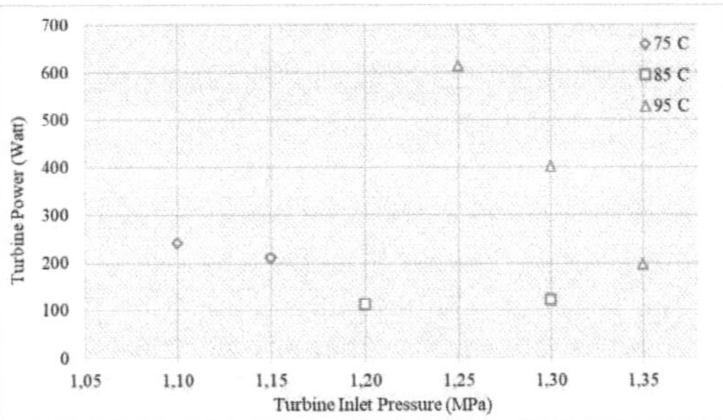

Figure 5. Turbine pressure effect on turbine theoretical power

Figure 5 displays the effect of turbine inlet pressure on the turbine power of the ORC system. From the experimental data results with variations of the heat source temperatures, turbine inlet pressure increases with time. However, it makes the power generated by the turbine to decrease. This phenomenon is caused by turbine power that is theoretically also determined by the pressure and temperature of the steam after passing the turbine. During the experiment, the temperature of the working fluid entering the turbine continues to increase, but the turbine is unable to reduce the pressure and temperature after steam expansion, that caused steam pressure and temperature after exiting the turbine is still high which affects the enthalpy value generated. The difference in enthalpy becomes smaller if turbine temperature increases without turbine ability to decrease pressure, then the longer operation time of ORC system, the more reduce in turbine power will be.

The highest efficiency of experimental data was idealized using Cycle Tempo to determine the percentage of successful experimental results based on ideal conditions. The comparison results of these two conditions can be seen in Fig 6 below.

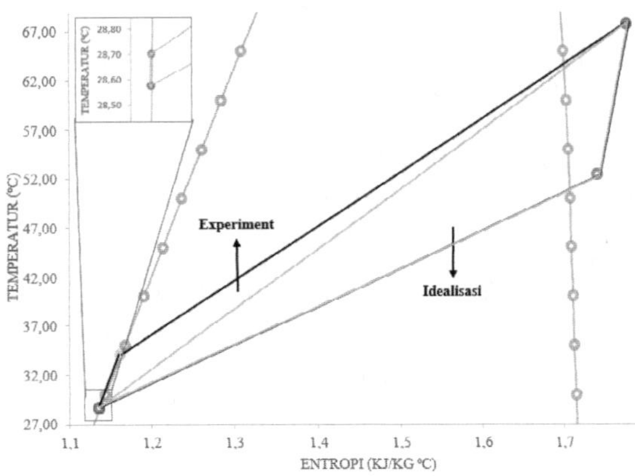

Figure 6. Comparison of experiment result performance and idealization result

Figure 6 shows the T-s diagram of the comparison of highest experimental results and idealized results using Cycle Tempo. It can be seen that the resulting performance idealization higher than the experimental results. After the idealization, it is known that the ORC system produces a thermal efficiency of 6.95%, which means that the best experimental results can only reach 52.1% of the ideal state. This is caused by the situation where the experimental data results showed an increase in fluid temperature through the pump which led to the required pump power to raise the fluid pressure. Some things that could be the cause of the rise in working fluid temperature is that the condensation process did not occur optimally in the condenser thus the pumped R-134a has a mixed quality of liquid and gas phase. Another factor that likely contributed to the experimental result is the ORC system that was designed using an ammonia pump. R-134a has smaller density from ammonia, thus the pump did not working optimally and requiring greater power to increase the pressure. In order to avoid losses due to the work of the pump, pump designed specifically for fluid refrigerant should be used.

4. Conclusion

Research on ORC system with low temperature utilization temperature using R134a working fluid has been conducted with temperature variation of heat source of 75°C, 85°C, and 95°C and cooling water temperature stabilized at 10°C. Turbine power is calculated theoretically based on the experimental data obtained because the turbine was not able to rotate. This experiment resulted the optimum efficiency and theoretical power of ORC are respectively 3.33% and 279.58 Watt, inlet pressure of turbine are 1.25 MPa using heat source temperature of 95°C. However, the manufactured ORC system produces stable power if the heat source temperature is 85°C, meaning that under those conditions R134a used for the ORC system can work optimally. The highest experimental results realized with Cycle Tempo are known to achieve only 52.1% efficiency of ideal condition efficiency.

References

[1] Martin A, Miswandi, Prayitno A, Kurniawan I, Romy 2016 International Journal of Technology5: 530-536
[2] World Coal Institut 2005 *A full review of coal resources* (www.worldcoal.org)
[3] Greenpeace Indonesia 2015 *Threat of deadly coal power plant* (greenpeace.org/seasia/id)
[4] Usman M, Imran M, Dong Hyun L, Byung Sik P 2015 *Energy Conversion and Management*103 (2015) 1089–1100
[5] Shahzad Baig K, Yousaf M 2017 *Journal of Earth Science & Climatic Change* Vol 8:7 1000404
[6] Ehsan Munawer M 2017 *Journal of Sustainable Miningxxx* (2017) 1-10
[7] Wakeyama T, Ehara S 2011 *World renewable enegy congress* (Linkoping, Sweden)
[8] Environmental Protection Agency (EPA) 2014 Vol 79 no 117
[9] Greenpeace 2015 *The High Cost of Coal Power: Internalizing Health Impacts and Cost from Coal-Fired Power Plants in Indonesia*(greenpeace.org)
[10] Rohmah N, Pikra G, Indra Praman R, Joko Purwanto A 2016 International Conference on Sustainable Energy Engineering and Application (ICSEEA)978-1-5090-2899-3/16
[11] Yuh-Ren L, Chi-Ron K, Chih-His L, Ben-Ran F, Jui-Ching H, Chi-Chuan W 2014 *Energies*7, 2436-2448
[12] Araya Nigusse H, M Ndiritu H, Kiplimo R 2014 *Int. Journal of Engineering Research and Applications* Vol. 4, Issue 8(Version 3) pp.141-147
[13] Tzu-Zhen H, Duen-Sheng L, Jaw-Ren L 2014 *International Journal of Photoenergy*239137
[14] Nur T B, Pane Z, Amin M N 2017*1st Annual Applied Science and Engineering Conference* 180 012035
[15] Bao J, Zhao L 2013 *RenewableandSustainableEnergyReviews*24325–342
[16] Eicke A, Smolen S 2015 *3rd International Seminar on ORC Power Systems* (Brussels, Belgium)
[17] Oralli E, Ali Tarique M, Zamfirescu C, Dincer I 2011 *International Journal of Low-Carbon Technologies* 6 200–206
[18] Douvartzides S and Karmalis I 2016*20th Innovative Manufacturing Engineering and Energy Conference (IManEE 2016)*IOP Cof. Series161 012087

IOP Conf. Series: Materials Science and Engineering **539** (2019) 012027 doi:10.1088/1757-899X/539/1/012027

Design and manufacturing of organic rankine cycle (orc) system using working fluid r-134a with helical evaporator and condenser

A Martin[1], Romy[1], D Agustina[1] and A M Ibra[1]

Laboratory of Energy Conversion, Department of Mechanical Engineering, Riau University, Pekanbaru, Indonesia

Email: awaludinmartin01@gmail.com

Abstract. Waste heat from industries such as steam power plants can be utilized to meet the world's electricity need. However, it cannot be converted efficiently to electric power by using conventional power generation methods as opposed to a technology called Organic Rankine Cycle (ORC). ORC is one of the power plant systems that is a modification of the Rankine Cycle using organic working fluid. In this research, ORC was designed and manufactured using R-134a as working fluid and helical heat exchanger as evaporator and condenser. The design of ORC systems based on simulation using Cycle Tempo 5.0 with evaporator temperature of 90°C, condenser temperature of 10°C, inlet pressure of pump of 0.55 MPa, and inlet pressure of turbine of 0.79 MPa. The results of this study showed the length of tube evaporator and condenser of 10.61 m and 10.56 m, respectively, with system efficiency achieved at 3.8%. Based on experiment, with evaporator temperature of 95°C, condenser temperature of 10°C, inlet pressure of pump of 0.48 MPa and inlet pressure of turbine of 0.52 MPa so resulted of system efficiency of 3.33%.

1. Introduction

Indonesia is one of the countries that has great geothermal resources potential, but a lot of the natural resources has been used to meet energy needs [1]. Human needs for energy will increase because of the high population growth, economic growth, industrial expansion and the advancement of technology used by humans [2]. It causes less energy on earth. Worldwide energy consumption has increased by more than 40% over the last few decades and causes significant environmental problems and economic value. The misused energy from fuel derived from engine exhaust is more than 30-40% and only about 12-25% can be converted for useful work [3].

Generally, energy and non-renewable energy sources. The fulfillment of energy needs so far has been focused on usage of non-renewable or fossil energy sources, such as petroleum, coal and natural gas which are currently in critical condition with limited reserves and were known had a negative effects on the environment such as the global warming potential, ozone depleting potential and acid rain [4][5]. Therefore, needed an effort to found alternative energy sources that can divert energy consumption from fossil energy which is more efficient and friendly to the environment [6].

Utilization of renewable energy sources such as wind energy, biomass, geothermal and waste heat is a way to solve the problems. In recent years, intensive research has been done on technology that can be converting for low temperature and low-pressure to produce energy. The heat source can be obtained from various thermal processes, one of them is a waste heat recovery [7].

Waste heat recovery from industries such as steam power plants has the potential to reduce dependence on fossil energy and to meet the needs of electricity in the world. However, the low temperature of the source could not be converted efficiently into electricity when using conventional power generation methods, so in this study, the heat converting technology can be realized in an Organic Rankine Cycle system [8][9].

Organic Rankine Cycle (ORC) was accepted as a technology to converting low-temperature to electrical power to ease the operation, low maintenance, and good operating pressure. The vapor

IOP Conf. Series: Materials Science and Engineering **539** (2019) 012027 doi:10.1088/1757-899X/539/1/012027

produced by the working fluid could be used to drive the turbine [10]. Generally, grade of the temperature is classified by temperature range, there are three classifications of it, such as high temperature (>650°C), medium temperature (230-650°C) and low-temperature (<230°C) [10]. Thus, the ORC system will be operated under the 230°C temperature.

Organic Rankine Cycle (ORC) is one of the power generation systems that is a modification of Rankine Cycle using organic working fluid (refrigerant and Hydrocarbon) [11]. Organic Rankine Cycle (ORC) has four major components, they are evaporator, turbine, condenser, and pump. The working fluid is pumped into the evaporator to be heated to generate the steam, the vapor is used to drive the steam turbine, the turbine spin result will be used to drive the generator and generate the power source, the turbine expansion vapor is condensed and flowed by the pump back to the evaporator [1][12].

There are much varieties of organic working fluids, which has the advantages and disadvantages respectively. Classification of working fluid based on the thermodynamics, environmental and safety criteria also availability in research region [11][12].

With this study, it is hoped that it can be used as one of the references for advanced research or other research that want to design this ORC system to produce bigger electric energy capacity.

2. Energy Analysis of Rankine Cycle System

Figure 1a shows the schematic diagram of basic Rankine Cycle. In ORC, a boiler is replaced by heat exchanger or evaporator. In this study, heat exchanger that used is shell-and-tube evaporator and condenser with helical tube.

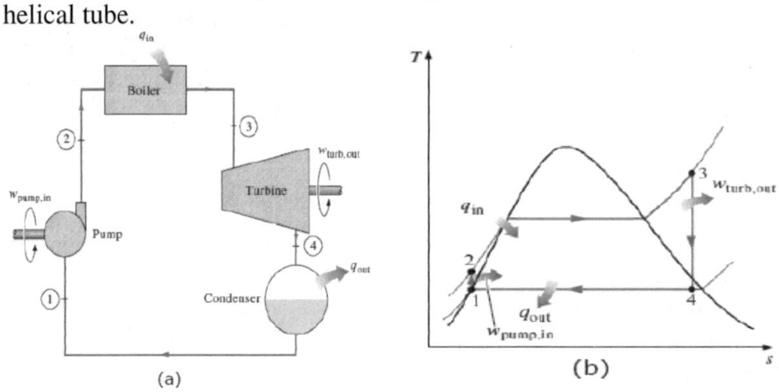

Figure 1. Schematic diagram of the Rankine Cycle (a); T-s diagram of ORC (b)

The ORC system diagram is shown in Figure 1b, the process of ORC starts from the fluid is pumped into the evaporator and occur the heat transfer process between fluid and hot water as the heater, the fluid will be changed to vapor condition, the steam will drive the turbine, with isentropic expansion that means the temperature and pressure will decrease, the fluid flows to the condenser, in the condenser the steam will be change to liquid condition because of the heat transfer process between fluid and cold water. Process of ORC is described as follow:

1. Process 1-2: Isentropic compression in a pump.
2. Process 2-3: heat addition in a evaporator.
3. Process 3-4: isentropic expansion in a turbine.
4. Process 4-1: heat rejection in a condenser

IOP Conf. Series: Materials Science and Engineering **539** (2019) 012027 doi:10.1088/1757-899X/539/1/012027

Energy analysis of the ORC will be describe as follow:

$$\eta_{th} = \frac{w_{net}}{q_{in}} = \frac{w_t - w_p}{q_{in}} = \frac{\dot{m}(h_3 - h_4) - (h_2 - h_1)}{\dot{m}\,(h_3 - h_2)} \tag{1}$$

3. Working Fluid Selection

The selection of working fluid is an important decision criteria for the performance and reliability of the design procedure of an Organic Rankine Cycle power plant. The characteristics of the selected Organic Rankine Cycle (ORC) working fluid must be compatible with thermodynamic characteristics, high temperatures stability, non-toxic, non-corrosive, appropriate pressure (evaporators and condensers) and their environmental impact (related to ODP and GWP) [11]–[13].

In general, working fluids are classified into three categories, namely wet fluids, isentropic fluids and dry fluids. The classification is based on the value of E which is the opposite of the slope of the steam saturation curve. The fluids having an E>0 are classified as wet fluids. Fluid having E<0 are classified as dry fluids. The fluid having the value E ≈ 0 is classified as isentropic fluid [11]–[13].

$$E = \frac{C_p}{T_H} - \frac{\frac{nT_{rH}}{1-T_{rH}} + 1}{T_H^{\,2}} \Delta H_{vap} \tag{2}$$

where:

E = ds/Dt
Cp = specific heat of the fluid at constant pressure (kJ/kg)
Th = evaporation temperature of the fluid (˚C)
n = 0.375 or 0.385
Trh = decreased of evaporation temperature (˚C)
ΔHvap = latent heat (˚C)

In addition, the classification of working fluids based on safety to the environment is as follows:

1. Based on Ozone Depleting Potential (ODP)
   a. Medium, ODP <0.1
   b. High, ODP> 0.1
2. Based on Global Warming Potential (GWP)
   a. Low, GWP <150
   b. Medium, GWP between 150 - 2500
   c. High, GWP> 2500
3. Based on ASHRAE Safety 34
   a. A, non-toxic
   b. B, toxic
   c. 1, not flammable
   d. 2, potentially flammable
   e. 3, flammable
   f. L, flammable

Based on Karmalis [11], we classified 10 candidate working fluids that would choose to use in this study as shown in table 1. This classification based on the thermodynamical criteria of working fluid.

Table 1. Thermodynamics Criteria

ASHRAE code	Tcr (° C)	Pcr (MPa)	Type of fluid
R123	183,381	3,6618	Dry
R124	122,245	3,6242	Wet
R134a	101,21	4,059	Isentropic
R152a	113,41	4,516	Wet
R21	178,48	5,181	Wet
R227ea	101,76	2,926	Wet
R245fa	154,2	3,64	Dry
R290	96,825	4,2476	Isentropic
R32	78,25	5,782	Wet
R600	152,125	3,796	Dry

Classification based on the characteristics of the working fluid that was described in Quoilin [12], such as:

1. Efficiency of output power should be higher than heat source.
2. Working fluid inlet turbine in positive or isentropic saturation vapor curve.
3. High environmental safety, it means value of ODP, GWP, the toxicity, and the flammability should be as low as possible
4. Good availability and low cost.

Based on the type of fluid, working fluid of R123, R134a, R245fa, R290, and R600 can be used in Organic Rankine Cycle system because they had positive and infinitive vapor saturation curve. In addition, working fluid based on environmental and safety criteria classifications shown in Table 2:

Table 2. Environmental and Safety Criteria

ASHRAE code	ODP	GWP	Safety
R123	0,02	77	B1
R124	0,022	609	A1
R134a	0,055	1430	A1
R152a	0	124	A2
R21	0,04	151	B1
R227ea	0	2900	B1
R245fa	0	1030	B1
R290	0	3,3	A3
R32	0	675	A2L
R600	0	4	A3

From the table above, R290 and R600 has a low value of ODP and GWP, however they had high environmental impact. From the safety criteria, R-134a and R124 can be used in ORC system. We have been selected 1 of 10 working fluid and selected the working fluid is R-134a, cause of availability in market.

4. Methodology

In this study, we used the method of design and manufacturing that performed for the design, manufacture, and experiment of the ORC systems. The schematic of systems described in Figure 2. In this figure, there are two cycles namely ORC shown by red lines and Refrigeration cycle shown by blue lines. In this ORC system, refrigeration cycles was used only to hold isothermal condition of water in condenser. Each step of the systems was equipped with pressure gauge and thermocouple to measure pressure and temperature of fluid in a tube when system operates. In addition, a flow meter was used to measure the mass flow rate of the fluid.

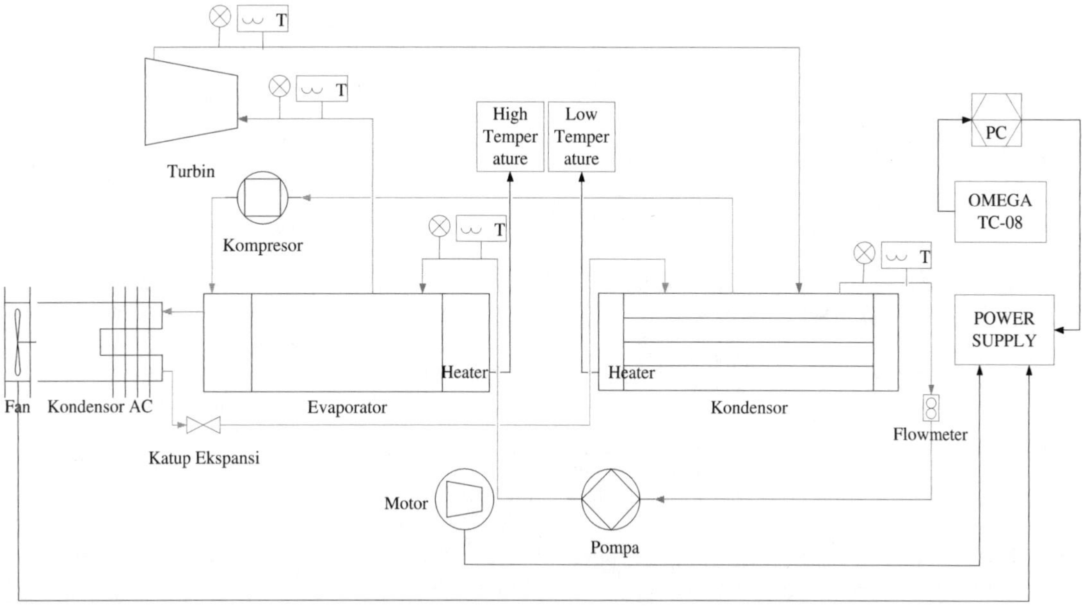

Figure 2. Schematic of ORC System

5. Design Parameter

The design of evaporator and condenser was calculated using the equation that described previously by Cengel [14] as follows:

$$Q_h = Q_c = \dot{m}.\Delta h = U.A_s.\Delta T_{lm} = U.A_s.\frac{\Delta T_1 - \Delta T_2}{\ln\left(\frac{\Delta T_1}{\Delta T_2}\right)} \quad (3)$$

To determine the rate of heat transfer for laminar flow is

$$h = \frac{k}{D}Nu = \frac{k}{D}\left(3{,}66 + \frac{0{,}065\left(\frac{D}{L}\right)Re\,Pr}{1+0{,}04\left[\left(\frac{D}{L}\right)Re\,Pr\right]^{\frac{2}{3}}}\right) \quad (4)$$

and for the turbulent flow is

$$h = \frac{k}{D}Nu = \frac{k}{D}\left(0.023 \times Re^{0.8}Pr^{1/3}\right) \quad (5)$$

IOP Conf. Series: Materials Science and Engineering **539** (2019) 012027 doi:10.1088/1757-899X/539/1/012027

6. Result and Discussion

Design simulation in Cycle Tempo 5.1 was described in Figure 3:

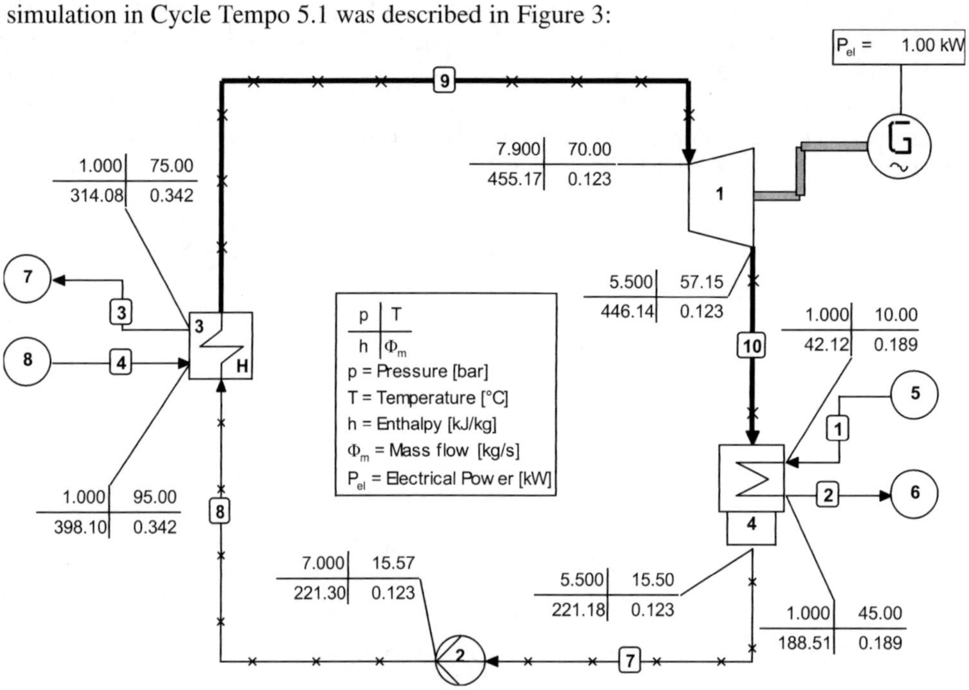

Figure 3. Design Simulation of ORC

Based on the data obtained from Figure 3, then to determine the design of evaporator and condenser was used design parameter equation as Eq. 3, Eq. 4, and Eq. 5 with other parameters as follow:

1. Power generator : 1 kW
2. Do tube : 0.009525 m
3. Di tube : 0.008407 m
4. U : 300 W/m^2.°C
5. Rf,i : 0.0002 m^2.°C /W
6. Rf,o : 0.0004 m^2.°C /W
7. K copper : 385 W/m.°C
8. Flow type : Parallel-flow
9. HE type : Shell-and-tube
10. Type of tube : Helical tube

The result of design shown in Table 3, helical tube of evaporator and helical tube of condenser shown in Figure 4 as follow:

Table 3. Data from the design

Tube	Length (m)	Helical diameter (m)	Pitch (m)	Number of Coil	Height of helical (m)
Evaporator	10.61	0.25	0.0222	13.5	0.3
Condenser	10.56	0.25	0.0222	13.5	0.3

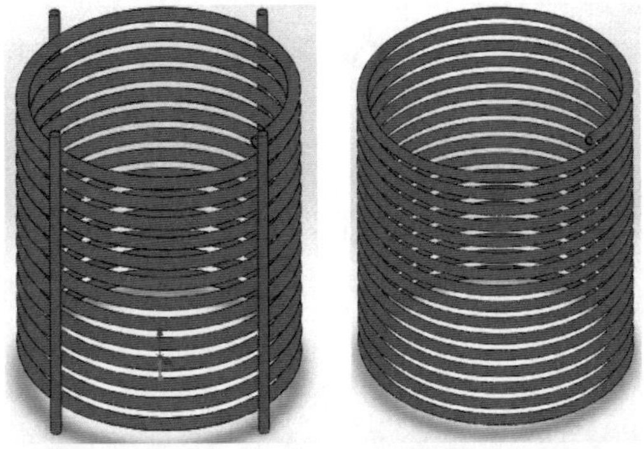

Figure 4. Helical Evaporator (a); Helical Condenser (b)

Thus, final result of this ORC system design shown in Figure 5 and Table 4 as follow:

Figure 5. ORC system

Where:

1. Pump
2. Motor
3. Condenser
4. Temperature Controller
5. Pressure Gauge
6. Thermocouple
7. Compressor as turbine
8. Evaporator
9. Flowmeter
10. Compressor for refrigeration cycle
11. Refrigeration condenser

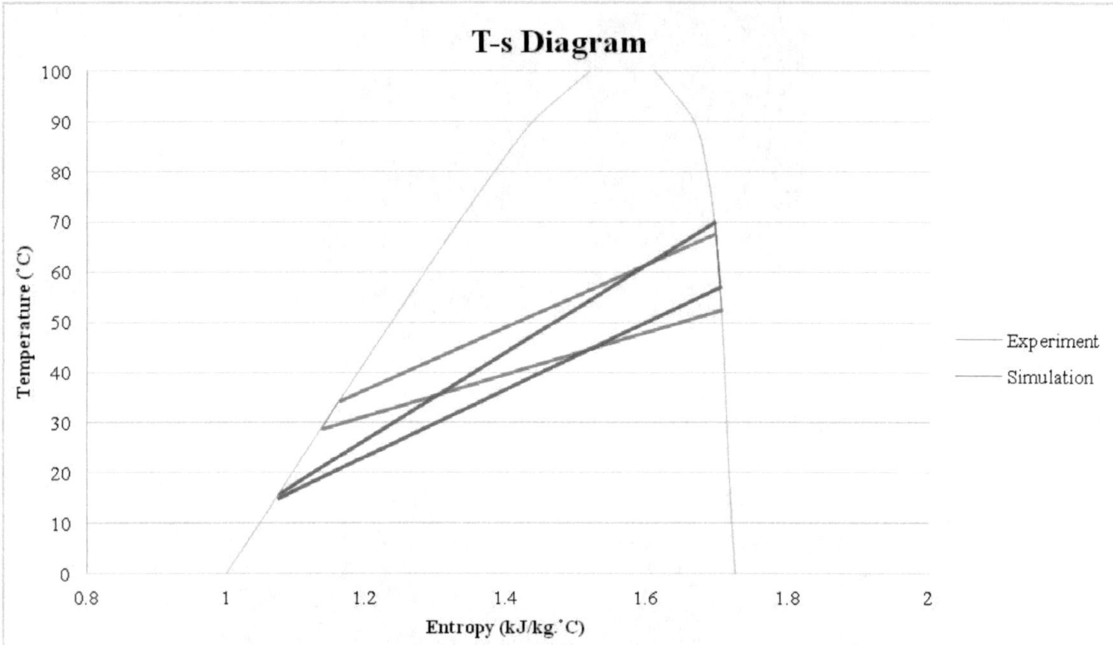

Figure 6. T-s Diagram of R134a

There is a difference between the design and experiment results, where in the design, the inlet pressure of pump of 0.55 MPa and the inlet pressure of turbine of 0.79 MPa was resulted the efficiency of 3.8%, whereas in experiment, inlet pressure of pump of 1 MPa and inlet pressure of turbine of 1.25 MPa, the efficiency of 3.33%. We can see in the figure, the blue line describe cycle of R-134a of experiment, red line describe cycle of R-134a of design. The difference of inlet pressure both of components can affect the power generated of turbine. High inlet pressure of turbine will be generate the turbine well. From the figure, we can see that curve that has a larger area has a higher efficiency, so curve which has greater efficiency is simulation shown in red line. Otherwise, the design of system was made in ideal conditions, whereas in the experiment that there are many energy losses in each process. This is causes the difference efficiency between design and experiment.

Table 4. Comparation of Design and Experiment Result

	W,t (kJ/s)	W,p (kJ/s)	Q,in (kJ/kg)	W,net(kJ/kg)	Efficiency (%)
Design	1.11	0.12	233.87	8.91	3.8
Experiment	0.614 0.310	0.334 0.169	197.89	6.59	3.33

The conclusion of this study is Organic Rankine Cycle system has been successfully designed and manufactured using R-134a working fluid with helical tube of evaporator and helical tube of condenser. Where, the length of helical evaporator is 10.61 m and helical condenser is 10.56 m. Based on the design, this study had resulted power turbine of 1110 Watt and efficiency of system of 3.8%, whereas result of the experiment is power turbine of 614 Watt and efficiency of system of 3.33%

References

[1] O. Nurhilal, C. Mulyana, and N. Suhendi, 'Disain Pembangkit Listrik Tenaga Uap Siklus Rankine Organik 100 kW dengan Fluida Kerja R-123', no. April, pp. 136–139, 2014.
[2] A. Martin, Miswandi, A. Prayitno, I. Kurniawan, and Romy, 'Exergy Analysis of Gas Turbine Power Plant 20 MW in Pekanbaru - Indonesia', *Int. J. Technol.*, pp. 921–927, 2016.
[3] S. Saadon and A. R. A. Talib, 'An analytical study on the performance of the organic Rankine cycle for turbofan engine exhaust heat recovery', *IOP Conf. Ser. Mater. Sci. Eng.*, vol. 152, p. 012011, 2016.
[4] T. B. Nur, Z. Pane, and M. N. Amin, 'Application of Biomass from Palm Oil Mill for Organic Rankine Cycle to Generate Power in North Sumatera Indonesia', *IOP Conf. Ser. Mater. Sci. Eng.*, no. 1, pp. 1–7, 2017.
[5] C. Liu, C. He, H. Gao, X. Xu, and J. Xu, 'The Optimal Evaporation Temperature of Subcritical ORC Based on Second Law Efficiency for Waste Heat Recovery', pp. 491–504, 2012.
[6] F. Vélez, 'Selecting working fluids in an organic Rankine cycle for power generation from low temperature heat sources Seleccionando fluidos de trabajo en ciclos Rankine para generación de energía a partir de fuentes de calor de baja temperatura', 2014.
[7] M. A. Siddiqi and B. Atakan, 'Investigation of the Criteria for Fluid Selection in Rankine Cycles for Waste Heat Recovery *', vol. 14, no. 3, pp. 117–123, 2011.
[8] K. S. Rao and C. L. Sindhuja, 'Working fluids for Organic Rankine Cycle – A Review', vol. 3, no. 8, pp. 447–451, 2017.
[9] B. Saadatfar, R. Fakhrai, and T. Fransson, 'The Journal of MacroTrends in Energy and Sustainability Waste heat recovery Organic Rankine cycles in sustainable energy conversion: A state-of-the-art review', *Jmes*, vol. 1, no. 1, pp. 161–188, 2013.
[10] U. Muhammad, M. Imran, D. H. Lee, and B. S. Park, 'Design and experimental investigation of a 1kW organic Rankine cycle system using R245fa as working fluid for low-grade waste heat recovery from steam', *Energy Convers. Manag.*, vol. 103, no. October, pp. 1089–1100, 2015.
[11] S. Douvartzides and I. Karmalis, 'Working fluid selection for the Organic Rankine Cycle (ORC) exhaust heat recovery of an internal combustion engine power plant', *IOP Conf. Ser. Mater. Sci. Eng.*, vol. 161, no. 1, 2016.
[12] S. Quoilin and V. Lemort, 'Technological and Economical Survey of Organic Rankine Cycle Systems'.
[13] F. Calise, C. Capuozzo, and L. Vanoli, 'Design And Parametric Optimization Of An Organic', vol. 6, no. 2, pp. 178–204, 2014.
[14] Y. A. Cengel and M. Boles, *Thermodynamics An Engineering Aproach*, 5th Intern. McGraw-Hill, 2006.

The effect of fiber volume fraction on tensile and impact properties of *eleusine indica* grass reinforced polypropylene bio composite

I P Lokantara[1,2], N P G Suardana[2], I W Surata[2], and I N S Winaya[2]
[1] Doctoral Study Program of Engineering Science, Faculty of Engineering, Udayana University, Kampus Sudirman, Denpasar-Bali, Indonesia
[2] Mechanical Engineering Department of Udayana University, Street of Bukit Jimbaran, Badung, Bali, Indonesia

Email: lokantara_santri@yahoo.com

Abstract. The purpose of this research was to investigate the tensile strength and impact strength of eleusine indica grass fiber-reinforced polypropylene recycle composite. In this study, the fiber was treated with alkali at 5% NaOH for 2 hours. The eleusine indica grass fiber cut with a length of 3 cm and the matrix used recycle polypropylene. In this research, the volume fraction of eleusine indica grass fiber varied by 20%, 25%, and 30% respectively. The composite molded with a hot press at temperature 200^{0}C with a holding time for 2 hours. Tensile test used ASTM D3039 and impact test used ASTM D256. The highest tensile and impact strength results on fiber volume fraction 30%.

1. Introduction

For the past decade, most manufacturing industries have been interested in developing eco-friendly products. The material of the developed product must be lightweight, widely available, easily biodegradable in nature, contain no hazardous and non-toxic materials. In the field of industry, natural fibers are being used to replace glass fiber. The advantages of natural fiber include easy to extract, light, renewable natural resources, decompose in the environment [1]. In particular, there is a great need for natural fiber in the automotive sector where the use of natural fibers in the vehicle body will reduce vehicle weight thereby impacting better fuel efficiency without sacrificing performance. There has been considerable research on the use of natural fibers as reinforcement in polymer matrix composites to replace synthetic fibers [2, 3, 4]. Synthetic fibers have been widely used in the industrial world, but synthetic fibers have the disadvantages; expensive, does not decompose in nature, high energy consumption, causes problems such as environmental pollution, skin irritation, and abrasion of equipment during processing. The increasing need for natural fiber composites required new natural fiber research as a reinforcement material for use in various applications and also as a hybrid composite. The natural fibers of the extracted plant generally grow in varying regions and environments. Soil conditions, climate, age, plant parts (roots, stems, leaves, fruits) determine the chemical content and other important properties of the fiber [5]. However, the characteristic of the most important amplifier is the adhesion property between fiber and matrix. Therefore, it is necessary to identify new natural fibers that have good physical and mechanical characteristics. Adhesion between fiber and matrix determines the quality of the composite. Natural fibers contain layers of lignin, hemoselulosa, waxes that wrap cellulose, this layer which inhibits the bond between the matrixes with the fiber. To reduce the layers it is necessary to do chemical treatment so that the surface structure of the fibers becomes coarser so that it has good adhesion properties between the fiber and matrix. Chemical treatments are mostly done by some researchers is to use a solution of NaOH. In general, natural fiber treatment process using NaOH is easy to process, low cost and good result. Chemical treatment of each fiber varies, specific to each fiber. The percentage of NaOH varies from 1% to 20%, as well as the duration of immersion varies from 15 minutes to 24 hours [6-8]

The matrix used in biocomposites can be of thermoplastic or thermosetting polymers. The type of thermosetting polymer that is often used is epoxy. The disadvantage of polymer thermosetting is that it cannot be recycled, has a bad impact on the environment, because plastic waste cannot be degraded, causing CO_2 emissions and producing toxic gases when the material is burned. The type of thermoplastic polymer that is often used as a product in the industrial world is polypropylene. One of them is a plastic glass of mineral water. Some researchers have conducted research using polypropylene type plastic waste combined with natural fiber. Polyprophelene is used as a matrix in composites because it can accept various types of reinforcing materials, such as glass fibers, glass balls, talc, asbestos, mica, calcium carbonate, and silica. Polypropylene also has a structure that can melt and be reformed several times.

Several studies have been conducted, among others, the use of polypropylene plastic waste combined with various kinds of natural fibers to be used as biocomposites, and the mechanical properties of the composites produced are close to the mechanical properties of glass fiber composites [9]. Other studies on the characteristics and mechanical properties of polypropylene plastic waste composites reinforced with coco fiber [10], palm fiber [11], sisal fiber [12], hemp fiber [13], and kenaf fiber [14]. Coconut fiber filters [15], corn husk [16]. In this study, *Eleusine indica* fiber was used as a reinforcement for polypropylene recycle composites.

2. Method

Eleusine indica grass is soaked with the water retting process for 7 days, then the fiber is extracted manually. The fiber is dried for 2 days in the open air, then oven for 2 hours to reduce the moisture content. The initial treatment of fiber uses 5% alkaline solution for 2 hours, then the fiber is rinsed with running water. *Eleusine indica* fiber is dried in an oven, until the moisture content is stable. Fiber is cut to 3 cm in size. Polypropylene recycle used comes from a drink with a certain brand. Polypropylene recycle is cut into small pieces, then cleaned with running water. The next process is drying for 1 day. Fiber density is calculated using a picnometer method. The volume fraction of fiber used is 20%, 25%, 30% respectively. The process using a hot press, with a temperature of 200^0 C with a holding time of 2 hours. Testing of tensile strength using ASTM D-3039 and impact strength testing using ASTM D256. SEM is needed to see morphology microstructure of fiber bonds with matrix.

3. Results and discussion

3.1. Tensile Strength

The results of the composite tensile strength testing are illustrated in Figure 1. The tensile strength of the composite at volume fractions of 20%, 25%, 30% is 13,465 Mpa, 16.01 MPa, 18.5 MPa respectively. The tensile strengths at 30% V_f were 37% greater than those of the materials prepared with 20% V_f. From the graph it can be seen that the trend of tensile strength increases with increasing volume fraction of fiber. In the 30% fiber fraction, the adhesion bond between the fibers with the best matrix, the polypropylene matrix fills the cavity between fibers and forms a good adhesion bond. In the volume fraction of 20% the lowest tensile strength fiber, polypropylene as a filler is too much compared to fiber, thus allowing the gap between fibers. This results in reduced composite ability to accept tensile loads.

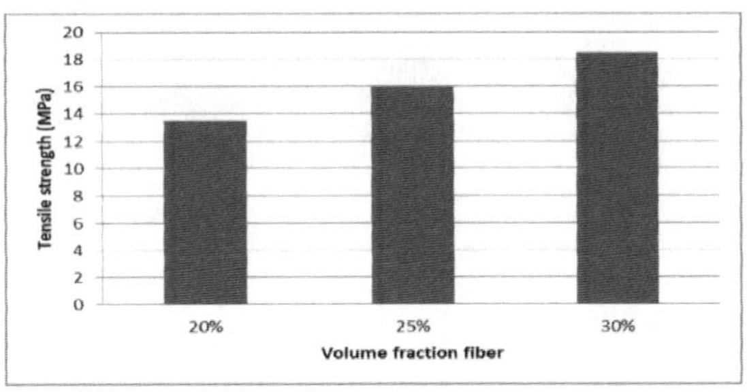

Figure 1. Tensile strength of composite

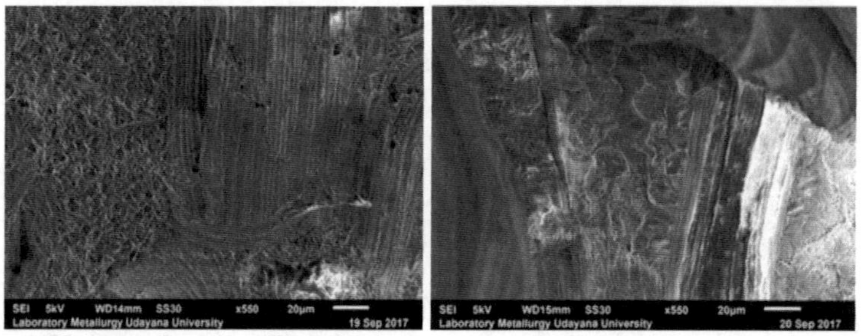

a. Volume fraction fiber 20% **b.** Volume fraction fiber 30%

Figure 2. Morphology microstructure of composite

Morphology of microstructure using SEM shows the same facts. In Figure 2.a it appears that the fault is darker, indicating that the fracture is uneven because the bond between the fiber and the matrix is uneven. In Figure 2.b it can be seen that the fracture is brighter or the fracture surface is more even. This happens because fiber bond with matrix are better.

3.2. Impact Strength

The results of the composite impact strength testing are illustrated in Figure 3. Impact strength composite at volume fractions of 20%, 25%, 30% at 0.005 Nm/mm^2, 0.029 Nm/mm^2, 0.041 Nm/mm^2 respectively. The increase in impact strength is 666% from a volume fraction of 20% to a volume fraction of 30%. The adhesion bond is very good at a 30% volume fraction, the polypropylene matrix binds fibers well so that the impact load is well distributed to the fiber. At a volume fraction of 20% the impact strength is very low because there is a cavity between fibers, as a result the load cannot be passed to the fiber. From the SEM photos, microstructure of composite morphology in the volume fraction of 20% contains many cavities, this is because the adhesion bond is less than perfect between the fiber and the matrix. At a 30% volume fraction, there is a fracture pattern with little cavity, a sign that the adhesion bond between fiber and matrix is very good.

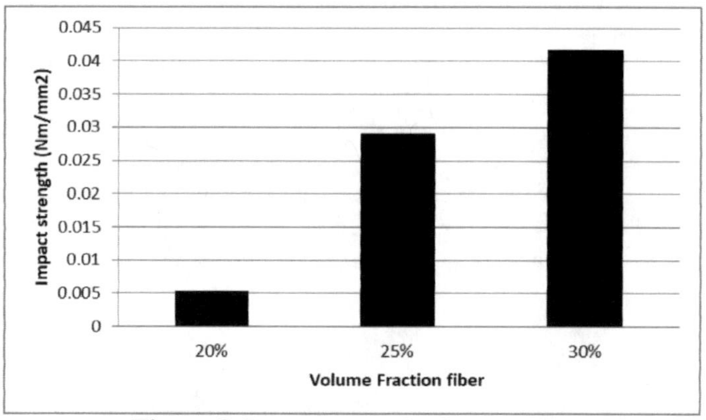

Figure 3. Impact strength of composite

IOP Conf. Series: Materials Science and Engineering **539** (2019) 012028 doi:10.1088/1757-899X/539/1/012028

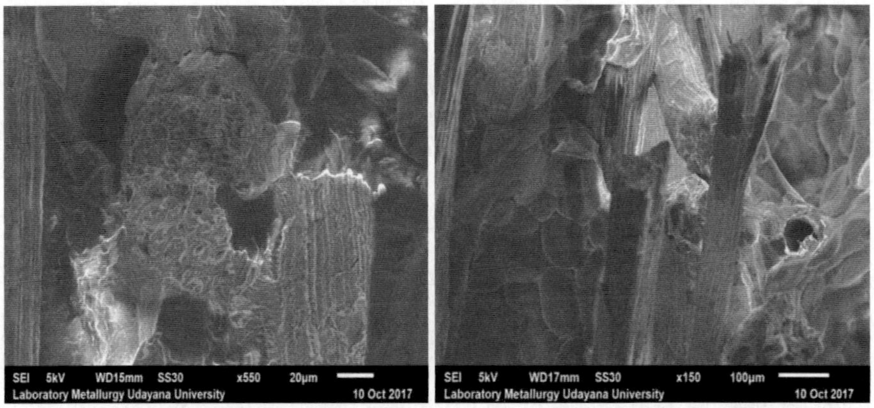

a. Volume fraction fiber 20% b. Volume fraction fiber 30%

Figure 4. Morphology microstructure of composite

4. Conclusions

The higher the fiber volume fraction increases the tensile strength and composite impact strength. In the 30% volume fraction, the adhesion bond between the fiber and the matrix is the most optimal, tensile load and impact load is well distributed to the fiber. From microstructure of composite morphology using SEM, visually visible fracture surfaces are more evenly distributed at 30% volume fraction. The bond between fiber and matrix is very good, with fewer cavities than the fracture surface at a volume fraction of 20%

References

[1] Jawaid, M., and H. P. S. Abdul Khalil. 2011. Cellulosic/synthetic fiber reinforced polymer hybrid composites: A review. Carbohydrate Polym. 86: 1– 18.

[2] Senthamaraikannan, P, Saravanakumar. S.S, Arthanarieswaran. V.P and Sugumaran.P, 2016, Physicochemical properties of new cellulosic fibers from bark of Acacia planifrons, Int. J. Polym. Anal. Charact. 21(3): 207-213.

[3] Mayandi. K, Rajini. N, Pitchipoo. P, Winowlin Jappes. J. T and Varada Rajulu. A, 2016, Extraction and characterization of new natural lingo cellulosic fiber Cyperus Pangorei**.** *Int. J. Polym. Anal. Charact*. 21: 175–183.

[4] R. Gopinath., K. Ganesan., S.S. Saravanakumar and R. Poopathi, 2016, Characterization of new cellulosic fiber from the stem of Sida rhombifolia.*Int. J.Polym. Anal. Charact*. 21(2): 123-129.

[5] C. E. Maepa, J. Jayaramudu, J. O. Okonkwo, S. S. Ray, E. R. Sadiku, and J.Ramontja. 2015. Extraction and Characterization of Natural Cellulose Fibers from Maize Tassel. *Int. J.Polym. Anal. Charact*.20: 99–109.

[6] Ridzuan, M.J.M., Majid, M.S.A., Afendi, M., Azduwin, K., Kanafiah, S.N.A, 2015. The Effects of the Alkaline Treatment's Soaking Exposure on the Tensile Strength of Napier Fibre. *Procedia Manufacturing* 2: 353-358.

[7] K. J. Nagarajan & A. N. Balaji. 2016. Extraction and Characterization of Alkali Treated Red Coconut Empty Fruit Bunch Fiber. *International Journal of Polymer Analysis and Characterization* 21(1): 18-28.

[8] Mohammad Hosein Pouriman, Alvin R. Caparanga, Majid Ebrahimi & Ali Dahresobh, 2017, Characterization of Untreated and AlkalineTreated Salago Fibers (Genus Wikstroemia Spp.), *Journal of Natural Fibers*.

[9] Quazi T. H. Shubhra, A. K. M. M. Alam and M. A. Quaiyyum. 2011, Mechanical Properties of Polypropylene Composites: A Review. *Journal of Thermoplastic Composite Materials* I (30)

[10] Chun, Koay Seong, Salmah Husseinsyah, Fatin Nasihah Azizi. 2013, Characterization and Properties of Recycled Polypropylene/Coconut Shell Powder Composites: Effect of Sodium Dodecyl Sulfate Modification. *Polymer-Plastics Technology and Engineering* 52, no. 3 : 287-94

[11] Tshwafo E. Motaung, Linda Z. Linganiso, Maya John, Rajesh D. Anandjiwala. 2015. The Effect of Silane Treated Sugar Cane Bagasse on Mechanical, Thermal and Crystallization Studies of Recycled Polypropylene. *Materials Sciences and Applications, 6, 724-733.*

[12] Arun Kumar Gupta, Manoranjan Biswal, S.Mohanty, and S.K.Nayak. 2012. Mechanical,Thermal Degradation, and Flammability Studies on Surface Modified Sisal Fiber Reinforced Recycled Polypropylene Composites. *Advances in Mechanical Engineering, Article ID 418031, 13 pages.*

[13] A. Bourmaud, A. Le Duigou, C. Baley, 2011, What is the technical and environmental interest in reusing a recycled polypropylene-hemp fibre composite?**,** *Polymer Degradation and Stability 96 1732-1739*

[14] Neng Sri Suharty, Hanafi Ismail, Kuncoro Diharjo, Desi Suci Handayani, and Maulid Firdausa. 2016, Effect of Kenaf Fiber as a Reinforcement on the Tensile, Flexural Strength and Impact Toughness Properties of Recycled Polypropylene/Halloysite Composites, *Procedia Chemistry 19 253 – 258*

[15] N.P.G. Suardana, I. Putu Lokantara, Y.J. Piao, J.K. Lim. 2013, Water Absorption and Tensile Strength of Coconut Filter Fibers/Polypropylene Composites, *Advanced Materials Research*, Vol. 702, pp. 207-212

[16] Sari N H, Wardana I N G, Irawan Y S, Siswanto E 2018 Characterization of the Chemical, Physical, and Mechanical Properties of NaOH–treated Natural Cellulosic Fibers from Corn Husks, *Journal of Natural Fibers* **15**(4) 545–558. doi: 10.1080/15440478.2017.1349707

IOP Conf. Series: Materials Science and Engineering **539** (2019) 012029 doi:10.1088/1757-899X/539/1/012029

Smoldering combustion propagation of subtropic peat; case study on pelalawan peat

A Martin[1], H Orion[1], A Sandhyavitri[2], P Palamba[3] and Y S Nugroho[3]

[1]Department of Mechanical Engineering, Faculty of Engineering Universitas Riau , Pekanbaru 28293, Indonesia

[2]Department of Civil Engineering, Faculty of Engineering Universitas Riau , Pekanbaru 28293, Indonesia

[3] Department of Mechanical Engineering, Faculty of Engineering Universitas Indonesia, Depok 16424, Indonesia

Email: awaludinmartin01@gmail.com

Abstract. Smoldering combustion was categorized as a very dangerous combustion and considered as a main cause to haze problem during wildland fires in Sumatera, Indonesia. This study aims to determine the characterized of the spread of peat fire, so the mitigation can be done. This study was used experimental methods. In this case, peat from Riau were analyzed and the samples to be used were peat that has been dried in the oven during 24 hours at 100 °C of temperature, moisture content after interaction with air was 7.09%. This study was resulted the maximum combustion temperature of 508.72 °C. The rate of propagation of peat fires was determine from the difference of time when the combustion temperature reaches 300°C and 365°C. The rate of smoldering propagation at temperature of 300°C of 14.3 cm/h on the vertical direction and 6.31 cm/h on horizontal direction. The rate of smoldering propagation at temperature of 365°C of 19.76 cm/h on the vertical direction and 5.17 cm/h on horizontal direction.

1. Introduction

Fire on peatlands is a very dangerous fire, which produced high toxic emission. Most of peat fires were caused by large-scale of land reveal and canal establishment of peatlands, which can damage the physical and chemical properties of the peat and affect the decrease in moisture content, until the peat becomes dry, especially in the dry season. Dry peat will be highly flammable and peatland fires can be dominated by self-sustained combustion, slow-moving of propagation and low temperatures [1]. Indonesia has the largest peatland among tropical countries. Fires in Indonesia in 2006 was 2.79 million hectares of peatland had been emission 98.38-180.38 million tons of CO2 [2]. In 2015 Badan Nasional Penanggulangan Bencana (BNPB) estimates that total of burned areas in Indonesia of 2,089,911 ha and 618,574 ha from peatlands. It can hamper national action plans to reduce greenhouse gas emissions to 26% in 2020, if peatlands continue to burn on a large scale [3].

Indonesia has the largest peatland among tropical countries. The area of peatland in Indonesia reaches 20.6 million ha or 10.8% of the land in Indonesia. The largest peatland was located in the island of Sumatra which is 6.4 million hectares, which is spread in some Provinces. One of the largest peatlands in Sumatera is Province of Riau, where 3.8 million hectares were scattered in some districts, where Pelalawan Regency is the region with the third largest peatland after Indragiri Hilir and Bengkalis [4]. The extent of peatlands in Riau, especially in Pelalawan District, is at risk for peat fires, and can lead to greenhouse gases, disrupt the balance of ecosystems, and contribute to financial losses, especially in Riau.

Initial research regarding smoldering combustion was first published in 1957. Palmer, conducted an experiment of smoldering combustion in dust and fibrous materials [5]. Smoldering combustion occurs at low-temperature, and generates incomplete combustion. Therefore, smoldering combustion produced toxic gases and particulates at a higher rate than flaming combustion. Palamba, P., was conducted a study of smoldering combustion analysis on dry peat in two different locations. Samples were taken from Ogan Komering ilir (OKI), South Sumatera Province, and from Kampung Bagaiserwar, Sarmi Regency, and Papua Province. In general, there are three main stages of smoldering combustion namely, heating, combustion (consisting of pyrolysis and oxidation), and charcoal or ash formation [6]. This study aims to determine the comparison of the characteristics of burning of peat from Ogan Komering Ilir (OKI), South Sumatera Province and Kampung Bagaiserwar, Sarmi Regency, Papua Province.

The purpose of this study is to find out the propagation characteristics of peat fires, so that mitigation of peat fires can be done.

2. Methodology

In this study, we used the experimental method. The sample that used was peat from province of Riau. Sampling technique using a systematic random. The method used for sample collection is an uninterrupted method, so the nature of the sample does not change and it was not affected by microbial activity until the experimental work time.

2.1 Experimental Setup

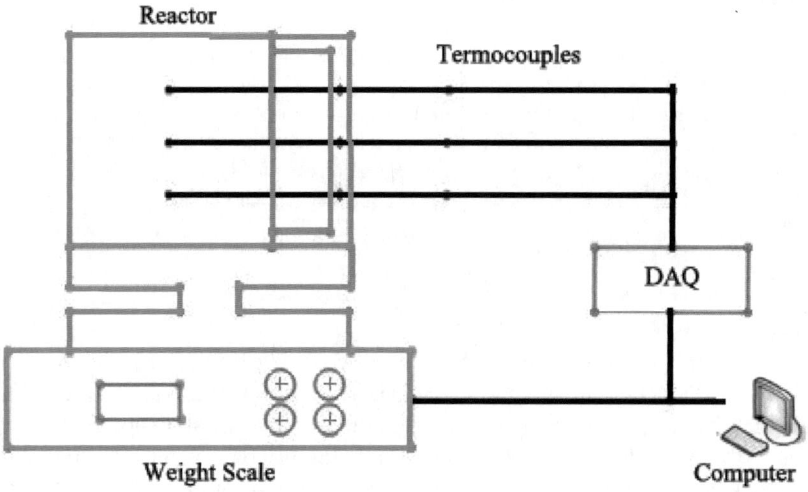

Figure 1. Schematic Diagram of the Experimental Setup

2.2 Experimental Preparation

The experimental apparatus that was used in this experiment can be shown in Figure 2, whereas reactor of 100x100mm was used for sample and digital weighing were equipped to measure mass loss of peat sample [7]. To reduce the waste heat that generated from combustion, this experiment equipment was designed using heat resistant calci boards. In other side, the reactor was perforated to insert nine thermocouples as a temperature gauge, thus the overall distribution in the combustion chamber will be measurable and will be converted in figures by a DAQ system connected to personal computer. Each thermocouple was set of 2.5 cm to another. So, the speed of propagation of combustion can be calculated by comparing the distance between the thermocouple with the difference of time between the thermocouple, when the burning temperature reaches 300 °C and 365 °C.

Figure 2. Experimental Apparatus

2.3. Moisture Content

The moisture content on a dry basis is a percentage equivalent to the ratio of water weight to dry material weight [8]. Water content (moisture content) on dry basis, obtained by using the following formula:

$$MC = \frac{m_1 - m_2}{m_2} x100\%$$

Dimana :

MC : *Moisture Content* (%)

m_1 : initial mass (gram)

m_2 : final mass (gram)

After testing and processing the data, moisture content from peat can be seen in Table 2.

Table 1. Moisture Content

m_2	m_1	*Moiture Content (%, dry base)*
0.202	2.002	891.089

By following sample preparation and sample drying procedures, the average moisture content of Pelalawan Regency was reached 891,089%.

2.4 Experimental Procedure

The peat samples were dried by using an electric oven, in accordance to ASTM standard [9] peat was dried at 100°C during 24 hours, the heating coil was connected to the reactor. Data acquisition systems and mass-change readings connected to laptop. Then the thermocouple was calibrated before the experiment was done and press the zero button on the scales, before the peat was filled up in the reactor to measure the initial mass and the final mass during the experiment. The heater (coil heater) was switched on for 45 minutes and considered heat heating energy was able to sustain peat burning until the peat can burn by itself.

2.5. Data Collecting

2.5.1. Physical Characteristics

Physical characteristics of Pelalawan peat are shown by Table 1. Pelalawan peatlands can be classified (ASTM D4227-13) as peat with moderate ash content (5-15%) and Hemic (33-67%) or Semi Fibrous.

Table 2. Physical Characteristics of Pelalawan Peat

Physical Characteristics	Unit	Value
Density	Kg/m^3	1.5
Wet Volume weight	ton/m^3	1.1
Dry volume weight	ton/m^3	0.1
Porosity	%	92
Ash Content	%	6.3
Organic Content	%	94
Fibre Content	%	44
Electric Resistance	Ohm meter	12

2.5.2. Higher Heating Value

The calorific value is sought to determine the amount of heat present in peat per gram. Heat value obtained from the test using the Digital Bomb Calorimeter C-200 tool, where the data obtained in the form of temperature rise. Temperature rise from Pelalawan peat samples is 1.71 with 1 gram peat mass. Until the HHV value is obtained by using the formula as follows:

$$HHV = \frac{\text{Temperature rise x 2332}}{\text{weight}}$$

HHV= 3987.72 caloric/gram

3. Result and discussion

3.1 Result

3.1.1 Propagation Smoldering Combustion of Peat

a. Smoldering Propagation Rate on Vertical Direction

Smoldering combustion propagation on vertical direction got from the distance between the thermocouple divided by the time difference in each thermocouple when the combustion temperature reaches 300ºC and 365ºC. smoldering combustion propagation of themocouple 1 (Th1) to thermocouple 4 (Th4) is v1, and v2 is the burning burning of themocouple 4 (Th4) to thermocouple 7 (Th7), and so on which can be seen in the burning propagation scheme Figure 3.

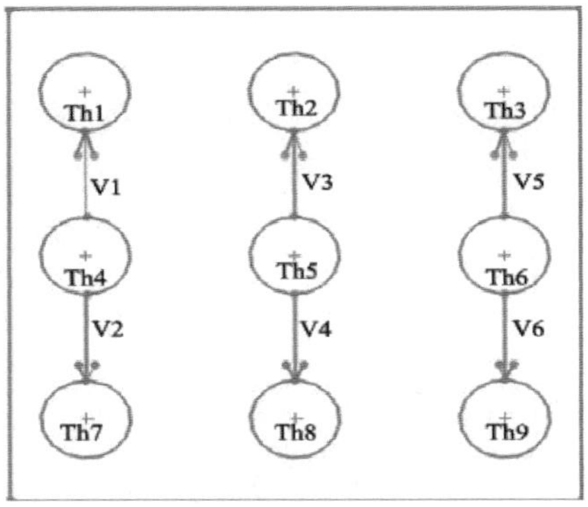

Figure 3. Vertical Direction Propagation Scheme

From the time difference obtained between the thermocouples when the combustion temperature reaches 300°C, the velocity of smoldering combustion on the vertical direction in each thermocouple, which can be seen in Table 3.

Table 3. Smoldering Propagation Rate

v1 (cm/jam)	v2 (cm/jam)	v3 (cm/jam)	v4 (cm/jam)	v5 (cm/jam)	v6 (cm/jam)
1.3	26.6	15.6	3.9	14.1	24.2

The mean speed of smoldering combustion propagation in the vertical direction is as follows:

$$v_{average} = \frac{v_1 + v_2 + v_3 + v_4 + v_5 + v_6}{6}$$

$$v_{average} = 14.3 cm / h$$

From the time difference obtained between the thermocouples when the combustion temperature reaches 365°C, the velocity of smoldering combustion on the vertical direction in each thermocouple, which can be seen in Table 4.

Table 4. Smoldering Propagation Rate

v1 (cm/jam)	v2 (cm/jam)	v3 (cm/jam)	v4 (cm/jam)	v5 (cm/jam)	v6 (cm/jam)
57.325	8.867	16.886	1.061	23.316	11.097

The mean speed of smoldering combustion propagation in the vertical direction is as follows:

$$v_{average} = \frac{v_1 + v_2 + v_3 + v_4 + v_5 + v_6}{6}$$

$$v_{average} = 19.76 cm / h$$

b. Smoldering Propagation Rate On Horizontal Direction
Smoldering combustion propagation on vertical direction is get from the distance between the thermocouple divided by the time difference in each thermocouple when the combustion temperature reaches 300°C and 365°C. smoldering combustion propagation of themocouple 1 (Th1) to thermocouple 2 (Th2) is v1, and v2 is the burning burning of themocouple 2 (Th2) to thermocouple 3 (Th3), and so on which can be seen in the burning propagation scheme of Figure 4.

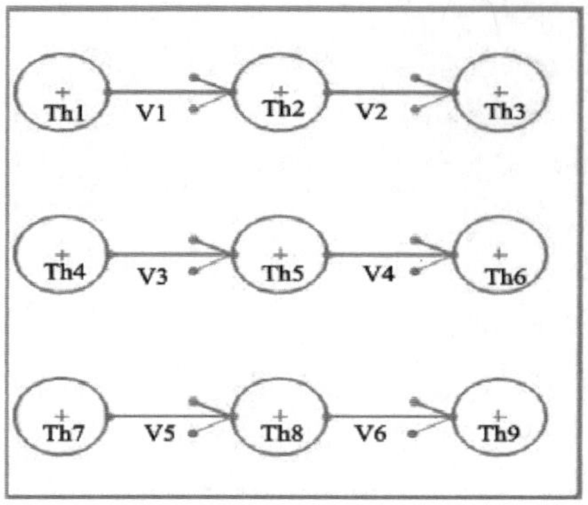

Figure 4. Vertical Direction Propagation Scheme

From the time difference obtained between the thermocouples when the combustion temperature reaches 300°C, the velocity of smoldering combustion on the vertical direction in each thermocouple, which can be seen in Table 5.

Table 5. Smoldering Propagation Rate

v1	v2	v3	v4	v5	v6
0.81	14.8	2.03	10.6	1.48	8.09

The mean speed of smoldering combustion propagation in the vertical direction is as follows.

$$v_{average} = \frac{v_1 + v_2 + v_3 + v_4 + v_5 + v_6}{6}$$

$$v_{average} = 6.31 cm / jam$$

From the time difference obtained between the thermocouples when the combustion temperature reaches 365°C, the velocity of smoldering combustion on the vertical direction in each thermocouple, which can be seen in Table 6.

Table 6 Smoldering Propagation Rate

v1 (cm/jam)	v2 (cm/jam)	v3 (cm/jam)	v4 (cm/jam)	v5 (cm/jam)	v6 (cm/jam)
1.211	5.806	1.046	8.427	0.539	13.997

The mean speed of smoldering combustion propagation in the vertical direction is as follows.

$$v_{average} = \frac{v_1 + v_2 + v_3 + v_4 + v_5 + v_6}{6}$$

$$v_{average} = 5.17 cm / jam$$

3.1.2 Mass Loss Rate

Before getting mass loss rate, then first look for mass loss from result of burning that have been done. Mass loss is obtained from the following formula:

IOP Conf. Series: Materials Science and Engineering **539** (2019) 012029 doi:10.1088/1757-899X/539/1/012029

Mass loss = initial mass - final mass

= 217 gram

Burning time up to extinguish for 16 hours. Until the mass loss rate (m˙) is obtained:

$$\dot{m} = \frac{\text{mass loss}}{\text{time}}$$

$$= 13.56 \text{ gram} / \text{h}$$

3.1.3 Emissions from Peat Burning

In this research, to get the emission of gas from the result of peat burning using technomotor test equipment. The average exhaust emissions from combustion in peat obtained from technomotor, with CO emissions of 0.078% and CO2 of 0.214%. CO and CO2 emissions test is carried out for 106 seconds, then scaled during combustion, ie for 16 hours and multiplied by mass loss rate, then CO gas percentage is obtained from 1.06% and CO2 gas 2.9%.

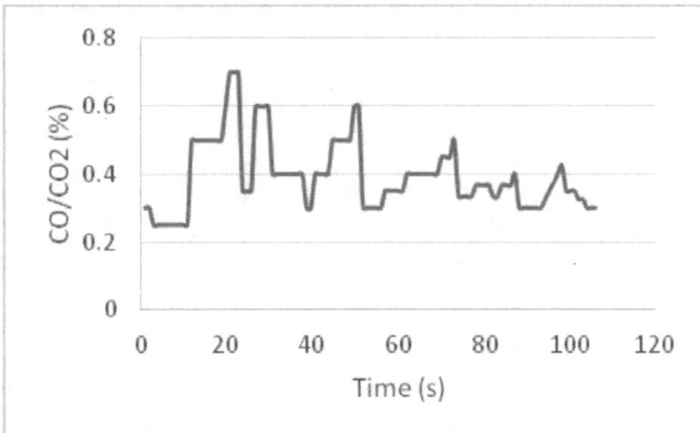

Figure 5. Graph of CO and CO2 Emissions on Peat Fires

3.2 Discussion

From Figure 6 can be seen changes in fluctuating temperature, when the heating coil is turned off then the temperature will decrease drastically, but eventually the temperature will increase again. This is because the smoldering propagation occurs outside the boundary of the reading system of the thermocouple and has not yet reached the limit of the other thermocouple readings. Combustion experiment in this study is a shallow burning, since combustion occurs only <2m [10].

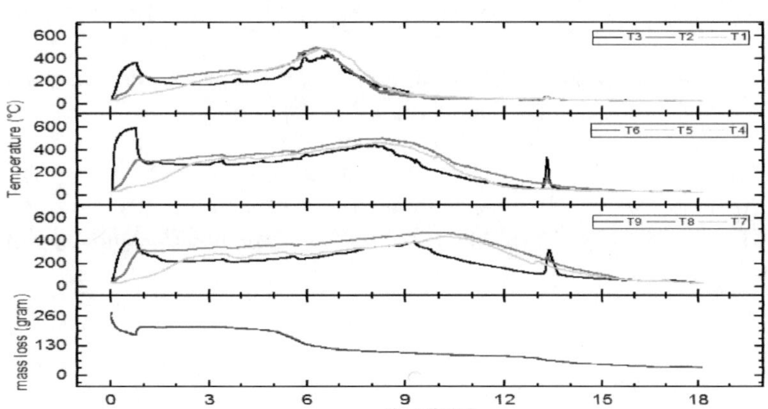

Figure 6. Profile of Temperature Changes and Mass Loss from Peat.

IOP Conf. Series: Materials Science and Engineering **539** (2019) 012029 doi:10.1088/1757-899X/539/1/012029

The maximum temperature of smoldering combustion is 508.72°C, when burning lasted for 8.3 hours. The combustion takes place for 16 hours, where on the heating coil is placed close to thermocouple 6 (T6) with its maximum temperature reaching 600°C. The initial propagation is more dominantly upward by seeing the high temperature in thermocouple 3 (T6) is 462 °C compared to thermocouple 9 is 400 °C at 45 minutes burning time, this is because more oxygen or air in the upper peat (thermocouples 1,2 and 3) than the bottom peat (thermocouples 7, 8 and 9). The upper peat more rapidly loses mass, which can be seen in Figure 6. Where the combustion time for 8.8 hours the average temperature on thermocouples 1, 2 and 3 reaches 72°C, which is close to unburned temperatures. This is due to the subsidence of vacancies from the center and bottom of the burning peat.

4. Conclusion

The rate of smoldering propagation on a temperature of 300°C is 14.3 cm/h on the vertical direction and 6.31 cm/h on horizontal direction. The rate of smoldering propagation on a temperature of 365°C is 19.76 cm/h on the vertical direction and 5.17 cm/h on horizontal direction. The rate of propagation of combustion on the vertical and horizontal combustion direction is constantly changing. This is caused by an ever-changing burning temperature and will automatically affect the burning speed. The higher the combustion temperature then the burning propagation rate will increase.

References

[1] Huang. Xinyan, Rein. Guillermo. 2017. Downward spread of smouldering peat fire: the role of moisture, density and oxygen supply. International Journal of Wildland Fire 2017, 26, 907–918.

[2] Ballhorn. U, Siegert. F, Mason. M, Limin. S. 2009. Derivation of burn scar depths and estimation of carbon emissions with LIDAR in Indonesian peatlands. Proceedings of the National Academy of Sciences of the United States of America.

[3] Nugroho, S. 2015. Analisis Luas Hutan dan Lahan Terbakar di Indonesia 2015. BNPB. (Accessed: 2018. http://www.disasterchannel.co/wp-content/uploads/2015/10/media_Analisis-luas-hutan-dan-lahan-terbakar-2015).

[4] Wahyunto, Ritung. S, dkk. 2005. Sebaran Gambut dan Kandungan Karbon di Sumatera dan Kalimantan 2004. Bogor. Wetland Indonesia.

[5] Palmer, K. Smoldering combustion in dusts and fibrous materials. Combustion and Flame, 1957. 1(2): p. 129-154.

[6] Palamba. P, Ramadhan. dkk. 2017. Investigation of Smouldering Combustion Propagation of Dried Peat. Dapertment of Mechanical Engineering, Faculty of Engineering Universitas Indonesia, Depok 16424, Indonesia.

[7] Rein. G, Cleaver. N, Ashton. C. 2008. The Severity of Smouldering Peat Fires and Demage to the Forest Soil. BRE Centre for Fire Safety Engineering University of Edinburgh, UK. Standard, A., D2216-10.

[8] Rein, G. Cohen, Simon. 2009. Carbon Emissions from Smoldering Peat in Shallow and Strong Fronts. Proceedings of the Combustion Institute 32:2489-2496.

[9] Standard, A., D2216-10: Standard Test Methods for Laboratory Determination of Water (Moisture) Content of Soil and Rock by Mass. West Conshohocken, PA: ASTM International (Accessed: 2014. http://www. astm. org/Standards/D6913. htm), 2010.

[10] Rein, G. 2013. Smouldering Fires and Natural Fuels. An Interdisciplinary Approach to Fire Science, pp. 15–34.

IOP Conf. Series: Materials Science and Engineering **539** (2019) 012030 doi:10.1088/1757-899X/539/1/012030

Analysis of transient elastohydrodynamic lubrication of point contact subjected to sinusoidal dynamic loads

D R P Cupu[1,2], A Stratmann[2] and G Jacobs[2]
[1]Department of Mechanical Engineering, University of Riau
28293 Pekanbaru, Riau, Indonesia

[2]Institut für Maschinenelemente und Systementwicklung, RWTH Aachen University
52062 Aachen, Germany

Email: dedi.cupu@lecturer.unri.ac.id

Abstract. Vibration is inevitable from all machine elements operations since the load, surface velocities or contact radii change in time. In the case of a ball bearing application, the load changes as the ball elements rotate around the track. This change will be more rapidly when the external load changes. The vibrations induce a sinusoidal dynamic load working on the contact. On the other hand, the imperfect waviness on the contacting surfaces causes the vibration effects on the film thickness to become more pronounced. This paper numerically investigates the influence of the sinusoidal varying loads due to vibrations on the film thickness and pressure profile of the isothermal, elastohydrodynamically lubricated point contact with a wavy surface. For that purpose, a single contact between ball element and inner raceway is modeled as a contact between a spherical ball and a flat plate. The results reveal that the external sinusoidal dynamic load induces the modulations on the film thickness and pressure profile. The waviness on one contacting surface causes the changes in the EHL behaviors to become more pronounced.

1. Introduction

A lubricated contact between non-conformal surfaces in machine elements such as gears, rolling bearings, and cam and followers are commonly identified as elastohydrodynamic lubrication (EHL) regime in which the contact generates a very high pressure over a very small contact area. The behaviours of EHL, such as the pressure and film thickness distribution, explain the important considerations in the analyses of machine elements contacts. The analysis of the generated pressure in the EHL contacts is crucial to investigate the subsurface stress tensor and to predict the failure of initial crack whereas the study of the film thickness formed between the lubricated contacts improves the performance of machine elements. However, it is generally known that machine elements operate almost always under time-dependent conditions. In the case of ball bearing applications, the contact between rolling elements and raceway in the loaded zone changes in time due to varying orbital position of the rolling elements during rotation. This such condition causes a harmonic change in the load applied to the contact and is now known as vibration. The change in the external load applied to the contact will cause the vibration to become more pronounced. In addition, manufacturing processes of ball bearing components cannot guarantee that the surfaces will be perfectly smooth, there will be imperfections on the surfaces, such as waviness. This imperfect surfaces may cause the film thickness changes and the pressure as well. Hence, a transient effect needs to consider during the analysis of the EHL contact.

A realistic condition of the contacting surfaces of EHL point contact has been numerically considered in the past four decades, for example, indentations [1, 2], ridges [3], and waviness [4]. The results observed that the imperfect surfaces induce a change in the film thickness and pressure profile. However, the solvers were still in the assumption of the constant load conditions. Furthermore, Wijnant [5] presented the sinusoidal dynamic load effects in the field of EHL point contact and the results revealed that although the surfaces are perfectly smooth, the film thickness shape and pressure profile in the contact area are no longer flattened as shown in the case of the static load condition for the smooth contacting surfaces. In this paper, the combination of the sinusoidal dynamic load effect due to the harmonic change in the external load as well as the waviness effect on the contacting surface is considered by accounting for both squeeze effect in the Reynolds equation and the sinusoidally varying loads in the equation of motion.

2. Governing Equations

A contact of ball-raceway in a ball bearing application can be modelled as a contact between a spherical ball and a flat plate. First, the spherical ball subjected to the static load as an initial condition. Then, the load alters as the sinusoidal dynamic loads along the *z*-direction in the coordinate system. The surface of the inner raceway is wavy whereas the surface of the ball is assumed to be smooth.
In order to analyze the behaviors of the transient EHL point contact, some equations must be solved simultaneously, i.e. Reynolds, film thickness, lubricant properties, and load balance equations. The dimensionless Reynolds equation for the transient EHL point contact can be written as:

$$\frac{\partial}{\partial X}\left(\epsilon\frac{\partial P}{\partial X}\right)+\frac{\partial}{\partial Y}\left(\epsilon\frac{\partial P}{\partial Y}\right)-\frac{\partial(\bar{\rho}H)}{\partial X}-\frac{\partial(\bar{\rho}H)}{\partial T}=0 \tag{1}$$

The boundary conditions are $P(X_a,Y,T) = P(X_b,Y,T) = P(X,Y_a,T) = P(X,Y_b,T) = 0$, where X_a, X_b, Y_a, and Y_b are the boundaries of the domain. It should be noted that the cavitation occurs in the contact, and hence the condition of $P(X_a,Y,T) \geq 0$ must be satisfied. ϵ is defined according to:

$$\epsilon=\frac{\bar{\rho}H^3}{\bar{\eta}\lambda} \tag{2}$$

where $\bar{\rho}$, $\bar{\eta}$, and H denote the dimensionless density, dimensionless viscosity and dimensionless film thickness, respectively. λ can be calculated as:

$$\lambda=\frac{6\eta_0 u_s R_x^2}{a^3 p_h} \tag{3}$$

In this paper, the density and the viscosity of the lubricant are assumed to depend on the pressure, and hence the density-pressure relation proposed by Dowson and Higginson [6] as well as the Roelands pressure-viscosity relation [7] are employed. The dimensionless film thickness reads:

$$H(X,Y,T)=H_0(T)-\mathcal{R}(X,Y,T)+\frac{X^2}{2}+\frac{Y^2}{2}+\frac{2}{\pi^2}\int_{-\infty}^{+\infty}\int_{-\infty}^{+\infty}\frac{P(X',Y')dX'dY'}{\sqrt{(X-X')^2+(Y-Y')^2}} \tag{4}$$

where $H_0(T)$ is an integration constant and determined by the dimensionless force balance equation:

$$\int_{-\infty}^{+\infty}\int_{-\infty}^{+\infty}P(X,Y,T)dXdY=\frac{2\pi}{3} \tag{5}$$

Under dynamic load condition, Equation (5) changes to be [5]:

$$\sum_{i=1}^{M}\sum_{j=1}^{N}P(X,Y,T)=\frac{2\pi}{3}(1+\bar{A}\sin(\Omega_e T)) \tag{6}$$

where $\bar{A}$ and Ω_e denote the dimensionless amplitude and dimensionless frequency of vibration, respectively. $\mathcal{R}(X,Y,T)$ in Equation (4) stands for the undeformed geometry of the lower body at dimensionless time *T*. This undeformed geometry can be calculated as [4]:

$$\mathcal{R}(X,Y,T)=\begin{cases}0 & , X\geq X_d\\ \bar{A}_n\sin\left(2\pi\frac{X_d-X}{\bar{W}_n}\right), & X<X_d\end{cases} \tag{7}$$

where $\bar{A}_n$ and $\bar{W}_n$ denote the dimensionless amplitude and the dimensionless wavelength of the waviness (see Nomenclature), respectively. X_d is the dimensionless location of the first wave passing into the contact and defined as:

$$X_d = X_s + \frac{u_2}{u_s} T \tag{8}$$

3. Numerical Simulation

First, the stationary condition was solved using the initial static load applied to the contact. To solve the Reynolds equation for stationary problem, the pressure distribution of the Hertzian dry contact was employed as the initial pressure using Jacobi relaxation scheme, and along with the Half-space theory for solving the film thickness equation as well as the lubricant properties relations, the new pressure profile was obtained. Then, the film thickness, the density as well as the viscosity of the lubricant were updated based on the new pressure. Subsequently, the stationary result was used to solve the transient EHL point contact by accounting for the squeeze-term effect in the Reynolds equation and the sinusoidal dynamic load effect in the force balance equation. The same procedure as the stationary case was employed for each time step simulation. It should be noted that in order to ensure the whole waviness effect has been included in this investigation, the effect of the sinusoidal load was calculated after the waviness left the contact, i.e. $X_d = 1.5$.

4. Results and Discussion

4.1 Stationary result

Table 1. Different parameters and their values for stationary and sinusoidally varying loads cases

Parameter	**Value**
Load, *F* (N)	20
Rolling speed, u_m (m/s)	1
Slide-to-roll ratio, *SRR*	0
Viscosity at ambient pressure, η_0	0.0835
Viscosity index	1.95×10^{-8} Pa^{-1}
Material parameter, *G*	2279
Dimensionless oscillation amplitude, $\bar{A}$	0.1
Dimensionless excitation frequency, Ω_e	2π

In the present paper, parameters for stationary case, i.e. operating conditions and lubricant properties are shown in Table 1. This table also shows the sinusoidal dynamic load parameters for the case of dynamic load condition. Load of 20 N and rolling speed of 1 m/s result in the maximum Hertzian dry contact of 0.53 GPa with 0.135 mm radius contact. Slide-to-roll ratio $SRR = 0$ means both surfaces move with the same velocity of 1 m/s, and the sliding motion is absent in this condition, i.e. pure rolling condition. The numerical domain for this solver is $-2.5 \leq X \leq 1.5$ and $-2.0 \leq Y \leq 2.0$ with a single grid solution consists of 129×129 points. The spatial mesh was $\Delta X = \Delta Y = 0.03125$ and the increment time was set to be equal to the spatial mesh, $\Delta T = 0.03125$. This corresponds to the 0.0042 milliseconds. Figure 1 shows the stationary result of film thickness for the smooth contacting surfaces by means of the reverse three-dimension plot of film thickness together with the centerline film thickness and pressure profile along *X*-direction. From the plots, it can be observed that all familiar features of EHL point contact for the medium load are displayed, such as the pressure distribution is almost semi-ellipsoid and the pressure spike occurs at the exit region as well as the well-known horse shoe shape in the contact. The central film thickness at $Y = 0$ is nearly flattened, and the minimum film thickness is found in the side lobes of the horse shoe shape near $X = 0$, $Y = \pm 1$.

In order to investigate the effect of wavy surface on the EHL behaviors, the quasi-transient analysis was performed by accounting for the squeeze-term in the Reynolds equation whereas the load is still assumed to be constant. Parameters of waviness were used in this simulation, i.e. $\bar{A}_n = 0.05$ and $\bar{W}_n = 2.0$ which give the waviness on the surface with 0.024 μm amplitude and 0.27 mm wavelength. This wavelength causes the waviness with one wave contacting the smooth surface. Figure 2 shows the influence of waviness on the film thickness and pressure profile. From the plots, it can be seen that the central film thickness shape is no longer flattened. However, the waviness on one contacting surface is almost completely deformed in the contact area, due to the high pressure in this area. The change in the pressure profile is also observed due to waviness in which the pressure is no longer semi-ellipsoid in the contact area.

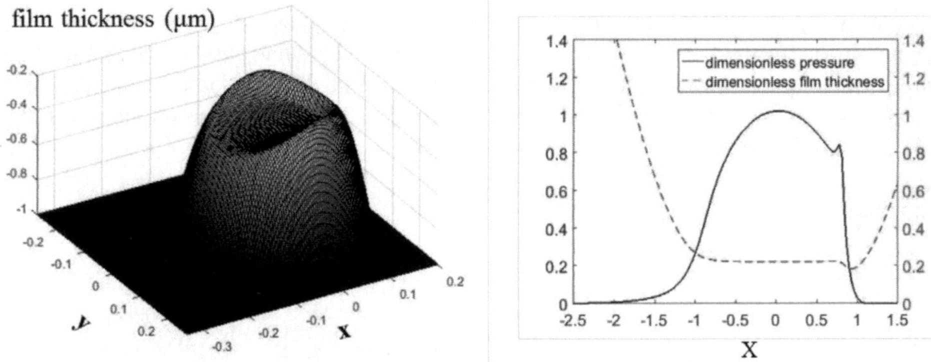

Figure 1. Reverse three-dimension plot of film thickness shape (left) and centerline pressure and film thickness along *X*-direction for the smooth contacting surfaces

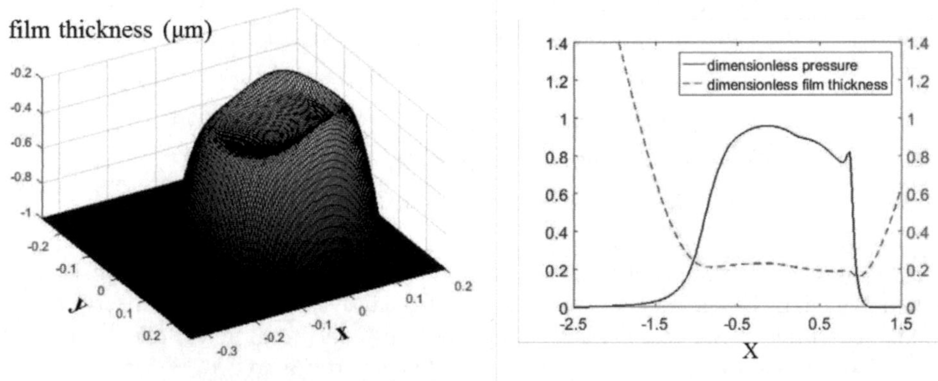

Figure 2. Reverse three-dimension plot of film thickness shape (left) and centerline pressure and film thickness along *X*-direction for the contact between wavy and smooth surfaces.

4.2 Transient result

The effect of changes in the applied load due to external sinusoidal dynamic load on the film thickness and pressure profile can be clearly seen in Figure 3 for the case of smooth contacting surfaces. Using the dimensionless frequency excitation of 2π and dimensionless oscillation amplitude of 0.1, Figure 3 captures the centerline dimensionless pressure and film thickness at the different momentarily time steps. From these plots, one can observe that although the contacting surfaces are assumed to be perfectly smooth, there are modulations in the film thickness and pressure profile due to dynamic load. Moreover, since the external dynamic load applied to the contact changes sinusoidally, the changes in the film thickness as well as the pressure profile are periodic. This periodicity starts after the initially induced film thickness changes have propagated through the contact, for example, in this case the periodicity start after $T \geq 2$. Taking a closer look at Figure 3, film thickness and pressure profile at T = 2.0 are coincident with those of at T = 3.0, and so do to the EHL behaviors at T = 2.5 which is similar to the EHL behaviors at T = 3.5. These phenomena are in good agreement with result shown in [5].

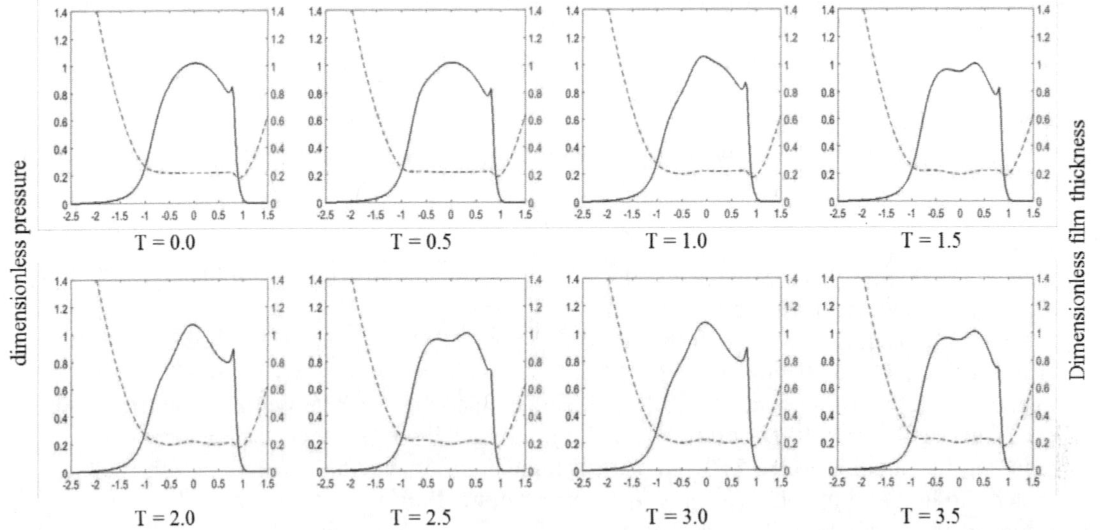

Figure 3. Centerline dimensionless pressure (left labels) and film thickness (right labels) for $\Omega_B = 2\pi$ and $\bar{A} = 0.1$ at different momentarily dimensionless time for the smooth contacting surfaces

To illustrate the combined effects of wavy surface and the sinusoidal dynamic load from external vibration applied to the lubricated contact, Figure 4 shows captures of the changes in the centreline film thickness and pressure profile along the X-direction at different times. Similar to the smooth contacting surfaces case, the film thickness and pressure profile modulate as the load applied to the contact changes sinusoidally. However, due to the waviness onto one surface, the modulations of film thickness and pressure profile are no longer periodic.

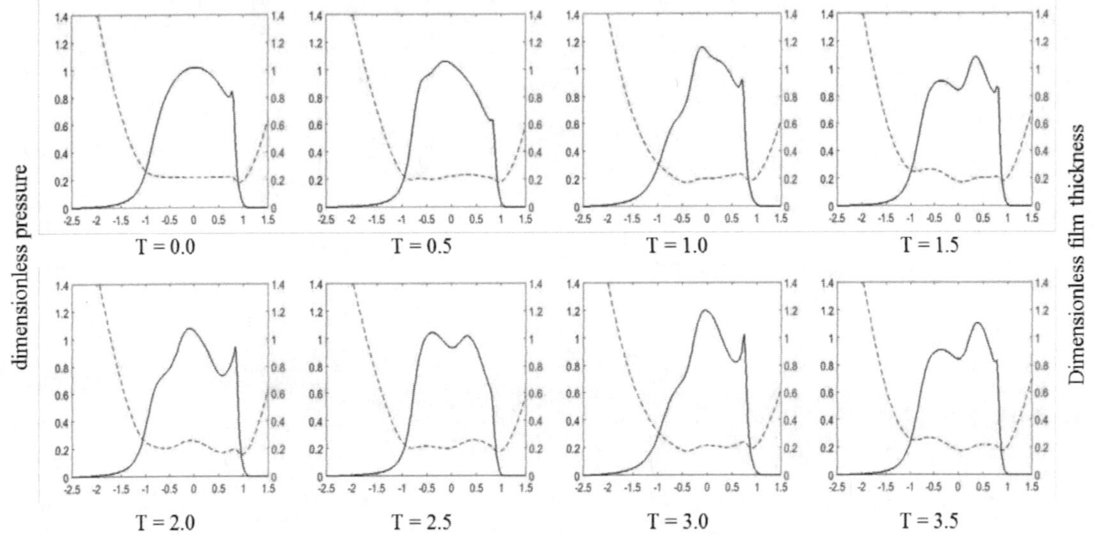

Figure 4. Centerline dimensionless pressure (left labels) and film thickness (right labels) for $\Omega_B = 2\pi$ and $\bar{A} = 0.1$ at different momentarily dimensionless time for the wavy contacting surface

5. Conclusion
In this work, a simple transient EHL point contact solver has been developed to investigate the influences of the waviness and sinusoidally varying load on the EHL behaviors. The simulation results reveal that the external dynamic load induces modulations on the film thickness and pressure profile in the lubricated contact. The waviness on one contacting surface causes the changes in the film thickness and pressure profile to become more pronounced.

Acknowledgement
This research is financially supported by a fellowship of the Indonesian Endowment Fund for Education (LPDP).

References
[1] Venner, C. H. Multilevel Solution of the EHL Line and Point Contact Problems. PhD Thesis, University of Twente, Enschede, The Netherland. 1991.
[2] Ai, X., and Cheng, H. S. The Influence of Moving Dent on Point EHL Contacts. Tribology Transactions. 1994; 37(2); pp. 323-335.
[3] Venner, C. H., and Lubrecht, A. A. Numerical Simulation of a Transverse Ridge in a Circular EHL Contact Under Rolling/Sliding. ASME Journal of Tribology. 1994; 116; pp. 751-761.
[4] Venner, C. H., and Lubrecht, A. A. Numerical Simulation of Waviness in a Circular EHL Contact, Under Rolling/Sliding. Lubricants and Lubrication: Proceedings of the 21th Leeds-Lyon Symposium on Tribology (Ed. Dowson et. al). 1995; 30; pp.259-272.
[5] Wijnant, Y. H. Contact Dynamics in the Field of Elastohydrodynamic Lubrication. Ph.D. Thesis, University of Twente, Enschede, The Netherland. 1998.
[6] Dowson, D., and Higginson, G. R. Elastohydrodynamic Lubrication. The Fundamentals of Roller and Gear Lubrication. Pergamon Press, Oxford, Great Britain. 1996.
[7] Roelands, C. J. A. Correlation Aspects of the Viscosity-Temperature-Pressure Relationship of Lubricating Oils. Ph.D. Thesis, Technical University Delft, Delft, The Netherland. 1966.

The characterization tensile and thermal properties of hibiscus tiliaceus cellulose fibers

N H Sari[1] and Y A Padang[1]
[1]Department of Mechanical Engineering, Faculty of Engineering, Mataram University, West Nusa Tenggara, Indonesia. Jalan Majapahit No 62. Mataram, West Nusa Tenggara, Indonesia.

Email: n.herlinasari@unram.ac.id

Abstract. In our study, hibiscus tiliaceus (HT) fibers were treated with potassium hydroxide (KOH) 8% for two hours at ambient temperature. The tensile strength and thermal properties of the treated fibers were examined by using tensile test and Thermogravimetric analysis (TGA). The surface structure of the hibiscus tiliaceus fibers were analysed using Scanning Electronic Microscopy (SEM). Experimental results revealed that under KOH treatment, the tensile strength increased up to 5144.9 MPa, resistance thermal and cellulose of the fibers become increased; due to hemicellulose content were reduced of the fibers. SEM images show that after alkali, the hibiscus tiliaceus fibers surface becomes clear, fibril and rougher.

1. Introduction

One source of fiber that is cheap, eco–friendly, abundant, and not harmful to human health is fiber from plants such as corn husks, Jute, Pandannus Amarylliofius, and Kapok [1,2]. Natural fibers have been used as a filler for composite resin [3–6], and they have been applied in the field of building construction, and automotive [4].

However, the use of natural fiber as filler in polymer materials is less preferred due to a number of hydroxyl groups and impurities attached to the surface of cellulose fibers, which causes them to be incompatible with resin [7,2]. Surface modification of natural fibers using chemical and physical treatments is a method to produce surface roughness and the reactive hydroxyl groups to form bonds with polymeric materials [1,7,8].

Many researchers investigated the effect of alkali treatment using sodium hydroxide (NaOH) on the surface of natural cellulose fibers. Most of the reported works focused on the mechanical properties of the fibers and its composites [1,9]. After NaOH treatment, the elongation, Young' moduli and tensile strength of natural fiber become increased [10,11,1]. On the contrary, investigation related to the effect of potassium hydroxide (KOH) on natural fiber properties still limited. Abdul Khalil et al. [12] analyzed the effect of activated carbon from oil palm empty fruit bunch, bamboo stem and coconut shell treated by potassium hydroxide and phosphoric acid on the flexural properties of epoxy–nanocomposites. They reported that potassium hydroxide (KOH) activated carbon-based epoxy nanocomposites had a higher flexural strength value compared to phosphoric–acid activated carbon. Furthermore, Pan et al. [13] reported the effect of $Ca(OH)_2$, KOH and NaOH as alkalinity on deacetylation of a konjac–glucomannan. These previous studies have shown that the influences of KOH treatment on the properties of the natural fibers still had not investigated.

Therefore, this paper has been done to provide comprehensive data and discussion on the thermal, and tensile properties of hibiscus tiliaceus fibers treated with potassium hydroxide (KOH). SEM, TGA and tensile strength analysis of the fibers were performed to expand an understanding of the single fiber constituent behaviors. These constituent behaviors were used to systematically analyze treatment effects on fiber tensile strength and thermal properties. The study suggests that KOH treated hibiscus tiliaceus (HT) fibers can be as an alternative reinforce in polymer composites fabrication.

2. Material and Methods

2.1 Raw materials

Hibiscus tiliaceus (HT) skins were obtained from around West Lombok, West Nusa Tenggara, Indonesia. For alkalization, a commercial grade of Potassium hydroxide (KOH) with concentrate 8% was used.

2.2 Extraction of fibers

HT skins were soaked in water for one week to undergo microbacterial. Then, they were combed with a wood brush to retained a bundle of fibers and dried through wind convection.

2.3 KOH treatment of fibers

In this process, HT fibers were immersed in 8% KOH solutions at ambient temperature (32°C) for 2 hours. After KOH treatment, the fibers were thoroughly washed six times with distilled water in order to remove the KOH adhering to the surface of the HT fibers and dried in an oven maintained at 70 °C. The physical properties of fiber as shown in Table 1.

Table 1. Physical properties of untreated and KOH treated hibiscus tiliaceus (HT) fibers.

Fiber types	Diameter (cm)	Density gr/cm^3	Moisture content (%)
Untreated	0.41 ± 0.02	0.76 ± 0.01	11.32 ± 0.01
Treated KOH	0.27 ± 0.02	0.68 ± 0.01	7.49 ± 0.01

2.4 Characterization

Tensile strength, the morphology and thermogravimetric (TGA) of the surface of untreated and KOH treated HT fibers were characterized.

2.4.1. Tensile strength

Individually fibers (untreated and treated fibers) prepared. All the samples were stored in a chamber at 21 °C and 63 % relative humidity for 24 h for proper conditioning as determined in ASTM D 1776–04 [14] before testing. The maximum load of 20 N was applied and a crosshead speed of 5 mm/min with a gage length of samples of 25 mm. In each treatment, there are five fibers samples were tested.

2.4.2 Morphology of fiber Surface

The surfaces of KOH treated and untreated HT fibers were evaluated using scanning electron microscopy, SEM (FEI Inspect S50) and has been compared with untreated fibers. The samples were coated with gold using a plasma sputtering–apparatus and accelerating voltage 10 kV.

3. Results and discussion

3.1. Fiber diameter

The diameter of HT fibers was decreased by 34 ± 0.2% upon 8% KOH treatment at 32°C (see in Table 1). After KOH treatment, the diameter reduction was caused by the deletion of lignin and hemicellulose in the fiber. These results indicate that 8% KOH of the alkaline treatment can eliminate noncellulosic materials (hemicellulose and lignin) more effective than using 8% NaOH as reported by Sari et al.[1]. They reported a 22% reduction of the cornhusk fibers diameter upon treatment with 8% NaOH at 31°C for 2 h.

3.2. Chemical analysis

The content of fiber constituents of treated fibers is presented in Table 2. From Table 2 it can be noted that after 8% KOH treatment, the hemicellulose, and lignin in the fiber were decreased compared to the untreated HT fibers (0% KOH); consequently, the cellulose content was increased because KOH treatment has no effect on cellulose molecules. These results indicated that the treatment of KOH dissolved most of the hemicellulose and lignin from the fiber.

Table 2. The chemical content of untreated and treated hibiscus tiliaceus fibers.

Fiber types	Cellulose (%)	Hemicellulose (%)	Lignin (%)
Untreated	64.0	27.1	17.3
Treated KOH	71.4	20.2	9.8

3.3. Tensile strength

Under 8% KOH treatment at 32°C, the tensile strength of HT fibers was observed to increase by 700% (5144.9 MPa) compared to untreated HT fibers (650 ± 110 MPa) (Figures 1 and 2) and the elongation was noted to increase by 61.3% (Figures 3a and 3b). The increase in strength was related to the decrease in diameter and roughness surface on HT fibers; due to lignin, or hemicellulose from the more effective interfibrillar region removed.

This result is significantly higher than the value obtained by NaOH treated cornhusk fibers developed by Sari et al. [1], which showed that the tensile strength of the 8% NaOH treated CHFs was 368.25 ± 78.97 MPa. It proves that the KOH–treatment on fiber surface can be a good alternative to the filler in the composite material.

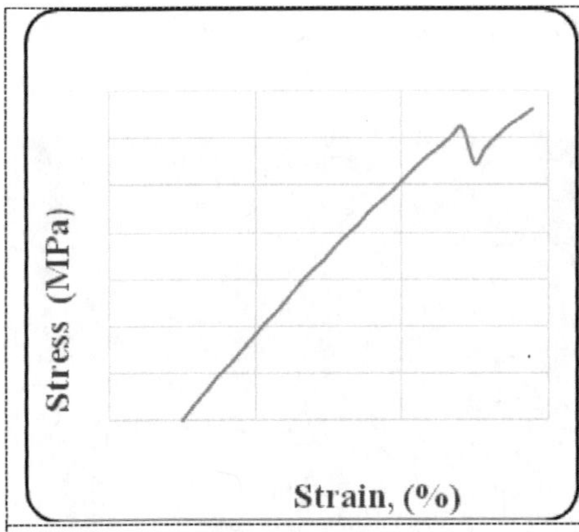

Figure 1. Stress vs strain of untreated hibiscus tiliaceus fibers.

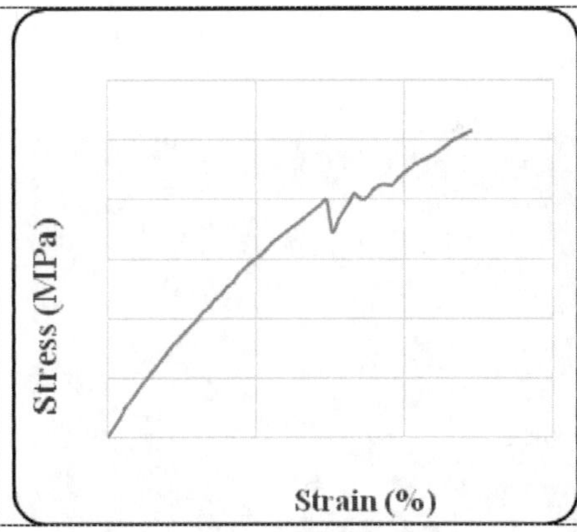

Figure 2. Stress vs strain of KOH 8% treated hibiscus tiliaceus fiber.

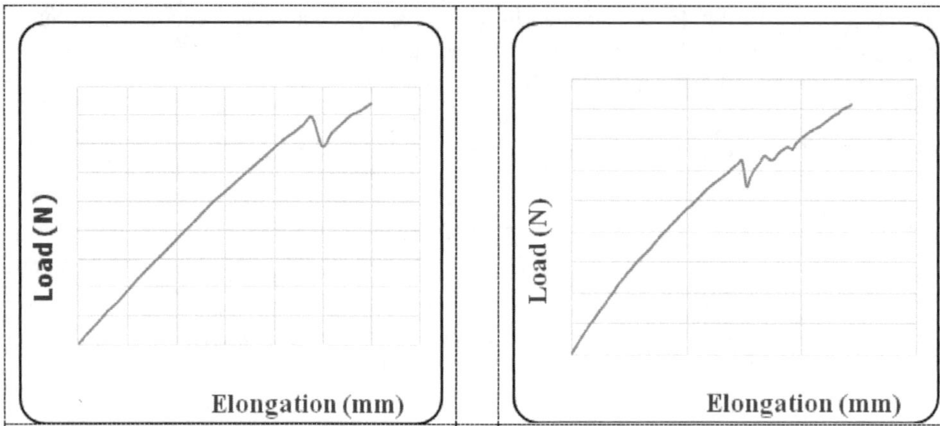

(a) **(b)**

Figure 3. Load vs elongation of hibiscus tiliaceus fibers, (a). Untreated, (b). Treated KOH 8%.

3.4. Surface morphology
The SEM photo exhibited that after 8% KOH treatment, the surface structure of the HT fiber become rougher and fibril. It is attributed to the removal of noncellulosic (hemicellulose and lignin) substances and impurities on the HT fibers (Figures 4a and 4b). These investigations indicated that the strength improvement of the HT fibers related to defibrillation. When the fibers were used as a filler in composite resins, the surface of chemically treated fiber allowed them to form chemical bonds between the fibers/matrix [2].

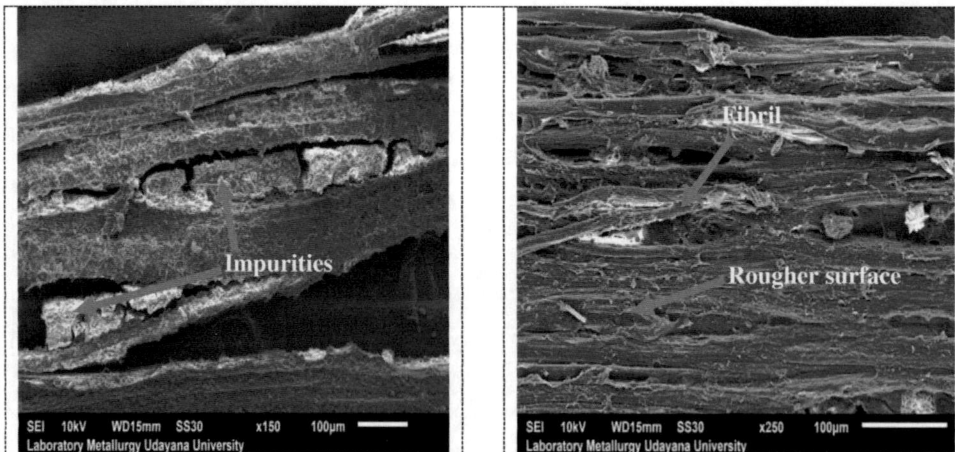

(a) **(b)**

Figure 4. SEM image of hibiscus tiliaceus fibers surfaces, (a). untreated, (b). Treated KOH 8%

3.5 Thermogravimetric Analysis
From Figure 5 it can be observed that at about 220 °C weight loss was very small. The large weight loss occurred at temperature ranges 280–370 °C. Kabir et al. (2013) stated that the cellulose was decomposed (250–350 °C) caused molecular structures of cellulose broken down. Then, Placet

(2009)[16] stated that cellulose has a crystalline structure containing strong intermolecular hydrogen bonding and intramolecular, so it takes a lot of energy to break the molecular structure of cellulose [16–18].

The decomposition of hemicellulose presented three steps of decomposition. The first step (25 –180 °C) exhibited the loss of moisture from hemicellulose. In the second step (180 – 280 °C), the majority of the hemicellulose constituents decomposed with fast weight losses (water evaporated). Then, at the last step (280 – 600 °C), they decomposed leaving a residue of charcoal.

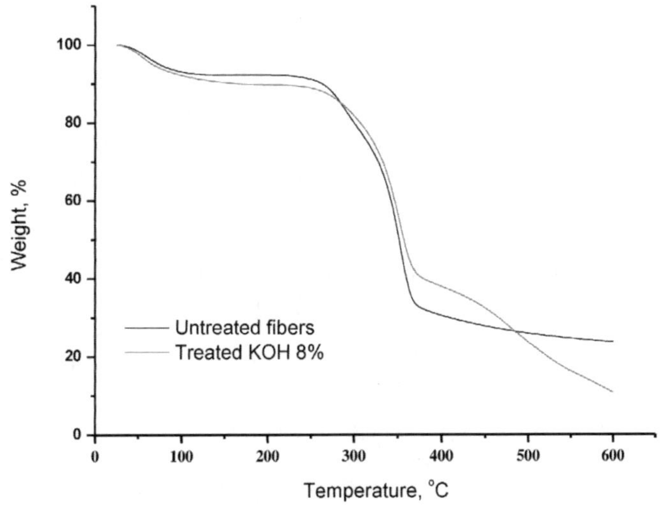

Figure 5. Thermogravimetric (TGA) of hibiscus tiliaceus fibers.

Thermal decomposition of lignin constituents occurred at 150–550 °C. At the heating temperature, lignin formed aromatic hydrocarbons hydroxyphenolics, and syringyl [19]. At higher temperatures, all components of lignin which contain –OH groups became unstable; this is because they experience a radical reaction between themselves which results in the structural rearrangement. The existence of these molecular activities causes degradation of lignin at prolonged temperatures and a smaller weight loss.

Figure 5 also shows alkalize fiber samples. Under alkali treatment, the decomposition temperature of the HT fiber is higher than untreated fibers. In the range of 250–350°C, weight reduction of KOH treated HT fibers is smaller than untreated fibers; due to the reduction of hemicellulose and lignin constituents. Consequently, the decomposition process occurs in cellulose, so that the degradation temperature of the KOH treated HT fiber increases.

4. Conclusions

On KOH treatment, the tensile strength of the HT fiber increased by up to 700%. The increase of elongation of 4.75 mm exhibited that the KOH treated fibers maintained their flexibility. Tensile and thermal characterization of HT fibers noted that the increase in strength and thermal resistance were due to the removal of hemicelluloses. This treatment has a potential and viable structural application.

References

[1]. Sari N H, Wardana I N G, Irawan Y S, Siswanto E 2018 Characterization of the Chemical, Physical, and Mechanical Properties of NaOH–treated Natural Cellulosic Fibers from Corn Husks. *Journal of Natural Fibers* **15**(4) 545–558. doi: 10.1080/15440478.2017.1349707.

[2]. Mwaikambo L, and Ansell M 2001 Chemical Modification of Hemp, Sisal, Jute, and kapok FibersbyAlkalization. *Journal of Applied Polymer Science* **84** 2222–2234.
[3]. Syafri E, Kasim A, Abral H, Sudirman, Sulungbudi G T, Sanjay M R, Sari N H 2018 Synthesis and characterization of cellulose nanofibers (CNF) ramie reinforced cassava starch hybrid composites. *International Journal of Biological Macromolec*ules **120** 578–586. https://doi.org/10.1016/j.ijbiomac.2018.08.134.
[4]. Sari N H, Wardana I N G, Irawan Y S, Siswanto E 2017 Corn Husk Fiber–Polyester Composites as Sound Absorber: Nonacoustical and Acoustical Properties. *Advances in Acoustics and Vibration* Volume 2017. https://doi.org/10.1155/2017/4319389.
[5]. Sari N H, Yudhyadi I G N K 2013 Karakteristik Kekuatan Bending Kayu Komposit Polyester Diperkuat Serat Pandan Wangi dengan Filler Serbuk Gergaji Kayu. *Jurnal Energi Dan Manufaktur* **6**(2): 157–164.
[6]. Sastra I P K A, Sari N H, Sujita 2013 Analisis Uji Penyerapan Air Dan Struktur Mikro Komposit Laminate Hybrid Serat Sisal Dan Batang Pisang Dengan Matrik Epoxy. *Dinamika Teknik Mesin* **3** (1) 41–49. http://dinamika.unram.ac.id/index.php/DTM/article/view/87/82
[7]. Sari N H, Wardana I N G, Irawan Y S, Siswanto E 2017 The Effect of Sodium hydroxide on Chemical and Mechanical Properties of Corn Husk Fiber. *Oriental Journal of Chemistry* **33**(6) 3037–3042.
[8]. Saha P, Manna S, Chowdhurya S R, Sen R, Royc D, Adhikari B 2010 Enhancement of tensile strength of lignocellulosic jute fibers by alkali–steam treatment. *Bioresource Technology* **101** 3182 – 3187.
[9]. Kabir M M, Wanga H, Lau K T, Cardona F 2013 Effects of chemical treatments on hemp fibre structure. *Applied Surface Science* **276** 13– 23.
[10].Bilba K, Arsene M A 2008 Silane treatment of bagasse fiber for reinforcement of cementitious composites. *Composites: Part A* **39** 1488–1495.
[11].Sari N H, Wardana I N G, Irawan Y S, Siswanto E 2016 Physical and acoustical properties of corn husk fiber panels. *Advances in Acoustics and Vibration* Article ID 2265971814, 8 pages. https://doi.org/10.1155/2017/4319389.
[12].Abdul Khalil H P S, Jawaid M, Firoozian P, Alothman O Y, Paridah M T, Zainudin E S 2014 Flexural Properties of Activated Carbon filled Epoxy Nanocomposites. *The Malaysian Journal of Analytical Sciences* **18**(2) 391 – 397.
[13].Pan Z, Meng J Wang Y 2001 Effect of alkalis on deacetylation of konjac glucomannan in mechanochemical treatment. *Particuology* **9** 265–269.
[14].ASTM 2004. *ASTM D1776–04 Standard Practice for Conditioning and Testing Textiles. American Society for Testing and Materials*, West Conshohocken, Pennsylvania, USA.
[15].Han G, Chenga W, Deng J, Daic C, Zhan S, Wud Q 2009 Effect of pressurized steam treatment on selected properties of wheat straws *Industrial Crops and Products* **30** 48–53.
[16].Placet V 2009 Characterization of the thermo–mechanical behavior of Hemp fibres intended for the manufacturing of high–performance composites. *Composites Part A: Applied Science and Manufacturing* **40**(8) 1111–1118.
[17].Tajeddin B, Abdul Rahman R, Chuah L, et al 2009 Thermal properties of low–density polyethylene–filled kenaf cellulose composites. *European Journal of Scientific Research* **32**(2) 223–230.
[18].Thakur V K, Singha A S 2010 Physico–chemical and mechanical characterization of natural fibre reinforced polymer composites, *Iranian Polymer Journal* **19**(1) 3–16.
[19].Brebu M, Vasile C 2010 Thermal degradation of lignin – a review, *Cellulose Chemistry & Technology* **44**(9) 353.

IOP Conf. Series: Materials Science and Engineering **539** (2019) 012032 doi:10.1088/1757-899X/539/1/012032

Performance of vertical axis Savonius wind turbines related to the fin number on the blade

Ridwan[1], I Setyawan[1], and Setiyono[2]
[1]Mechanical Engineering Department, Gunadarma University Indonesia
[2]Electrical Engineering Department, Gunadarma University

Email: ridwan@staff.gunadarma.ac.id

Abstract. Model and the number of blades on the wind turbine greatly affect its performance. The purpose of this research is to know the effect of the fin on Savonius turbine U type 4 number blade. Simulation using Solidworks (Flow Simulation) software based on Finite Element Analysis (FEA) Method, will show pressure distribution and velocity distribution of blade rotor savonius wind turbines. The simulations to compare on the variation of the addition of 1 fin, 2 fins, and blade without fin (standard). The wind speed variations applied to each of the blades is 5 m/s, and 7 m/s respectively. The simulation results showed that the performance of pressure distribution and velocity distribution on each blade is higher and more widespread and evenly distributed, occurring on each blade with the addition of two fins compared to a single fin turbine and standard blade (without the addition of fin). The higher wind speed applied to the turbine blades will have a significant effect on the value of pressure distribution and velocity distribution on the blade. The effectiveness of pressure distributions and velocity distributions that occurs on the turbine blades will be significant to effect for the rotation on the turbine shaft, so as to have a significant impact on the performance the turbines.

1. Introduction

Turbines are classified into two categories, namely horizontal axis wind turbines (HAWTs) and vertical axis wind turbines (VAWTs). Where, HAWT has a horizontal rotation axis, while VAWT has a vertical rotation axis. The advantages of VAWT, is able to produce power at relatively low wind speeds compared to HAWTs. In addition, VAWTs are also easy to install [1]. VAWT has a simpler structure and installation than HAWT. VAWT generators are placed at the bottom of the central shaft on the ground so it does not require to be supported by the tower. Turbines are useful in different wind speeds and directions [2]. However, VAWTs have low pressure coefficients, so the scope for research, especially in VAWT rotors is still wide open which is intended to improve its performance. VAWT rotors have different types, such as Savonius, Darriues, and H-Darrieus rotor rotors [4]. Greenpeace predicted that wind power is able to reach nearly 2000 GW by 2030 that supplying 16.7% to 18.8% of global electricity, helping save over 3 billion tons of CO2 emissions annually [5,6].

As a simplest turbine, Savonius wind turbine works due to the difference of forces exerted on each blade. The concave part to the wind direction caught the air wind and forces the blade to rotate around its central vertical shaft. Otherwise, the convex part hits the air wind and causes the blade to be deflected sideway around the shaft. The blades curvature has less drag force when moving against the wind or *Fconvex* than the blades moving with the wind or *concave* as seen in Fig. 3. Hence, concave blades with more drag force than the other half cylinder will force the rotor to rotate. [3].

IOP Conf. Series: Materials Science and Engineering **539** (2019) 012032 doi:10.1088/1757-899X/539/1/012032

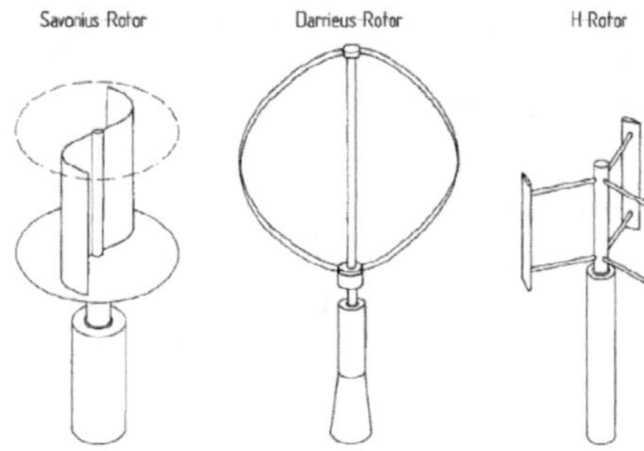

Figure 1. Types of vertical axis wind turbines [6]

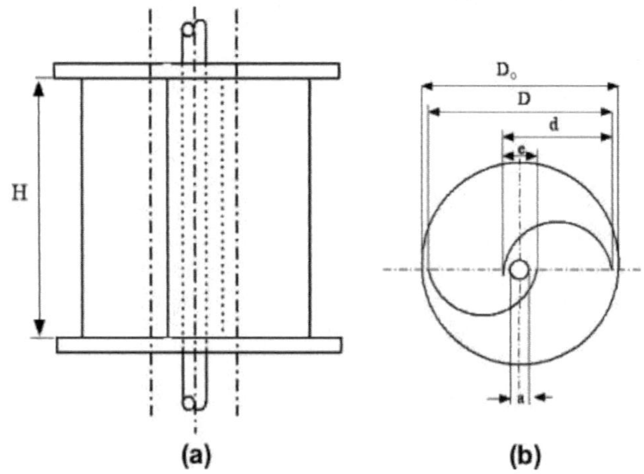

Figure2. Savonius wind rotor[4,7]: (a) Front view ; (b). Top view

As for the characteristics of the type of VAWT wind turbine has a variety of advantages [5]

1. Blade has a basic design that is simpler construction than other types of VAWT
2. Able to leverage the flow of the wind from all directions and thus no yaw mechanism construction as found on HAWT
3. The construction is simple and the manufacturing costs are also more affordable than HAWT
4. The treatment process easier because of the simple construction and the number of components that are above the soil surface
5. Being able to perform self start at relatively low wind velocitys
6. Able to produce higher torque from low wind velocity range to high wind velocitys.

2. Methodology

In this study, a VAWT was developed. Where the model of the wind turbine was built with overlap ratio (e : d) equal to 0.15; aspect ratio (D: h) equal to 1.0 and end plate parameter (D0 : D) equal to 1.1. The model of blades we described in Figure 3. In this figure, the diameter of the blade (d) = 200 mm, rotor height = 370 mm, and thickness of blades and end plates = 2 mm. [3,6]

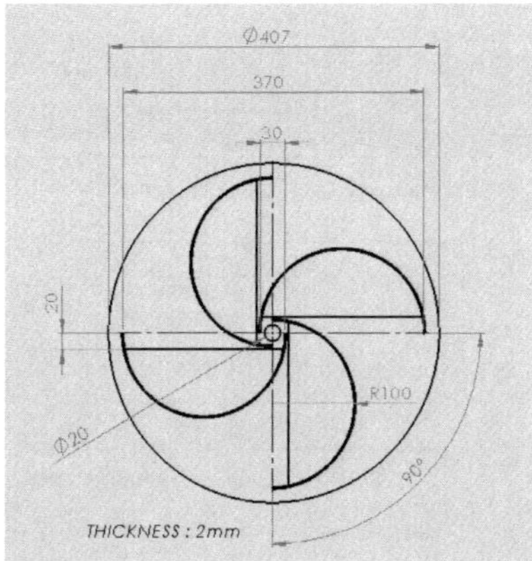

(a)

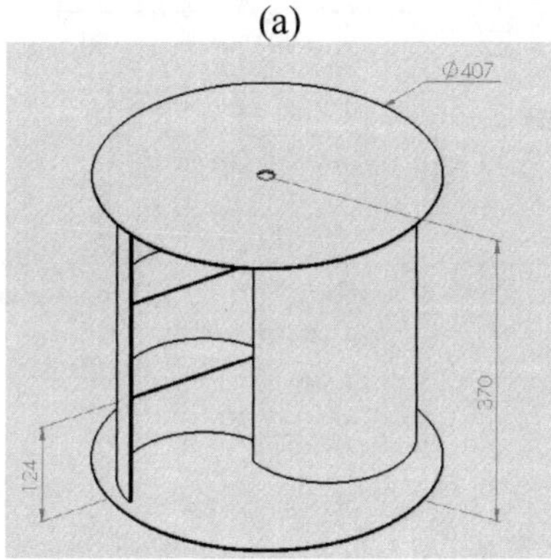

(b)

Figure 3. , Blade dimentions. (a)Top view, (b) cross section

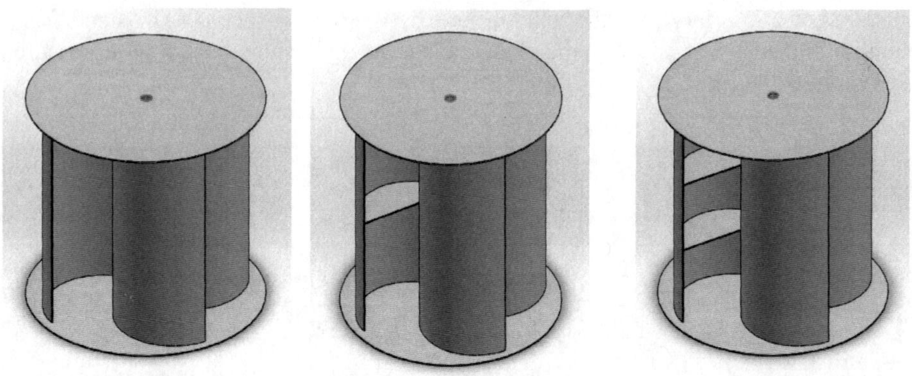

(a) (b) (c)

Figure 4. Blade rotor U Type. (a) Blade without fin, (b) blade with 1 fin, (c) Blade with 2 fin .

The flowchart for running simulations using Solidworks software, The flow Simulations based on Finite Element Analysis (FEA) Method, as shown in figure 5. [9].

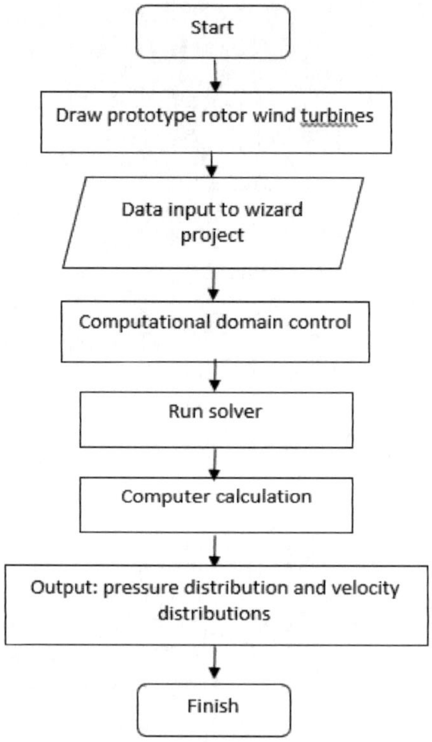

Figure 5. Flowchart Simulation

3. Result and Discussion

Results of simulations indicated the relationships between wind speeds 5 m / s and 7 m / s as well as the number of the fin was added to the blades. Fig.6, to fig.9, show the cut plot pressure distribution and velocity distribution.

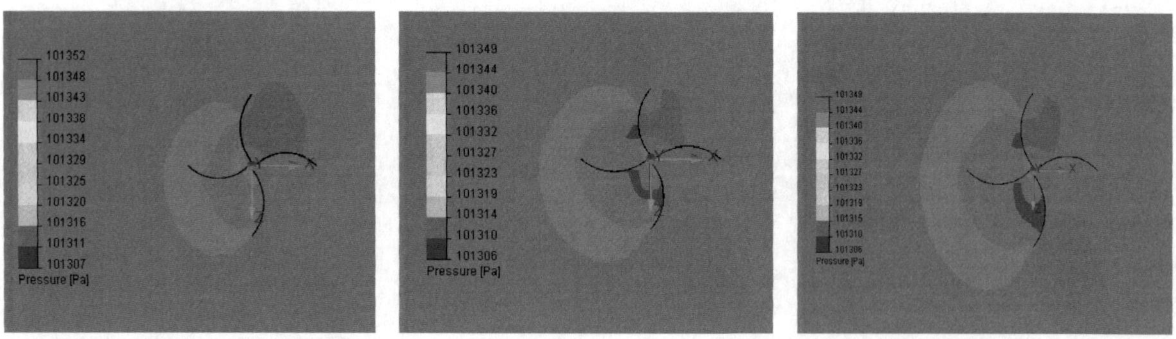

(a) (b) (c)

Figure 6. Cut Plot Pressure at wind speed 5 m/s. (a) 0 Fin (b) 1 Fin (c) 2 Fin.

Figure **6** shows the cut plot pressure of the U-type blade with a wind speed of 5 m/s. In Figure 6 (a), can be seen that the pressure area for the U-blade without the fin, 0 fin where the largest Pressure is 101,343 Pa and the smallest Pressure is 101,311 Pa. Fig.6(b) for the U-type blade with 1 Fin, can be seen that the largest Pressure is 101,344 Pa and the smallest Pressure is 101,310 Pa. Furthermore, figure 6(c) shows the U-type blade with 2 Fin where the largest Pressure is 101,344 Pa and the smallest Pressure is 101,310 Pa. The area of high pressure increases with the addition of fin, where Figure 6 (c) is the largest area with a Pressure 101,344 Pa.

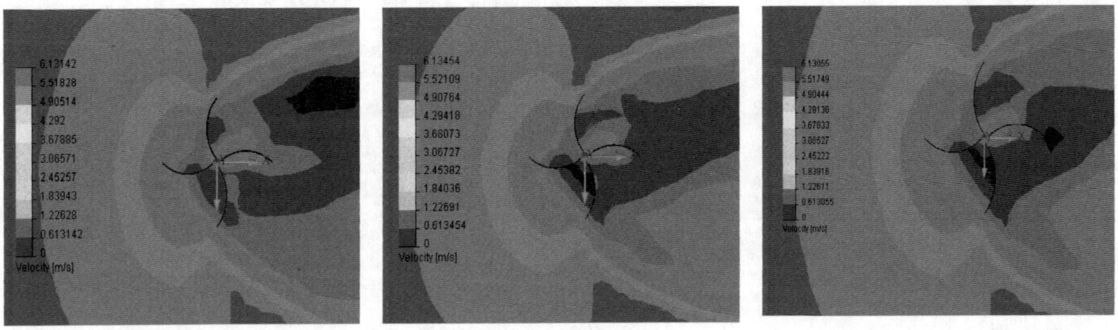

(a) (b) (c)

Figure 7. Cut Plot Velocity at wind speed 5 m/s. (a) 0 Fin (b) 1 Fin (c) 2 Fin.

Figure 7 shows the cut plot velocity of a U-type blade with a wind speed of 5 m/s. In Figure 7 (a), can be seen that the area of wind velocity for U-type blades without fin/ blade standard has the highest velocity of 1.839 m/s and the lowest velocity of 0.613 m/s. In figure 7 (b) which is the U-type blade with 1 fin, has the highest velocity of 1.840 m/s and the lowest velocity of 0 m/s. Furthermore, in figure 7 (c) which is a U-type blade with 2 fins has the highest velocity of 1,839 m/s and the lowest velocity of 0 m / s. The widest area velocity is 1,839 m/s.

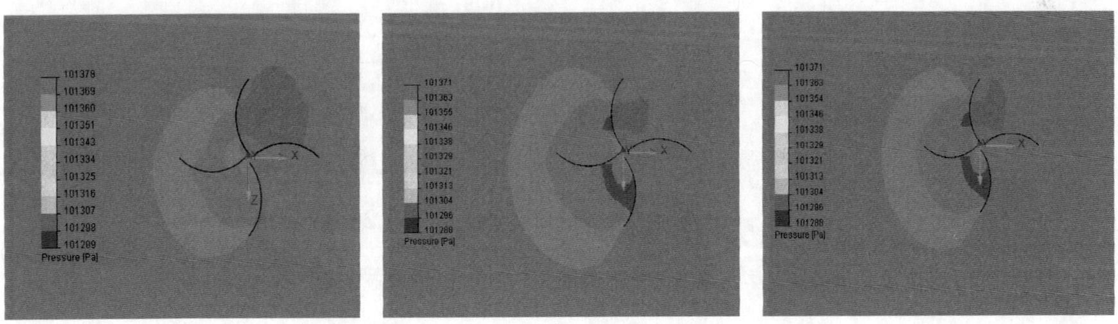

(a) (b) (c)

Figure 8. Cut Plot Pressure at wind speed 7 m/s . (a) 0 (without) Fin (b) 1 Fin (c) 2 Fin.

Figure 8 shows the cut plot pressure of the U-type blade with a wind speed of 7 m/s. In In Figure 8 (a), described the pressure area by a U-type with four of the blade without modification (standard blade) wherein the largest pressure is 101,360 Pa and the smallest Pressure is 101.298 Pa. In Figure. 8 (b), illustrated the pressure of the U-type blade with 1 fin where the largest Pressure is 101.363 Pa and the smallest Pressure is 101.298 Pa. In Figure. 8 (c), for the type U blade with 2 fins, can be seen the plot where the largest Pressure is 101,363 Pa and the smallest pressure is 101.298 Pa. The area of

IOP Conf. Series: Materials Science and Engineering **539** (2019) 012032 doi:10.1088/1757-899X/539/1/012032

pressure is increasing with the addition of the fin, in which for both Figure 8 (b) and Figure 8(c) have a pressure area equal to 101,363 Pa.

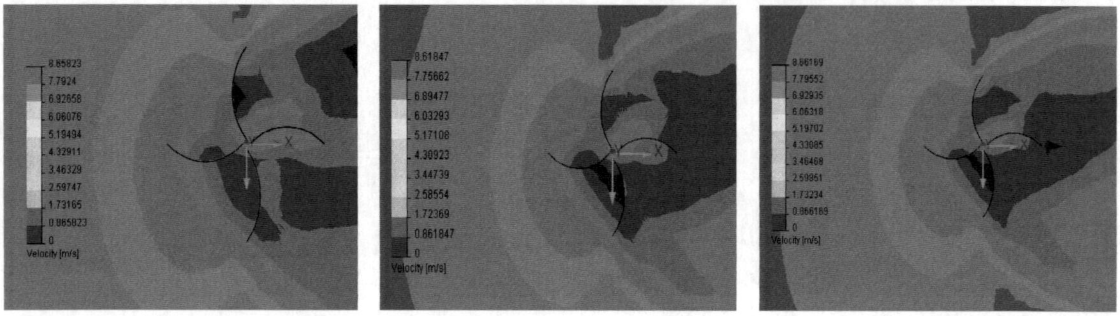

(a) (b) (c)

Figure 9. Cut Plot Velocity at wind speed 7 m/s . (a) 0 Fin (b) 1 Fin (c) 2 Fin.

Figure 9 shows the cut plot velocity of the U-type blade with a wind speed of 7 m / s. In figure 9 (a), displayed the area of wind velocity for U-type blades without fin where the highest velocity is 2.597 m/s and the lowest velocity is 0 m/s. Figure 9 (b) for the U-type blade with 1 fin produce the highest velocity of 2.585 m / s and the lowest velocity of 0 m / s. Furthermore, In Figure. 9 (c) which is a U-type blade with 2 fins, generate the highest velocity of 2.598 m/s and the lowest velocity of 0 m/s. The largest area with the largest velocity in Figure 9 (c) with the velocity of 2,598 m/s.

Table 1. Drag coefficient/drag force of the blade

Number Fin	**F_D (N)**	**Wind speed (m/s)**	**C_D**
0	2.775592686	5	0.001724402
1	4.138989592	5	0.002571444
2	4.146832679	5	0.002576317
0	5.408223959	7	0.001714278
1	8.122327823	7	0.002574585
2	8.136160157	7	0.002578969

Table 1. shows the value of the drag coefficient/drag force. From the table, can be seen the increase of the both with the addition of the fin. The addition of 2 fins give the higher drag coefficient and drag force compared to1 fin and without fin at the same wind speed. This is clearly visible for the wind speed of 5 m/s and 7 m/s.

4. Conclusion

Some conclusions on the simulations study in savonius type wind turbines are the following:

1. The fin number on the blade will influence the value of the pressure distributions and velocity distribution. Where two fins on the blade produced the higher pressure distribution and velocity distribution.
2. Drag coefficient was increasing with addition of fin. Where two fins produced the higher drag coefficient/drag force compared as blade 0 fin and blade 1 fin.
3. Pressure on the concave blades is higher than the convex blades.

References

[1] M.Islam, DSK. Ting, A. Fartaj. Aerodynamic models for Darrius type straight bladed vertikal axis wind turbines, Renew, sust. Energy rev. 12 (4) 2008 pp. 1087-1109.
[2] Martinus, M. Dyan Susila E.S.,dan Martinus Budiyono, 2011. Analisis Fenomena Penampang Alir Vertical Axis Wind Turbine (VAWT) Tipe Heliks Terhadap Kecepatan Angin Sebagai Pembangkit Listrik Alternatif Berskala Rumah Tangga, Department of Mechanical Engineering University of Lampung, Lampung. Journal of Mechanical, September Vol. 2 No.2.
[3] F. Wenehenubun, A. Saputra, and H. Sutanto, "An experimental study on the performance of Savonius wind turbines related with the number of blades " Energy Procedia, vol. 68, p. 7, 2015
[4] N. H. Mahmoud, et al, "An experimental study on improvement of Savonius rotor performance," Alexandria Engineering Journal, vol. 51, pp. 19-25, 2012.
[5] Ali, MH. Experimental Comparison Study for Savonius Wind Turbine of Two and Three Blades at Low Wind Speed. *Int. J. Modern Eng.Research*, Vol.3 issue 5, 2013, p. 2978-2986.
[6] GWEC, Global wind energy outlook, Greenpeace & Global Wind Energy Council, 2014.
[7] Hau, Eric, 2006. Wind Turbines Fundamentals, Technologies, Applications, Economics, 2nd Edition, Springer, Berlin.
[8] Altan, BD., Atilgan, M. and Ozdamar, A. An Experimental Study on Improvement of a Savonius Rotor Performance with Curtaining.Experimental Thermal and Fluid Science 32, 2008, p. 1673-1678.
[9] Dassault, SolidWorks Essential solver modeling guide, USA, 2012.

IOP Conf. Series: Materials Science and Engineering **539** (2019) 012033 doi:10.1088/1757-899X/539/1/012033

Application of stiffened plate technique to reduce crack propagation in spring plate of girth gear connector in rotary cement kiln

H Dahlan[1], M Rusli [1] and R Muchtar[2]

[1] Mechanical Engineering Department, Faculty of Engineering, Andalas University, Kampus Limau Manis, Padang 25163, Indonesia
Phone: +6275172586; Fax: +6275172566
[2] Engineering Department, PT. Semen Padang, Jln. Raya Indarung, Padang 25237, Indonesia

Email: meifal@eng.unand.ac.id

Abstract. This paper shows the effect of stiffened plate to the stress concentration and the stress intensity factor of crack on the spring plate. The spring plate has cracked at its base after several years of operation in a kiln in the Semen Padang cement factory. In order to obtain the cause of crack appearing and effect of stiffened plate on spring plate, the finite element model is used to analyze the stress distribution and the stress intensity factor of the crack. In conclusion, the additional stiffened plate is able significantly to reduce the stress concentration and stress intensity factor of crack for both models. The stress concentration can be reduced from 50 to 57 percent with addition of stiffened plate. Meanwhile, the stress intensity factor can be reduced from 63 to 67 percent when spring plate using the addition of stiffened plate. It reveals that the implementation of additional stiffened plate on spring plate has an important role to avoid the crack growth.

1. Introduction

Kiln plays an important role during the cement production process in a cement plant. A rotary cement kiln in general consists of some main components, kiln shell with fire brick inside, kiln tires and tire-bearing to support shell sections, girth gear and pinion for driving system inlet and outlet segment system. The kiln system rotates about three to five rpm constantly with 80 to 100 thousand tons of total weight depending on dimension and capacity of the kiln, and inside temperature excess 1500°C.

Damage occurring in kiln components can cause large losses in the factory, either in terms of loss of time and in terms of material. Therefore, all efforts must be done to make the essential components of this cement plant work well. Many machine elements, particularly the large-sized and heavy weight ones fail not because of being erratically designed or improperly operated but often because of unavoidable defects such as shrinkage pinholes, material porosity, non-metallic inclusions that become an initial crack then by dynamic load its propagates into catastrophic fatigue failure [1]. Some type of fatigue failure that commonly found in rotary kiln components are in tire ring [1], some welded joint component [2], and fretting fatigue fracture of the supporting shaft [3].

Failures in machine components have been caused by a fatigue crack initiated from points having a stress concentration. If the stress concentration level is higher than a critical value, so crack will grow and propagate resulting failure in the machine components. Therefore, it is a useful idea to reduce a stress concentration and then arrest the crack growth by some method before the failure occurring.

A number of researches have been conducted to retard the crack growth of defective machine components such as stop-hole technique by drilling a hole in the crack tip to remove the crack tip singularity [4-6], metal crack stitching, Vee and weld, hammer peening [7], and stiffened plate [8]. Stop whole technique is one of the most popular techniques used to repair method to fatigue crack. These techniques require a hole of sufficient diameter to be drilled at the crack tip, and it completely reduces the stress concentration on the tip of the crack for successful arrest of the crack propagation.

Girth gear is a gear that is attached to the kiln as one part of the kiln drive system. In general, the kiln drive system consists of a drive motor, pinion and gearth gear as shown in Figure 1. Girth gear is attached to the wall of the kiln using a spring plate. The stiffness of this spring plate should be stiff enough to avoid major deformation of the kiln during operation.

Figure 1. Kiln, supporting and and driving system

This paper will discuss the technique to stop the crack that found in spring plate that supporting the girth gear in a kiln at semen Padang cement factory. The spring plate has cracked at its base after several years of operation like shown in. The shape and position of the crack can be seen in Figure 2. The stiffened bar will be used to reduce the crack propagation by reduce stress concentration and stress intensity factor at the spring plate. Analysis will be done using finite element software.

Figure 2. Crack in the spring plate

2. Modeling and Load Analysis

Mechanical loads that working in kiln structure is a summation of gravitational load of kiln shell, fire brick, coating material, clinker, girth gear, live ring, and torsional load from the driver [9, 10]. The load distribution of a kiln is described on figure 3. The kiln is rotated three rpm by 1200 kW electric motor power. In general, the consuming power is about 40% of the total power of electric motor.

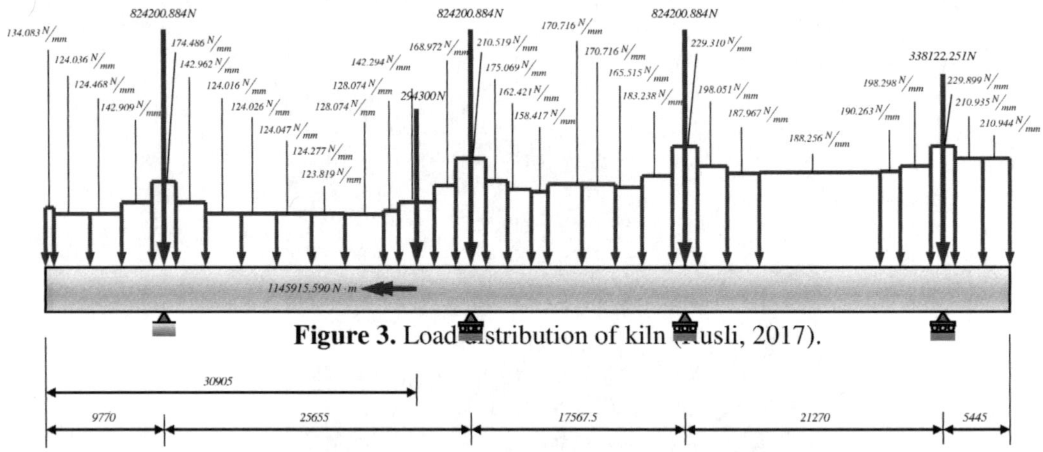

Figure 3. Load distribution of kiln (Rusli, 2017).

Therefore, to connect girth gear and kiln shell and to transfer the torsional power, there are 12 leaf springs made from mild steel. The leaf springs transfer the torque from the driver and girth gear weight and gear contact force. The load model of the leaf spring is shown in figure 4. The finite element model of the spring plate is shown in Figure 5. The load in each leaf spring is depending upon the position the spring. The load distribution according to the spring position is shown in table 1. It shows that the spring load is fluctuated along the rotation of the kiln from 66 kN to 105.5 kN in axial (F_a) and 41 kN to 58.2 kN in tangential (F_t) for maximum design, or 14.5 to 54 kN and 11.3 to 28.4 kN for 40% operational load respectively.

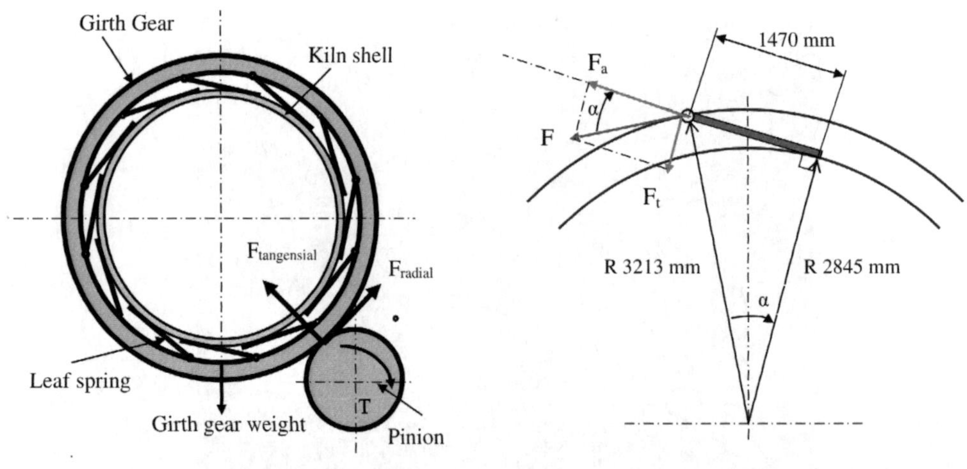

Figure 4. Load model in driving system (left) and load in a leaf spring (right).

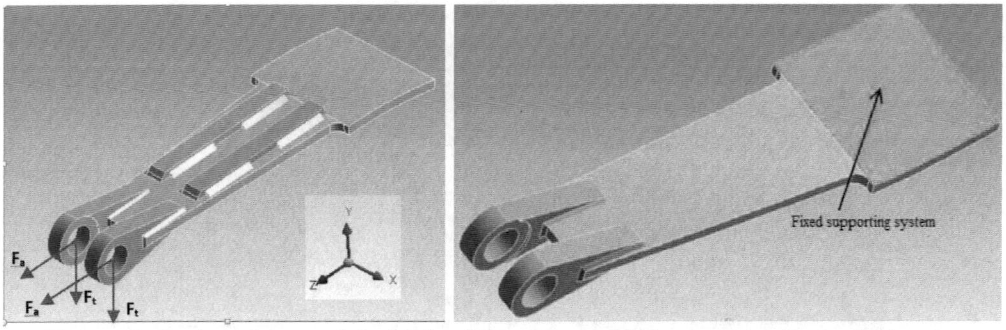

Figure 5. Spring plate model with loading system and fixed supporting in the bottom

Table 1. Estimation of load distribution in different position of the leap spring.

No spring	angular position	Full load		40% load (operational load)	
		Axial load (N)	Tangential load (N)	Axial load (N)	Tangential load (N)
1	0	85,779.8	49,680.4	34,311.9	19,881.9
2	30	90,712.8	58,213.8	39,245.0	28,415.3
3	60	100,569.5	58,213.5	49,101.6	28,415.0
4	90	105,502.1	49,679.9	54,034.2	19,881.4
5	120	100,582.6	41,138.7	49,114.7	11,340.2
6	150	90,726.0	41,123.3	39,258.1	11,324.8
7	180	85,779.8	49,649.0	34,311.9	19,850.5
8	210	80,873.9	41,130.3	29,406.0	11,331.8
9	240	71,017.3	41,099.2	19,549.4	11,300.7
10	270	66,057.5	49,617.1	14,589.7	19,818.6
11	300	70,949.8	58,173.9	19,481.9	28,375.4
12	330	80,806.3	58,220.7	29,338.4	28,422.2

Analysis will be done by calculating the stress intensity factor (SIF) and working stress at the spring plate fillet that generates stress concentration. Crack propagation will be reduce using stiffened plate technique. Two models of stiffened plate will be used in the analysis.

3. Result and Discussion

3.1. Stress Analysis on the spring palate without stiffened plate

In order to obtain the cause of crack appearing as mention early, so the finite element model is used to analyze the stress distribution on spring plate. The spring plate are subjected to loading and supporting system as shown in Figure 5. Figure 6 shows stress concentration at the crack area for 40 percent of operational loading for spring plate without stiffened plate. As seen from figure, the highest stress concentration at area is 235.5 MPa. Furthermore, the highest stress concentration at area increase up to 436.31 MPa for 100 percent of loading (full loading) on spring plate as shown in Figure 7. A large and fluctuating distribution of stress in the fillet area can cause crack growth and propagation on spring plate.

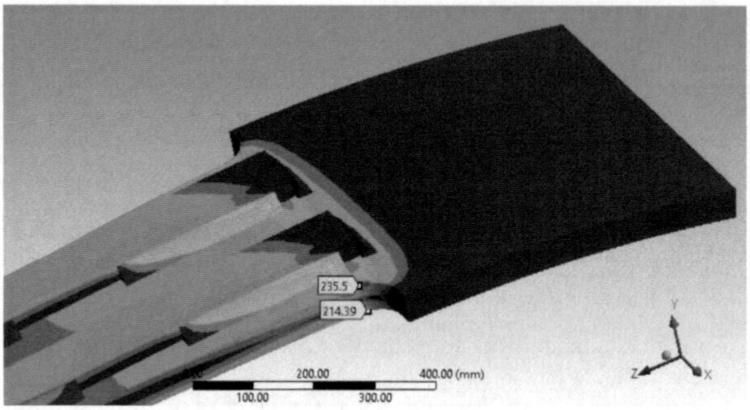

Figure 6. Stress distribution of spring plate for 40 percent of loading

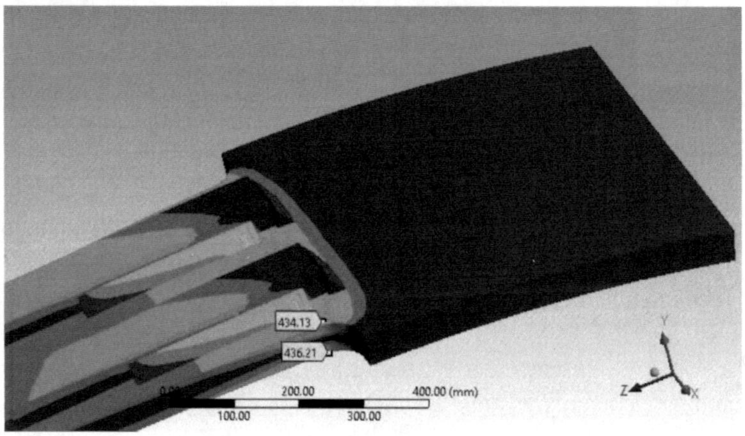

Figure 7. Stress distribution of spring plate for 100 percent of loading

3.2. Stress Analysis on the spring palate with stiffened plate

The one of techniques to avoid the crack occurring is additional stiffener. In this study, two kind of modification with the addition of the stiff plate length are carried out. The first model uses two stiffened plates (Figure 8) and the second model with one wider stiffened plate with a length to the base of 150 mm as shown in Figure 9.

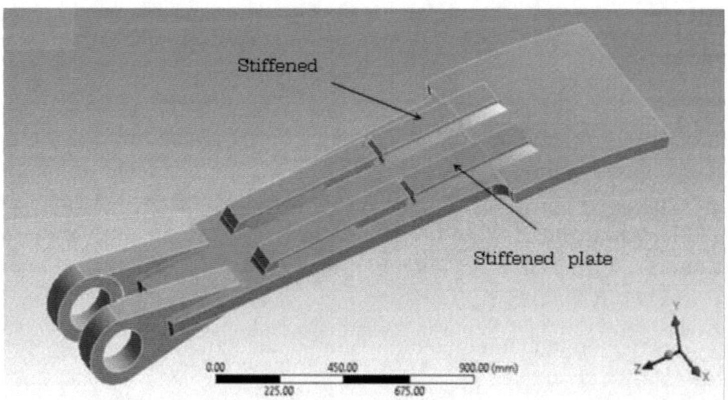

Figure 8. Spring plate model with two stiffened plate

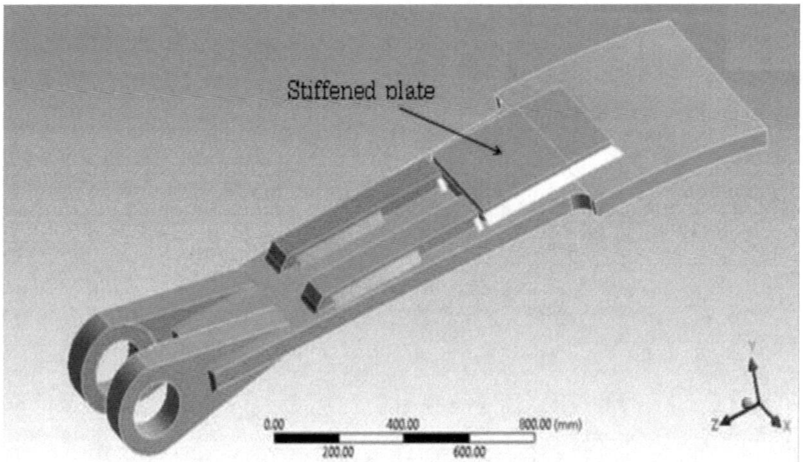

Figure 9. Spring plate model with one stiffened plate

Figure 10 and Figure 11 show the stress concentration at the crack area for 40 percent and 100 percent of loading for spring plate with two stiffened plate (first model). As seen from figure, the stress concentration decrease significantly to 114 MPa for 40 percent of loading as shown in Figure 10. Furthermore, the stress concentration also decreases significantly to 218.4 MPa for 100 percent of loading as shown in Figure 11. The stress concentration reduces about 50 percent for both type of loading.

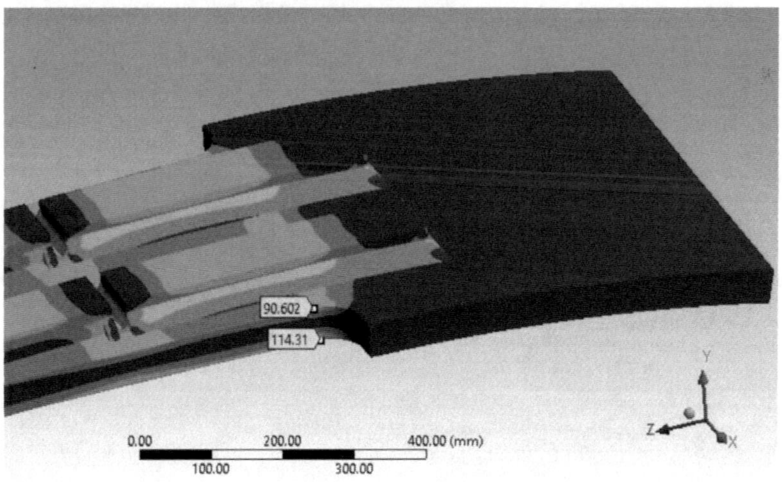

Figure 10. The stress distribution of spring plate with two stiffened plate for 40 percent of loading

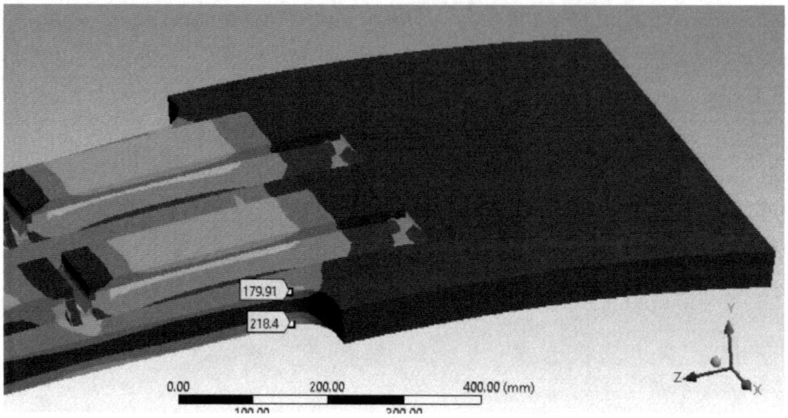

Figure 11. The stress distribution of spring plate with two stiffened plate for 100 percent of loading

Figure 12 and Figure 13 show the stress concentration at the crack area for 40 percent and 100 percent of loading for spring plate with one stiffened plate (the second model). As seen from figure, the stress concentration decrease significantly to 101 MPa for 40 percent of loading as shown in Figure 12. It can be seen that the stress concentration reduces about 57 percent. Furthermore, the stress concentration also decreases significantly to 217 MPa for 100 percent of loading as shown in Figure 13. It can be seen that the stress concentration also reduces about 50 percent.

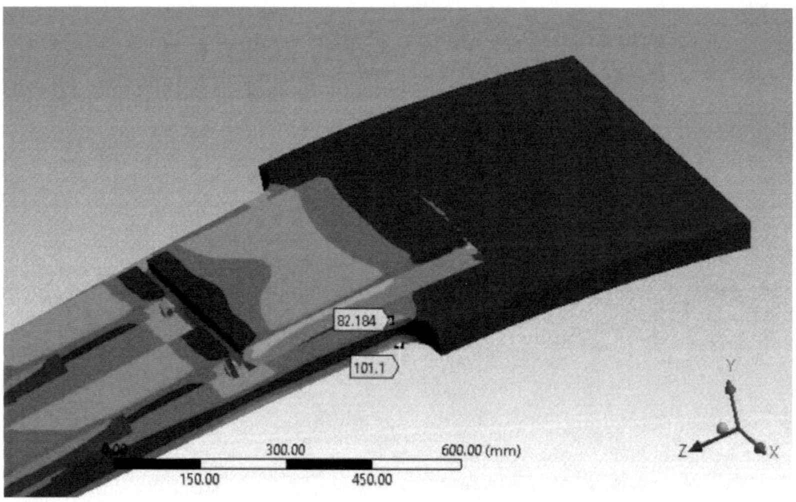

Figure 12. The stress distribution of spring plate with one stiffened plate for 40 percent of loading

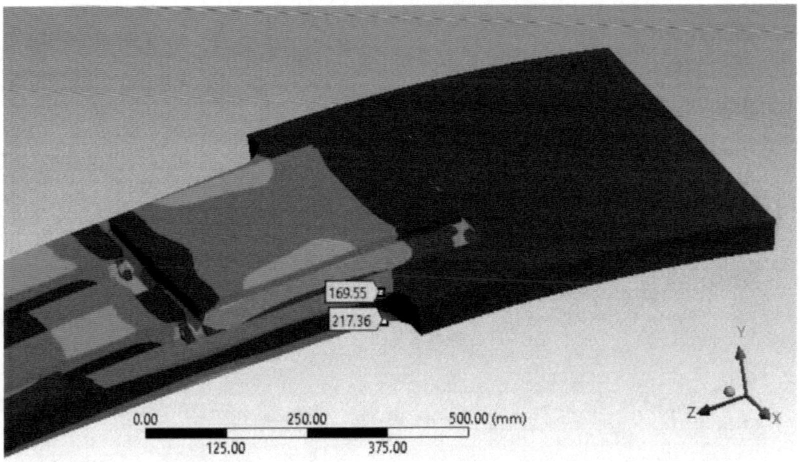

Figure 13. The stress distribution of spring plate with one stiffened plate for 100 percent of loading

3.3. Analysis of Stress Intensity Factor (SIF) on the spring plate

In order to obtain the effect of stiffener on stress intensity factor of crack, so the crack modelling is used in the finite element analysis the stress on spring plate without stiffener as shown in Figure 14. The crack modelling used is semi-elliptical with 20 mm of major and minor radius. Figure 14 shows that the stress intensity factor is 886.89 $\sqrt{\mathrm{MPa.\,mm}}$ for 40 percent of loading. Moreover, Figure 15 shows that stress intensity factor reduce significantly to 326.44 $\sqrt{\mathrm{MPa.\,mm}}$ for spring plate with two stiffened plate (the first model). It reveals that the stress intensity factor decrease about 63 percent. Meanwhile, stress intensity factor decrease significantly to 293.95 $\sqrt{\mathrm{MPa.\,mm}}$ for spring plate with one stiffened plate (the second model) where the stress intensity factor decrease about 67 percent as shown in Figure 16.

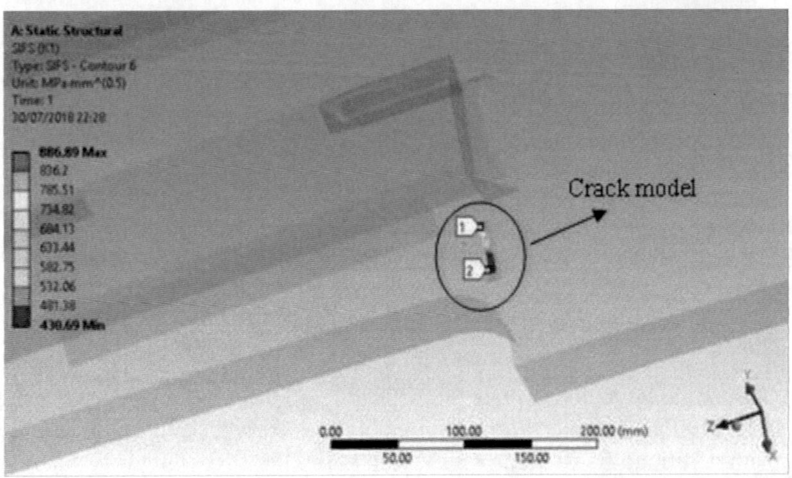

Figure 14. The stress intensity factor of the crack on spring plate without stiffener

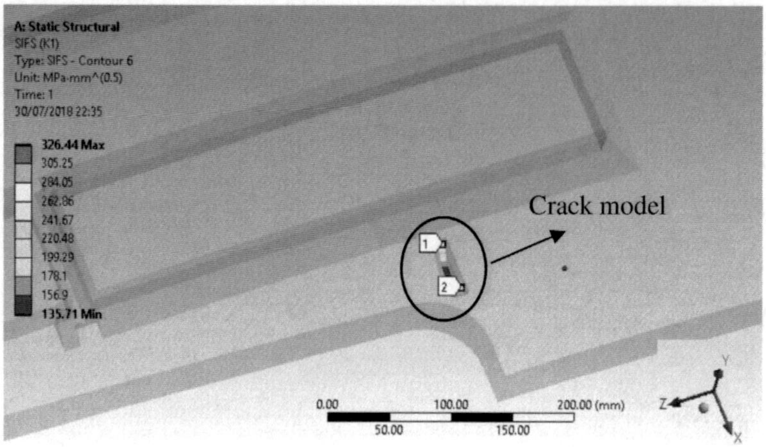

Figure 15. The stress intensity factor of crack on spring plate with two stiffened plate

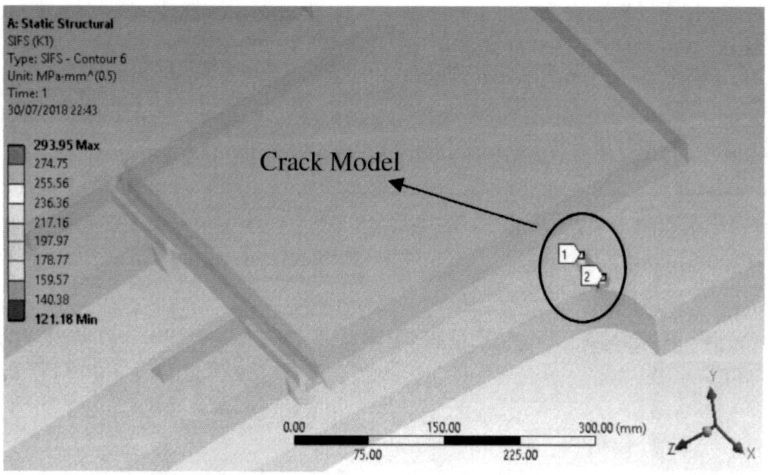

Figure 16. The stress intensity factor of crack on spring plate with one stiffened plate

4. Conclusion

In conclusion, the additional stiffened plate has been able significantly to reduce the stress concentration and the stress intensity factor of crack for both model. The stress concentration can be reduced from 50 to 57 percent with addition of stiffened plate. Meanwhile, the stress intensity factor can be decreased from 63 to 67 percent when spring plate using the addition of stiffened plate for 40 percent of loading. It reveals that the additional stiffened plate on spring plate gives effect significantly to avoid the crack growth.

Acknowledgement

The authors gratefully acknowledge the financial support from International publication grant from Mechanical Engineering Department, Engineering Faculty - Andalas University. We also would like to express profound gratitude to PT Semen Padang for supporting data and analysis.

IOP Conf. Series: Materials Science and Engineering **539** (2019) 012033 doi:10.1088/1757-899X/539/1/012033

Reference

[1] Maziarz M, Tasak E. Case study of fatigue failures in the support rings of a cement kiln. International Journal of Fatigue 1992; 14(2): 84-90.

[2] Shen YP, Bin GF, Wang SL. Fatigue propagation lifetime prediction of rotary kiln weld shell. Applied Mechanics and Materials 2010; 34-35: 1724-1729.

[3] Shi J, Ding N, Xu N, Guo W, Liu L, Chen L, Wu X, Zang Q, Wu CL, Fretting fatigue fracture of the supporting shaft in a rotary kiln. Engineering Failure Analysis 2017; 82 : 474-480.

[4] Song PS, Shieh YS. Stop drilling procedure for fatigue life improvement. International Journal of Fatigue 2004; 26 (12): 1333-1339.

[5] Murdani A, Makabe C, Saimoto A, Kondou R. A crack-growth arresting technique in aluminum alloy. Engineering Failure Analysis 2008; 15 4): 302-310.

[6] Nateche T, Meliani MH, Matvienko YG, Pluvinage G. Drilling Repair Index (DRI) based on two-parameter fracture mechanics for crack arrest holes Engineering Failure Analysis 2016; 59: 99-110.

[7] Marazani T, Madyira DM, Akinlabi ET, Repair of Cracks in Metals: A Review, Procedia Manufacturing 2017; 8: 673-679.

[8] Huang X, Liu Y, Huang X, Dai Y. Crack arrest behavior of central-cracked stiffened plates under uniform tension. International Journal of Mechanical Sciences 2017. 133 : 704-719.

[9] Diaz JJC, Mazon RD, Nieto PJG. Design and finite element analysis of a wet cycle cement rotary kiln. Finite Elements in Analysis and Design 2002; 39 :17–42.

[10] Rusli M, Hendra AH, Yunirsyam D, Bur M. Shell Thickness Reduction Effect on Working Stress of Cement Kiln with Mechanics and Thermal Loads, Proceeding Seminar Nasional Teknik Mesin 2017; 16: PMT 16.

IOP Conf. Series: Materials Science and Engineering **539** (2019) 012034 doi:10.1088/1757-899X/539/1/012034

The using of sound signal and simple microphone to detect damages in induction motor

F Huda[1], A Anggriawan[1] and M Rusli[2]

[1] Mechanical Engineering Department, Faculty of Engineering, University of Riau, Kampus Bina Widya, Pekanbaru 28293, Indonesia
Phone: +62761566786; Fax: +62761566786
[2] Mechanical Engineering Department, Faculty of Engineering, Andalas University, Kampus Limau Manis, Padang 25163, Indonesia

Email: feblil.huda@eng.unri.ac.id

Abstract. In this work, the induction motor damage detection based on sound signal was developed by using simple microphone and sound card as sound data grabber to reduce the cost of instrument. For that purpose, the induction motor apparatus was constructed, and the artificial damages were applied to induction motor to make the conditions of unbalance, broken rotor, and bearing fault. The Fast Fourier Transform (FFT) and wavelet transform were then used to process sound signal, to show the signature of each type of damage. The results show that the unbalance exhibited very clear large wavelet coefficient in the frequency of 1x rotational speed. In the broken rotor bar damage, the strong red color of wavelet graph appeared in the sideband frequencies associated with the broken rotor bar (f_{BB}). Further more, the bearing fault case showed the different color contour of wavelet graph in the ball-pass outer raceway frequency (f_{BPOF}). It can be concluded that the using of sound signal and simple microphone were powerful tools to detect unbalance, broken rotor and bearing fault in induction motor.

1. Introduction

Induction motor is an electromechanical equipment used in various applications to convert electrical power into mechanical power [1]. In industrial world, induction motor becomes one of the most important components, which can be found as a drive in conveyor, pump, lathe, press machine, elevator and etc. Long time operation will let the induction motor undergoes aging that tends to have damages. The type of damages that often occur in induction motor are mechanical unbalance, bearing fault and broken rotor. Because of such crucial function of induction motor, damage detection of induction motor has became an important research area in mechanical engineering.

Several methods have been developed to detect damages in induction motor. The most popular method to detect damage in rotating machine is using vibration measurement as part of predictive maintenance [2]. In this method, accelerometers are usually used as sensor which have to be mounted in the area of measurement. This condition is inconvenient and may endanger the measurement operator in narrow, and restricted area.

In order to overcome the deficiency of measurement instrument, the non-contact measurement method need to be applied. One of the popular method of non-contact measurement is using the sound signal. The sound data can be grabbed by using microphone [3-5], however most of microphones used in Structural Health Monitoring are spending much budget, so the simple microphone with less expenses need to be investigated to find the solution of expensive instrument.

In this work, the simple microphone and sound card used as sensor to get the sound signal from induction motor with artificial damages (unbalance, broken rotor bar and bearing faults). The sound signals were then processed by using Fast Fourier Transform (FFT) and Wavelet transform. From this work, it is hoped that the potency of simple microphone and soundcard to detect damage in induction motor can be shown as the solution to get low budget instrument for induction motor damage detection.

IOP Conf. Series: Materials Science and Engineering **539** (2019) 012034 doi:10.1088/1757-899X/539/1/012034

2. Methodology

2.1. Experimental Set-Up

In this research, damages detection in induction motor was conducted by using simple microphone and sound card as sound signal grabber which connected to a PC. The *microphone* used in this setup was model of Zm-360 with frequency response 70-20KHz and the sensitivity of -20dB below 1 Khz, which usually used as a part of sound system. In induction motor, the artificial damages (unbalance, broken rotor bar and bearing fault) were applied. The experimental set-up can be seen in Figure 1.

The sound data of each type of damage were grabbed with microphone and PC sound card. The sound data were then processed in the PC. By using FFT and wavelet transform with Mexican heat mother wavelet [6], the sign of every damages could be displayed in the PC. The motor speed would be set in steady speed of 2850 rpm corresponds to 47.5 Hz.

2.2. Experiment in Normal Condition

Induction motor testing in normal (without damage) conditions was conducted to show the sound signal of induction motor without any damages applied. Normal condition data will be used as a reference of the induction motor testing condition which will be compared with damage condition.

Before testing the induction motor in normal condition, the induction motor and rotor as shown in the experimental setup need to be balanced. The balancing process was applied the induction motor apparatus to make sure the rotor attached on motor well balanced. The balancing was conducted based on sound signal.

2.3. Experiment in damage condition

The experiment in damage condition were conducted by applying artificial damages to the induction motor. The damages were bearing defect, broken rotor bar, and unbalance. The given artificial damages can be seen in Figure 2. The bearing defect was applied by devastating the outer race of the motor bearing. The broken rotor bar was produced by making holes in the rotor bar using drilling machine. The unbalance condition was applied by installing unbalance disk to the motor setup.

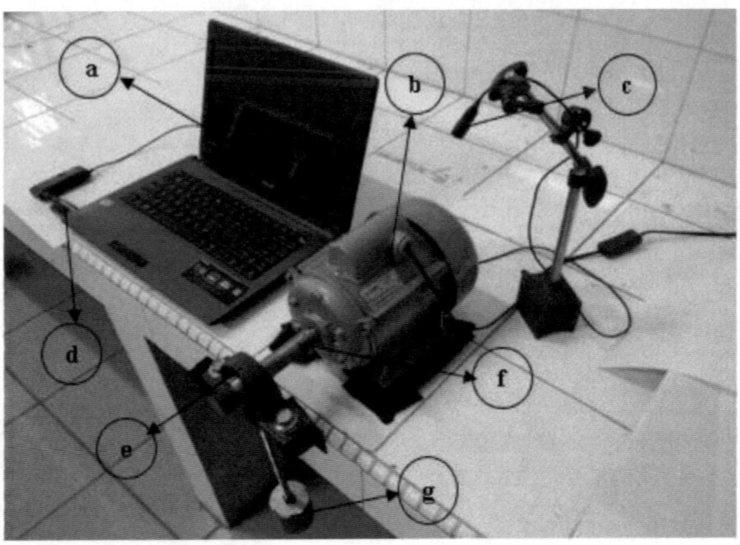

a. PC
b. Induction motor
c. Microphone
d. Soundcard
e. Bearing of Load
f. Shaft
g. Load 0.8 kg

Figure 1. Experimental setup of induction motor damage detection

In early stage, the experiments were conducted for every single of damage, and continued with combination of two types of damages. Finally, the experiment was completed by combining all type of artificial damages. The sound signal data were then processed and analyzed by using FFT and wavelet transform-Mexican head mother wavelet, so that the signatures of damage in motor induction from sound signal could be determined.

(a) (b) (c)

Figure 2. Artificial damages in Induction motor (a) unbalance, (b) Broken rotor (c) Bearing fault

3. Result and Discussion

3.1. Induction motor signal in normal condition

Figure 3 shows the sound signal of induction motor for normal condition in form of time domain and frequency domain. In the time and frequency domain signals, the high amplitude does not appear. It means that there are no any damages, even though there is a little bit high amplitude in around operational speed frequency, but it is still normal.

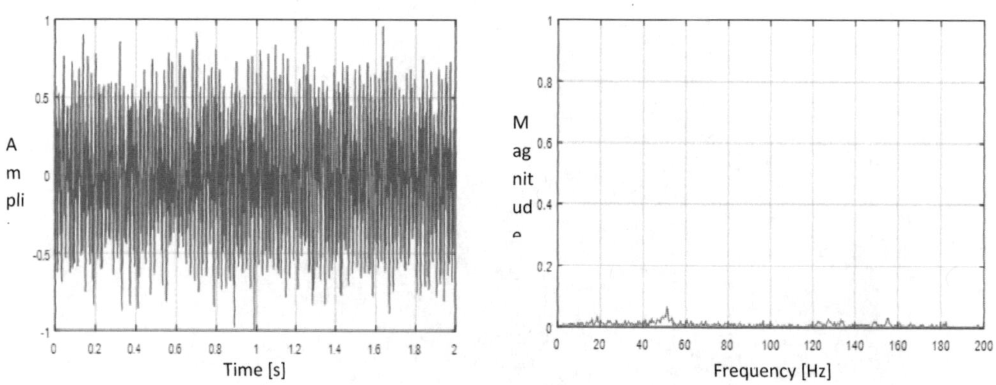

(a) (b)

Figure 3. The sound signal of induction motor in normal condition (a) Time domain (b) Frequency domain

3.2. Induction motor signal in unbalance condition

The signature of unbalance in a vibration signal normally has the form of increased amplitude along the rotor frequency, being located in this case at 49.5 Hz. The same tendency will appear in sound

IOP Conf. Series: Materials Science and Engineering **539** (2019) 012034 doi:10.1088/1757-899X/539/1/012034

signal. To make sure that the rotor in unbalance condition, the rotor would be balanced by the process of balancing. After balancing, the unbalance mass was added to the rotor to give unbalance condition. 7 gram mass was attached to the rotor. Figure 4 shows the sound data signal for unbalance condition in form of frequency domain (Figure 4a), scalogram of wavelet transform (Figure 4b), 3-D wavelet coefficient (Figure 4c), and the view of time-frequency (Figure 4d).

Induction motor was operated in the speed of 2970 rpm corresponds to 49.5 Hz. The sound signal in frequency domain shows the high amplitude in the frequency of motor speed. In the scalogram, the yellow contour appears straight in the frequency of 49.5 Hz where the contour becomes wider as the scale raise up, and in time-frequency wavelet red-yellow and blue contours form row in the frequency 49.5. It could be analyzed that the unbalance problem is detected very well by using proposed method.

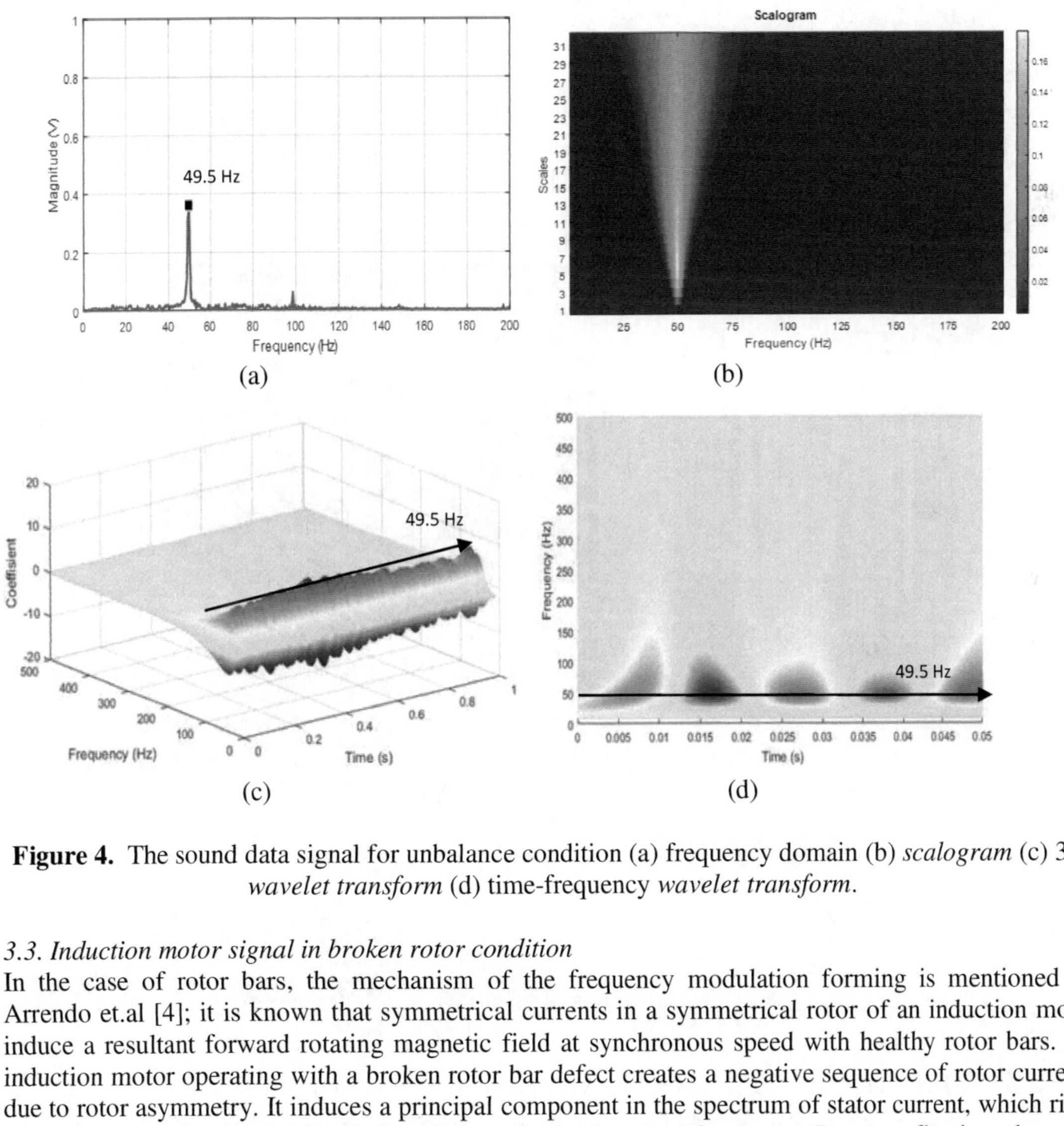

Figure 4. The sound data signal for unbalance condition (a) frequency domain (b) *scalogram* (c) 3-d *wavelet transform* (d) time-frequency *wavelet transform*.

3.3. Induction motor signal in broken rotor condition

In the case of rotor bars, the mechanism of the frequency modulation forming is mentioned by Arrendo et.al [4]; it is known that symmetrical currents in a symmetrical rotor of an induction motor induce a resultant forward rotating magnetic field at synchronous speed with healthy rotor bars. An induction motor operating with a broken rotor bar defect creates a negative sequence of rotor currents due to rotor asymmetry. It induces a principal component in the spectrum of stator current, which rises up to the frequency $(1-2s)\,f_s$ and it is close to the fundamental frequency. Due to reflection, the rotor asymmetry frequencies are $-2ksf_s$, where $k = 1, 2, 3, \ldots$ is any positive integer and s is the slip. Thus, in the case of broken rotor bars, there is a speed oscillation, which acts as a frequency modulation on rotation frequency and two side band frequencies appear around f_r in the vibration spectrum [7] that are given by:

$$f_{BB} = f_r \pm 2ksf_s \tag{1}$$

Where f_{BB} are the sideband frequencies associated with the broken rotor bar. The slip s is defined as the relative mechanical speed of the motor n_m with respect to the motor synchronous speed n_s as follows:

$$s = \frac{n_s - n_m}{n_s} \tag{2}$$

The motor speed was measured by *tachometer*, and it is found that the motor rotated in 2865 rpm corresponding to 47.75 Hz. The values of f_{BB} were calculated to be 39 HZ and 59 Hz.

The sound signal of the induction motor in the condition of broken rotor bar can be seen in the Figure 5. The frequency domain signal shows the high amplitude in the frequency of 78.5 Hz. This value is close to $2 \times f_{BB}$ (harmonic of f_{BB}), which reflect the signature of broken rotor bar. The difference of 0.5 Hz caused by fluctuation of motor speed.

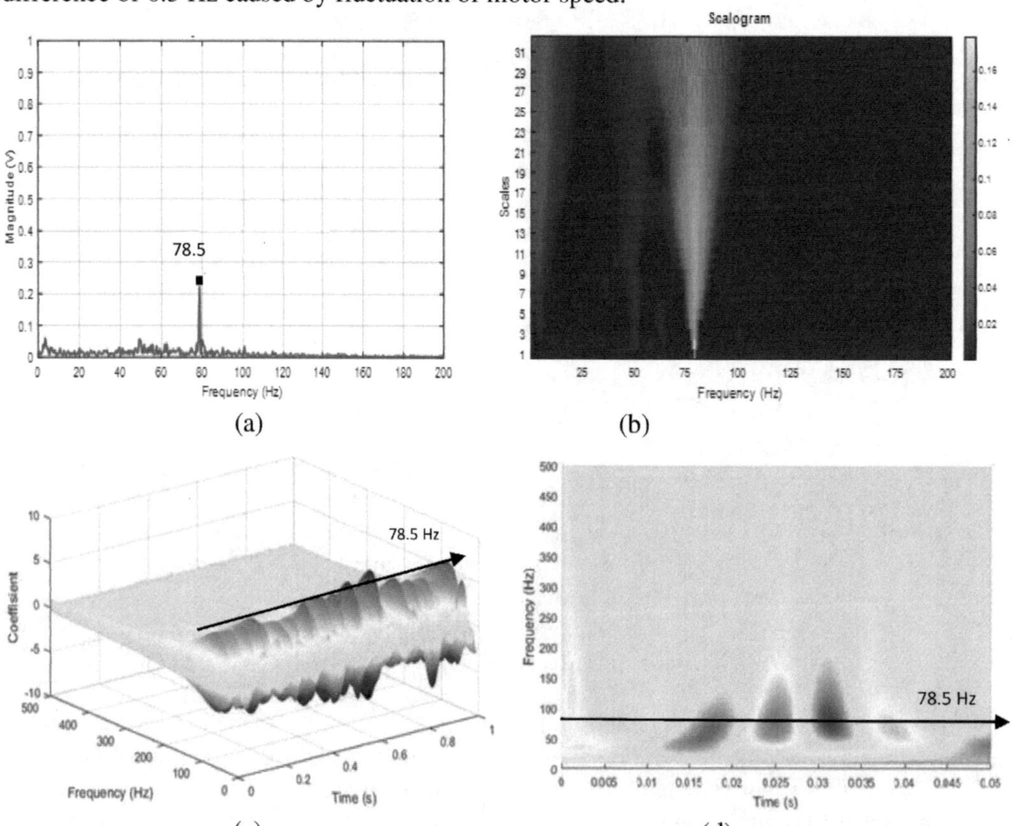

Figure 5. The sound data signal for broken rotor bar condition (a) frequency domain signal (b) *Scalogram* (c) 3-d *wavelet transform* (d) time-frequency *wavelet transform.*

In the scalogram, the yellow contour is shown clearly in line of 78.5 Hz. In the 3-D wavelet transform, very clear red contour appears in the same frequency line. In the time-frequency *wavelet transform,* the blue-yellow-red color form a line of 78.5 Hz. These results indicate that the simple microphone has a good ability to detect the induction motor sound.

3.4. Induction motor signal in bearing fault condition

The artificial damage which applied to induction motor's bearing was the outer race fault. As the bearing rotates, impacts occur at the ball-pass outer raceway frequency (f_{BPOF}) as given by [8]:

$$f_{BPOF} = \frac{N_B}{2} f_r \left(1 - \frac{D_B}{D_c} Cos\theta\right) \quad (3)$$

Where θ is the contact angle between the bearing surfaces, D_C is the cage diameter of the bearing and is measured from a ball center to the opposite ball center, D_B is the ball diameter, and N_B is the number of balls in the bearing. By calculating f_{BPOF} from the data of induction motor in the experimental setup, the ball pass outer race frequency defect was found to be 144.5 Hz.

Figure 6 shows the sound data signal for bearing fault condition in form of frequency domain (Figure 6a), scalogram of wavelet transform (Figure 6b), 3-D wavelet coefficient (Figure 6c), and the view of time-frequency (Figure 6d). It can be seen that the signature of damage caused by bearing fault f_{BPOF} in the frequency of 144.5 Hz. In the frequency domain, the high amplitude appears in that frequency. The wavelet transform were then used to show the other aspect of damage detection. By applying the wavelet transform, this signature is also visible clearly. In the scalogram, in line of 144.5 Hz, the yellow contours widen as the scale increases. The 3-D wavelet coefficient in Figure 6c shows the consistent high value of wavelet coefficient in the frequency of 144.5 Hz. In the view of Time-Frequency the red and blue contours have the centers in the line of $f_{BPOF.}$ This result show that by using sound signal and simple microphone, the bearing fault in induction motor can be detected well.

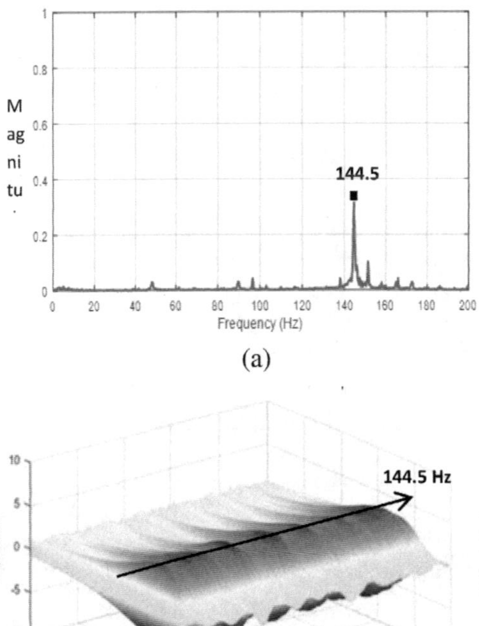

(a)

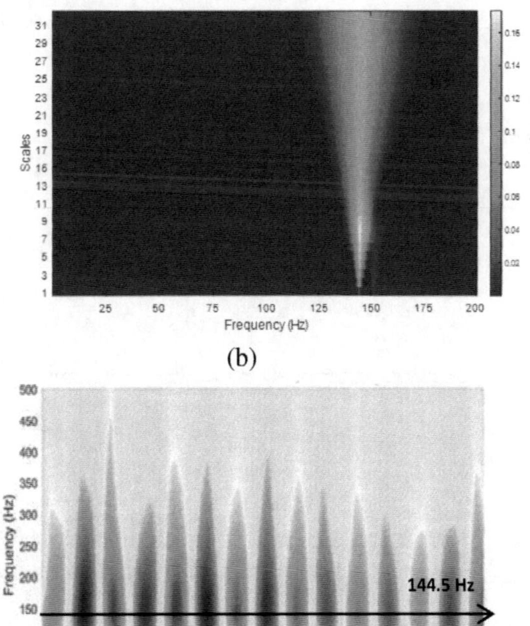

(b)

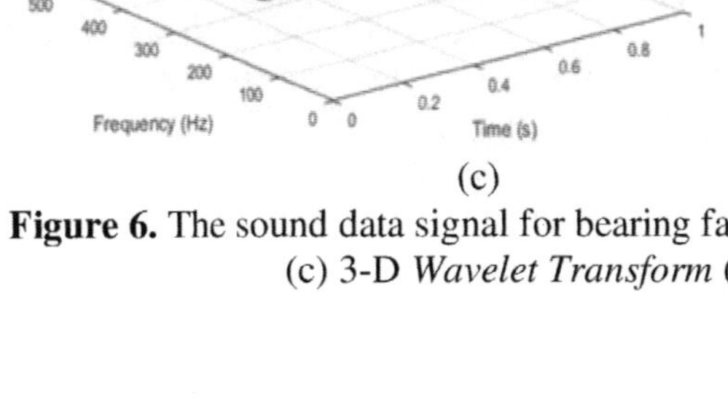

(c)

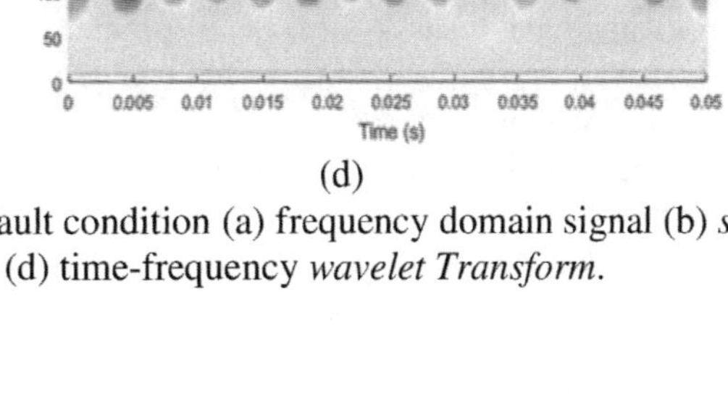

(d)

Figure 6. The sound data signal for bearing fault condition (a) frequency domain signal (b) *scalogram* (c) 3-D *Wavelet Transform* (d) time-frequency *wavelet Transform.*

IOP Conf. Series: Materials Science and Engineering **539** (2019) 012034 doi:10.1088/1757-899X/539/1/012034

3.5. Induction motor signal in multiple damage condition

For multiple damage condition, the unbalance, broken rotor and bearing fault were applied to induction motor apparatus simultaneously. Figure 7 shows induction motor sound signal in multiple damage condition. In the frequency domain data, there are three dominant peaks which show the signature of unbalance, broken rotor and bearing fault respectively. The frequency value of damage signature in multiple damage are slightly different from the frequency value of single damage condition, which caused by the interplay between each type of damage. In the scalogram (Figure 7b) there are three contours in the signature of damage frequencies, but the contour in the frequency of 146.5 Hz as the signature of bearing defect is not clear. This condition is caused by the low amplitude at that frequency compared with other damage frequency signature, so it is overwhelmed. In the 3-D wavelet coefficient, so many peaks appear and it mixes all of the pattern. The same tendency is found in time-frequency *wavelet transform.*

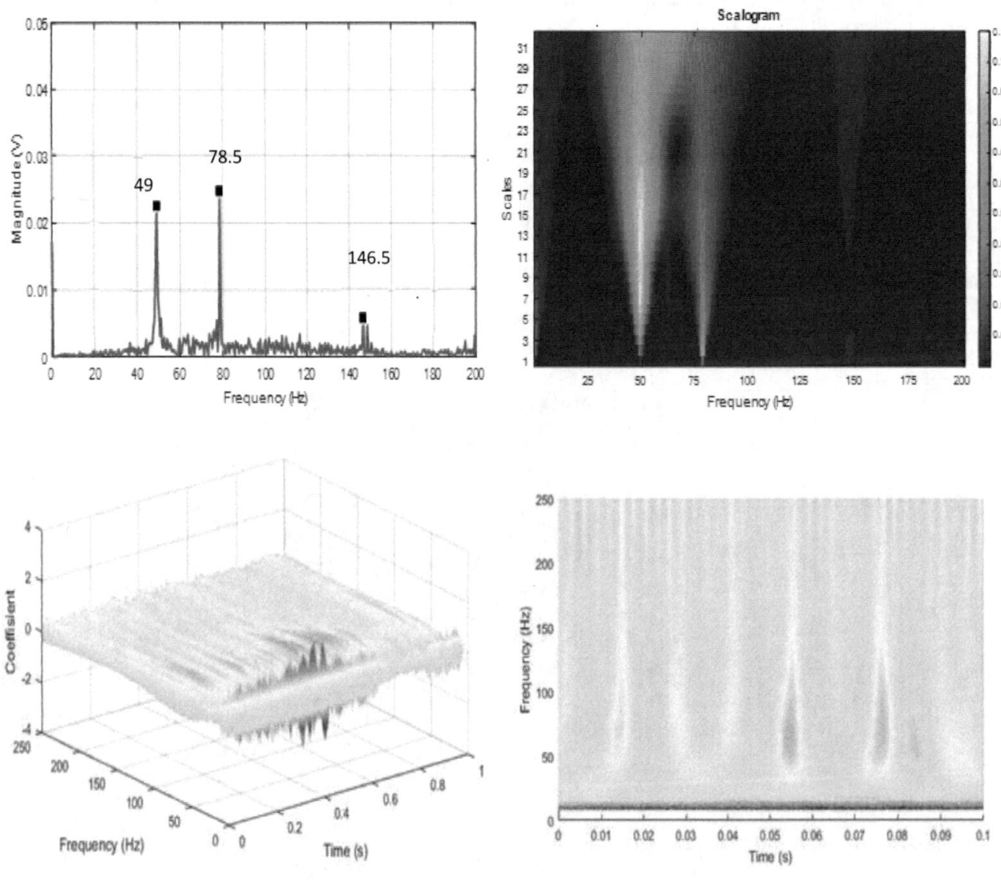

Figure 7. The sound data signal for multiple damage condition (a) frequency domain (b) *scalogram* (c) 3-d *wavelet transform* (d) time-frequency *wavelet transform.*

4. Conclusion

From the works that have been done, it is shown that the sound signal and simple microphone can be used to detect induction motor damage well, where the FFT and wavelet transform were applied as signal processor. By using this method, the unbalance condition, broken rotor bar and bearing fault could be detected well by recognizing their own signature in frequency domain sound signal and wavelet transformation for the single damage. For multiple damage, all of damage signature frequency appear clearly in frequency domain signal. In the scalogram, the high amplitude signature would close the lowest one, so the lowest amplitude signature was not so clear but it still could be detected. For future work it is really recommended to apply this method in industry, so the useful this method in the real application.

References

[1] Andrzej M. Trzynadlowski, Control of induction motors a volume in engineering, Academic press; 2001.

[2] Mulyadi B, Feblil H, Jhon M, Masaki O. Experimental study of vibration on rotor system due to unbalance mass and misalignment, Jurnal poros; 2010: Vol. 8 No. 4.

[3] Huda F, Mario D, Nazaruddin. Analisis Suara pada Rotordinamik akibat Unbalance, Misalignment, dan Looseness. In: Proceeding Seminar Nasional Tahunan Teknik Mesin XIV (SNTTM XIV); 2015.

[4] Arrendo D, Sotelo M, Methodology for fault detection in induction motors via sound and vibration signals. Mechanical system and signal processing, 2017; (83): 568-589.

[5] Yumi O, Yoshifumi O, Takafumi K, Soichiro T, Osamu H. Anomaly detection of motors with feature emphasis using only normal sounds. Proceedings of the 2013 IEEE International Conference on Acoustics Speech and Signal Processing. IEEE; 2013: 2800–2804.

[6] Mertins A, Signal analysis: wavelets, filter banks, time-frequency transform and applications. In University of Wollongong, Australia; 1999.

[7] Alireza S, Mohammad E, Esmaeil R. A new approach for induction motor broken bar diagnosis by using vibration spectrum. Proceedings of the 2006 SICE-ICASE International Joint Conference. 2006: 4715–4720.

[8] Zeki K, Hira K. Simulation and analysis of vibration signals generated by rolling element bearing with defects. Tribology International; 36(2003) 667–678.

IOP Conf. Series: Materials Science and Engineering **539** (2019) 012035 doi:10.1088/1757-899X/539/1/012035

Remaining strength assessment of dented pipe based on depth and strain based criteria

M Akbar[1], F Huda[1] and D K Dewi[2]
[1]Department of Mechanical Engineering, Universitas Riau, Pekanbaru, Indonesia
[2]Center of Technology for Machinery Industry, Agency for Assessment and Application of Technology, Tangerang, Indonesia

E-mail: akbarmust@eng.unri.ac.id

Abstract. There are currently two methods which are used in feasibility operation of dented pipeline systems, namely depth based and strain based. In this research, both methods will be compared to determine conservatism degree of each method and as a guideline and considerations in the integrity assessment of pipeline system. The finite element method with including nonlinearities of material properties and dimension will be used as a tool in the analysis. Static indentation was held with using half ball indenter. Variations of applied load are 600 N, 1000 N, 1500 N, 2000 N, and 2500 N. The results of indentation are dent depth ratio (d/D) with value 1,93 %, 3,78 %, 5,53 %, 7,08 %, 10,63 %, and plastic strain with value 4,24 %, 5,87 %, 8,36 %, 11,43 %, and 12,93 %, respectively. Based on the Code, depth based criteria use d/D<6 % and strain $(\varepsilon) < 6$ % as an acceptance criteria. Based on this research, it can be concluded that strain based method is more conservative rather than depth based method. Then, five model of defected pipe will be simulated in the condition of operation pressure applied to the pipe. Operation pressure was varied with (P/Py) 0.2, 0.4, 0.6, 0.8, and 1. The results show that depth of dent will decrease with variation of strain value.

1. Introduction

Piping system as a fluid transportation system plays a very important role in an industry. In the oil and gas industry, piping systems are the most important equipment which became focus and main concern in the design process, installation, operation and maintenance. In the oil and gas industry, losses will be experienced by the company in case of problems in the piping system resulting in cessation of oil or gas transportation. In fact, in a country whose main income comes from the oil and gas industry, problems in pipeline networks may cause a financial deficit for the country. Seeing the importance of the piping system, integrity analysis is needed to ensure sustainability of pipeline operations. Integrity analysis generally refers to predefined and generally accepted Standards, e.g. API 579 / ASME FFS-1 2007, BS 7910, and ASME B31.G. Those standards include integrity analysis for various defect modes of mechanical equipment, including piping systems.

Dent is one of the defects in the pipe material due to applied load, whether static or impact load, resulting in emergence of a local curve or dent, generally inward of the main axis of the pipe. In piping systems, external applied load may be caused by many things, such as due to rock pressure around the pipe, falling objects of pipe material, or other object impacts during pipe installation, maintenance and operation. Dent in the pipe causes the emergence of stresses and residual strain around the pipe material that suffers the defect. Such conditions may lead to decreased fatigue life of the pipeline, increased risk of leakage, and more fatal effects may result in pipeline rupture.

Various previous studies have much to discuss about cases of dent defects, either on the pipe or on the pressure vessel. Theoretically, Ying Wu et al. [1] have formulated an analytical equation for calculating the effect of dent defects on pipe structures in the form of a shell or tube. The equation was derived by considering the effect of operating pressure, either from internal or external pressure. Such research can be useful as a guide in the treatment of piping systems. In 2010, Noronha et al. [2] construct a mathematical formulation to calculate the size of the strain on the dent defect. The geometry of the dent defect is obtained based on the B-Spline curve which is interpolated from the measurement result using the In-Line Inspection (ILI) instrument. The formulation then was proposed to be used in ASME B.31G Code. If referring to the ASME Code B.31.G-2000 edition [3], the pipe is feasible to operate if depth ratio of defect (d) to the outer diameter of the tube (D) is small than 6%. The method was finally revised in the 2009 edition by incorporating the latest criteria based on the value of strain occurring around the dent defect area i.e. the pipeline was said to be feasible to operate if the strain value is small than 6%. Ramezani and Neitzert [4] undertook research to develop and understand the geometry and strain distribution characteristics of pipe materials with dent defects. In the study, the analysis was performed using finite element method which was validated with the result of theoretical calculation. The conclusion is that by pressing the larger pipe operation the force required to produce the dent defect is also greater. Likewise, the strain on the pipe increases with increasing depth of dent and internal pressure. Other researchers, Oliviera et.al. [5] use finite element analysis to obtain the value of the stress intensity factor around the dent defect area. The analysis was conducted using two-dimensional model with indenter of cylinder geometry. The stress concentration factor was calculated based on the ratio between the maximum stress and the nominal stress at the center of the dent. It was concluded that the value of stress intensity factor increased with decreasing of indenter diameter. Rachman and Musthafa [6] was generated numerical data of cracked cylindrical structure in order to simplified process of integrity assessment. In their research, failure assessment diagram as an acceptance criteria was constructed based on numerical data an compared to diagram which was available in most commonly used Code or Standard.

Some of the above-mentioned studies are still using two-dimensional models to save computational costs, so that no observation of dent deformities in both longitudinal and circumferential directions is possible. In addition, the variations performed are limited to the size and geometry of the indenter and the depth of the defect, not to the size of the pipe. In this study, pipes and indenters are modeled using three-dimensional elements to obtain a more in-depth analysis of strain, stress, and displacement distributions at each coordinate around the defect area. In addition, investigations were also conducted to look at the behavior of the API 5L-X42 pipeline material when supplied with the induced pressure that caused the dent defect and when the defective pipe was operated with a variety of internal pressures.

2. Methodology

Finite element analysis was performed using three-dimensional model of a pipe that a dent deformed on the surface of pipe material. The pipe was modeled intact with a length of 150 mm, a diameter of 40 mm, and a thickness of 1 mm. The determination of the length of the pipe shall be based on the reference to the crack analysis where the ratio of pipe length to defect diameter shall be greater than 10 ($L/D \geq 10$) [7]. In this case, the indenter is a half-sphere with a diameter of 10 mm so that the maximum diameter of the resulting defect is not expected to be much different from indenter's diameter.

The pipe was supported on both free ends by partitioning both ends of the pipe at a width of 10 mm and a holding angle of 90 °. Both free ends of the pipe are held in motion in the x-axis and y-axis direction and to the z-axis rotation ($Ux = Uy = Rz = 0$, $Uz = Rx = Ry =$ free). The emphasis was carried out precisely at the center of the length of the pipe by providing the boundary conditions of free movement of the x-axis indenter, held in the y-axis direction, the z-axis, and all rotational directions ($Ux =$ free, $Uy = Uz = Rx = Ry = Rz = 0$). In finite element analysis three dimensional elements are used with three dimensional brick elements with eight nodal on each element [8]. In order to save computing time and to get more accurate results, meshing strategy was applied by dividing the model into multiple meshing zones. The main zone is the contact area between the indenter and the pipe. In this zone, 3480 elements with size of 0.5 mm were used. For the whole model of pipe and indenter used as many as 5721 elements. The contact line coefficient between the pipe and the indenter is assumed to be 0.3. Interaction and discretization model used are surface to surface with indenter as master surface and the pipe as slave surface.

In finite element analysis, the indenter material is assumed to be very rigid while the pipe material is API 5L-X42 with mechanical properties as given in Table 1. The indenter was assumed as a rigid body to prevent deformation of the indenter at the time of compression and to save computation time. Emphasis will cause plastic deformation of the material, so that the plasticity model of the material must be defined. In this study, elastic-plastic model of material were constructed using the Ramberg-Osgood formula [9]. The relationship between stress and strain on the Ramberg-Osgood formula was given in equation (1). The value of the Ramberg-Osgood constant (n) was obtained from calculations using equations (2) and (3). The mechanical properties of the API 5L-X42 pipeline material are taken from the API 5L-2007 Specification for Line Pipe data issued by the API Institute and as shown in Table 1. The pipe specification data obtained then converted to stress-strain curves using non-linear elastic plastic equation that was given by Ramberg-Osgood, the stress-strain curve of the material was showed in Figure 1.

$$\varepsilon = \frac{\sigma}{E} + 0.002 \cdot \left(\frac{\sigma}{\sigma_y} \right)^n \tag{1}$$

$$n = \frac{Ln(\varepsilon_{us} / 0.2)}{Ln(\sigma_u / \sigma_y)} \tag{2}$$

$$\varepsilon_{us} = 100 \cdot \left(\varepsilon_r - \frac{\sigma_u}{E} \right) \tag{3}$$

Table 1. Mechanical properties of API 5L-X42 pipe material. [10]

No	Mechanical properties	Minimum value
1	Yield strength	290 Mpa
2	Tensile strength	415 MPa
3	Strain	26,5 % *
4	Elastic moduli	210 GPa
5	Poisson ratio	0,30

*based on test specimen with section area equal to 0,44 in^2

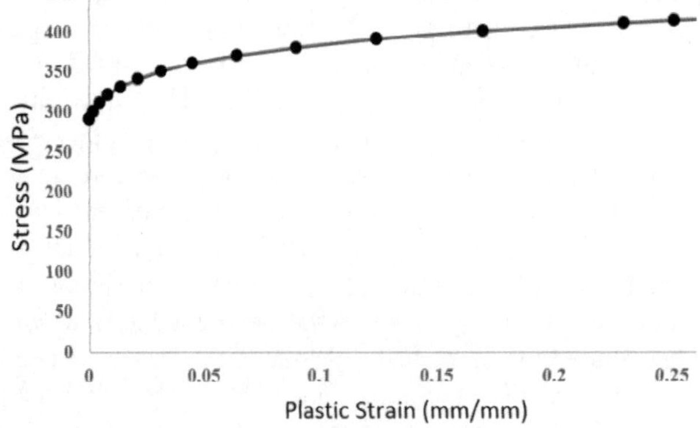

Figure 1. Stress-strain curve of API 5L-X42 pipe material which was generated using Ramberg-Osgood equation.

3. Results and discussion

The indentation was made on API 5L-X42 pipes using spherical indenters with variations of load are: 600 N, 1000 N, 1500 N, 2000 N, and 2500 N. At pressing, the maximum depth of the dent occurs precisely at the center of the compression by the indenter with the value 2.13 mm, 3.04 mm, 5.6 mm, 7.09 mm, and 9.45 mm, respectively. The final depth of the dent defect was measured under conditions after the compression force was removed by a depth of values of 0.77 mm, 1.51 mm, 2.21 mm, 2.83 mm, and 4.25 mm, respectively.

To facilitate the analysis and ease of use for pipes of different dimensions, the dent depth data is presented in the percentage of depth dent to the outer diameter of the pipe (d / D). With an outer diameter of API 5L-X42 pipeline of 40 mm and variations of pressure force of 600 N, 1000 N, 1500 N, 2000 N, and 2500 N, the percentage of dent depth to outer diameter of pipe (d / D) 93%, 3.78%, 5.53%, 7.08%, and 10.63% (Table 1). The curve of the relationship between the indentation force (F) to the value of the depth ratio of the dent defect (d / D) can be seen in Figure 2. Based on the curve, it can be seen that the d /D value increases when pressurized and decreases as the compression force is slowly lowered. The final depth of the resulting dent defect is then used to analyze the integrity of the pipe when operated with certain internal pressures.

Table 2. Variation of indentation load to resulting dent defect with certain depth

No	Load (N)	Maks. dent depth when loading (mm)	Dent depth after unloading (mm)	d/D (%)
1	600	2,13	0,77	1,93
2	1000	3,04	1,51	3,78
3	1500	5,6	2,21	5,53
4	2000	7,09	2,83	7,08
5	2500	9,45	4,25	10,63

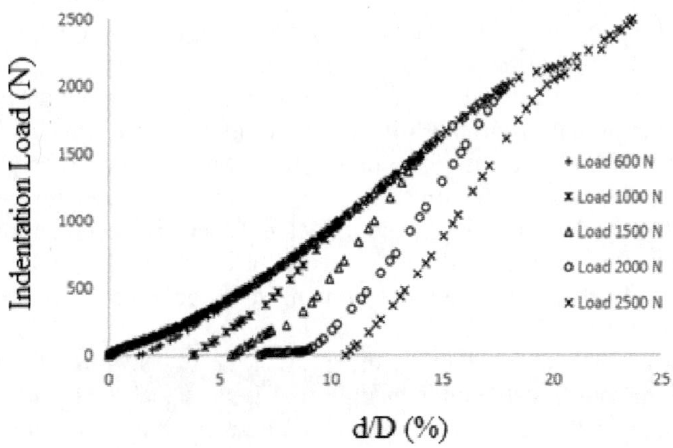

Figure 2. Loading-unloading curve of indentation load to resulting dent defect.

IOP Conf. Series: Materials Science and Engineering **539** (2019) 012035 doi:10.1088/1757-899X/539/1/012035

Besides the depth of the dent, indentation also causes the emergence of stress and residual strain on the pipe material, particularly in the compression area. As the indentation was made by adding the force slowly to its maximum, the maximum stress and strain appeared. Likewise, when the compression force is slowly lowered to 0 N or without pressure force, there will be tension and residual strain on the pipe material. The change in stress and strain values can be seen in Table 3 and Table 4. Distribution of stress on the pipe when it is suppressed with maximum force and when the force of force is removed there is a change in the value of the stress and the displacement from the position of the maximum stress initially concentrated at the center of the dent defect toward the circular direction (circumferential) of the pipe diameter.

Table 3. Stress on pipe materials during indentation (loading-unloading).

No	Load (N)	d/D (%)	Maximum stress (MPa)	Residual stress (MPa)
1	600	1,93	367,7	309,9
2	1000	3,78	377	322,2
3	1500	5,53	388	348,6
4	2000	7,08	395,6	382,4
5	2500	10,63	400,4	358,6

Table 4. Plastic strain on pipe materials during indentation process (loading-unloading).

No	Load (N)	d/D(%)	Maximum strain (%)	Residual strain (%)
1	600	1,93	4,24	4,24
2	1000	3,78	5,87	5,88
3	1500	5,53	8,36	8,75
4	2000	7,08	11,43	11.47
5	2500	10,63	12,93	13,56

In a cylindrical structure, for example in a pipe, the greatest stress occurs in the circumferential direction, which is twice the value when compared to the longitudinal direction. In this study, to observe the behavior of changes in the value of stress and strain of the pipe in the longitudinal or circular direction, numerical data was taken in two models, namely the pipe with indentation of 1500 N and 2500 N. In the longitudinal or circular direction, the observations for nodal-nodal distance 15 mm to the right or left from the midpoint of the dent defect. From Figure 3, Figure 4, Figure 5, and Figure 6 it can be concluded that at the time of compression, the maximum stress occurs at the midpoint of the dent defect or the midpoint of the indenter. When the compression force is lowered, there is maximum displacement of the residual stress, the maximum value not at the midpoint of the dent again. This mapping of values and stress positions needs to be done in relation to the analysis of the integrity of the piping system.

In addition to the above conclusions, from the mapping of the stress values it is seen that in general the value of the circular direction of the circular direction is not much different from the maximum stress when the emphasis is made. A different thing is seen in the longitudinal direction of the stress where the residual stress has a value quite different from the stress when it is suppressed.

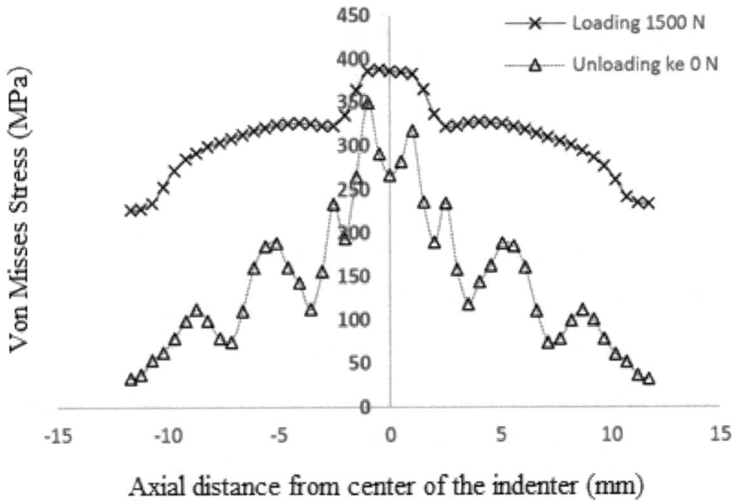

Figure 3. Variations of stress along longitudinal axis of dent defect with indentation load 1500 N.

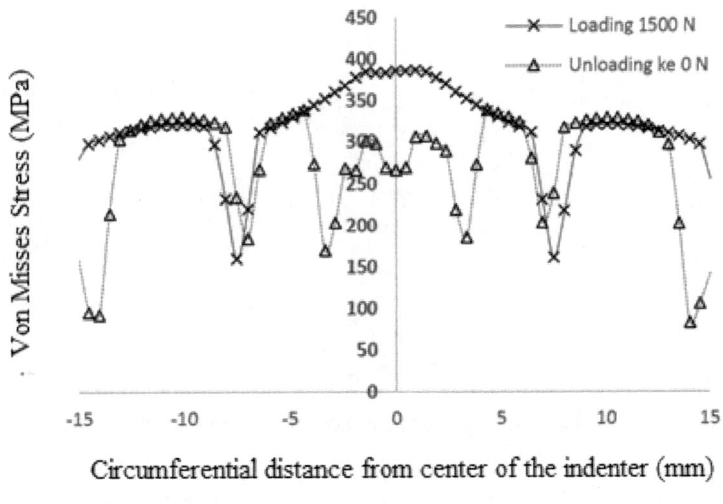

Figure 4. Variations of stress along circumferential direction of dent defect with indentation load 1500 N.

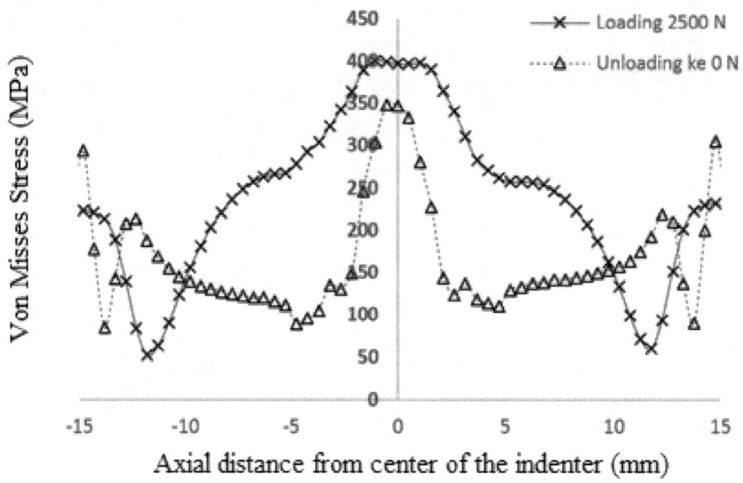

Figure 5. Variations of stress along longitudinal axis of dent defect with indentation load 2500 N.

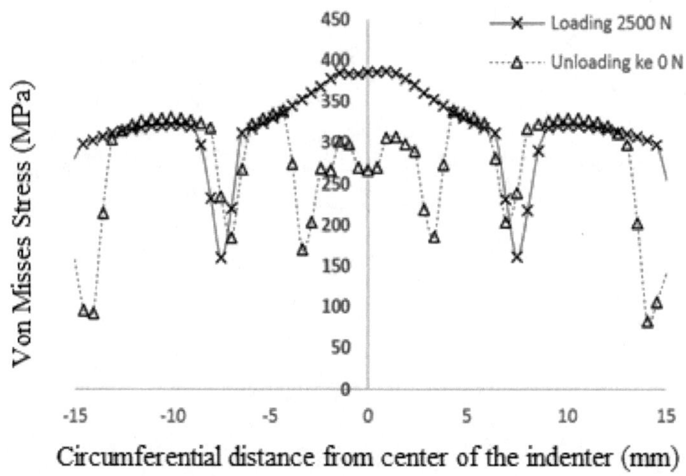

Figure 6. Variations of stress along circumferential direction of dent defect with indentation load 2500 N.

The next analysis is the value of the strain that arises when the indentation process. The strain contours in the stress areas generally show no change in position as well as significant strain values when emphasis or when the force of force is removed. The maximum strain value as well as the remaining strain of the elongated or circular direction is in an improper position at the center of the defect, but still in the area of the pipe section subject to the indenter. In Figure 7 to Figure 10 shows the value of strain in a longitudinal or circular direction from the midpoint of the dent.

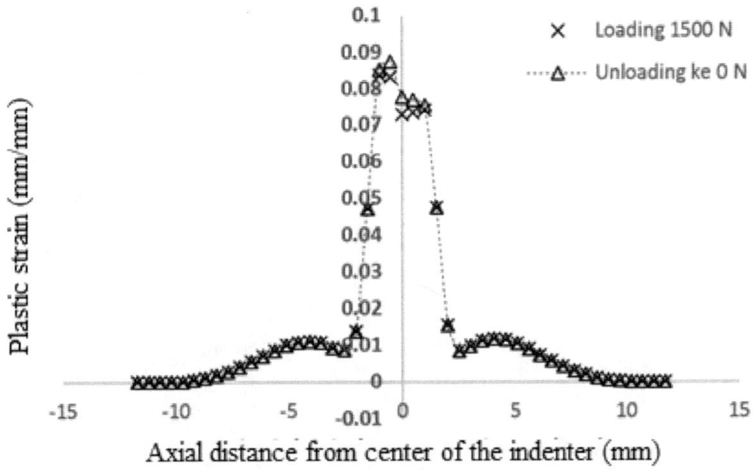

Figure 7. Variations of plastic strain along longitudinal axis of dent defect with indentation load 1500 N

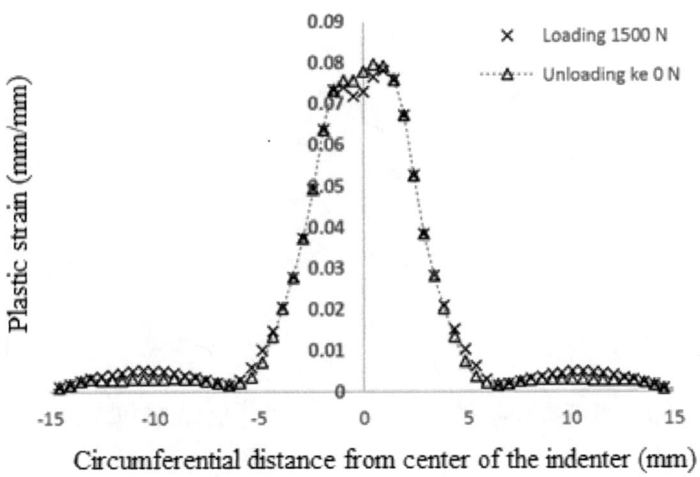

Figure 8. Variations of plastic strain along circumferential direction of dent defect with indentation load 1500 N.

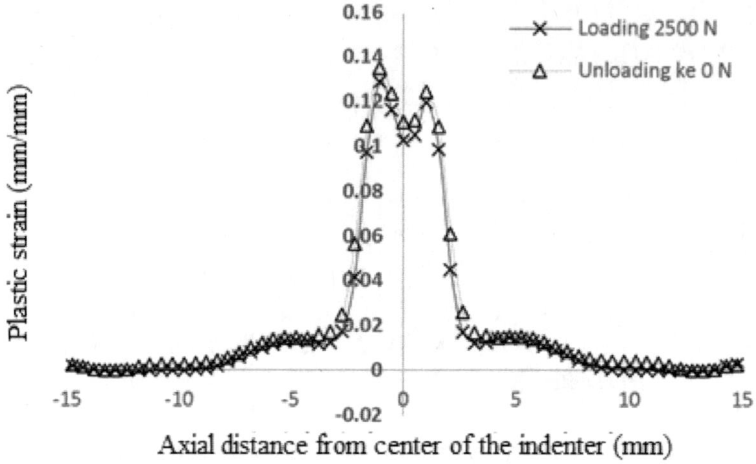

Figure 9. Variations of plastic strain along longitudinal axis of dent defect with indentation load 2500 N.

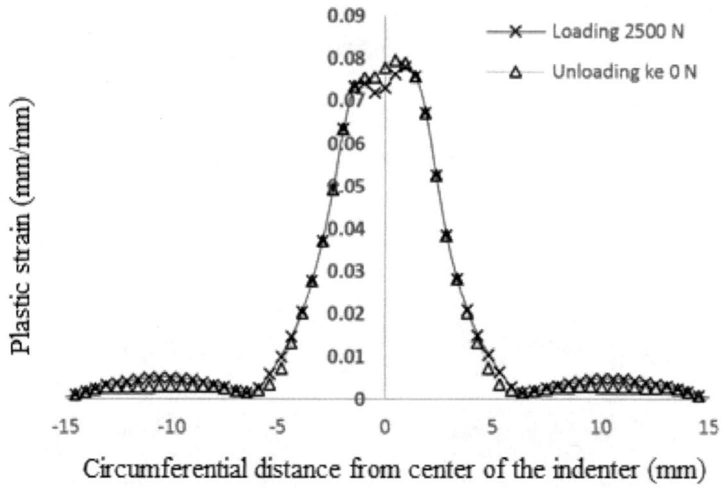

Figure 10. Variations of plastic strain along circumferential direction of dent defect with indentation load 2500 N.

The integrity analysis in this research was conducted on defect pipe that operated with certain operating pressure variation. The selection of operating pressure is carried out with reference to the ratio between actual operating pressures (P) to the operating pressure which may cause the tube to undergo plastic deformation (Py). Py value is derived from the theoretical equation of the maximum stress on the pipe, as shown in equation (4).

$$P_y = \frac{\sigma . t}{R} \tag{4}$$

Based on the above equation then obtained Py of 15.26 MPa. Furthermore, in the simulation used operating pressure of 2.3 MPa, 4.6 MPa, 7.6 MPa, 11.5 MPa and 15.26 MPa, or in the P / Py ratio of 0.15, 0.3, 0.5, 0.75, and 1. When the pipe having a dent defect is operated by providing internal pressure it will change the dimension of the defect. Figure 11 shows that in pipes with d / D = 5.53%, an increase in operating pressure causes a decrease in the depth of the dent defect until the P / Py value reaches 0.5. Pipe operation may also result in changes in the value of the plastic strain in the dent defect area. Figure 12 shows that for defects with d / D = 5.53% there is a decrease in residual strain value if P / Py is raised from 0.15 to 0.5.

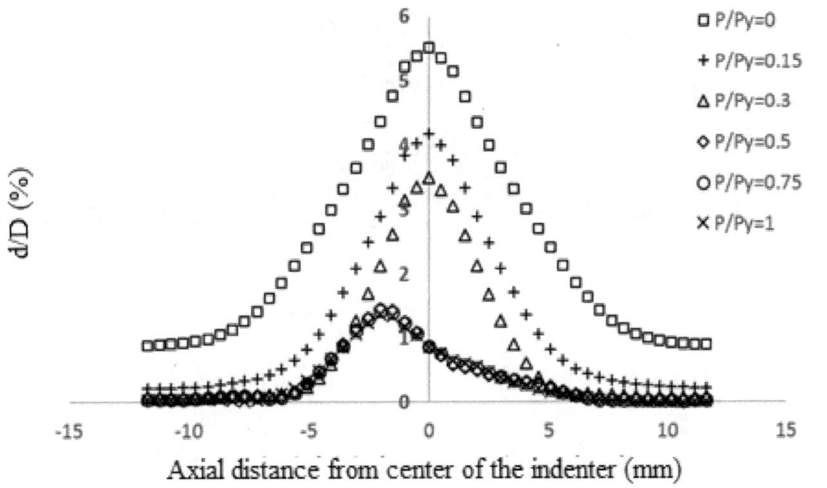

Figure 11. Depth of dent defect with d/D=5.53 % when operated under various internal pressure.

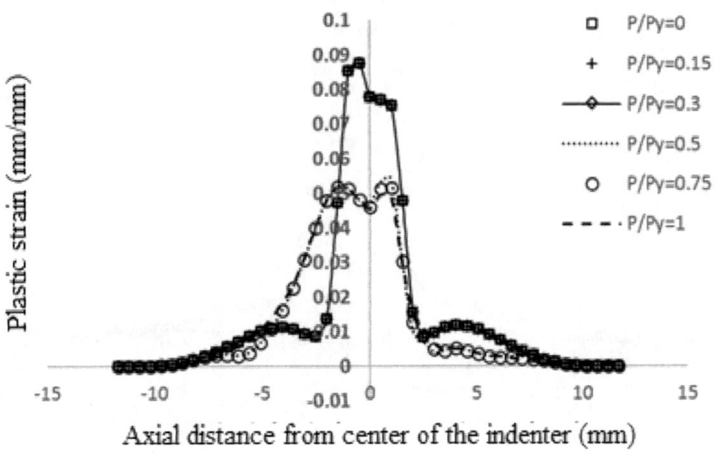

Figure 12. Plastic strain distribution of dent defect with d/D=5.53 % when operated under various internal pressure.

The same phenomenon can also be observed in Figures 13 and 14. In the case of defect with the ratio d/D = 10.63% there is a reduction in defect depth until the P/Py ratio reaches 0.75. The residual strain value on the pipe decreases in some positions and increases in other positions if the P/Py ratio is raised to 0.75.

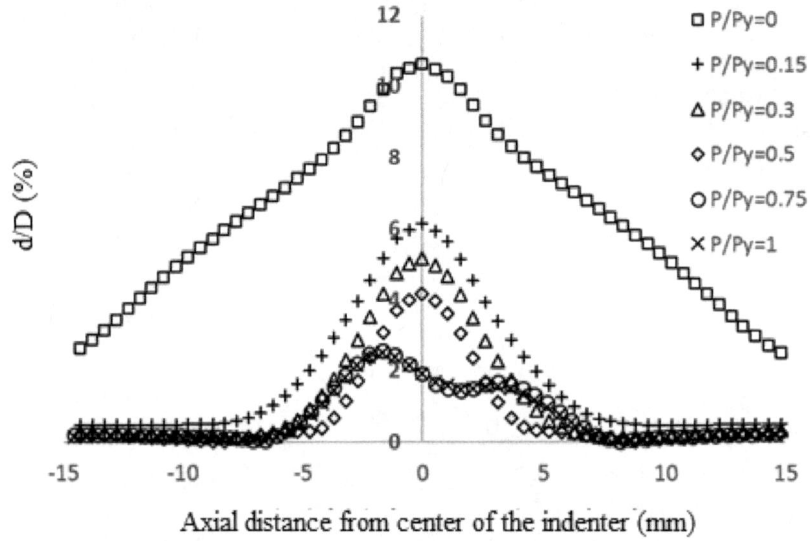

Figure 13. Depth of dent defect with d/D=10.63 % when operated under various internal pressure.

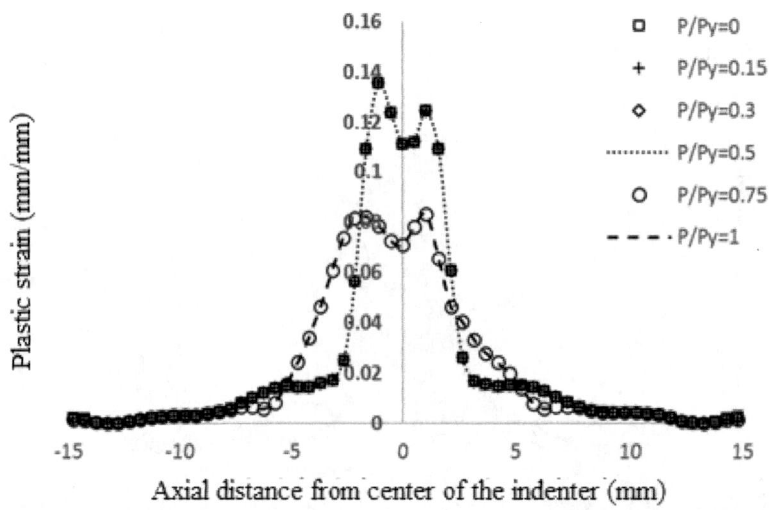

Figure 14. Plastic strain distribution of dent defect with d/D=10.63 % when operated under various internal pressure.

From the simulation performed on 5 models of dent defects with different defect depths, it can be concluded that the pipe has decreased the defect depth due to the increase of operating pressure, in this case in the form of P/Py ratio (Figure 15). Figure 16 shows that the stress value of the defect increases with the addition of operating pressure, while the residual strain tends to decrease (Figure 18).

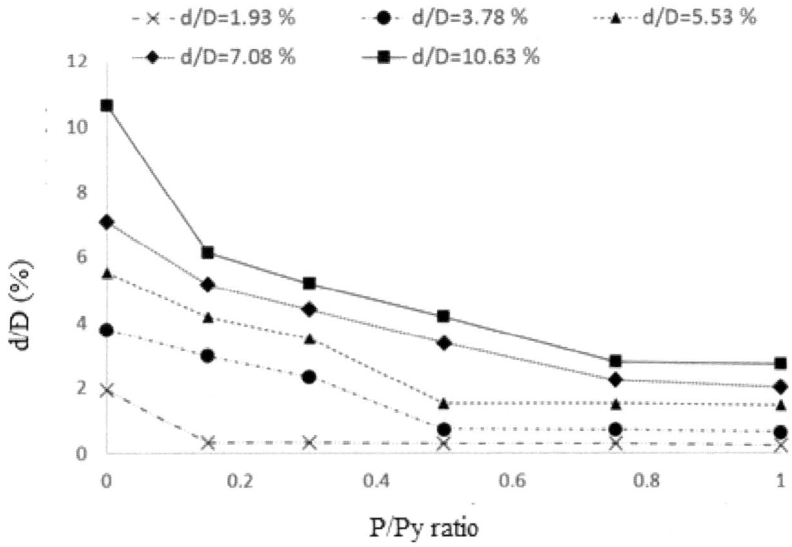

Figure 15. Effect of increasing internal pressure to depth of dent on pipe material.

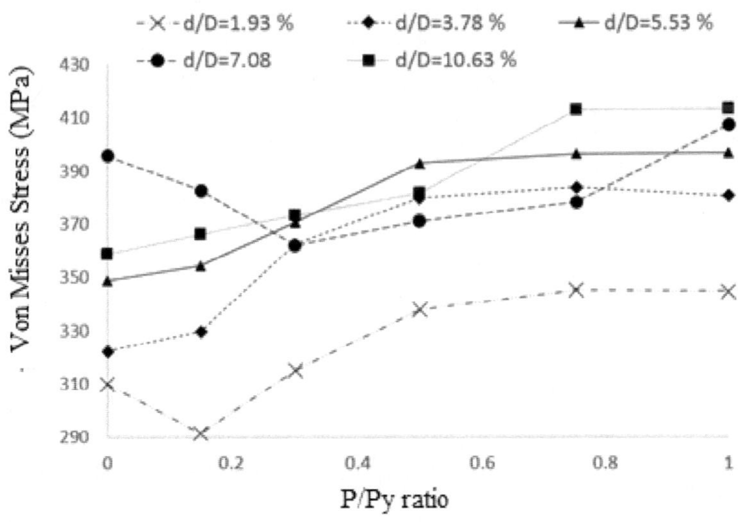

Figure 16. Effect of increasing internal pressure to stress value on pipe material.

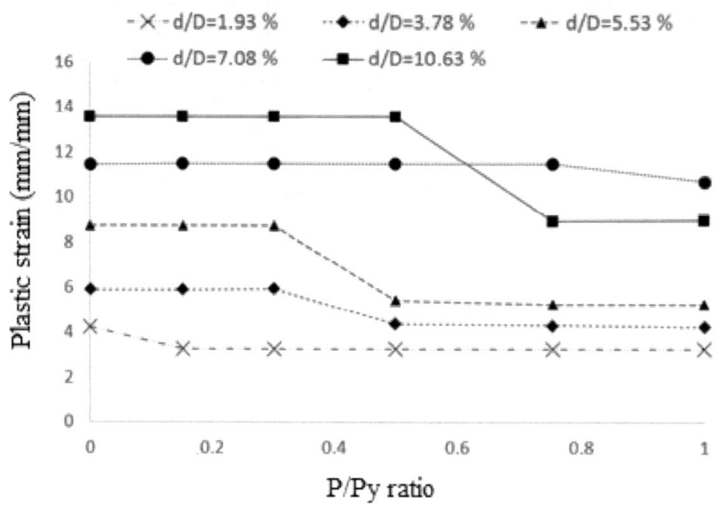

Figure 17. Effect of increasing internal pressure to plastic strain on pipe material.

4. Conclusions

Finite element analysis can be used to perform integrity analysis of pipeline systems that experienced a dent defect. There are two criteria which were commonly used in pipeline integrity analysis, namely based on depth based and strain based. The strain values obtained from finite element analysis can be used as acceptance criteria by referring to the existing Code. Some results that can be inferred from the research are:

1. Dent defects that occur due to indentation 2000 N and 2500 N produce dent with d/D 7.08% and 10.63%. Both models are declared FAIL or not feasible to operate if they refer to the depth based criteria.
2. Dent defects that occur due to indentation of 1500 N, 2000 N and 2500 N produce dent with a strain value of 8.36%, 11.43%, and 12.93%. The three pipeline models are declared FAIL or not feasible to operate if referring to strain based criteria.
3. It was concluded that strain based methods are more conservative (easier to assume to fail) than depth based methods.
4. Defective pipes operated by varying internal stresses may cause a decrease in the depth of the defect as well as the strain value.

References

[1] Y. Wu, P. Jin, and P. Zhang, "Theoretical analysis of pipeline with type I dent under the external force," vol. 9, no. 5, pp. 1–8, 2017.

[2] D. B. Noronha, R. R. Martins, B. P. Jacob, and E. de Souza, "Procedures for the strain based assessment of pipeline dents," *Int. J. Press. Vessel. Pip.*, vol. 87, no. 5, pp. 254–265, 2010.

[3] P. Piping, "Manual for Determining the Remaining Strength of Corroded Pipelines," vol. 2009, 2009.

[4] M. Ramezani and T. Neitzert, "Strain Based Evaluation of Dents in Pressurized Pipes," vol. 7, no. 6, pp. 974–979, 2013.

[5] S. C. O. Jr, E. P. Deus, and A. M. M. Alverne, "Finite Element Analysis of Pipes Considering The Effects of Stress Concentration Factors Due to Dents," 2010.

[6] R. Setiawan and M. Akbar, "Numerical data used for producing failure assessment diagram in integrity analysis of cylindrical shell with crack," *Int. J. Struct. Integr.*, vol. 8, no. 5, pp. 530–

543, Aug. 2017.

[7] J. Newman J. C. and I. S. Raju, "Stress-Intensity Factors for Internal Surface Cracks in Cylindrical Pressure Vessels," *J. Press. Vessel Technol.*, vol. 102, no. 4, pp. 342–346, Nov. 1980.

[8] Hibbit, Karlsson, and Sorensen, *ABAQUS/Standard Analysis User's Manual*. USA: Hibbit, Karlsson, Sorensen Inc., 2007.

[9] W. Ramberg, W. R. Osgood, U. States., and N. A. C. for Aeronautics., *Description of stress-strain curves by three parameters.* Washington D.C.: National Advisory Committee for Aeronautics, 1943.

[10] A. P. Institute, "API 5L Specification for line pipe," *Api Spec 5L*, vol. Forty Four, pp. 1–40, 2007.

Thermal resistance analysis of central processing unit cooling system based on cascade straight heat pipe

W N Septiadi [1,2], W A W Ula [2,3], I G A A D Wulandari[2,3] , I A Tnunay[4] and M R Murti[1,2]

[1]Department of Mechanical Engineering Udayana University, Kampus Bukit Jimbaran Badung-Bali, Indonesia

[2]Heat Transfer Laboratory Department of Mechanical Engineering Udayana University

[3]Bachelor Programme Student of Department of Mechanical Engineering, Udayana University

[4]Master Programme Student of Department of Mechanical Engineering, Udayana University

Email: wayan.nata@gmail.com

Abstract. The development of electronic device is moving toward Smart Technologies, But, the idea of Smart Technologies brings a new problem to heat flux management because the device produces more heat to be banished, especially CPU. So to solve those problems, heat pipe was chosen to be an alternative solution. Heat pipe was hoped to be a passive cooling system technology, but further research shows that goal can't be achieved, due to the high condenser temperature, and the cooling system still needs fan support to banish the heat quickly from the device. So, this paper will discuss a research about new cascade system for heat pipe with variations of heat load and amount of condenser. The results were compared to non-cascade heat pipe to know its thermal performance, especially in thermal resistance. The results show that application of cascade straight heat pipe is effective to solve high condenser temperature problem. In thermal resistance, non-cascade heat pipe still better, but the differences only around 0.3-0.37 °C/W. For the comparison between single and double condenser, double condenser shows better thermal performance. In the end, this cascade heat pipe technology is effective to be applied for central processing unit cooling system technology.

1. Introduction

The development of electronic device's technologies has been developed toward the trend of Smart Technologies, which is the technologies become smaller in dimension, but have to be followed by some improvements in performance [1]. One of the technologies that affected by this trend is a computer, specifically its Central Processing Unit (CPU).

Smart technologies are a nice idea, but it brings a new problem to the heat flux management of electronic devices and CPU specifically. Smart technologies increasing the heat that needs to be banished from the CPU system to maintain its performance and lifetime. So, to solve this problem, the CPU system needs a cooling system with a high performance, slow dimension, and may works passively without any electrical energy supply [2,3].

If we see from all cooling system's properties that have been mentioned before, the heat pipe can be chosen to be an alternative solution for cooling system problems, because heat pipe applies the concept of natural convection phenomenon as its working method. So, the heat pipe can work passively and also have a better performance compared to the previous method that exists before.

The heat pipe is a heat exchanger device which generally made of a copper pipe combined with wick and coolant to move the heat from evaporator side toward condenser side. The performance of a heat pipe is affected by some parameter, such as heat load, heat pipe's material, wick's porosity, and permeability, coolant's amount and type, and the geometry of heat pipe itself [4].

The working method of a heat pipe starts from evaporator receiving some heats from the heat source. And then, the heat will evaporate coolant of the heat pipe. Based on the theory of convection, the vapor will flow naturally to the condenser part of the heat pipe. In the condenser, the heat saved in coolant's vapor will be banished to the surround and then the coolant will be back to the liquid phase. The last phase, the coolant will be flowing back to the evaporator helped by the wick. And the cycle will continue. The simple illustration of heat pipe mechanism can be seen in Fig. 1.

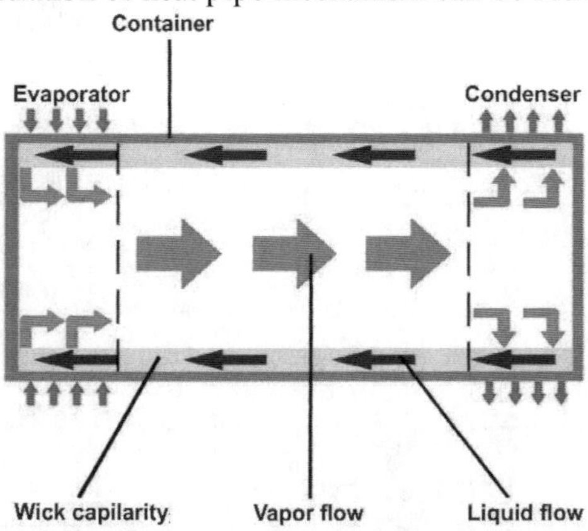

Figure 1. Schematic of a heat pipe [5].

Further, Putra, et al. [6] has been conducted an experimental study on sintered powder wick loop heat pipe. Putra, et al. [4] done a research on thermal resistance of a sintered powder wick heat pipe with the variation of concentration and loading fluid of nanofluids. Septiadi, et al. [7] has done a characterization of screen mesh wick heat pipe which loaded with nanofluid as a passive cooling system. But, all those research find out that heatpipe still produce a high outlet temperature from condenser which needs to be banished fastly from the system to maintain its performance and lifetime. Hence, a fan still used to banish the heat from condenser faster. So, the main purpose of heat pipe application has not been achieved yet.

But later, Putra et al. [8] develop a cascade system for heat pipe by combining two loop heat pipes for CPU cooling system. Then, it was reported that the use of a cascade system for heat pipe was succeeded to reduce the outlet temperature at the condenser and reduce its thermal resistance up to 17.6%.

So, this research tries to develop a new heat pipe's cascade system by combining two straight heat pipes as a CPU cooling system technology. This research conducted to gain more information about the thermal performance of a cascade straight heat pipe by analyse its thermal resistance. So, a better evaluation of its thermal performance can be achieved.

2. Methodology

Cascade straight heat pipe was designed by combining two straight heat pipe, and it is divided into two level, they are level I heat pipe consist of first evaporator and condenser and level II heat pipe consist of second evaporator and condenser, and both of them are connected at first condenser and the second evaporator. The heat pipe is made of copper, water was used as its coolant, and it used sintered powder wick. Fig. 2 shows the design of cascade straight heat pipe.

The data was taken with two types of cascade straight heat pipe, they are cascade straight heat pipe with a single condenser and double condenser. CPU Core i5 2.90 GHz was used as the heat source for this experiment. The data was taken under two condition of the CPU, they are an idle condition with 10W heat load and maximum condition with the 48W heat load. The data of distributed temperature was taken by connecting 4, 5, and 8 type-K thermocouples to not cascade heat pipe, cascade straight heat pipe single condenser, and double condenser, respectively. For the nation cascade heat pipe, a thermocouple was connected to processor's surface, condenser, heatsink, and environment. Meanwhile,

IOP Conf. Series: Materials Science and Engineering **539** (2019) 012036 doi:10.1088/1757-899X/539/1/012036

for the cascade heat pipe, a thermocouple was connected to processor's surface, first and second evaporator part, first and second condenser part, and environment. Schematic diagram of the experiment can be seen in Fig. 3.

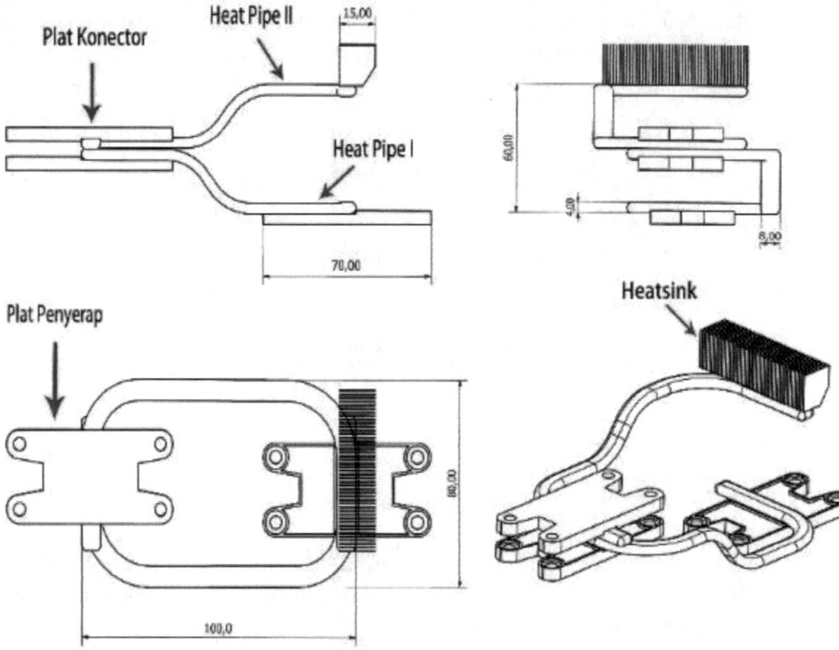

Figure 2. Design of cascade straight heat pipe.

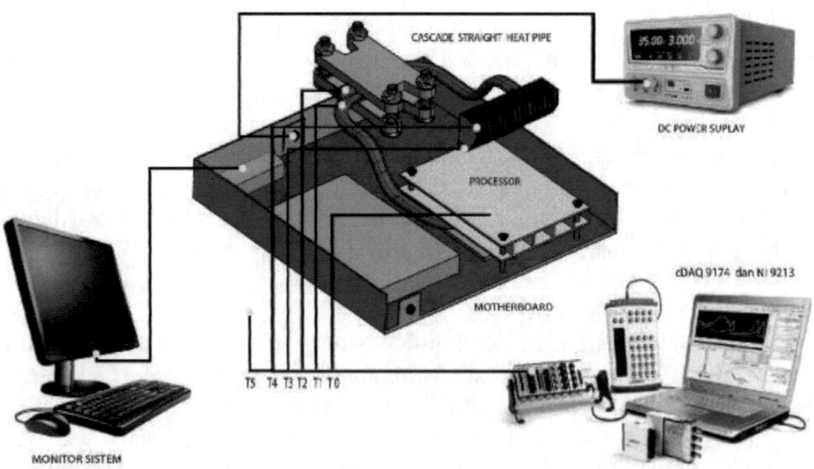

Figure 3. Schematic diagram of the experiment.

Thermal resistance calculation in this research was done for each of heat load (10W and 48W). The calculation was done in three steps. The first one was thermal resistance for the whole level I heat pipe system, followed by calculation for whole level II heat pipe system, and the last one was the calculation of the whole system of cascade straight heat pipe. The equation for each calculation is presented in equation (1), (2), (3), (4), and (5).

$$R_{hps} = \frac{T_e - T_c}{Q} \quad (1)$$

$$R_{hpI} = \frac{T_{eI} - T_{cI}}{Q} \quad (2)$$

$$R_{hpII} = \frac{T_{eII} - T_{cII}}{Q} \quad (3)$$

$$R_{hp,csc} = \frac{T_{eI} - T_{cII}}{Q} \quad (4)$$

$$R_{hp,cdc} = \frac{T_{eI} - (T_{cII} - T_{cIII}/2)}{Q} \quad (5)$$

3. Results and discussion

In order to analyze the thermal resistance of cascade straight heat pipe. The temperature data from four, five, and eight measure points for non cascade heat pipe and both cascade straight heat pipe single and double condenser respectively has been taken. All data can be seen in Table 1, 2, and 3, respectively. Looking at table 1, 2, and 3, we can say that the purpose of cascade straight heat pipe system to reduce the outlet temperature at the condenser are succeed. The outlet temperature difference is quite good, it is up to 19.6% for cascade straight heat pipe single condenser and 21.42% for the double condenser. It can happen because by using a cascade system, the heat is taken from processor going through two-step of the cooling process, in level I and II heat pipe.

Table 1. Temperature data for non-cascade heat pipe

Heat load (Watt)	T_{ev-nc}	T_{c-nc}	T_{hs-nc}	T_{env-nc}
Idle 10 Watt	69.10	59.99	53.32	26.13
Max 48 Watt	69.68	63.06	60.85	25.64

Table 2. Temperature data for cascade straight heat pipe single condenser

Heat load (Watt)	T_{ev1-cs}	T_{c1-cs}	T_{ev2-cs}	T_{c2-cs}	T_{hs-cs}	T_{env-cs}
Idle 10 Watt	66.36	53.03	45.32	40.39	38.38	25.91
Max 48 Watt	68.19	46.25	45.48	43.66	39.73	26.23

Table 3. Temperature data for cascade straight heat pipe double condenser

Heat load (Watt)	T_{ev1-cd}	T_{c1-cd}	T_{ev2-cd}	T_{c2-cd}	T_{ev3-cd}	T_{c3-cd}	T_{hs-cd}	T_{env-cd}
Idle 10 Watt	65.21	54.00	49.57	45.64	42.3	38.57	36.20	24.30
Max 48 Watt	65.44	54.35	49.74	48.20	45.8	39.49	36.69	24.39

By using equation number (1), (2), and (3). Thermal resistance calculation for non-cascade heat pipe, level I, and level II cascade straight heat pipe was done. Based on Chen et al. [9], the highest temperature was produced by temperature at the evaporator and then followed by temperature at the condenser. That is the reason why the temperature used for calculating thermal resistance of cascade is temperature at the first evaporator and the temperature at the last condensor. The comparison result for non-cascade heat pipes, level I's, and level II's cascade straight heat pipe is showed in Fig. 4.

As we can see in Fig. 4. (a) the trends of the data are the same, the more heat load was supplied to the heat pipe, the thermal resistance of the heat pipe was decreased. This has been mentioned before in

Chen et al. [9] research, they also found that thermal resistance of heat pipe was decrease as the heat load supply increasing. But it will only happen until heat pipe reach its maximum heat load, after that the thermal resistance was increased as the heat load supply increased. It could be happened due to the effect of dry out. We also can see in the figure that thermal resistance value for level II heat pipe when Q_{in} 10W is so small compared to thermal resistance of non-cascade heat pipe and level I cascade straight heat pipe, it happened because when heat load received by heat pipes in the first evaporator is 10W, actually the heat load received by level II smaller than 10W. Hence, level II heat pipe haven't received the heat maximally. So, all the coolant inside level II heat pipe hasn't fully evaporated yet. It makes level II heat pipe wasn't work in its real performance.

The differences between thermal resistances of level I heat pipe and non cascade heat pipe at 10W and 48W heat load are between 0.3-0.4 °C/W. Meanwhile, the differences between thermal resistance of level II heat pipe and non cascade heat pipe at 10W and 48W heat load are between 0.11-0.17 °C/W. And when thermal resistance for level I and level II heat pipe compared to each other, the difference are bigger, specially at 48W heat load, the differences are between 0.43-1.26 °C/W. Author's considered that thermal resistance data for level II heat pipe are too small, it can be happened because Q used in level II heat pipe thermal resistance's calculation was considered the same with level I heat pipe. Meanwhile, actually heat load received by level II heat pipe of course smaller than the first received by level I heat pipe.

To see the effectiveness of the whole cascade heat pipe system. A calculation was done by following equation (4) and (5). The result of the calculation was compared with thermal resistance calculation for the non-cascade straight heat pipe. And the comparison is showed in Fig. 4 (b).

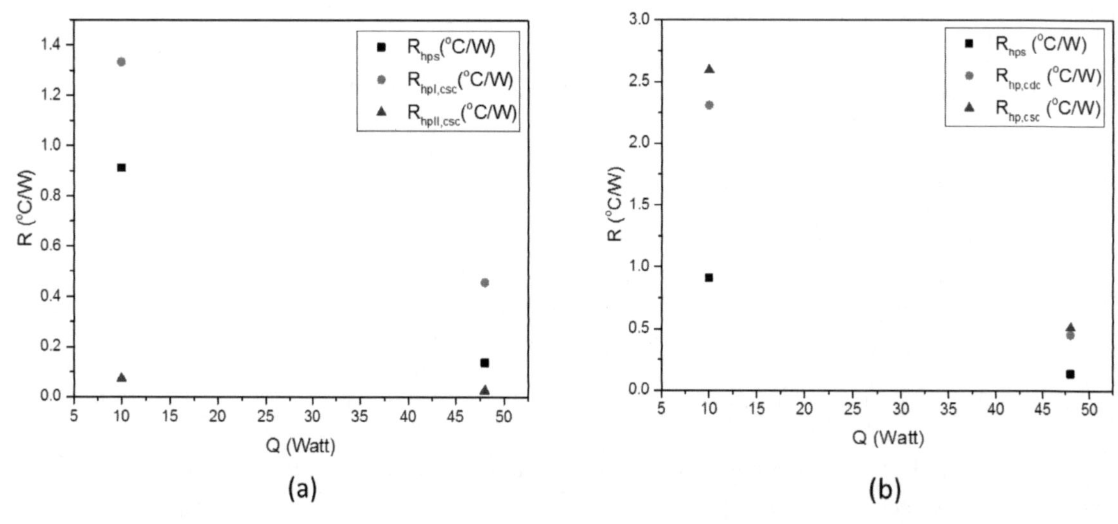

Figure 4. (a) Thermal resistance comparison of non-cascade heat pipe, first level, and second level cascade straight heat pipe, (b) Thermal resistance comparison of the non-cascade straight heat pipe, cascade straight heat pipe single condenser, and cascade straight heat pipe double condenser.

We can see in Fig. 5 that thermal resistance for non-cascade heat pipe was smaller than thermal resistance for the cascade system heat pipe. It was normal because there are more resistances element in the cascade system heat pipe compared to the non-cascade heat pipe. In non-cascade heat pipe, the heat received by heat pipe only get through three-steps of heat transfer. The first one in evaporator from the heat source by conduction, move the the coolant by convection and then move to the condenser surface by conduction. Meanwhile in cascade straight heat pipe, the heat received by heat pipe need to get through twice-steps of heat transfer compared to non-cascade heat pipe. The thermal resistance network comparison for non-cascade and cascade heat pipe can be seen in Fig. 5 (a) and 5 (b), respectively.

From Fig. 4 we also can see that when heat pipe receives 10 W of heat load, thermal resistance for cascade heat pipe and non-cascade heat pipe have a quite big difference, the difference are ±1.7 °C/W for cascade heat pipe with single condenser and ±1.4 °C/W for cascade heat pipe with double condenser. It happened because when cascade heat pipe receives only 10W of heat load, it hasn't fully worked yet, especially for the level II heat pipe because the coolant used in the heat pipe hasn't fully evaporated yet. So, at that time heat transfer that happened in level II heat pipe wasn't in its true performance, and it also affected the performance of whole cascade heat pipe systems.

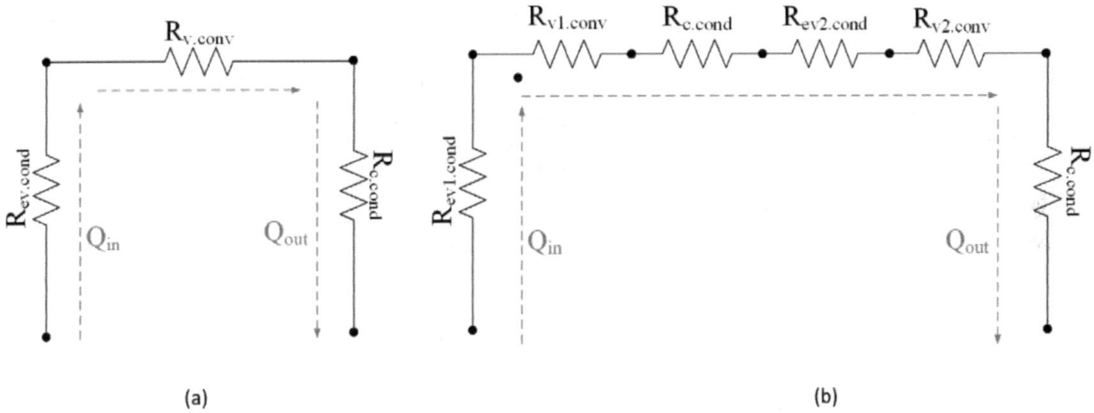

Figure 5. (a) Thermal resistance network for non-cascade heat pipe, (b) thermal resistance network for cascade straight heat pipe.

The thermal resistance of cascade straight heat pipe double condenser in 10W heat load was quite better than the cascade straight heat pipe single condenser, the difference is ±0.287 °C/W. Meanwhile, in 48W heat load, cascade straight heat pipe with double condenser only slightly better than the one with single condenser, the difference is only ± 0.062 °C/W. It can be happened because of the effect in using two heat pipes as its level II heat pipes, so the heat has to get through more cooling process than the one with a single condenser.

4. Conclusion

From the research, it was found:the cascade straight heat pipes are effective to reduce the outlet temperature at condenser part until range between 19-25 °C. Even the differences are not so big, but we can say that based on the experimental result, cascade straight heat pipe double condenser has better cooling performance than the one with one condenser only.

Acknowledgments
Equations Thank you to the Ministry of Technology and Higher Education and the Udayana Institute for Research and Community Service for financial support through the 2018 Higher Education Primary Research Grant (PTUPT) scheme with Contract Number 171.123/ UN14.4.A / LT/ 2018..

References

[1] Q. Chen and Y. Huang, "Scale effects on evaporative heat transfer in carbon nanotube wick in heat pipes," *Int. J. Heat Mass Transf.*, vol. 111, pp. 852–859, 2017.
[2] D. Liu, F.-Y. Zhao, H.-X. Yang, and G.-F. Tang, "Thermoelectric mini cooler coupled with micro thermosiphon for CPU cooling system," *Energy*, vol. 83, pp. 29–36, 2015.

[3] S. Wang, J. Chen, Y. Hu, and W. Zhang, "Effect of evaporation section and condensation section length on thermal performance of flat plate heat pipe," *Appl. Therm. Eng.*, vol. 31, no. 14, pp. 2367–2373, 2011.

[4] N. Putra, W. N. Septiadi, and R. Irwansyah, "Effect of Concentration and Loading Fluid of Nanofluids on the Thermal Resistance of Sintered Powder Wick Heat Pipe," *Adv. Mater. Res.*, vol. 651, pp. 728–735, Jan. 2013.

[5] N. Putra and W. N. Septiadi, *Teknologi Pipa Kalor Teori, Desain, dan Aplikasi*. Jakarta: Universitas Indonesia, 2014.

[6] N. Putra, Saputra, M. I. Bimo, R. Irwansyah, and S. N. Wayan, "Experimental study on sintered powder wick loop heat pipe," 2012, pp. 612–620.

[7] W. N. Septiadi, N. Putra, M. Juarsa, I. P. . Putra, R. Sahmura, and R. Sahmura, "Characteristics of Screen Mesh Wick Heat Pipe with Nano-fluid as Passive Cooling System," *Atom Indones.*, vol. 39, no. 1, p. 24, Jun. 2013.

[8] N. Putra, A. Duanovsah, and K. Haliansyah, "Investigation of Cascade Loop Heat Pipes," *Int. J. Mech. Aerospace, Ind. Mechatron. Manuf. Eng. Vol9*, vol. 9, no. 10, pp. 1868–1872, 2015.

[9] J. S. Chen and J. H. Chou, "Cooling performance of flat plate heat pipes with different liquid filling ratios," *Int. J. Heat Mass Transf.*, vol. 77, pp. 874–882, 2014.

IOP Conf. Series: Materials Science and Engineering **539** (2019) 012037 doi:10.1088/1757-899X/539/1/012037

Experimental study of thermoacoustic cooling with parallel-plate stack in different distances

Amirin[1], Triyono[2] and M Yulianto[2]
[1]Master of Mechanical Engineering Trisakti University, Jakarta
[2]Department of Mechanical Engineering Trisakti University, Jakarta

Email: amirin_king@yahoo.co.id

Abstract. Thermo-acoustic is one of Non-Vapor compression cooling technologies also environmentally friendly technology developed rapidly to the present time due to its only use the sound wave to get the magnitude of cooling temperature with the oscillatory motion of the working gas pass through the *stack* inside a *resonator* is made in such manner, not use moving part such as the compressor. No lubrication system, low electricity consumption, and without the use of any chemical refrigerants so that the constructions, it will be more simplified and cheaper. The Stack is considered as the heart of the thermo-acoustic system which is the important heat transfer processes will be occur in this area. The variation of the shape, stack distances, kind of material stack applied on the system will be influenced the output of temperature as well. The writer will build a thermo-acoustic refrigerator with the parallel plate stack that is made from PVC Rigid sheets. This study will be done to see how the stack performance with temperature different aspect that would be achieved in the stack, the expectation on this experiment that the performance will have optimum result. The Writer also tried to conduct simulation with the DeltaEC's Software.

1. Introduction

The development of cooling technology over time has experienced many innovations that are oriented towards the achievement of cost aspects, simple forms and high efficiency. Currently conventional cooling systems that use vapor compression technology are still widely used in commercially as human needs for air conditioning are increasing, this vapor compression system uses compressors and chemicals refrigerant such as CFCs which are very detrimental to environment also the consumption of electrical energy to drive the compressor is consider still high. Nowadays there are many cooling systems being studied that use non-vapor compression HVAC technologies such as one of the institutions in America, the American Energy Department [6] makes a research target and ranks about the potential of HVAC technology options as describes in Figure 1 below. Thermoacoustic as a one of the choice of cooling method has many advantages if compared to conventional systems, including: no moving components, no use of chemical refrigerants and lubrication, simpler use of components, low electricity consumption and most importantly environmentally friendly [10][9]. Thermoacoustic can be a cooling tool with low cost and effective alternatives when compared to the current conventional system. Thermoacoustic is a field related to physical phenomena, where temperature differences can generate sound waves, and conversely sound waves can produce temperature differences [11], this cooling does not require a relatively expensive cooling fluid such as Freon or other chemical refrigerants but can use helium or free air as a sound wave propagation medium, this medium will produce differences in temperature when the acoustic waves is generated at certain frequencies.

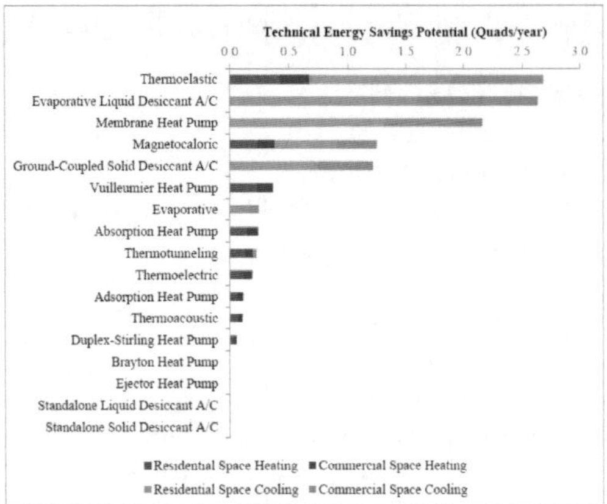

Figure 1. Technical Energy Saving Potensial [6]

A sound wave (acoustic) in a working fluid or gas is usually only seen as pressure and motion oscillations, actually the temperature oscillation is always the case. When the sound propagates in small canals, oscillating heat will also will flow to and from the canal walls [11]. The combination of all these oscillations will produce thermoacoustic phenomena or thermoacoustic effects, or in a easier understandable that the sound waves can cause temperature differences due to the air passing through small channels in the propagation. The thermoacoustic effect of the pressure and temperature changes due to sound waves on the resonator has been analyzed with reasearch by Tijani and his colleagues [18]. This phenomenon or effect can then be used as a refrigeration mechanism. The components used in thermo-acoustic refrigeration can be seen schematically in Figure 3 [22].

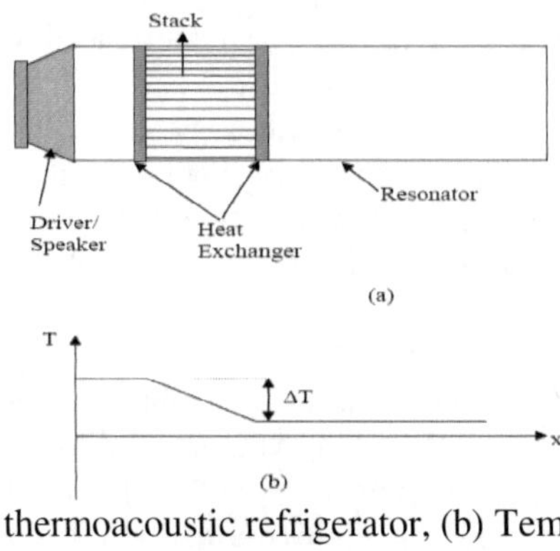

Figure 2. (a) Schematic of a thermoacoustic refrigerator, (b) Temperature variation across the resonance tube

The component consists of acoustic wave generators which usually use loudspeakers, resonators which are usually cylindrical (tube) and stack which usually use parallel plates. The thermoacoustic effect itself has been thoroughly examined for a long time with two very well-known tools, namely "Soundhauss tube" in 1850 and "Rijke tube" in 1858 [37] [41], however the theoretical explanation of the. Wheatley, Swift and others [28] [29][30] have developed a relationship between thermoacoustic components in general from the point of view of thermodynamics. The Thermoacoustic heat engine is further categorized into two types, namely "Standing wave engine" and the "Traveling-Wave engine". Currently the Traveling wave engine is better known as the Stirling engine [17] [5], while the discussion of thermoacoustic cooling engines generally refers to "Standing wave engine".

Many thermoacoustic studies have been carried out with various variations in geometry, major component modifications and engineering of the main parameters in an effort to improve the performance and efficiency of thermoacoustic machines. M.E.H Tijani [19] in his thesis and papers has made a lot of research on thermoacoustic and has become a reference by other researchers in developing this thermoacoustic technology. The procedure and thermoacoustic refrigerator design optimization strategy using thermoacoustic linear theory has been introduced and described by M.E.H Tijani et al. [12], also the procedure for making and testing experimental thermoacoustic refrigerator performance has been carried out by Tijani et al. 65°C [13], optimization of stack design with parallel plate type has also been tested by looking for the optimal inter-plate distance (2yo) by Tijani, et al. and Stack Parallel plate form for cylindrical stack and resonator is the best form for thermoacoustic machine performance [14]. Further analysis of the improvement of thermoacoustic performance was developed by looking at the potential characteristics of the working fluid such as seeing the effect of the prandtl number on the thermoacoustic refrigerator as described in the paper Tijani, H. Zeeger and M.D. Waele [16] and concludes the COP machine will be high with a low prandtl value. Russell & Weibul [42] have been able to realize and operate a thermoacoustic cooling machine with a temperature difference between the hot tendon and cold tendon of 15°C within four minutes of the start of operation. Mathematic studies have also been analyzed by many mathematical equations like those conducted by PHM Wihelmus [25] as well as the general formulas outlined by Arnott et al. In his paper [1], Weismen & Diana [43] also outline many mathematical modeling in optimization of thermoacoustic components, making it easier for other researchers to determine design parameters.

In Indonesia, too many studies have been conducted, for example by Iksan Setiawan, et al. [11] who have successfully designed and made thermoacoustic devices as refrigerants and heaters simultaneously, Iksan setiawan [10] also made an experiment about the porosity effect of the stack. Anastasia et al. [4] have analyzed the effect of the frequency of operation and stack length experimentally using a porous stack from gambas material, which concluded that the optimal frequency is at 103Hz with the maximum temperature difference between the heat reservoir and the cold tendon of 4.7°C while the stack length optimal around 6 cm. Edy Hartulistiyoso and Muhamad Yulianto [9] also made research on the potential of Bamboo to be used as a resonator tube material. Analysis of the effect of hydraulic spacing from the stack was also carried out by Prastowo et al. [23] concluded that the greater the hydraulic track value, the temperature difference on onset tends to be smaller and the stack length also affects the difference in the temperature of the onset and the time of the sound. With the potentials and advantages of the thermoacoustic system when compared to the current conventional cooling technology system, it is deemed necessary for the advancement and development of cooling technology to conduct further studies and research on thermoacoustic systems, especially in the hope that this technology can be applied to rural technology. , application in the field of cooling of transportation equipment, cooling of electronic devices.

2. Basic Principle of Thermoacoustic

The word "Thermoacoustic" comes from a combination of 2 words, namely "Thermo" or heat, and "acoustic" or sound. Thermoacoustic is the interaction between temperature, density and variations in sound wave pressure. Thermoacoustic works based on the principle that sound waves will also cause wave pressure, while sound waves (acoustics) in a working fluid or gas are usually only viewed as pressure and motion oscillations, whereas temperature oscillations always occur. When the sound propagates in small canals, oscillating heat will also flow to and from the canal walls [11]. The combination of all these oscillations will produce thermoacoustic effects, or in a more understandable sense are sound waves that can cause temperature differences because of the air through small channels in the propagation, this principle is used in the thermacoustic refrigerator. This system rely on two main principles of thermodynamics.

First; the temperature of the fluid will rise if compressed and down when expanding on a fixed volume.

Second; when two substances or objects are attached directly, the heat will flow from substances/objects that are hotter to cooler objects.

There are 2 types of thermoacoustic devices namely thermoacoustic engine and thermoacoustic refrigerator. In the thermoacoustic engine heat is converted into sound wave energy and this energy is able to be used for work, while thermoacoustic refrigerator is the opposite of the process that occurs that uses sound waves to absorb low temperature heat and remove heat in high temperature media. A thermoacoustic device consists of a loud speaker, where the sound is directed to the resonator tube which has been filled with working gas for example air, helium or other gases, and a stack is made with a thin layer of plate where the thickness and distance between plates are determined, there is also a pore circular stack, the stack equipped with heat exchanger at the ends of the stack to utilize the cold temperature produced can then be used for various things. A simple illustration of the thermoacoustic system in a standing wave can be seen in Figure 2.

Stack (also called regenerator) is a place where a heat pump process consists of small canals in such a way that the plates is placed parallel to the resonator which is used for the purpose of producing a temperature gradient along the layers of the plates.

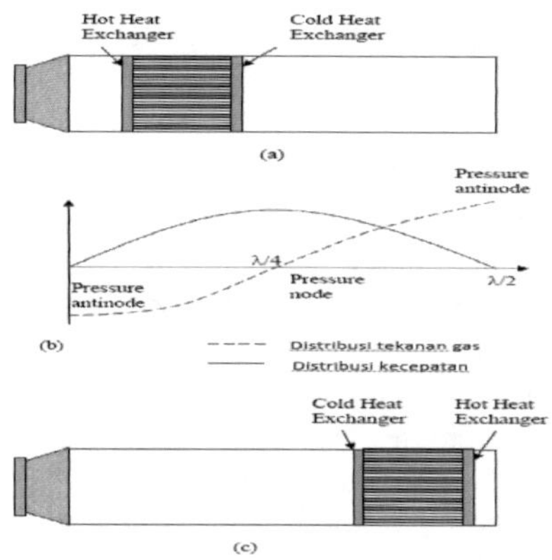

Figure 3. (a) Stack position for λ/4, (b) The Temperature & pressure distribution along the resonance tube, (c) Stack position for λ/2

3. Thermoacoustic Cooling Cycle & Heat Transfer Processes

As explained earlier that the stack can be said to be the heart of the thermoacoustic system because the heat transfer process occurs here, then if we take a small portion of the inter-plate channel image for our analysis from the example of the parallel stack plate, then the process cycle can be described as shown 4.

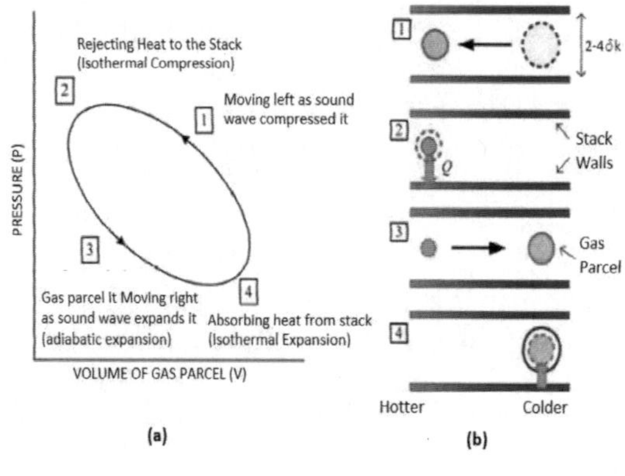

Figure 4. (a) Thermoacoustic cycle (b) Heat transfer processes in the stack [42] [11]

The working principle of thermoacoustic cooling consists of 4 stages that occur repeatedly. In Figure 4 shows the working stages of the system based on conditions of increase and decrease in temperature and pressure as follows [11]:

- Adiabatic Compression (1-2)
  When the acoustic wave causes the gas package between the plate channel to move to the left, the compressed gas and the pressure increase
- Isothermal Compression (2-3)
  The compressed gas package gets hotter than the stack wall, causing heat transfer from the gas to the stack wall because the stack temperature is lower and the volume of the gas parcel begins to shrink
- Adiabatic Expansion (3-4)
  When the upright wave continues the cycle the gas package returns to the right direction, it will expand the air so the pressure drops again. The gas becomes more tenuous, this pressure drop will simultaneously reduce the air temperature so that the gas temperature is lower than the plate temperature
- Isothermal Expansion (4-1)
  At this stage, heat moves from the stack to the gas because the gas absorbs heat from the stack wall and the gas package expands so that the temperature and pressure return to the initial state of the cycle.

From this stage it can be seen that the working principle of thermoacoustic refrigeration is different from the working principle of vapor compression refrigeration system.

4. Design Strategy

Many design procedures were carried out by several researchers such as Swift (1988), Wetzel, Herman (1996) and M.E.H Tijani (2001). The design method approach as done by Tijani in the Thermoacoustic refrigerator planning process and will be used as a reference for this research.

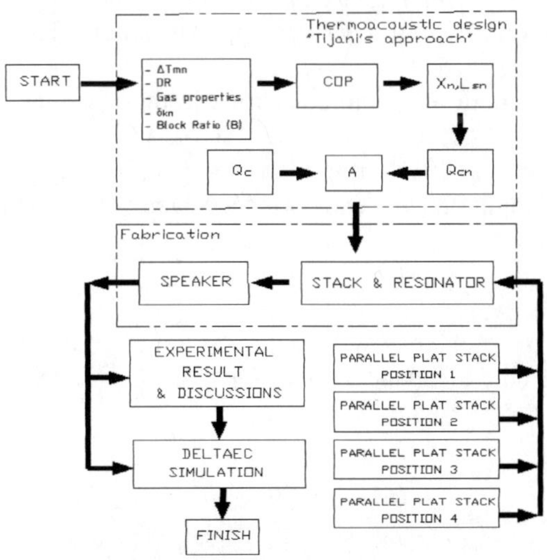

Figure 5. Design Strategy flow

This method will be used to determine design calculations in order to obtain the optimal dimensions and geometry of the test equipment, after which data analysis will be carried out and the test data will be compared with DeltaEC software simulation. Broadly speaking, the overall method of this research

IOP Conf. Series: Materials Science and Engineering **539** (2019) 012037 doi:10.1088/1757-899X/539/1/012037

will be carried out in accordance with the scheme as shown in Figure 5, because the limitations in the fabrication of heat exchanger devices in this study were not used.

5. Design Parameters

In determining the thermacoustic refrigerator as done by MEH Tijani there are several parameters that must be determined in advance, namely determining the amount of temperature difference (ΔTm) and desired cooling / cooling power (Qc) as MEH Tijani uses in designing thermoacoustic [12][13][14], so that we can determine the function of thermoacoustic to be applied to what cooling is suitable, from this parameter it will have an impact on the shape and size of the thermoacoustic refrigerator. ΔT_m=75 °K; T_m=27 °C (300 °K); Q_c=40 Watt (Cooling).
The selection of several operating parameters, working fluid and material from the stack and resonator are determined as follows.

5.1. Frequency & Resonator Length

The frequency will be determined at 245 Hz, where the frequency magnitude affects the wavelength value, while the wavelength (λ) is the distance between two identical points that are identical in waves and repeat. At certain sound wave frequencies, the waves resulting from interference will produce standing waves. This event is called resonance. The condition for the occurrence of resonance is given by

$$L = (2n-1)\,\lambda / 4 \quad (1)$$

Where L is the tube length and n = 1, 2, 3, ... is the harmonic or basic frequency resonance order, the frequency of the first, second and so on, while *l* is the wavelength. This wave is related to the length of the resonator tube So that the minimum L resonator length is generally for the closed tube side is λ / 4. The shape and length of the resonance is determined by the resonance frequency and minimal losses on the wall of the resonator tube. Resonator tube length usually varies with the acoustic wavelength produced, in previous studies determined the length of the resonator is λ/4, or λ/2 as done for example by Collard [9], because the frequency will determine the wavelength. This wavelength is equivalent to the length of the resonator that we need.

Determination of the wavelength to be more accurate in the resonance of upright sound waves as described by Taylor, et al (2011) that end correction factors must be used in the calculation of 0.33 times the diameter of the resonator [49]. Furthermore, the length of the resonator is determined equal to the magnitude of the wavelength plus the end correction factor that is given by

$$\lambda/4 + (0.33\emptyset) \quad (2)$$

According to Tijani for the λ/4 resonator can be optimized by reducing the diameter of the right side with a large diameter ratio of $\emptyset_1$ and small diameter $\emptyset_2$ of 0.54. The speed of sound in air at a certain temperature is obtained from the following equation:

$$a = \sqrt{\frac{\gamma P_m}{\rho}} = \sqrt{\gamma RT} \quad (3)$$

$$RT = P_m/\rho \quad (4)$$

5.2. Thermal penetration depth

There are more than 2 parameters defined in the perpendicular direction of gas movement known as the thermal penetration depth (δk) and the viscosity penetration depth (δv) see figure 6.

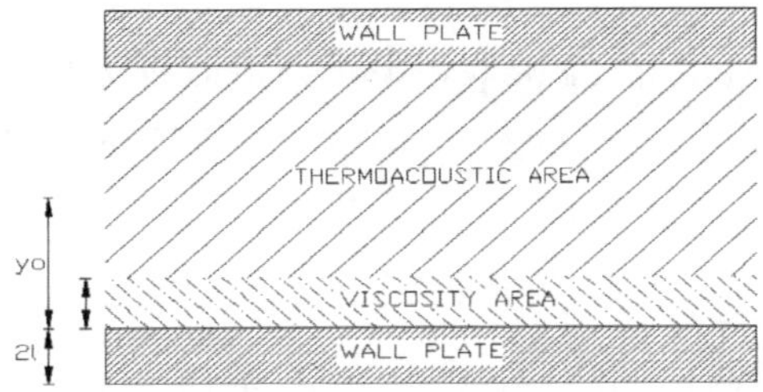

Figure 6. Illustration of the area δk and δv on the stack channel [40]

The thermal penetration depth is an estimate of the distance at which thermal can blend through the working fluid over a time interval of $1/\pi f\,(2/\omega)$ which corresponds to the period (T) of the sound wave, remember that the period T is inversely proportional to frequency *f*.

$$T = 1/f. \tag{5}$$

- Gas thermal penetration depth (δk)

$$\delta_k = \sqrt{\frac{2k}{\omega.\rho_m.c_p}} = \sqrt{\frac{2K}{\omega}} \tag{6}$$

- Viscosity penetration depth δv

$$\delta_v = \sqrt{\frac{2\mu}{\omega\rho_m}} = \sqrt{\frac{2v}{\omega}} \tag{7}$$

- Solid Thermal penetration depth δs

$$\delta_s = \sqrt{\frac{2ks}{\omega.\rho_m.c_s}} = \sqrt{\frac{2Ks}{\omega}} \tag{8}$$

The wall plate thickness (2l) should be have $2\delta s < 2l$; while the angular frequency (ω) is determined from equation 8 as follows:

$$\omega = 2\pi f \tag{9}$$

5.3. Average Pressure (P_m)

The average pressure that occurs inside the resonator is better made as large as possible because it is directly proportional to the amount of acoustic power and to the mechanical strength of the resonator material, also the greater average pressure will get the smaller thermal penetration depth. Because it does not use inert gas but only free air, the mean pressure in the resonator tube is 3.0 Bar.

5.4. Dynamic Pressure (Po)

Tijani said that to avoid nonlinear effects, the value of M (Mach number) must be limited to $M \approx 0.1$, which was also suggested by Swift (1988) so that air movement keep remains linear.

$$M = P_o / \rho_m.a^2 \tag{10}$$

IOP Conf. Series: Materials Science and Engineering **539** (2019) 012037 doi:10.1088/1757-899X/539/1/012037

5.5. Drive ratio (D)

Drive Ratio (D) is the dynamic pressure ratio (Po) to the average pressure (Pm) or can be written below the equation

$$D = \frac{Po}{Pm} \tag{11}$$

So we get the Po value = 0.42 bar. While the amount of the Drive ratio (D) according to MEHTijani must be less than 3%, $D \leq 0.03$ (Tijani, 2002), but in the calculation to make it a more realistic with the amount of pressure in the resonator tube, so this value is not used in terms of equation 11. It is obtained D value as follows:

$D = Po/P_m = 0.14$ (dimensional)

5.6. Working Gas

The working gas is one of the main parameter that must be considered because it will also affect the performance of the thermoacoustic system, the gas selection uses considerations such as has a high sound velocity, because it is important to remember that the speed of sound passing through different fluids will have a specific value in the speed of sound as well, other indicators are the gasses must have a low value of thermal conductivity and prandtl number [16].In this research will be used the air as a working gas.

Table 1. Gas properties (at P_{atm}, 27 °C)

Design Operating
Thermal conductivity (k) = 0.0257 W/m.K
Sound Velocity (a) = 597,61 m/s (3 bar,300K)
Spesific Heat ratio fo air γ = 1.4
Heat capacity (Cp) = 1007 J/Kg K
Dynamic Viscosity (μ) = 1.87×10^{-5} Kg/s.m
Prandtl number (σpr) = 0.72904 at 300K
Air density (ρ) = 1.176 Kg/m3
Constant (R) = 0.287 kJ/Kg.K (kPa-m3/Kg.K)
Gas Universal (Ru) = 8.314 kJ/Kmol.K
Molar Mass (M_o) = 28.970 Kg/Kmol
Molar Volume (V_m) = 0.0883 m3/Kmol
Thermal Difusiviy (K) = 2.17×10^{-5} (m^2/s)

6. Material Stack

Material stack must have a value of low thermal conductivity and heat capacity must be greater with a heat capacity of the gas work., MEH Tijani (2002) using a plate material Mylar as a stack on his research [12] also TJ Hofler (1986) using the material of camera roll films as a material in the form of a spiral stack [33]. Previous researchers in the early stages of development, for example GW swift, et al. (1990) still used stainless steel plate stack [29], JC Wheatley (1985) used tungsten as a plate material on the stack [37], but the use of stack using metal material was reduced currently because it refers to optimizing the performance of the engine the results of the years after that research that non-metal materials with low thermal conductivity values will increase efficiency. This research will use PVC Rigid Sheet or some known as Mica Film which is used for binding the book.

Table 2. PVC Rigid Properties

Thermal Conductivity (k_s) = 0.15 W/mK
Heat capacity (C_p) = 1260 J/Kg K
Solid Density (ρ_s) = 1400 Kg/m3 (1.4 g/cm3)

There are many types of stacks such as circular, parallel plate, triangle pores, etc. In this research will use parallel plate stack type according to Rott and MEH Tijani that this is the best shape. Parallel plate parameter illustration can be seen in figure 7 with the symbol definition; 2l is the thickness of the plate, 2yo is the distance of the air channel or the distance between the plates, Ls is the length of the stack, Π is the plate periphery

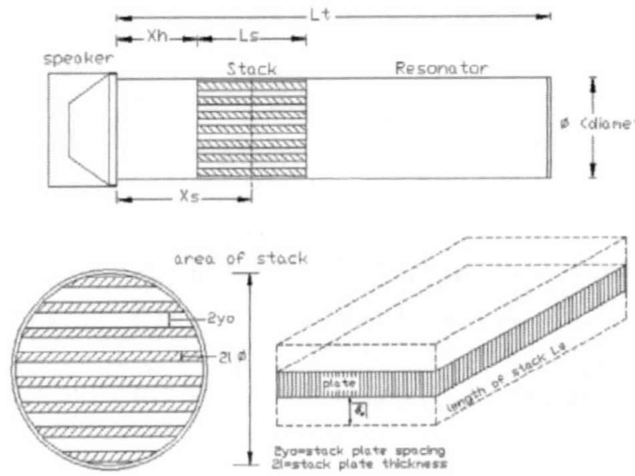

Figure 7. Parallel Plate Stack Parameters

According to Fig.7 parameter Π is given by

$$\Pi = \frac{A}{yo + 1} \tag{12}$$

Another important parameter is the Blokage ratio or porosity stack, with the equation:

$$B = \frac{y_o}{y_o + 1} \tag{13}$$

This parameter explains the presentation of the total area between solid and gas area, if the value of B is reduced then the thermoacoustic area will decrease but will increase the number of gas particles in the system.

Various choices of stack shapes that we might be able to use such as parallel plate, circular pore, array pin, triangle pore. The stack shape as illustrated in the Rott function diagram (see figure 8). Figure 8 shows the Real and imaginary parts of the Rott function (*f*k) of several shapes of the stack as a function of the ratio of the hydraulic radius (rh) and the thermal penetration depth (δk), from the visible pin shape and parallel plate shape is the most effective form used , the use of array pins in this study is not used because it is very difficult in fabrication so it will only use a stack with parallel plate, As we seen from Figure 8, for parallel plates the hydraulic radius value (rh = yo) so that the Im value (-*f*k) from Figure 8 below has a value for rh/δk = yo/δk.

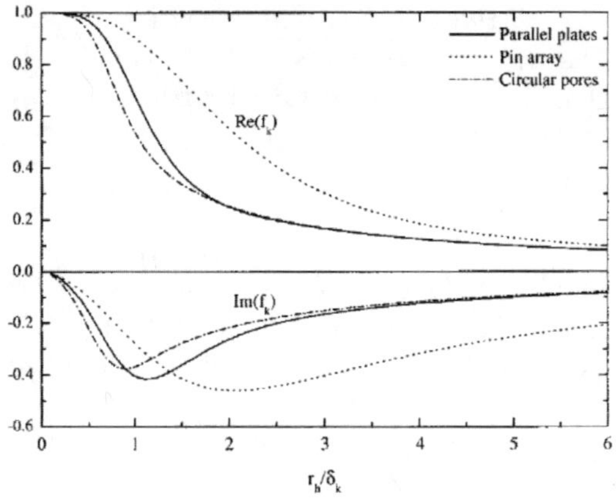

Figure 8. Rott's Function Diagram as a function of the hydraulic radius (rh) and thermal penetration depth (δk).

7. Stack Parameter Design

There are three stack design parameters namely Normalized stack position (Xn), Normalized stack length (Lsn) and the third is the cross section of the stack (A). See figure 7 as a reference notation for determining the stack and resonator dimensions. Generally the cross-sectional area (A) of the resonator tube is the same as the cross-sectional area of the stack. From the three parameters above can be made the following equation:

$$Xn = kv \, . \, Xs \tag{14}$$

$$Lsn = kv \, . \, Ls \tag{15}$$

While the Wave number is determined by the following equation:

$$kv = \frac{\omega}{a} \tag{16}$$

Some other parameters for stack design are

$$\Delta T_{mn} = \Delta T_m / T_m \tag{17}$$

Usually the average Temperature (Tm) works at room temperature of around 27°C or 300°K,

a.Parallel plate Stack type

Referring to figure 7, the dimensions of the plate thickness (2l) and the distance between plates (2yo) will be determined. The relationship between 2yo and 2l is determined by equation 13 so that the value 2l becomes:

$$2l = 2y_o \left(\frac{1}{B} - 1\right) \tag{18}$$

Blockage Ratio value itself according to Iksan, et al (Iksan, 2013) will improve performance with increasing values, taken in this study by 0.778, whereas when referring to the graph of Rott's function as illustrated in Figure 8, for parallel stack plate selection types, has a maximum Im (*-fk*) value with rh/ δk =1.1, and with a hydraulic radius *(rh)* = *yo*, then from the maximum point graph the parallel plate stack type is rh/δk = yo/ δk= 1.1, because (rh) Radial radius is the middle distance from the free area for gas media, so for stack type parallel the distance between plates is equal to 2yo,

So that the theoretical value 2yo = 1.1 x 2 = 2.2 δk. This is also according to the optimum distance suggestion based on Arnott (1991) which is about 2.2 δk, while Tijani (2002) suggests 2yo = 3δk, while Swift (1998) mentions about 4δk for 2yo, because the reason for uncertainty is that Arnot is only based on theory, while Tijani & swift are based on experimental test results, but this value is still proportional to the range (2 to 4) δk and this is also in accordance with the advice of Wheatley (1985). In this study the author uses 2.1δk according to the graph due to availability the plate spacing.

b. Stack Length (Ls) & Stack Diameter

M.E.H Tijani (2002) uses Figure 9 to explain the performance of the stack calculation as the Normalized function of the stack length (LSn) and the Normalized Distance Stack (Xn).

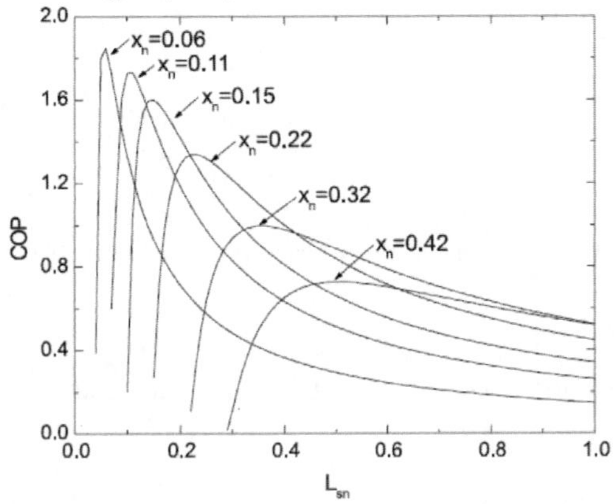

Figure 9. Performance calculation stack, as a function of the Normalized Stack Length (Lsn) & Normalized Distance Stack (Xn) [12]

Refer to figure 9, if the position at Xn = 0; defining the position of the loud speaker as (antinode pressure), in other words the smaller the Lsn value means the closer to the speaker, then based on the number of experimental cases the COP value shows the maximum limit that can be achieved from each stack position optimally, as if the normal value stack position (Xn) increases which means the position of the stack distance is also large, while the COP value will decrease. We will use Xn=0.11, and Lsn=0.135.

8. Performance and Efficiency

Dimensional Determination of the thermoacoustic heat rate (Qcn) that occurs through the gas medium and stack material around the stack area, and the dimensions of the acoustic power (Wn) can be determined using the equation as used by MEH Tijani (Tijani, 2001) and (Hao, 2006). Dimensional Thermoacoustic heating rate / Normalized Cooling power (Qcn).

$$Q_{cn} = -\frac{\delta_{kn} D^2 \sin((2x_n))}{8\gamma(1+\sigma)\Lambda}\left(\frac{\Delta T_{mn} \tan(x_n)}{(\gamma-1)BL_{sn}}\frac{1+\sqrt{\sigma}+\sigma}{1+\sqrt{\sigma}} - (1 + \sqrt{\sigma} - \sqrt{\sigma}\delta_{kn})\right) \quad (19)$$

While the values of Λ and δkn have the following equation:

$$\delta_{kn} = \delta_k / y_o \quad (20)$$

$$\Lambda = 1 - \sqrt{\sigma}\delta_{kn} + \frac{1}{2}\sigma\delta_{kn}^2 \quad (21)$$

IOP Conf. Series: Materials Science and Engineering **539** (2019) 012037 doi:10.1088/1757-899X/539/1/012037

where, y_o = half the distance between the stack plate walls. Normalized acoustic power (Wn) is given by

$$W_n = \frac{\delta_{kn}.L_{sn}D^2}{4\gamma}(\gamma-1)B\cos(X_n)^2\left[\frac{\Delta T_{mn}\tan(X_n)}{BL_{sn}(\gamma-1)(1+\sqrt{\sigma})\Lambda}-1\right] - \frac{\delta_{kn}L_{sn}D^2\sqrt{\sigma}\sin(X_n)^2}{4\gamma\,B\Lambda} \quad (22)$$

Note: All trigonometri function in radian not degree. Relationship between Q_{cn} and Wn

$$Q_{cn} = Q_c / P_m.a.A \quad (23)$$

$$Wn = W / P_m.a.A \quad (24)$$

With the substitution of Q_{cn} and Wn values, we will get the stack cross-sectional area (A). The performance of the thermoacoustic refrigerator is determined by the COP value or the Coefficient of performance, the maximum COP that can be produced is given in equation 25, namely;

$$COP = \frac{T_c}{T_h - T_c} \quad (25)$$

COP for refrigeration systems can also be determined by the equation;

$$COP = \frac{Q_{cn}}{W_n} \quad (26)$$

With reference to Figure 10, the results of the calculation will be summarized as follows:

Table 3. Calculation results

a = Sound Velocity = 597,61 m/s
λ = Wave length = 2,439242 m
kv= Wave number = 2.57457
D = Drive ratio = 0.14
P_o = Dynamic pressure = 0,42 bar
P_m = Mean Pressure = 3.0 bar (3.0 x 10^5 Pa)
ΔT_{mn} = Normalized Temperature differ.= 0.25
ΔT_m = Temperature different = 75 °K
T_m = Mean Temperature = 300 °K
δ_k = Thermal penetration depth= 0.168 mm
δ_{ks} = Solid Thermal penetr.depth=0.0105mm
δ_v = Viscosity penetration Depth= 0.1436mm
ω = Angular frekuensi; $2\pi f$ = 1538.6 rps
δ_{kn} = Normalized thermal penetr.depth=0,961
Q_{cn} = 4.8176 x 10^{-5} (dimensional)
Wn= 5.5366 x 10^{-5} (dimensional)
COPtheory = 0.870141
Wtheory = Power speaker = 45,9696 Watt
Stack Parameter
A = Cross Section area of Stack = 0.0046 m^2
B = Blockage ratio = 0.778
Ø = Diameter of Stack & Large tube = 3 Inch
$Ø_2$= Diameter of small tube = 1 1/2 Inch
Lt = Resonator length= 0.61 m = 61 cm
Xn=0.11 ; Lsn = 0.135
Ls = Stack length = 0.52 meter ≈ 5,2 cm
2yo = Stack plate spacing = 0.35 mm
2*l* = Stack plate thickness = 0.1 mm

9. Fabrication & Experimental Setup

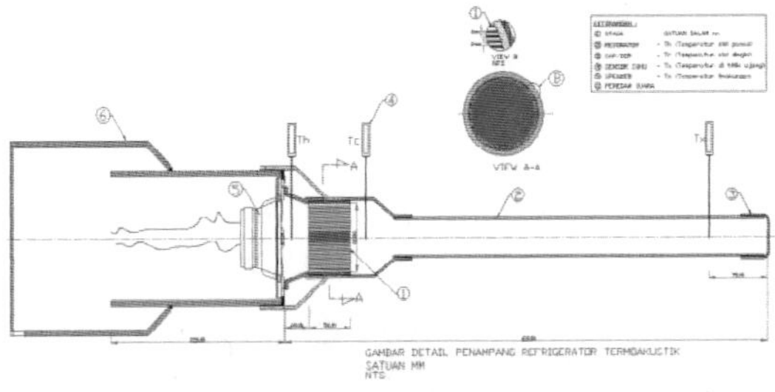

(a)

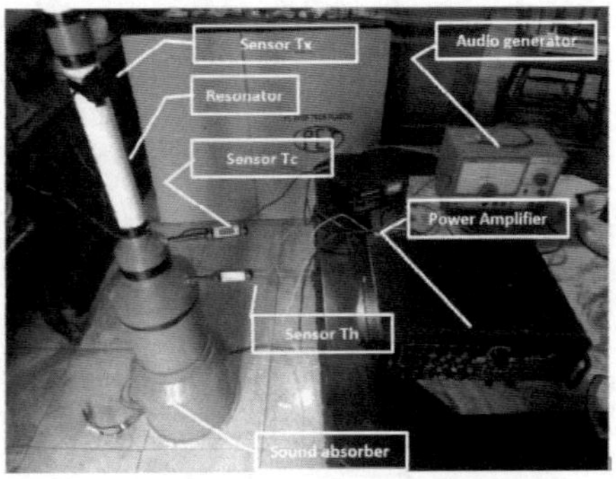

(b)

Figure 10. (a)Thermoacoustic Refrigerator detail, (b) Thermoacoustic devices

The Stack Posisition can be describe in figure 11, that the test will be devided to be fours at different distances also the sensor location was placed based on figure 10.According to Table 3 the parallel plate Stack was made as can be seen at figure 12.

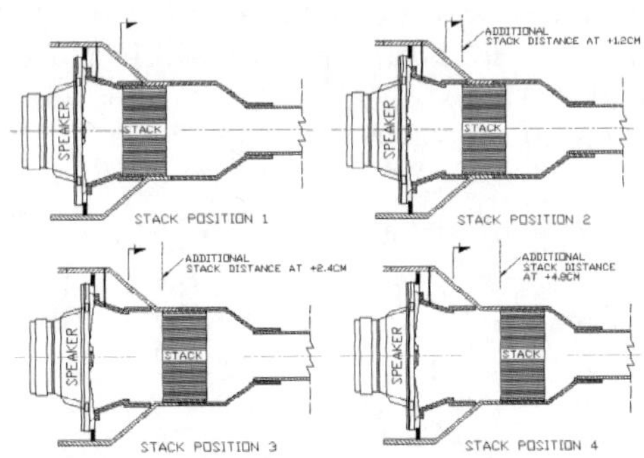

Figure 11. Stack Position

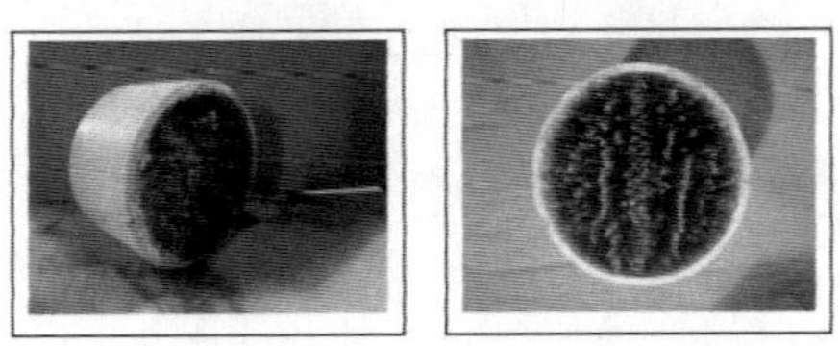

Figure 12. Stack Fabrication

10. Results and Discussion

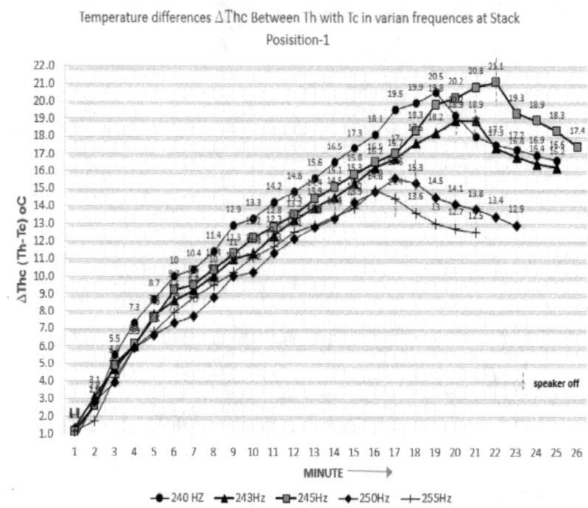

Figure 13. Temperature vs Minute at varian frequencies in Stack Position 1

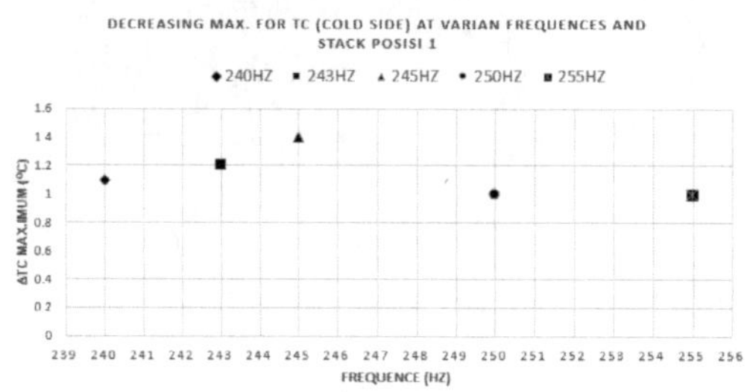

Figure 14. Decrease of temperature maximum Tc at varian frequencies in Stack Position 1

In fig. 13 and fig.14 the test has been conducted at stack position 1 with initial frequence 245Hz, also above and below from intial frequence are 240Hz, 243Hz, 250Hz and 255Hz respectively has tested. The results can be seen that the 245Hz has a value with optimum result. So the next it will be used to be initial frequence to applied to other stack position testing.

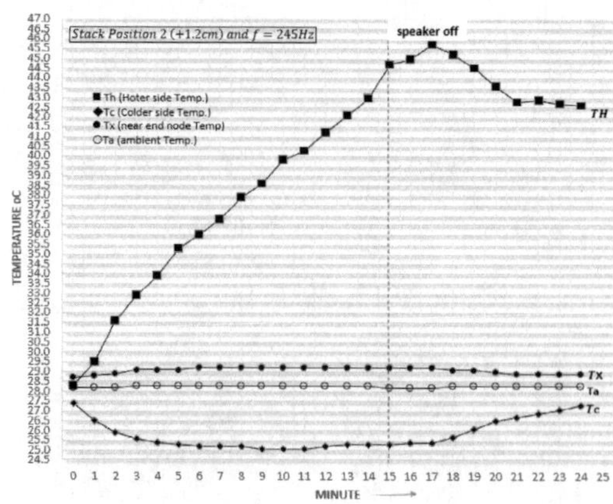

Figure 15. Temperature vs Minute at f=245Hz in Stack Position 2

IOP Conf. Series: Materials Science and Engineering **539** (2019) 012037 doi:10.1088/1757-899X/539/1/012037

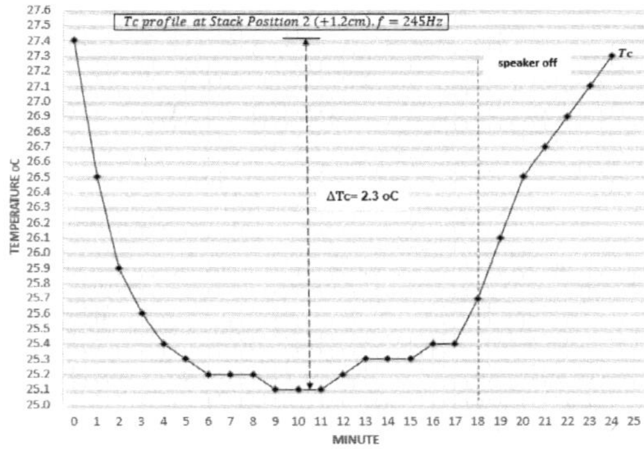

Figure 16. Temperature colder side (Tc) vs Minute at *f*=245Hz in Stack Position 2

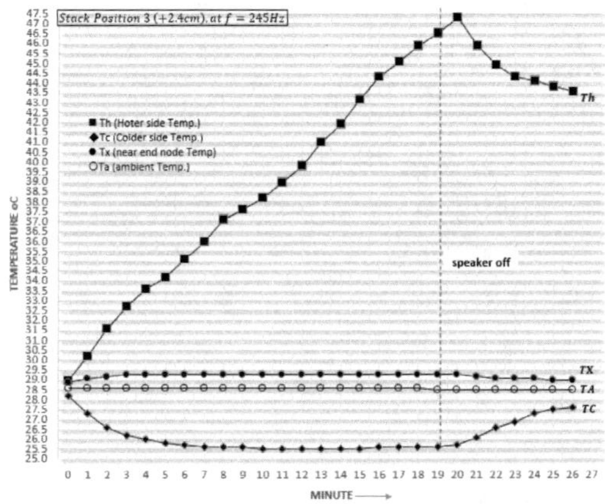

Figure 17.Temperature vs Minute at *f*=245Hz in Stack Position 3

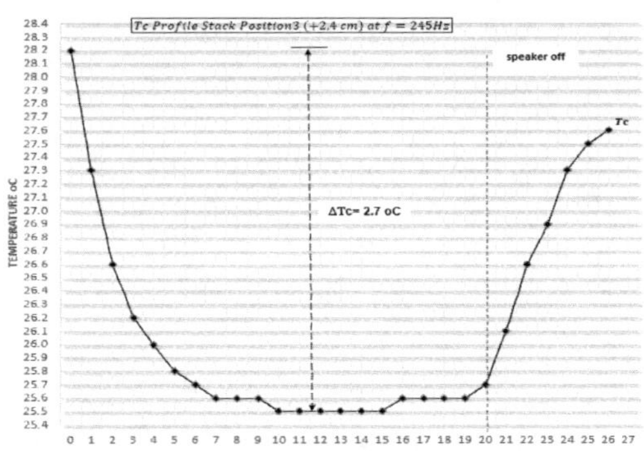

Figure 18. Temperature colder side (Tc) vs Minute at *f*=245Hz in Stack Position 3

Fig.17 to fig.19 and fig.21 has been tested and the results shows the optimum value that can be achieved on this experiment that the Stack position 3 is the best position. The value of decrease of temperature Tc (Colder side) maximum around 2.7°C for 10 minute.

IOP Conf. Series: Materials Science and Engineering **539** (2019) 012037 doi:10.1088/1757-899X/539/1/012037

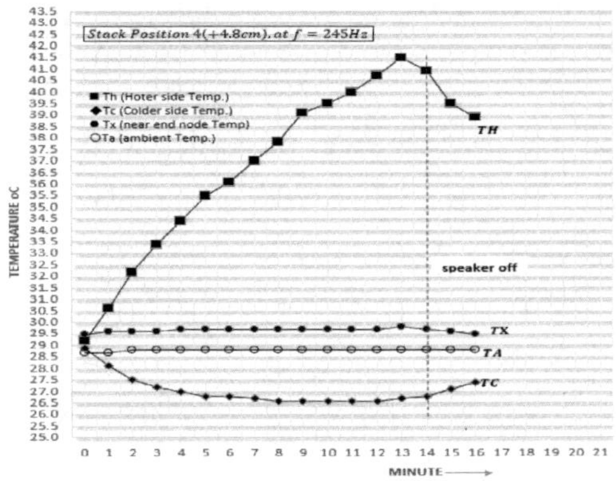

Figure 19.Temperature vs Minute at *f*=245Hz in Stack Position 4

Fig.21 has described that the position 3 is the best position than others, the value of temperature difference (Th-Tc) around 21.6°C in 20 minute, perhaps this value can be used for heatpump needed for other research.

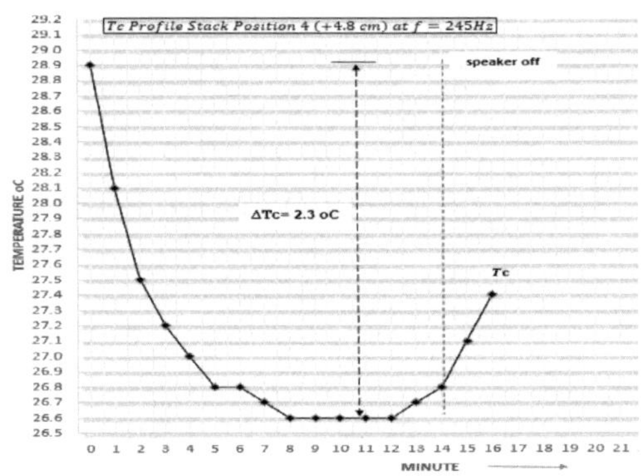

Figure 20.Temperature colder side (Tc) vs Minute at *f*=245Hz in Stack Position 4

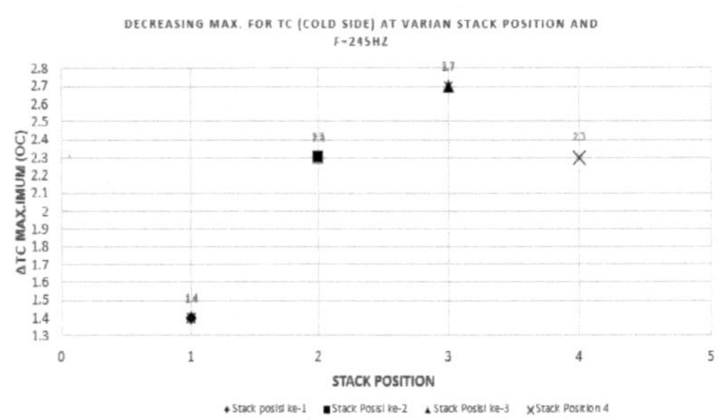

Figure 21. Temperature colder side (Tc) vs Minute at *f*=245Hz in Stack Position 1- 4

IOP Conf. Series: Materials Science and Engineering **539** (2019) 012037 doi:10.1088/1757-899X/539/1/012037

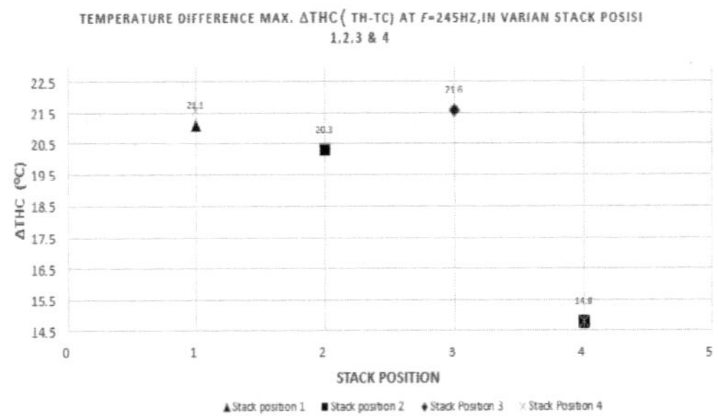

Figure 22.Temperature difference maximum (Th-Tc) at *f*=245Hz in Stack Position 1- 4

-DeltaEC Simulation

From the Stack position 3 with *f*=245Hz, the DeltaEC segment model is made and the geometry dimensions will be followed acoording to fig.10a.

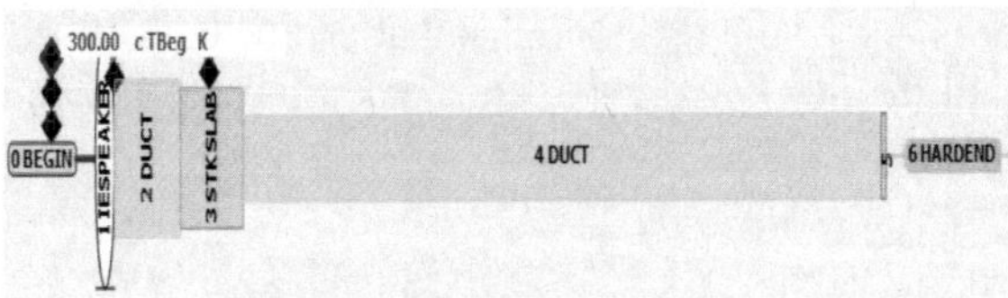

Figure 23.DeltaEC Geometry model

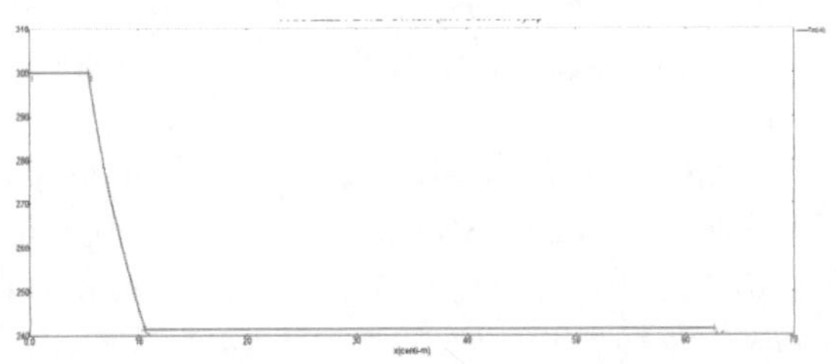

Figure 24. DeltaEC result for Temperature distribution along the resonance tube

In Fig 23. Shown that the distribution temperature on the stack it's going down, with the maximum Colder side Temperatur end of stack (Tc) 300 °K to be 241°K, so the simulation is given the large number value for decrease of temperature deltaTc around 59°K, if compared with the experiment result the deltaTc only 2.7°C, but need to know that on this software there is no explanation about how many time the result Tc end of stack will be taken., fig 24. Also described about the value of temperature (T) and pressure (P) distribution along the resonance tube, also the value of oscillatory volume flow rate (U) and Acoustic Power flow (E).

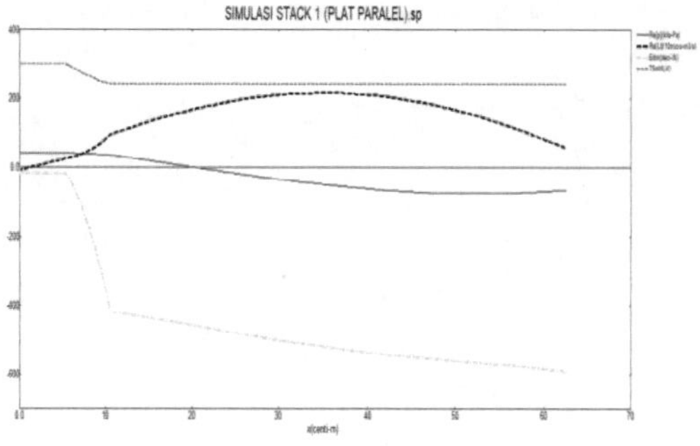

Figure 25.DeltaEC result for Temperature (T); Pressure (P);(U) oscillatory volume flow rate; E (acoustic power flow)

11. Conclusions

Thermoacoustic refrigerator is one of the technologies that can be considered as a "green technology" to the future. The precision of the Stack and resonator domensions will have huge impact of the result also the right position stack along the resonance tube.

References

[1] Arnott, W.P., Bass, H.E., and Raspet, R, 1991, "*General Formulation of Thermoacoustic for Stack Having Arbitrarily Shaped Pore Cross Sections*". J. Acoust. Soc. Am 90(6), 3228-3237

[2] Asmara P, Pebrianti,dkk, 2015, *"Pengembangan pendinginan termoakustik dengan menggunakan penukar kalor tambahan dalam regenerator",* Jurnal fisika Indonesia No.5 Vol XIX edisi November 2015 ISSN:1410-2994

[3] Aditya Nugraha,dkk, 2015, *"Studi eksperimental penggerak mula termoakustik piston air dengan diameter selang osilasi 1 inchi",* Prosiding SNST-6 2015, FT Univ. Wahid Hasyim, Semarang

[4] Anastasia F.Candraresita,dkk, 2015, *"Pengaruh frekuensi resonasi dan panjang stack pada kinerja pendingin termoakustik menggunakan stack berpori acak bahan organic (Gambas)",* Seminar Nasional XI SDM Teknologi Nuklir ISSN:1978-0176, Yogyakarta

[5] Backhaus,S. and Swift,G.W. 2002. "*New Varieties of Thermoacoustic Engines*" 9th International Congress on Sound and Vibration.

[6] Building Technologies office, 2014, *'Energy Savings Potential and RD&D Opportunities for Non-Vapor-Compression HVAC Technologies",* US Departement Of Energy, http://www.buildings.energy.gov

[7] Channarong.W,Kriengkrai.A, 2011, *"The impac of the resonance tube on performance at thermoacoustic",* Global digital central, ISSN:2151.8629

[8] Ibrahim girgin, Mehmet turker, 2012, *"Thermoacoustic system as an alternative to conventional coolers",* Journal of naval science and engineering, Vol.8 No.1,pp.14-32

[9] Edy hartulistiyoso, M.Yulianto, Irawan S, 2013, *"Potensi penggunaan bambu sebagai tabung resonator",* Jurnal keteknikan pertanian Vol.1 No.1 Oktober 2013

[10] Ikhsan Setiawan,et al, 2013, *"Experimental study on the influence of the porosity of parallel plate stack on the themperature decrease of a themoacoustic refrigerator",* Journalof physics : conference series 423 (2013)012035 doi:10.1088/1742-6596/423/1/012035

[11] Ikhsan Setiawan,dkk, 2007, *"Rancang bangun piranti termoakustik sebagai pemompa kalor"* SIGMA Vol.10 No.1 2007 :25-33, ISSN:1410-5888

[12] M.E.H Tijani, J.C.H Zeegers and A.T.A.M Waele,2002. "*Design of a Thermoacoustics Refrigerator*" Cryogenics 42 (2002) 49-57

[13] M.E.H Tijani, J.C.H Zeegers and A.T.A.M Waele,2002. "*Construction and Performance of a Thermoacoustics Refrigerator*" Cryogenics 42 (2002) 59-66

[14] M.E.H Tijani, J.C.H Zeegers and A.T.A.M Waele, 2002. "*The Optimal Stack Spacing for Thermoacoustic Refrigeration*" J. Acoustical Society of America 112(1)

[15] M.E.H Tijani, 2012. "*High Temperature thermoacoustic heat pump*" ,Presented at ICSV-19,Vilinius, Litouwen

[16] M. E. H. Tijani, J. C. H. Zeegers, A. T. A. M. de Waele. 2002. "*The Prandtl number and thermoacoustic refrigerators*", J. Acoust. Soc. Am. [submitted on 2002]

[17] M. E. H. Tijani, S.Spoelstra, G.A Poignand, 2008, *"Study of a thermoacoustic-stirling engine",* Energy research centre of the netherland (ECN), presented at acoustic'08-paris

[18] M. E. H. Tijani, S.Spoelstra, P.W.Bach, 2003, *"Thermal relaxation dissipation in thermoacoustic systems",* Journal of applied acoustic ECN-RX-03-054

[19] M. E. H. Tijani, 2001, *"Loud speaker driven thermo-acoustic refrigeration ",* Eindhoven: Technische Universiteit Eindhoven, *DOI:10.6100/IR547542*

[20] Mandhata yudev,2015. "*Effect of working gases and stack material on thermoacoustic refrigerator,* IJIRST Vol.1 issue 11 2015

[21] Normah M.G, et al, 2006, "Environmentally friendly refrigeration with thermoacoustic", Research Vote No.74166, Fakulti Kejuruteraan Mekanikal Universiti Teknologi Malaysia

[22] Nurudin B.H.M.A,et al, 2008, "Thermal performance of a thermoacoustic stack" Faculty of Mechanical Engineering Universiti Teknologi Malaysia

[23] Prastowo Murti,dkk, 2015,*"Pengaruh jejari hidrolik stack terhadap beda suhu onset pada primemover termoakustik gelombang berdiri",* SPEKTRA Vol.16 No.2 Okober 2015

[24] P. Merkli and H. Thomann, 1973. "*Thermoacoustic effects in a resonance tube*" J.Fluid Mech. (1975), vol 70, part1, pp. 161-177

[25] P.H.M.Wilhelmus, 2009, *"Mathematical aspects of thermoacoustics",* Eindhoven, The netherland ISBN 978-0-386-1862-3

[26] Rott N. 1980. "*Thermoacoustic*", Advances in Applied Mechanics, Volume 20 (135):135. doi:10.1016/S0065-2156(08)70233-3

[27] Richard Raspet,Henry EB, WP Arnott, 1991, *"Theoritical and experimental study of thermoacoustic engine",* Physical acoustic research laboratory, University of missipi

[28] Swift,G.W. 1988. "*Thermoacoustic Engines*" J. Acoustical Society of America. 84(4)

[29] Swift,G.W, 1990,*"Acoustic Cryocooler",* US Patent No.4953366 ,Monterey: US Patent and Trademark Office

[30] Swift G.W., J.C. Wheatley, T. Hoßer, and A. Migliori, 1983, *Ex-periments with an intrinsically irreversible thermoacoustic heat engine.,* Phys.Rev. Lett. 50, 499 (1983). https://doi.org/10.1103/PhysRevLett.50.499

[31] Swift,G.W, 2001. "*Thermoacoustics: A Unifying perspective for some engines and refrigerators*" Condensed Matter and Thermal Physics Group of Los Alamos National Laboratory

[32] S.Garret, 2016, *"Thermo-acoustic engines & refrigerators",* Penn State University, CFA 2016/VISHNO Le mans

[33] T.J. Hofler, 1986 "*Thermo-acoustic Refrigeration Design and Performance*" Ph.D. Thesis University of California San Diego.

[34] Tomas tisousky,2016, "*Design of theoretical optimal thermoacoustic cooling device*" EDP science.
[35] Wetzel,M. and Herman,C. 1997. "*Design Optimization of a Thermoacoustic Refrigerator*" Int J. Ref. Vol 20 No. 1, pp,s3-21.
[36] Wetzel,M. and Herman,C. 1996. "*Design Issues of a Thermoacoustic Refrigerator and its Heat Exchangers*" HTD-Vol 31, National Heat Transfer Conference Vol. 9.
[37] Wheatley,J., Hofler,T., Swift,G.W., Migliori,A.1985. "*Understanding Some Simple Phenomena in Thermoacoustics with Applications to Acoustical Heat Engines*" Acoustical Society of America. 74(1).
[38] Wheatley et al, 1984, *"Intrinsically Irreversible heat engine",* US Patent No.4489553 ,Monterey: US Patent and Trademark Office
[39] Swift G.W., J.C. Wheatley, and A. Migliori, 1983, *"Acoustical heat pumping engine"* US Patent No.4398398 ,Monterey: US Patent and Trademark Office
[40] Keolian,R.M, et al, 1995, *"Pin Stack array for thermoacoustic energy conversion"* US Patent No.5456082,Monterey: US Patent and Trademark Office
[41] Wheatley et al, 1989, *"Heat-driven acoustic cooling engine havin no moving parts",* US Patent No.4858441 ,Monterey: US Patent and Trademark Office
[42] Weibull .P & Russel D.A, 2002, "*Tabletop thermoacoustic refrigerator for demonstration*", Am J.Phys, 70.1231-1233
[43] Weisman C & Diana Baltean, 2014, *"Mathematical Modelling and numerical simulation in thermacosutics"*, Presented at PAMIR theme 9:Space Space Tripe summer School Intitut Francais, Riga Latvia
[44] Patcharin S, 2014, "Application of thermoacosutic technologies for meeting the refrigeration needs of remote and rural communities in developing countries", Department of Engineering University of Leicester.
[45] Valiyandi.S ,2016,*"Design & Fabrication of thermoacoustic refrigeration system",* India International centre, ICSTM-16 ISBN:978-81-932071-8-2
[46] Jithin george, 2016, "*Loud speaker driven thermoacoustic refrigerator",* International jurnal of scientific & engineering research", Vol.7 Issue 4
[47] http://www.physicsclassroom.com/class/sound/Lesson-5/Closed-End-Air-Columns

Central processing unit based on cascade straight heat pipe

W N Septiadi [1,2], I G A A D Wulandari[2,3], W A W Ula [2,3], I A Tnunay[4], I K G Wirawan[1], M R Murti[1,2] and N Putra[5]

[1]Department of Mechanical Engineering Udayana University, Kampus Bukit Jimbaran Badung-Bali, Indonesia
[2]Heat Transfer Laboratory Department of Mechanical Engineering Udayana University
[3]Bachelor Programme Student of Department of Mechanical Engineering, Udayana University
[4]Master Programme Student of Department of Mechanical Engineering, Udayana University
5Department of Mechanical Engineering University of Gunadarma, Depok-Indonesia

Email: wayan.nata@gmail.com

Abstract. Smart technologies are the scientific knowledge application that develops into computer technology. Central Processing Unit (CPU) is a computer hardware that carries out the computer program instructions. The CPU technology development is very fast with the impression of smaller dimensions yet maintain its performance and efficiency, this results in a heat flux or overheating that should be reduced and even removed from the CPU system to maintain the CPU performance and life because the overheating could potentially damage and slow down the computer work. To solve the CPU heat flux problem, we use heat pipes as the cooling system, because of its ability to transfer heat and contribute to the high outlet temperature in the condenser. The cascade heat pipe design is made by combining two heat pipes into one. The design is expected to be able to reduce the processor temperature by 2.74^{0}C at the idle condition, 1.48^{0}C at maximum conditions, and condenser section by 14.94^{0}C at the idle condition, and 21.12^{0}C at maximum condition. Cascade heat pipe with double condenser reduce processor temperature by 3.89^{0}C at the idle condition, 4.24^{0}C at the maximum condition, condenser section by 17.12^{0}C at the idle condition and 24.16^{0}C at maximum condition lower than the non-cascade cooling system.

1. Introduction

Central Processing Unit (CPU) is an important hardware of a computer which controls the instruction of a computer program. Nowadays, CPU technology has been developed towards Smart Technologies which has a better performance with smaller dimension [1]. The development of Smart Technologies affected the heat flux and overheating which are produced by CPU. Most electronic devices produced more than 100 W/cm^2 heat flux which needs to be removed from the CPU system to maintain its performance and life [2].

Some research has been done to solve the heat flux management problem, such as application of heat pipe using biomaterial wick as cooling system [3], application of heat pipe as cooling system integrated with sintered powder and screen mesh wick [4], potential of both nanofluids and hybrid nanofluids as heat pipe's coolant such as Al_2O_3-water, Al_2O_3-ethylene glycol, TiO_2-water, TiO_2-ethylene glycol, and ZnO-ethylene glycol [5] have been done to improve the performance of heat pipe to solve the CPU's heat flux management problem.

The heat pipe is a device used in transferring the heat from one place to another which is hollow, closed, evacuated pipe containing a working fluid (coolant). The heat pipe is divided into three sections: evaporator, adiabatic and condenser section [6]. The heat pipe is generally made by aluminum, cooper or nickel coated cooper. In the inner wall of the heat pipe contains capillary pipes (wick) which have a function as a track of the coolant from condenser back to the evaporator. After

IOP Conf. Series: Materials Science and Engineering **539** (2019) 012038 doi:10.1088/1757-899X/539/1/012038

the coolant evaporates in the evaporator, the vapor flows into the condenser and it condenses at this section, so the coolant flows back into the evaporator through the wick [7].

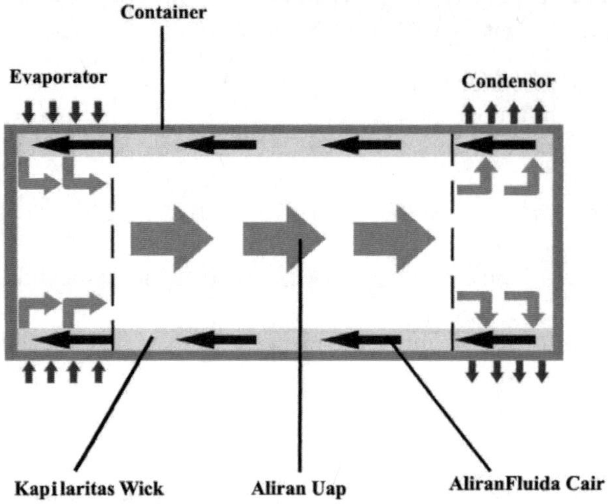

Figure 1. Heat pipe's work principal

However, a good heat transfer performance of a heat pipe affects the outlet temperature level at the condenser [8]. So, Putra et al. [9] have done a research to solve the high outlet temperature at the condenser by designing a cascade heat pipe by combining two loop heat pipes. It is reported that the application of cascade system was succeeded on reducing the temperature at the condenser and reduce thermal resistance up to 17.6%.

Cascade heat pipe consists of two evaporators (level I evaporator at the first heat pipe and level II evaporator at the second heat pipe) and two condensers (level I condenser at the first heat pipe which is connected to level II evaporator and level II condenser at the second heat pipe). The principle of cascade loop heat pipes are not so different with single heat pipe, where in single heat pipes, there is only one cooling process, but in cascade heat pipe, there are two cooling processes in level I condenser and level II condenser, so the outlet temperature at the last condenser will be smaller than the one in single heat pipe.

2. Methodology

This research was done by the experimental method through several steps. The first step was designing the cascade straight heat pipe model with a single and double condenser, then the next step was the experiment to find the temperature of the processor and condenser.

2.1. Heat pipe design

In this step, the design of cascade straight heat pipe with single and double condenser was made by a flat cooper pipe with 8 mm width, 4 mm thick and 100 mm length for the first heat pipe and the second heat pipe was a flat cooper pipe with the same width to the first heat pipe. The design size was made by considering the heat pipe work limits and the CPU and the processor's surrounding area. The first heat pipe was functioned as the evaporator which absorbed heat from the processor. One of the ends of the first heat pipe was equipped by a thermal contact plate made by Cooper with the dimension of 40 mm x 40 mm, the second heat pipe was equipped by fins.

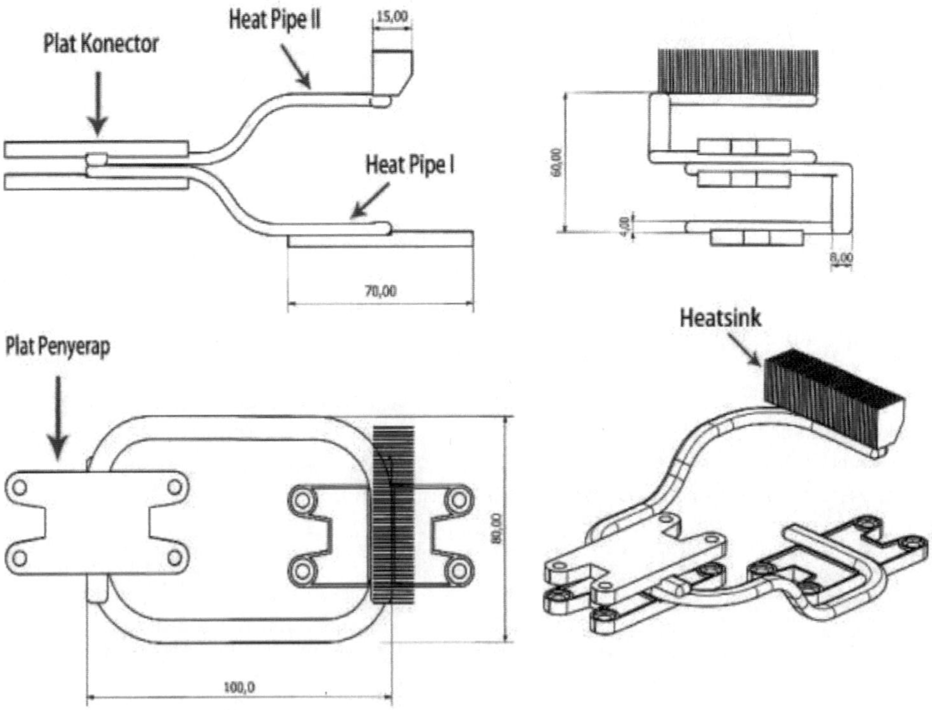

Figure 2. Cascade straight heat pipe design

2.2 Experimental setup

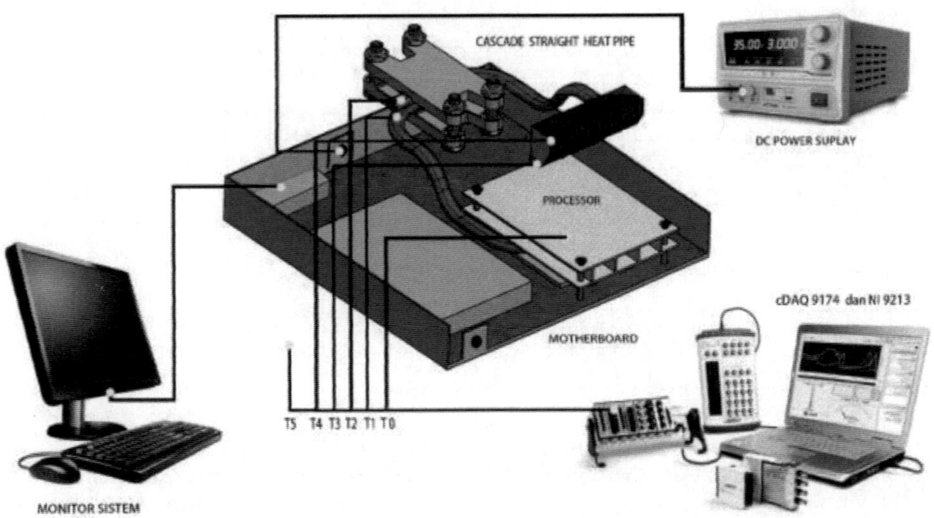

Figure 3. Schematic diagram of the experiment

In the step of the performance experiment of CPU cooling system based on cascade straight heat pipe was done on the Central Processing Unit (CPU) Core i5 2,90 GHz which is a new manufacturer and widely used, it also gave an idle load which was the CPU load condition without operation, and the maximum condition was when the CPU was being operated. The performance of the cascade straight

IOP Conf. Series: Materials Science and Engineering **539** (2019) 012038 doi:10.1088/1757-899X/539/1/012038

heat pipe cooling system was observed by putting 5 K-type thermocouples on the single condenser heat pipe and 8 K-type thermocouples on the double condenser heat pipe. Thermocouples were put on the surface of the processor, evaporator of the first and second heat pipe, and condenser of the first and second heat pipe.

3. Result and discussion

3.1 The temperature distribution of non-cascade CPU cooling system

Figure 4 (a) shows the temperature comparison of non-cascade at the 10 Watt idle condition. From the shown image, there is a dramatic rise of temperature from 0 seconds to 800 seconds, after that the temperature is steady, in a steady state, the processor temperature (Evaporator) reaches 69.10 ^{0}C and the outlet temperature of the condenser reaches 59.99 ^{0}C, and the temperature of the heat sink reaches 53.32 ^{0}C.

Non-cascade temperature distribution with a processor loading of 48 Watts or at maximum conditions can be specified in Figure 4 (b), the figure shows that the temperature rise drastically starts from 0 seconds to 800 seconds, after that the temperature is steady. In a steady state, the processor temperature (Evaporator) reaches 69.68 ^{0}C and the outlet temperature at the condenser reaches 63.06 ^{0}C, and the temperature at the heat sink reaches 60.85 ^{0}C.

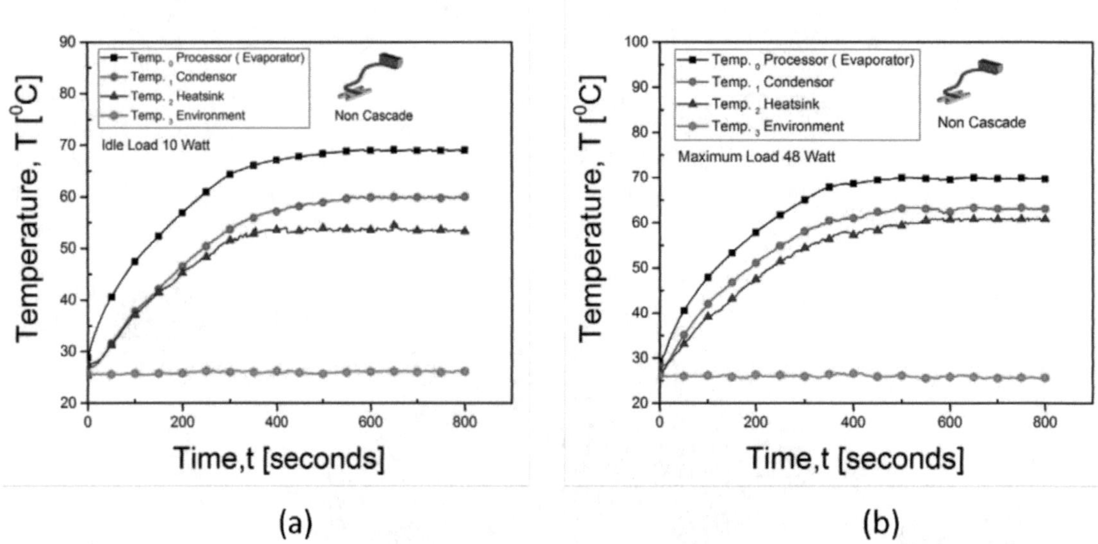

Figure 4. (a) Temperature distribution of non-cascade cooling system at the idle condition, (b) Temperature distribution of non-cascade cooling system at maximum condition

3.2 The temperature distribution of cascade straight heat pipe with a single condenser

Figure 5 (a) shows the temperature distribution of a cascade straight heat pipe with a single condenser at the 10 Watt idle conditions. From the shown image, it can be seen that the temperature rise drastically from 0 seconds to 800 seconds, after that the temperature is in the steady state, the processor temperature (Evaporator) reaches 66.36 ^{0}C and the outlet temperature at the condenser reaches 53.03 ^{0}C, and the temperature at the heat sink reaches 38.38 ^{0}C.

The temperature distribution of a cascade straight heat pipe with a single condenser at a maximum condition of 48 Watts can be seen in Figure 5. (b). The figure shows the temperature rise drastically starts from 0 seconds to 800 seconds, after that the temperature is steady. In the steady state, the processor temperature (Evaporator) reaches 68.19 ^{0}C and the outlet temperature of the condenser reaches 43.66 ^{0}C, and the temperature at the heat sink reaches 39.73 ^{0}C.

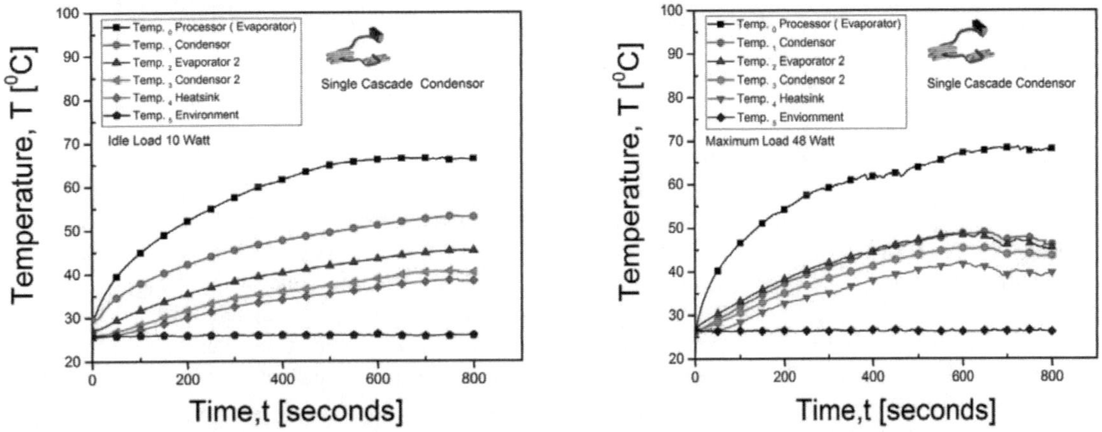

(a) (b)

Figure 5. (a) Temperature distribution of cascade straight heat pipe with a single condenser at the idle condition, (b) Temperature distribution of cascade straight heat pipe with a single condenser at maximum condition

3.3 The temperature distribution of cascade straight heat pipe with a double condenser

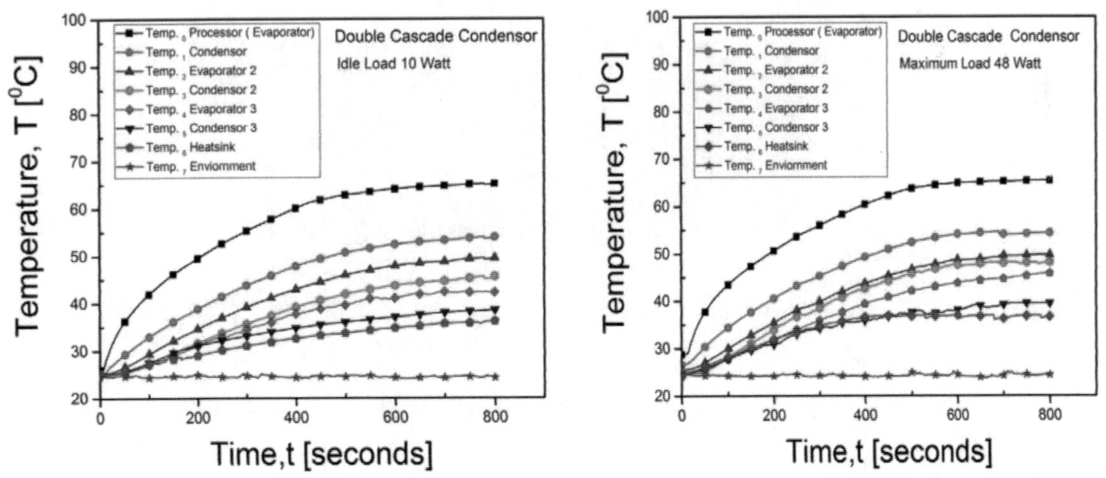

(a) (b)

Figure 6. (a) Temperature distribution of cascade straight heat pipe with a double condenser at the idle condition, (b) Temperature distribution of cascade straight heat pipe with a double condenser at maximum condition

Figure 6 (a) shows the temperature distribution of the cascade straight heat pipe with a double condenser at the idle condition of 10 Watts. The figure shows a dramatic rise in temperature from 0 seconds to 800 seconds, after that the temperature is steady. In the steady state, the processor temperature (Evaporator I) reached 65.21 ^{0}C and the outlet temperature of condenser 3 reached 38.57 ^{0}C, and the temperature of the heat sink reached 36.20 ^{0}C.

The temperature distribution of the cascade straight heat pipe with a double condenser at a maximum condition of 48 Watts is seen in Figure 6 (b). The figure shows that the temperature rise drastically starts from 0 seconds to 800 seconds, after which the temperature is steady. In the steady state, the processor temperature (Evaporator I) reaches 65.44 ^{0}C and the outlet temperature in condenser 3 reaches 39.49 ^{0}C, and the temperature at the heat sink reaches 36.69 ^{0}C.

3.4 Processor operational temperature of CPU cooling system

From the graph analysis that has been done then the calculation of the CPU cooling system design used will be done based on the operating temperature at idle and maximum loading, the condenser operating temperature is shown in the form of a diagram in Figure 7. CPU cooling system based on cascade straight heat pipe is able to reduce the operational temperature of the processor below 80 ^{0}C which is 66.36 ^{0}C for design with a single condenser and 65.21 ^{0}C for design with the double condenser. Cascade heat pipe with single condenser can reduce processor temperature to 2.74 ^{0}C at idle conditions, and 1.48 ^{0}C at maximum conditions lower than the non-cascade cooling systems. Cascade straight heat pipe with the double condenser is able to reduce processor temperature to 3.89 ^{0}C in idle conditions, and 4.24 ^{0}C at maximum conditions lower than the non-cascade cooling systems.

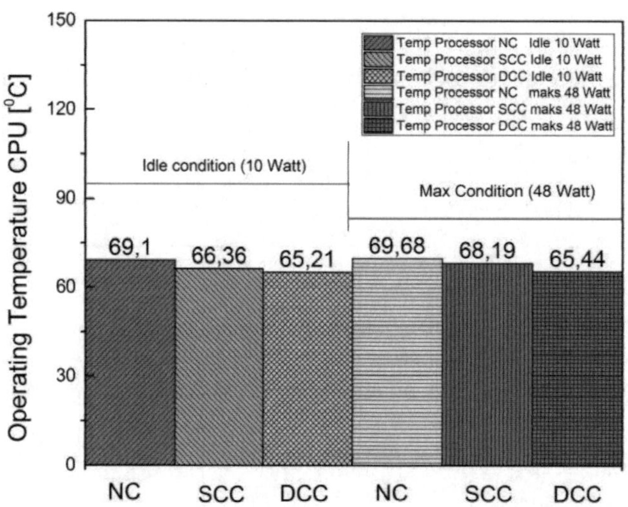

Figure 7. Comparison of non-cascade, single condenser and double condenser design processor operational temperature

The lower result of the processor temperature decrease happened because the cascade straight heat pipe with single and double condenser model absorb the processor heat twice, but the non-cascade model only absorb the heat once. According to the design of cascade straight heat pipe, the first level heat pipe functioned as the evaporator which absorbs heat from the processor, and the second level heat pipe absorbs the outlet heat from the first heat pipe and release it from its condenser, it makes the cooling system with cascade straight heat pipe is better than the non-cascade cooling system for its ability to remove more heat from the processor.

3.5 Condenser operational temperature of CPU cooling system

From the graph analysis that has been done, the calculation of the efficiency of the use of CPU cooling system design based on the operating temperature at idle and maximum loading, the condenser operating temperature is shown in the diagram shown in Figure 8. CPU cooling system based on cascade straight heat pipe is able to reduce the condenser operational temperature below 40 ^{0}C which is 38.38 ^{0}C for the design with a single condenser and 36.20 ^{0}C for design with the double condenser.

The condenser of cascade straight heat pipe with the single condenser is able to reduce the outlet temperature in the condenser section to 14.94 ^{0}C at idle conditions, and 21.12 ^{0}C at maximum conditions lower than the non-cascade cooling systems. The condenser of cascade straight heat pipe with the double condenser is able to reduce the outlet temperature in the condenser section to 17.12 ^{0}C at idle conditions, and 24.16 ^{0}C at maximum condition lower than the non-cascade cooling systems. Percentage of condenser temperature decrease at idle loading using CPU cooling system based on cascade straight heat pipe of each design to the temperature non-cascade condenser are 28% for cascade straight heat pipe with a single condenser and 32% for the double condenser. And the maximum loading of each design is 35% for design with a single condenser and 40% for the double condenser.

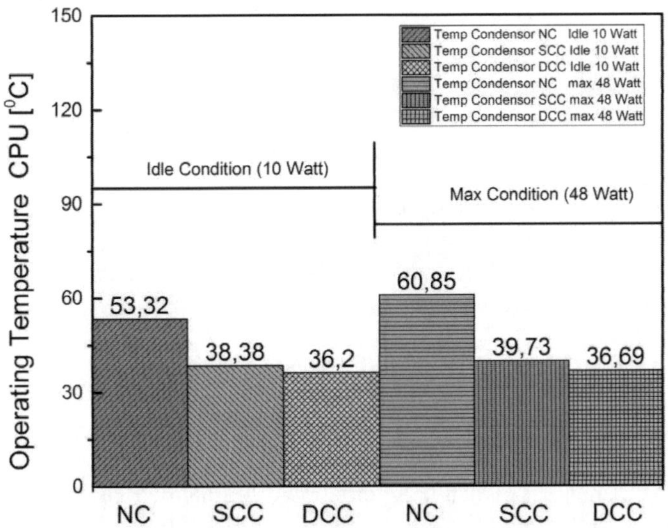

Figure 8. Comparison of non-cascade, single condenser and double condenser design condenser operational temperature

The lower result of the condenser outlet temperature happened because of the continued absorption of the cascade straight heat pipe with single and double condenser model where the second level heat pipe absorbs the heat form the first level heat pipe affects the lower temperature of the second heat pipe, while the non-cascade model which only absorbs the heat once affects the higher condenser temperature. This makes the cooling system with cascade straight heat pipe is better than the non-cascade cooling system.

4. Conclusions

CPU cooling system based on cascade straight heat pipe with single condenser can reduce the processor temperature at idle condition of 10 Watt by 2.74 ^{0}C and at a maximum loading condition of 48 Watts can reduce processor temperature by 1.48 ^{0}C, and reduce the condenser temperature at idle conditions of 10 Watts to 14.94 ^{0}C and at a maximum loading condition of 48 Watt can reduce the condenser temperature to 21.12 ^{0}C lower than the non-cascade cooling system. CPU cooling system based on cascade straight heat pipe with double condenser can reduce the processor temperature at idle condition of 10 Watt by 3.89 ^{0}C and at a maximum loading condition of 48 Watt can reduce the processor temperature by 4.24 ^{0}C, and reduce the condenser temperature at idle condition of 10 Watt to 17.12 ^{0}C and at a maximum loading condition of 48 Watt can reduce the processor temperature to 24.16 ^{0}C lower than the non-cascade cooling system. The use of cascade straight heat pipes is better than the non-cascade because it can reduce more temperature of the processor and the temperature of the heat sink. The use of the CPU cooling system on the processor using a cascade heat pipe must adjust to the size of the heat absorption area of the CPU, both in terms of processor size and CPU size. In the research action, the use of cascade heat pipes can be implemented by using a fan/cooling fan on the heat sink.

Acknowledgments

Equations Thank you to the Ministry of Technology and Higher Education and the Udayana Institute for Research and Community Service for financial support through the 2018 Higher Education Primary Research Grant (PTUPT) scheme with Contract Number 171.123/ UN14.4.A / LT/ 2018.

References

[1] Q. Chen and Y. Huang, "Scale effects on evaporative heat transfer in carbon nanotube wick in heat pipes," *Int. J. Heat Mass Transf.*, vol. 111, pp. 852–859, 2017.
[2] A. B. Solomon, K. Ramachandran, and B. C. Pillai, "Thermal performance of a heat pipe with nanoparticles coated wick," *Appl. Therm. Eng.*, vol. 36, pp. 106–112, Apr. 2012.
[3] N. Putra, R. Saleh, W. N. Septiadi, A. Okta, and Z. Hamid, "Thermal performance of biomaterial wick loop heat pipes with water-base Al2O3 nanofluids," *Int. J. Therm. Sci.*, vol. 76, pp. 128–136, 2014.
[4] N. Putra, Saputra, M. I. Bimo, R. Irwansyah, and S. N. Wayan, "Experimental study on sintered powder wick loop heat pipe," 2012, pp. 612–620.
[5] N. Putra, W. N. Septiadi, and R. Irwansyah, "Effect of Concentration and Loading Fluid of Nanofluids on the Thermal Resistance of Sintered Powder Wick Heat Pipe," vol. 651, pp. 728–735, 2013.
[6] N. K. Gupta, A. K. Tiwari, and S. K. Ghosh, "Heat transfer mechanisms in heat pipes using nanofluids - A review," *Exp. Therm. Fluid Sci.*, vol. 90, pp. 84–100, 2018.
[7] N. Putra and W. Nata Septiadi, *Teknologi Pipa Kalor Teori, Desain dan Aplikasi*. Jakarta: UI-PRESS, 2014.
[8] J. Xu, Y. Zou, M. Fan, and L. Cheng, "International Journal of Heat and Mass Transfer Effect of pore parameters on thermal conductivity of sintered LHP wicks," *Int. J. Heat Mass Transf.*, vol. 55, no. 9–10, pp. 2702–2706, 2012.
[9] N. Putra, A. Duanovsah, and K. Haliansyah, "Investigation of Cascade Loop Heat Pipes," *World Acad. Sci. Eng. Technol. Int. J. Mech. Aerospace, Ind. Mechatron. Manuf. Eng.*, vol. 9, no. 10, pp. 1868–1872, 2015.

IOP Conf. Series: Materials Science and Engineering **539** (2019) 012039 doi:10.1088/1757-899X/539/1/012039

Combustion characteristics of gas fuel from basic materials *arak bali*

I G K Sukadana[1] and I K G Wirawan[1]
[1] Mechanical Engineering Department of Udayana University, Badung-Bali

Email: igksukadana@unud.ac.id

Abstract. Analyze the characteristics of gas fuel from *arak Bali* like shape and flame speed. Testing characteristics such as methanol and ethanol content material gas from *arak Bali*, after it tested the gas fuel combustion characteristics of *arak Bali* such as the shape and speed of flame. Testing characteristics such as the content of methane and ethanol gas from *arak Bali* performed in the laboratory of Forensic, while testing the ignition characteristics of the shape and speed of fuel from evaporating *arak Bali* done using a helle-shaw cell combustion chamber model. Air mixture ratio variations with gas fuel from *arak Bali* is 24/1, 25/1, 26/1, 27/1, 28/1, 29/1, 30/1 and 31/1. The observed effect is the shape and speed of the premixed flame propagation in the helle-shaw cell combustion chamber model. The results of the study, the moisture content of the basic ingredients of gas fuel from *arak Bali* consisting of 40% methanol and 60% ethanol. To gas fuel from *arak Bali* has AFR stoichiometry on 30/1. Getting closer to the AFR stoichiometry, flame color changes from reddish color faded to red, reddish blue, blue and the last is bright blue. Maximum propagation speed of fire occurred in AFR stochiometry is 328.33 cm/sec.

1. Introduction

The Indonesia government in 2008 to make a policy on the management of national energy, about the use of ethanol, biodiesel and gasohol as an alternative energy in 2022. One source of alternative fuel in Bali is *arak Bali*. *Arak Bali* with quality above of 85% has an octane number about 108,6. *Arak Bali* are non-toxic and when used as an environmentally friendly fuel. *Arak Bali* fuel mixed with gasoline at a certain percentage can increase the octane number of gasoline. Increased octane rating will improve the quality of combustion, decreased residual gases of combustion and increased engine performance.

The use of *arak Bali* as an alternative energy, from 2006 has done research the use of *arak Bali* in liquid form as a fuel substitute for gasoline, with examining on the internal combustion engine carburetors and injection machines, with some testing variables such as rotation and compression ratio. Compared with gasoline, *arak Bali* fuel produces CO_2 emissions is a larger than gasoline, lower CO emissions, higher HC emissions and higher O_2. increased compression ratio to affects the increase of CO_2, decrease of CO and HC and the smaller increase in gas O_2, rotation engine is less steady, engine torque is decreas, engine power and engine torqy decrease at low rpm.

In this study using *arak Bali* fuel in gaseous form. To obtain the fuel gas from *arak Bali*, *arak Bali* requires heating processes. Perfect gas combustion is influenced by three things: the air fuel ratio, homogeneity of mixing and combustion temperatures. Comparison of fuel with air is less than perfect impact to perfection of combustion, which can be seen in the shape and velocity of the fire. Research using gas fuel combustion has been carried out such as: *Adi Winarta* in 2007 a study the mixture ratio of air and natural gas fuel to form and velocity of a fire, the results showed that the stoichiometry of the combustion air is in comparison with the fuel gas is 8/1 and the maximum flame propagation velocity is achieved in the mixture stoichiometry. Therefore we are interested in continuing research to determine the characteristics of the fuel gas content of *arak Bali* and its effect on combustion characteristics.

IOP Conf. Series: Materials Science and Engineering **539** (2019) 012039 doi:10.1088/1757-899X/539/1/012039

2. Literature Review

Nanda, Sukadana 2006, tested the mixture of *arak Bali* and gasoline by varying the fraction of the mixture to obtain physical properties closer on gasoline. And followed by *Artayana, Sukadana* on 2007 did test an addition of alcohol on gasoline to know the quality of the exhaust gas was tested on a motorcycle engine. This research produced, the greater percentage addition of alcohol causes the exhaust gases produced as hydrocarbon and oxygen is increased, the gasoline exhaust gas tends to be lower. The greater percentage of the engine rotation the exhaust gases produced is decreases.

Joni Artawan, Sukadana 2007, conducts research *arak Bali* as a substitute fuel motorcycle to the acceleration and fuel consumption. The results showed, at a compression ratio of 9.3/1 uses arak api can improve acceleration and can save fuel consumption, where are: in 1st gear at a speed of 0-20 km/h acceleration is 2.835 m/dt^2 with a fuel consumption is 0.091 lt/km, in 2nd gear at a speed of 20-40 km/h acceleration is 1.190 m/dt^2 with a fuel consumption is 0.102 lt/km, in 3rd gear at a speed of 40-60 km/h acceleration is 0.518 m/dt^2 with a fuel consumption is 0.117 lt/km and at the 4th gear speed of 60-70 km/h acceleration is 0.146 m/dt^2 with a fuel consumption is 0.183 lt/km.

Ervan, Sukadana 2007, doing research on the procession of *arak api* as a substitute fuel motorcycle to the content of the flue gas, the results of research by varying the concentration of ethanol as a fuel will greatly affects the content of the exhaust gases. With a higher concentration of exhaust gas produced such as carbon dioxide is increases. The greater concentration of ethanol produced CO, HC and O_2 emissions is decreases.

Sukadana, Tenaya 2008, researchs the use of *arak Bali* four stroke engine as a fuel with a compression ratio variations. The results obtained fuel produces *arak bali* to engine performance are better than gasoline, because it is able to work at a higher compression ratio so that the thermal efficiency in fuel use higher fire procession.

Sukadana, Bandem 2009 and 2010, conducting technical studies the performance of the continuous distillation and the use of wine as a fuel substitute for gasoline, by testing in conventional combustion engines in carburator, with several variable testing such as rotation variable, compression ratio variable to the engine performance such as emissions. Research results obtaineds, the higher temperature influence to the higher evaporation capacity wine products but inversely proportional to the product quality is getting lower. Compared with gasoline, fuel arak produce high CO_2 gas, low CO, high HC and high O_2. Increased compression ratio affect the increase in CO_2, reduced CO, HC emissions increase and the less gas O_2.

Sukadana, Tenaya 2011, conduct technical studies continuous distilator column-type to the capacity and quality production. The results from this study that: capacity and quality of the wine production is influenced by many number of levels distillation. The greater level number influence to the lower of distillate production capacity, the higher quality of production and production efficiency are lower.

Sukadana, Tenaya 2013, doing research "Improving the quality of *arak Bali* production as an alternative fuel with Continuous distillation method". In this research conducted a variable number of levels distillation of one level, two levels and three levels. With each variable has a setting different tempratures. Showed that the more level of distillation in used can be produced quality wine production higher but lower production capacity.

Tenaya, Sukadana 2014, conducts research to increase production capacity of alternative fuels *arak Bali* used forced fluid flow method. From research conducted showed that more level of continuous distillation influence to temperature distribution pattern, lower evaporation, lower production rate and the quality of production is increases.

Sukadana 2015 has been carrying out research application the method of forced condensation-type crossflow on alternative fuel of *arak Bali* to quality and capacity production process. The bigger cooling fluid flow rate to results the greater cooling rate to affect the rate of condensation is greater, the rate of production is increases. But the greater cooling fluid flow rate to affects the quality of production is low. *Arak Bali* for testing as a fuel in the carburator type engine gived the result is to increased concentration of *arak Bali* affects the specific fuel consumption is increasing, decreasing torque and engine power.

One of the alternative fuel are beginning to get special attention is ethanol from *arak Bali*, a clear liquid with no color and unique aroma. Because it is non-toxic, widely used as a solvent in the pharmaceutical, food and beverage industry. Ethanol mixture with water gives a sweet taste, but at high concentrations will give you a burning sensation. With the molecular formula Ethanol is C_2H_5OH, which is a group of chemical compounds containing hydroxyl compound-OH attached to a carbon.

In addition to its clean combustion, as well as easily produced or by another name is a green fuel. Ethanol is a fuel that is produced by some plants includes palm trees. The main result of a number of plant species among others is juice, the juice obtained by tapping virgin coconut tree that has not bloomed. A virgin can be tapped for 10-30 days with the results obtained from 0.5 to 1 liters of juice per day. Nira can be easy to experience because it contains wild yeast fermentation. Fermentation will end one day later after the sugars in the juice runs out are converted so that the sap will contain alcohol. The sap will be distilled to produce wine. Nira used is derived from the sap of palm trees are often called by the Balinese coconut palm wine. In the destilation process of coconut sap will do some fractional distillation process to obtain varying levels of ethanol.

Fuel gas is ideal for internal combustion engines, because the gas phase is homogeneously mixed with air. Analysis of gas fuel is based on the analysis of volume measured at standard temperature and pressure 60 ^{o}F, 30 inHg or 15, 56 ^{o}C, 1 atm. Characteristics of gas fuel include: ignition temperature is the temperature at which the fuel gas can ignite by it self. Ignition temperature can not be calculated exactly the same as specific gravity is obtained from the experimental results. Inflammability limits. The temperature of the mixture of fuel and air can ignite in air at certain mixed conditions. The mixture boundary conditions known as flammability limit both the lowest and the highest.

Good combustion is obtained the release of heat energy contained in the fuel, and suppres the amount of heat lost. Ideal Combustion is the burning of which produces only certain products. Ideal combustion is called stoichiometry, its combustion products of hydrocarbon fuels is CO_2, H_2O, and N_2. Combustion stoichiometry combustion says where all the atoms of the fuel and oxidizer react perfectly.

Combustion is classified into two: mechanics burning and natural burning. Premixed combustion is the process of burning fuel and air are mixed first mechanically and then burned. Combustion of this type can be seen in the process of burning gasoline vehical, carbide welding, burning on a rocket. In the premixed combustion wave propagation combustion occurs is called the front frame. Combustion wave propagates toward the reactants, behind a wave of arson formed combustion products.

The air fuel ratio is the ratio between the air with the fuel into the combustion chamber. The mixture contains enough air for complete combustion process called stoichiometry. The mixture contains excess fuel called rich mixture, a mixture containing less fuel called lean mixture. In this study was engineered at the determination of fire propagation speed. Flame propagation velocity is defined as the velocity of the gas unburned perpendicular to the surface as a gas combustion wave moves toward the burning fields. Variables that affect the speed of combustion are: the ratio of oxidizer and fuel mixture, the molecular structure of the fuel, Additive, pressure and temperature of the mixture, the temperature of the fire and thermal diffusivity.

IOP Conf. Series: Materials Science and Engineering **539** (2019) 012039 doi:10.1088/1757-899X/539/1/012039

3. Research Method.

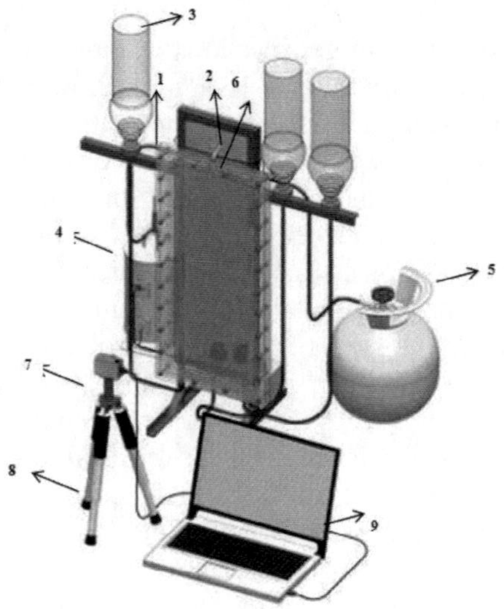

Figure 1. Research facility

Testing characteristics such as methane and ethanol content of the fuel gas from *arak Bali* conducted in Forensic laboratory. Testing characteristics of the shape and speed of ignition of the fuel gas from *arak Bali* performed by varying mixtures of air and fuel gas from *arak Bali*. Variables air fuel ratio is 24/1, 25/1, 26/1, 27/1, 28/1, 29/1, 30/1 and 31/1.

4. Results and Discussion.

In generaly, the propagation of flame shape is parabolic, from a small satellite dish near lighters steadily developed into a large satellite dish, even further away from the lighter originally parabolic horizontal and eventually broke up into several small satellite dish, as shown figure 2.

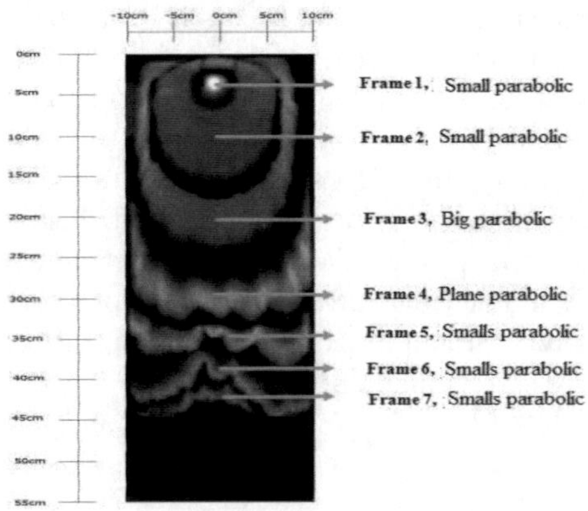

Figure 2. The shape of flame propagation

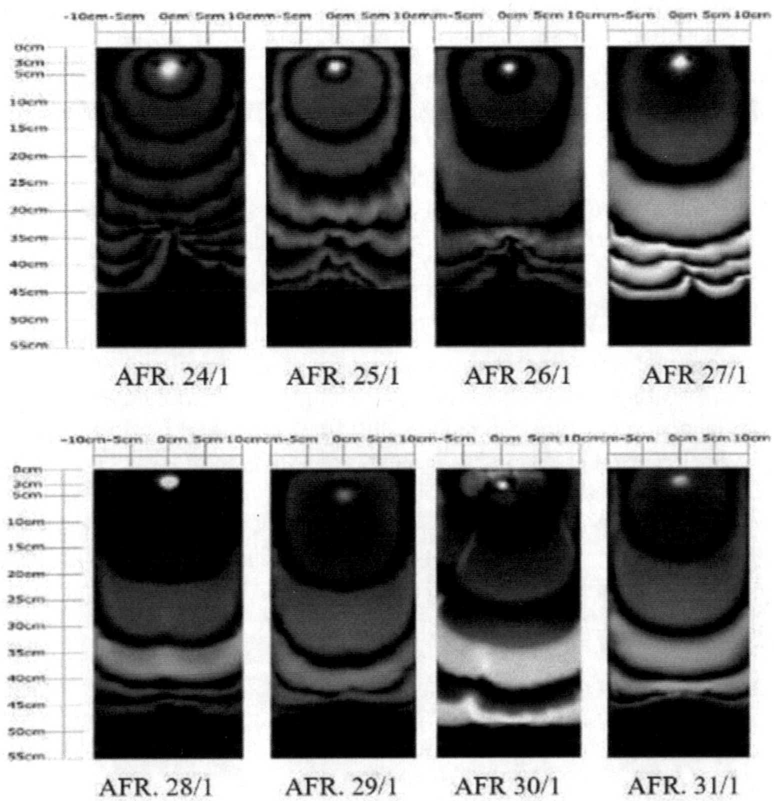

Figure 3. The flow patterns propagation of fire with the lighting from above on a variation of AFR.

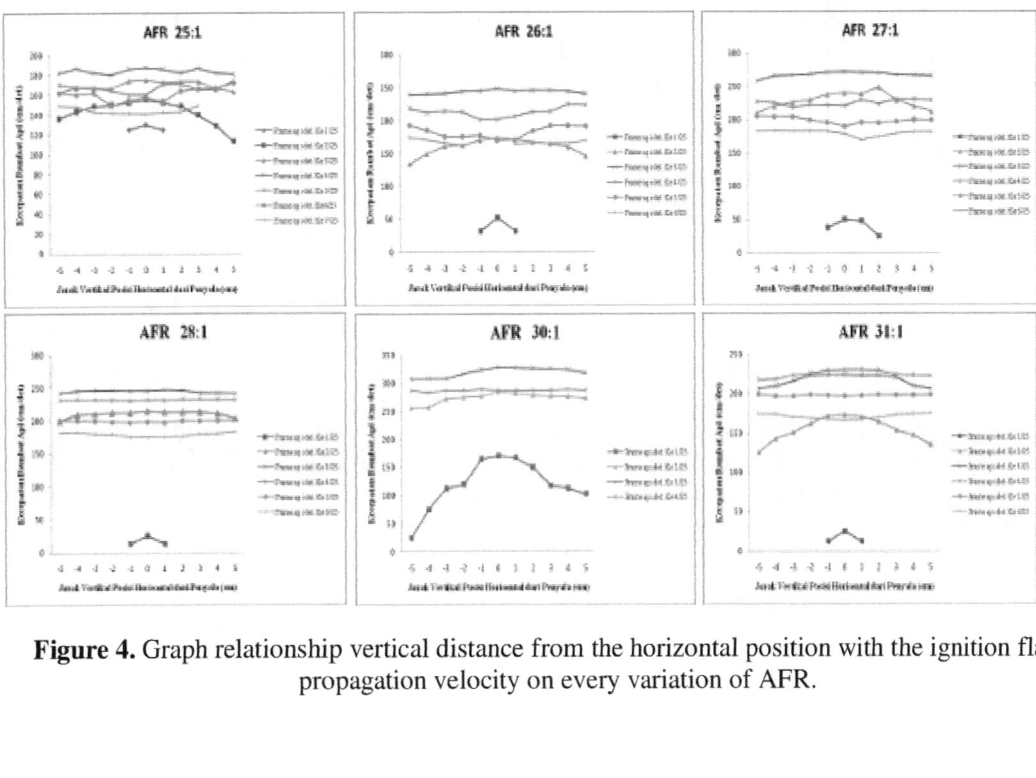

Figure 4. Graph relationship vertical distance from the horizontal position with the ignition flame propagation velocity on every variation of AFR.

From the graph in Figure 4 above shows that the AFR greatly affect the speed of propagation of fire. This phenomenon occurs because with a ratio of air and fuel gas from *arak Bali* making process more complete combustion. Comparison of air and fuel is getting closer comparison stoichiometry raise fire velocity. In AFR 24/1, 25/1, 26/1, 27/1, 28/1, 29/1, 30/1, and 31/1 maximum flame propagation velocity of each is 178.75 cm/sec, 187.50 cm/sec, 248.33 cm/sec, 271.67 cm/sec, 246.67 cm/sec, 268.33 cm/sec, 328.33 cm/sec, and 230.63 cm/sec. With the AFR stoichiometry is on 30/1 will generate the maximum flame propagation velocity, where the flame propagation velocity of 328.33 cm/sec.

5. Conclusions

Based on the results of research and discussion that has been done, it can be concluded as follows: the gas content of the basic ingredients of *arak Bali* consists of 40% methanol and 60% ethanol. For gas fuel from *arak Bali* has AFR stoichiometry occur at 30/1. Getting closer to the AFR stoichiometry the fire changed color from reddish color faded to red, reddish blue, blue, and bright blue last. Maximum speed of propagation of fire occurred in AFR stochiometry is 328.33 cm/sec.

Acknowledgements

The author thanks to the Ministry of Research Technology and Higher Education of the Republic of Indonesia and LPPM University of Udayana for supporting this research and paper through The Grants at 2016 - 2017.

Reference

[1] Sukadana 2009, “Pengaruh rasio kompresi terhadap unjuk kerja mesin empat langkah menggunakan arak bali sebagai bahan bakar”, Jurnal Ilmiah Teknik Mesin (CAKRAM), UNUD. ISSN: 1979-2468, April 2009, hal 26-32, Vol. 3 No. 1.

[2] Sukadana 2009, "Kajian Teknis Destilator Tipe Kontinu Penghasil Bahan Bakar Alternatif dari Bahan Dasar Arak Bali”, Jurnal Ilmiah Teknik Mesin (CAKRAM), UNUD. ISSN: 1979-2468, April 2009, hal 38-42, vol. 3 No. 1.

[3] Sukadana 2014, “Pengaruh Jumlah Tingkat Destilasi Kontinyu terhadap Kualitas dan Kapasitas Produksi Arak Bali sebagai Bahan Bakar Alternative”, Jurnal Energi dan Manufaktur, Th. 2014, hal. 211-216, ISSN: 2302-555, Volume 7, Nomor 2.

[4] Sukadana 2016, “Pengaruh Penggunaan Arak Bali Sebagai Bahan Bakar Pada Mesin Empat Langkah Dengan Rasio Kompresi Bervariasi”, *Flywheel* Jurnal Teknik Mesin Untirta, Volume II Nomor 1, April 2016.

[5] Sukadana 2016, “Ethanol Concentration Variations on Arak Bali influence to Torque, Power and Spesific Fuel Consumption of Engine “,The International Journal Of Engineering And Science (IJES), Volume || 5 || Issue || 8 || Pages || PP -35-39 || 2016 || ISSN (e): 2319 . 1813 ISSN (p): 2319. 1805

[6] Sukadana, 2009, 2010, "Kajian teknis destilator tipe continu penghasil bahan bakar alternatif berbahan dasar arak Bali”, Laporan Penelitian, Universitas Udayana.

[7] Sukadana, 2011, "Kajian teknis pemanfaatan arak Bali sebagai bahan bakar alternatif mesin pembakaran tipe injeksi”, Laporan Penelitian, Universitas Udayana.

[8] Sukadana, 2011, "Kajian teknis distilator kolom bertingkat tipe kontinu terhadap kapasitas dan kualitas produksi arak Bali”, Laporan Penelitian, Universitas Udayana.

[9] Sukadana, 2013, "Peningkatan kualitas produksi arak Bali sebagai bahan bakar alternative dengan metode distilasi Kontinyu bertingkat”, Laporan Penelitian, Universitas Udayana.

IOP Conf. Series: Materials Science and Engineering **539** (2019) 012040 doi:10.1088/1757-899X/539/1/012040

Solidification characteristic of cu-20%sn bronze alloys casting process by using sand and metal molds

I K G Sugita[1], I G N Priambadi[1], C I P K Kencanawati[1] and K Astawa[1]
[1] Mechanical Engineering Department Udayana University,Bali, Indonesia
Bukit Jimbaran 80 Badung-Bali Indonesia

Email: tutdegita@unud.ac.id

Abstract. Solidification is an essential part of casting process. This is a change process range from liquid phase into solid phase. Difference of solidification rate affects the needed time to change liquid phase into solid phase as well as the release of latent heat. This study aimed at reviewing solidification characteristic of Cu-20%Sn Bronze Alloys as Balinese gamelan materials during casting process by using sand and metal molds. Pure commercial copper (Cu) and white tin (Sn) were melted in crucible furnace until 1000° C. The melted metal was poured into sand and metal molds. History of solidification temperature was measured and recorded by using thermocouple type K which was directly connected to computer equipped with data acquisition system. The results of this study showed that the solidification rate of casting process in metal molds faster than sollidification rate on using sand molds. Finer dendrites with multiple branches are formed on casting on metal molds than compared that dendrite forms that produced on castings by sand molds

1. Introduction

Bronze is commonly called as tin bronze because the main alloys of this alloy are copper (Cu) and white tin (Sn). Tin bronze is widely used as engine bearing, pump blade, ring piston, bell, gear, and some engine elements. This alloy has good mechanical characteristic to hold the weight and wear, so it is always used in as-cast condition without heat treatment. If this alloy is used in as-cast condition so the properties material is merely controlled by casting process [1].

Tin bronze with composition Cu - 20% Sn is generally used as music instruments such as bell, *gamelan,* because this alloy has good mechanical characteristic, stable in room temperature, good acoustic characteristic, i.e. it can produce low damping vibration [2,3].

Casting has some strengths, i.e. it can produce complex thing, produce big and small products, produce net shape in mass number with low cost. With this strength, most of the production process is done by casting process [4].

One of important parameters in casting process is solidification process, i.e. change process from liquid phase into solid phase. Two main things happen in solidification process are nucleation and crystal growth [4]. Nucleation is started by forming solid particle in liquid form. This solid particle is formed when energy in lower particle from liquid energy. These small cores are formed independently in random spots. These cores will be the center of further crystallization process. Nucleation rate is depended on metal natural characteristic, impurity, and solidification rate.

In metal solidification process, solidification temperature gradient mostly happens in solidification area of liquid cooling. Solidification in liquid and solid area (mushy zone) releases latent heat; the temperature change gradient is relatively small. After there is no latent heat, the solid cooling starts which is followed by re-raising gradient temperature [5].

Solidification rate influences micro structure form such as grain size and dendrite arm spacing (DAS). The change of micro structure influences mechanical characteristic of the casting product. Some studies about solidification have been done previously such as solidification on alloy of binary system [6], solidification on bronze alloy Cu-8%Sn [7], fast solidification on tin solder alloy Sn-Cu [8], solidification rate on ductile cast iron [9], peritectic alloy Cu-Sn [10].

So far, a study about solidification of bronze casting with composition Cu -20%Sn on sand and metal molds has not been conducted yet. In this study, the effect of solidification rate on the formed micro structure and the mechanical characteristic as the result of casting process on both molds was reviewed.

2. Set - Up Experimental

Bronze alloy used in this study was alloy of copper (Cu) and white tin (Sn) with composition 80%Cu - 20%Sn. Commercial pure copper 99.99 wt.% and commercial pure tin 99.99 wt% were melted in crucible furnace until liquid temperature of 1000°C. This liquid metal was poured into sand and metal molds (permanent molding) with molding temperature of 400°C. the shape of the mold is cylinder with diameter 30 mm and height 80 mm. Measurement of solidification temperature used thermocouple type K which was directly connected to data acquisition system (National Instrument thermocouple system). Set-up of the solidification temperature measurement is showed in Figure 1.

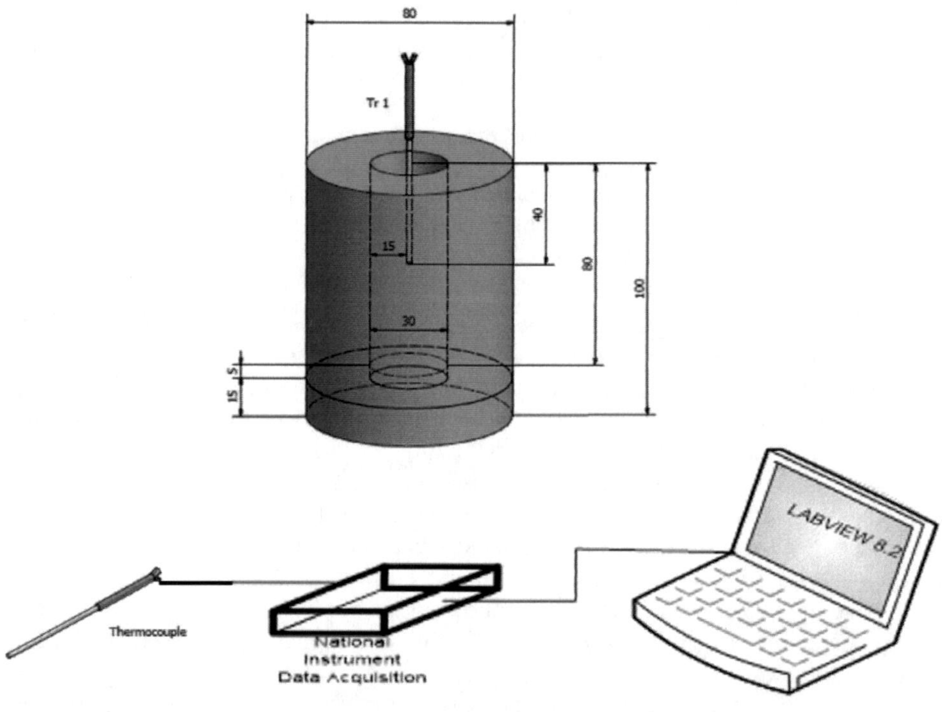

Figure 1. Set-up of the solidification temperature measurement

Characteristic of micro structure from the casting product was observed by using optical microscope. Specimen with diameter 30 mm and height 15 mm was prepared for micro structure investigation. The cut specimen was smoothed by using sandpaper respectively sized 240, 400, 600, 800, 1000, 2000, polished with diamond pasta, and finally processed with solution of 10% HNO3 + 90 % alcohol.
Tensile testing was done by using universal testing machine digital with capacity of 20 KN. The size of specimen followed the standard of JIS Z2201. No.7. Vickers hardness testing was used to test specimen hardness.

3. Results and Discussion

Figure 2 shows phase diagram and curve of solidification rate on casting process. From the solidification rate curve, it can be shown that liquid cooling happened along the A-B curve. In this area, the cooling rate is high, which is caused by high temperature difference between pouring temperature and molding temperature.

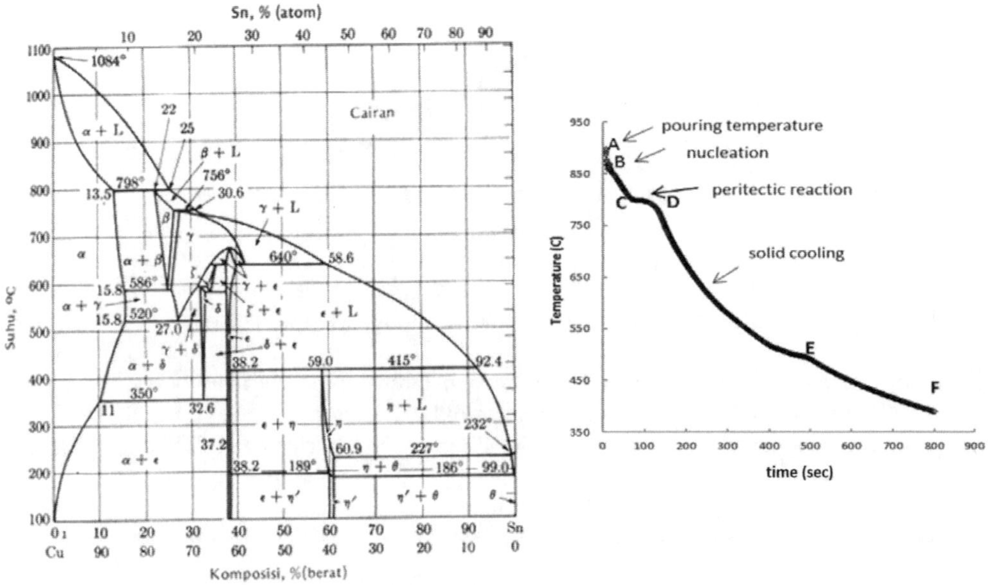

Figure 2. Phase diagram and Curve of solidification tin bronze alloy 80-20%Sn

Liquid metal which directly touched with the mold had earlier cooling, because the heat was absorbed by the mold which had lower temperature. Dendrite nucleation was started in 878°C (spot B) until temperature of 796.3° C in spot D. in this temperature range, there was alloy cooling (mushy zone), between liquid and solid phase α. Cooling process in this area was ended with peristatic reaction, i.e. reaction of phase change L + α → β, happened along C-D line. In the cooling process of mushy zone, there was release of latent heat of the alloy so the freezing was complete in spot D. After there was no latent heat, solid cooling was started accompanied by the decrease of high temperature gradient (D-F line), cooling rate in area D-F was lower than liquid area because in this condition the mold had started to heat as consequence of the heat transfer of liquid metal into the mold so it produced low temperature gradient between wall temperature of the mold and casting metal. Figure 3 shows cooling curve between sand and metal molds. Both graphs show relation of temperature with cooling time in the cooling process.

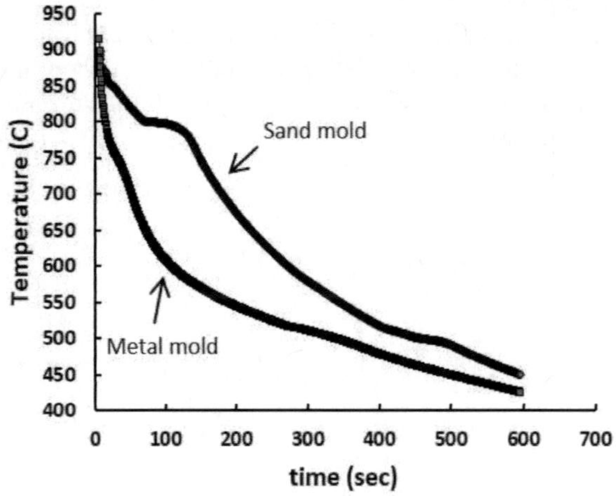

Figure 3. Curve of cooling rate on sand and metal molds casting process.

Characteristic of cooling rate of both graphs is highly different. Characteristic of cooling rate on sand mold was slower than cooling rate by using metal mold. Cooling rate on sand mold was 0.442^{o} C/s and cooling rate on metal mold was 7.725^{o} C/s. The difference of cooling rate influenced the total time needed in cooling process, which was measured since pouring temperature until complete cooling [11]. Total time of solidification in sand mold casting was higher than using metal mold. When solidification process happened, both molds were heating, cooling gradient (cooling solid) of both molds was almost the same.

Cooling rate was influenced by the amount of heat transfer which happened between liquid metal into the mold. The occurred heat transfer was depended on heat conductivity (k), density (ρ) and specific heat (C_p). Both molds in this casting process had different heat transfer characteristic so it produced different cooling rate curve. The ability of heat transfer on metal mold was higher than the ability of heat transfer on sand mold.

3.1. Microstructure Observation

Figure 4 and 5 show the micro structure of castings materials on sand and metal molds casting process. The form of DAS (dendrite arm spacing) was smooth with many branches on metal mold casting. Bigger dendrite structure with few branch was found in sand mold. These differences were formed by cooling rate on the casting process.

In the gradient of molding and high liquid temperature, heat transfer is faster and α dendrite primer became smoother and columnar with few branches (arms). Cooling time on sand mold is longer than on metal mold. Longer cooling time gave longer opportunity to diffusion process and crystal growth become bigger. Along the solidification process, the mold temperature raised so heat gradient decreased, solidification rate declined, primary dendrite grew, and DAS size raised up.

Variation of dendrite structure form was influenced by variation of solidification rate from the wall into the middle of the mold [12]. Metal solidification occurred on surface directly contacted with the mold and spread to the center of specimen diameter. Inner side of liquid metal had longer time for solidification compared to the liquid metal which directly contacted with the mold.

From Figure 3, it can be found that solidification rate on the metal mold casting is higher than on the sand mold. The effect of solidification rate is important to be found to design smoothing process from metal structure. It significantly influenced the mechanical characteristic.

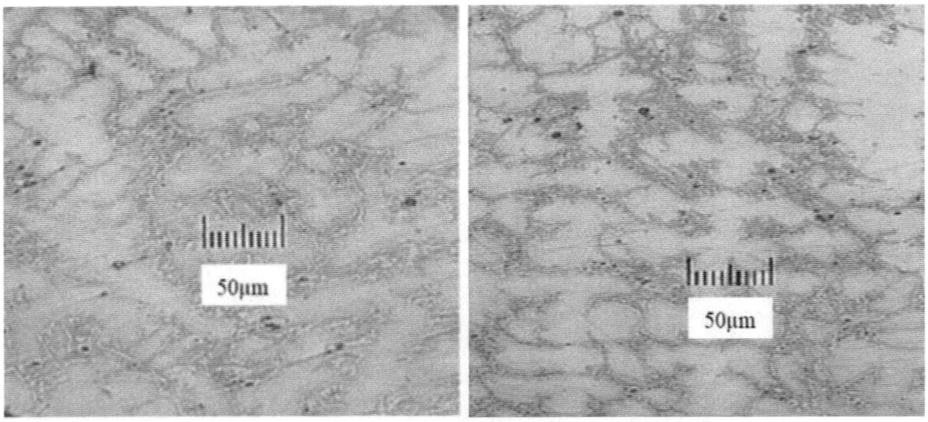

Figure 4. Micro structure of tin bronze alloy Cu-20%Sn
(Sand mold casting)

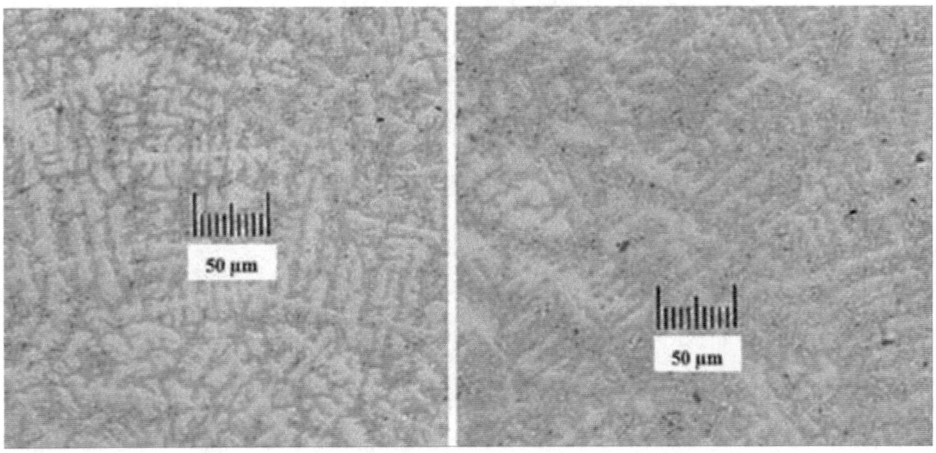

Figure 5. Micro structure of tin bronze alloy Cu-20%Sn
(Permanent mold casting)

In high solidification rate, the needed total time is short, results smoother structure form and augments solute solubility [11]. The higher cooling rate, the micro structure is getting smoother and dendrite tip radius will be getting pointed [13].

3.2. Mechanical Properties

The difference of solidification rate resulted by the different use of mold influenced the mechanical properties. The mechanical properties resulted by casting of sand and metal molds is shown by Figure 6. It shows that mechanical properties, i.e. tensile strength, hardness and impact toughness from the metal mold was higher than using sand mold. The increase of solidification rate improved mechanical properties of the casting product.

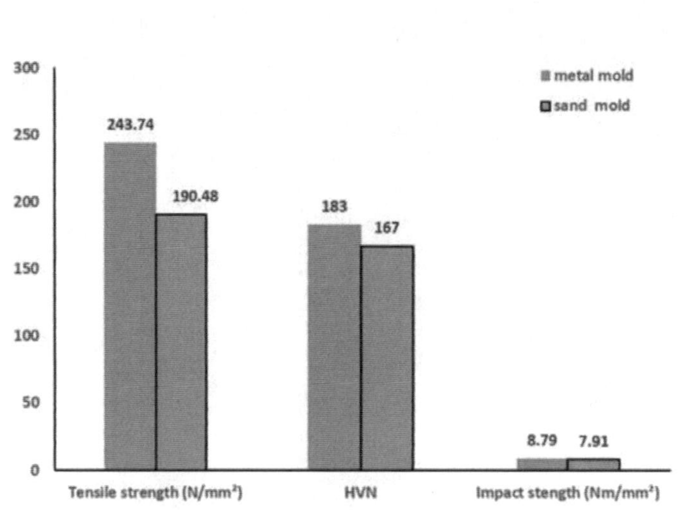

Figure 6. Mechanical properties of tin bronze alloy Cu-20%Sn

High solidification rate results better mechanical properties compared to low solidification rate. Smoother structure was formed in high solidification rate. These fine grains form many grain boundaries which will be able to prevent dislocation movement so it will improve material strength [11].

4. Conclusion

From the finding and data analysis of the casting using different molds, it shows that solidification rate influences micro structure and mechanical properties, i.e. tensile strength, hardness, and impact strength. Micro structure of the casting using metal mold is smother than using sand mold. Higher solidification rate results higher mechanical properties compared to lower solidification rate.

References

[1] Schmidt R.F., Schmidt D.G., 1993, Selection and Application of Copper Alloy Castings, ASM Handbook, Metals Handbook, Vol. 2, American Society of Metals, Cleveland, OH, p. 346.

[2] Lisovskii,V.A, Lisovskaya.O.B, Kochetkova,L.P, Favstov, Yu.K, 2007, Sparingly Alloyed Bell Bronze with Elevated Parameters of Mechanical Properties, Journal Metal Science an Heat treatment,vol 49, pp 232-235

[3] Hosford F. William, 2005, Mechanical Behaviour of Materials, Cambridge University Press

[4] Stefanescu Doru M. 2002, Science and Engineering of Casting Solidification, Kluwer Academic/Plenum Publisher, New York, Boston, Dondrecht, London, Moscow

[5] Slodicˇka Maria´n, Schepper Hennie De, 2005, Determination of the Heat-Transfer Coefficient during Solidification of Alloys, Comput. Methods Appl. Mech. Engrg. 194 pp 491-498

[6] Lee, S.L. and Tzong, R., Y., 1995, Latent Heat Method for Solidification Process of a Binary Alloy System, Journal of Heat and Mass Transfer, Vol. 38, No. 7, pp. 1237-1247.

[7] Martorano, M.,A. and Capocchi, J., D.,T., 2000, Heat Transfer Coefficient at the Metal-Mould Interface in the Unidirectional Solidification of Cu-8%Sn Alloys, Journal of Heat and Mass Transfer vol. 43 pp.2541-2552.

[8] Shen Jun, Liu Yong Chang, and Gao Houxiu, 2006, Rapid Directional Solidification in Sn-Cu Lead-Free solder Journal of University of Science and Technology Beijing Volume Z3

[9] Hemanth, J., 2000, Effect of Cooling Rate on Dendrite Arm Spacing (DAS), Eutectic Cell Count (ECC) and Ultimate Tensile Strength (UTS) of Austempered Chilled Ductile iron, Materials and Design 21 pp 1-8

[10] Kohler, F., Germond L, Wagniere J-D., Rappaz M., 2008, Peritectic Solidification of Cu–Sn Alloys: Microstructural Competition at Low Speed, Acta Materialia 57 pp 56–68.
[11] Askeland, D. R., 1984, The Science and Engineering of Materials, University of Misouri-Rolla, California, USA
[12] Kaplan, M. and Yieldiz, A.K., 2003, The Effects of Production Methods on Microstructures and Mechanical Properties of an Aluminum Bronze, Materials Letters, pp.4402-4411.
[13] Stefanescu Doru M. and R. Ruxanda, 2004, Fundamentals of Solidification, Metallography and Microstructures, Vol 9,ASM Handbook, p. 71–92

.

IOP Conf. Series: Materials Science and Engineering **539** (2019) 012041 doi:10.1088/1757-899X/539/1/012041

Experimental investigation on the effect of resonator shapes on the temperature characteristic of thermoacoustic cooling device

D Agustina[1], Sabri[2], and S Purnama[2]
[1] Department of Mechanical Engineering, University of Riau, Pekanbaru, Indonesia
[2] Department of Mechanical Engineering, Syiah Kuala University, Banda Aceh, Indonesia

Email: dinniagfa@lecturer.unri.ac.id

Abstract. Thermoacoustic refrigerator has emerged as one of new alternative ways of cooling. The use of this technology which converts sound energy to heat energy has not been popular despite its environmentally friendly and simple construction. A laboratory scale thermoacoustic cooling device has been constructed to conduct experiments on three different shapes of resonators that results in temperature difference occurring at its stack ends. Geometry of the device parts is one of the parameter that holds importance to its characteristic. The device applied a different shape of resonator for each experiment and record temperature data at stacks ends. The experiments were conducted with variations in resonator which are square tube, cylinder tube and tapered tube. The material of resonator is acrylic and its inner volume remains unchanged.

1. Introduction

The need for reliable and sustainable refrigeration technology is ongoing demand of today's world where one of daily living necessities is non-stop availability of cooling energy. The efforts to maintain supply of this energy conventionally relies on fossil fuel power which generate electricity. Even though this large need for refrigeration and freezing technology on different temperature levels has been served by vapour compression cycles and absorption cycles, thermoacoustic refrigeration has a potential as alternative cooling provider [1]. The use of massive conventional refrigeration technology development has a negative impact on the environment due to heavy air pollution that causes undesired impacts to the environment which is known as global warming. Refrigerant such as chlorofluorocarbons gas (CFCs) or hydrofluoric carbons (HFC)s are dangerous for the earth, because it can damage the ozone layer that results in the increase of temperature on earth.

Due to its tropical climate, and uncertain climate change, the existence of cooling systems is needed in Indonesia. Conventional cooling systems generally use Freon and CFC or HFC refrigerants as their working medium. The use of such substances has the potential to cause long-term hazards such as global warming. Therefore, an environmentally friendly non-CFC / HFC alternative cooling system such as thermo-acoustic cooling system is required. Specifically, thermoacoustic are sound waves that can cause temperature differences at the two ends of the stack because of the air through small channels in varying shapes, materials, stack designs and resonators [2].

Thermoacoustic devices have attracted the attention and interest of researchers in recent years because they are environmentally friendly, durable, and relatively have simple parts that can be installed and operated at a relatively low cost. On the other hand, thermo-acoustic refrigerators use a medium of working air or noble gases as an alternative to conventional cooling systems so as to be harmless to the environment [3]. In addition to its many advantages, the thermoacoustic refrigerator has its disadvantages, that it is generally very low in efficiency.

The effects of stack parameters and frequency variation on the thermal performance of thermoacoustic refrigerator by using acrylic resonator tubes of 4.0 cm and 5.3 cm diameter with stack thickness variation of 0.15mm, 0.5 cm and 1 mm has been investigated [4]. Based on some research that has been done by previous researchers, a continuous development is required by varying the geometry of the resonator tube and the stack form an important first step for the performance of thermo-acoustic cooling devices. The main problem in this research is how the effect of resonance frequency changes on the geometry variation of resonator tube, and the stack shape on thermo-acoustic performance.

Thermoacoustic is the event of heat transfer by sound waves or vice versa [5]. Thermoacoustic itself is a combination of two fields of science, which are acoustics and thermodynamics. Simple construction that typically produces thermoacoustic effects is commonly referred to as thermoacoustic devices. Figure 1 shows some of the major parts of thermoacoustic device as shown in Fig. 1.

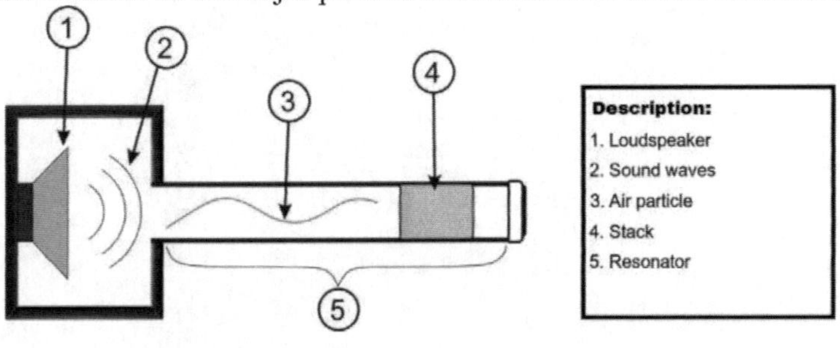

Figure 1. Main components of Thermoacoustic device

When working on the design of a thermoacoustic device, there are several things that must be considered, such as the dimensions of the resonator and of the stack. The frequency of input sound must be determined in order to be able to match the resonance frequency of a resonator tube.

While in designing the stack, the main limitation that is the width of the stack channel must be several times greater than the depth of the medium gas thermal penetration where the value of four in the ruler is the optimum distance value [6]. The thermal penetration depth (δ_K) is the distance of heat diffusion through the working fluid or gas in this case being at a time interval t = 1 / πf, calculated by the equation:

$$\delta_K = \sqrt{\frac{K}{\pi \rho f c_p}}$$

The basic concept of a thermoacoustic system utilizes sound waves generated by loudspeakers moving through a resonator tube that contains stack which act like a compressor in a vapor compression refrigeration. Air as a working fluid enters small canals located on the stack and compresses so that its temperature rises and releases heat to the material stack. The fluid further expands and decreases the temperature resulting in heat transfer resulting in temperature differences at the two ends of the stack. The transfer of heat from the air to the stack and vice versa is the result of acoustic waves that resonate in a tube that acts as an external work [7].

2. Research Methodology

The research was conducted at Thermal Engineering Laboratory of Faculty of Engineering, Syiah Kuala University and Thesis, Vibration and Acoustics Applied (PUSTVAT) University of Syiah Kuala Laboratory.

2.1. Material and Experimental set up

The material used for the resonator is acrylic, while the material for the stack is mica film. In this study first of all the calculated resonance frequency to be used in the research. To achieve this resonance frequency required a sound level meter measuring instrument placed at the end resonator. Sound waves generated from the function generator go through the amplifier and oscilloscope as a wave reader, then from the amplifier to the loudspeaker. From this loudspeaker is the release of sound in the form of a frequency that has been set on the function generator to the level of the sound level meter and then from the sound level meter. Frequency used ranging from 40 Hz to 200 Hz with filter 1/3 octave. Figure 4 shows the resonance frequency data capture. From the test results obtained resonance frequency value of resonance-based resonator made of cylinder tube is 140 Hz, and resonator of tapered tube have resonance frequency of 160 Hz, as well as square tube resonator form resonance of 180 Hz. The shape and length of the resonator affect the resonance frequency, although the volume of the resonator shape is unchanged.

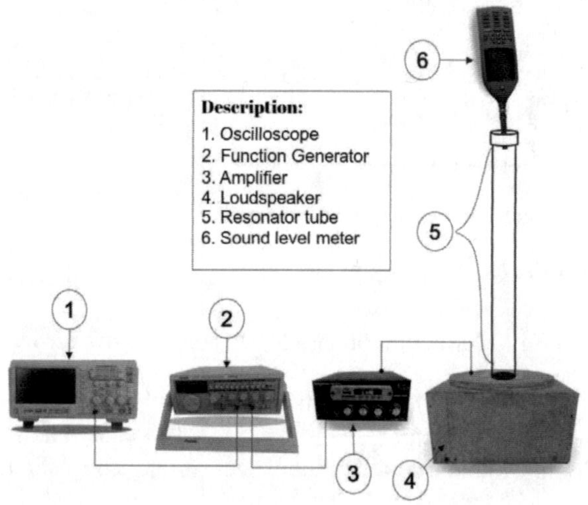

Figure 2. Experimental set up of the thermoacoustic device

Figure 2 shows the experimental setup of the thermoacoustic device. The stack was placed at 6 cm from the closed end. The frequency and wave amplitudes were measured by a condenser microphone which was located at pressure anti node of the resonator. The microphone was connected to an oscilloscope. The oscilloscope confirmed the form and frequency of the wave input. Type K thermocouples were inserted at the ends of the stack and the temperatures were recorded every 30 second with Phywe™ measurement system.

3. Results and Discussion

Figure 3 shows the temperature data at the ends of stacks inside the square tube resonator, named Th and Tc. Th is the temperatures at hot side of the stack while Tc is the temperatures at cold side of the stack. After 30 seconds, the temperature of T_h increased as much of 1,4 ^{0}C while T_c decreased by 0,4 ^{0}C. T_h experienced an increase in temperature up to 32.2°C at 540 seconds and a decrease of 26.1°C at 630 seconds. Above 630 seconds there is no increase and decrease in temperature. The ambient temperature was 29°C

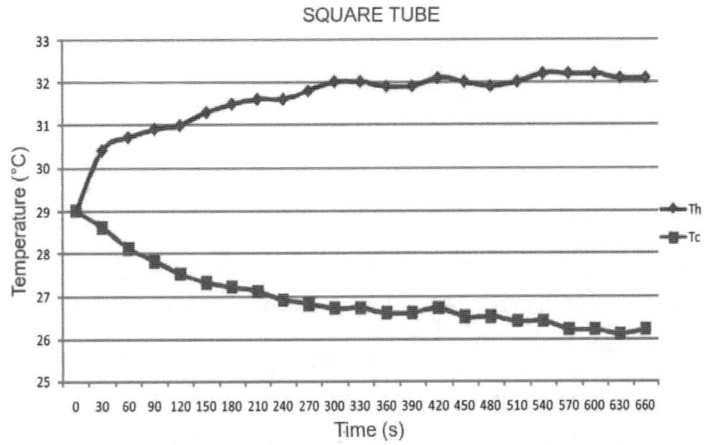

Figure 3. Graph of temperature of square tube resonator

Figure 4 shows the temperature data at the ends of stacks inside the tapered tube resonator. After 30 seconds, the temperature of T_h increased as much as 1,5 ^{0}C while T_c decreased by 0,8 ^{0}C. $T_h$experienced an increase in temperature up to 31.8°C at 540 seconds and a decrease of 25°C at 630 seconds. Above 630 seconds there is no increase and decrease in temperature. The ambient temperature was 29°C

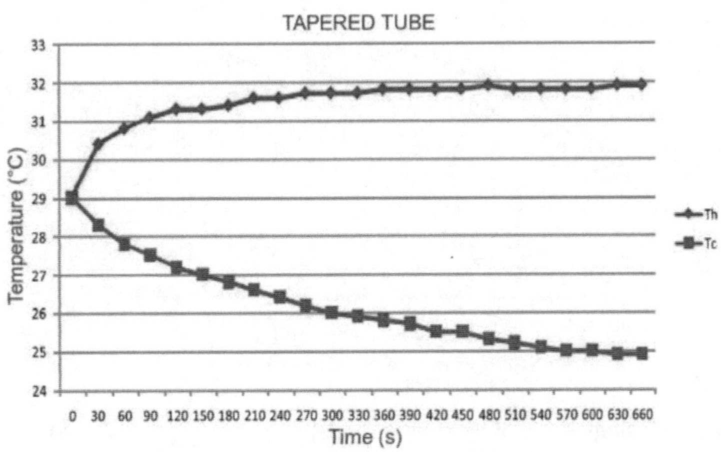

Figure 4. Graph of temperature of tapered tube resonator

Figure 5 shows the temperature data at the ends of stacks inside the cylinder tube resonator. After 30 seconds, the temperature of T_h increased as much as 1,5 ^{0}C while T_c decreased by 0,8 ^{0}C. $T_h$experienced an increase in temperature up to 32.2°C at 540 seconds and a decrease to 24.8°C at 630 seconds. Above 630 seconds there is no increase and decrease in temperature. The ambient temperature was 29°C

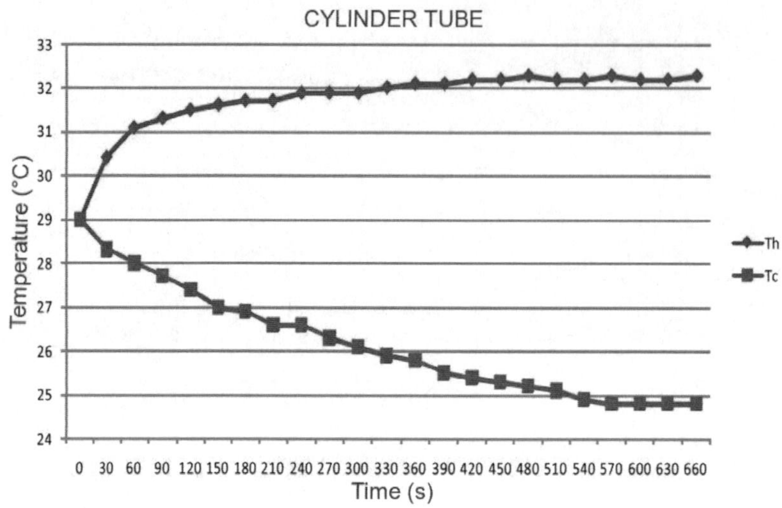

Figure 5. Graph of temperature of cylinder tube resonator

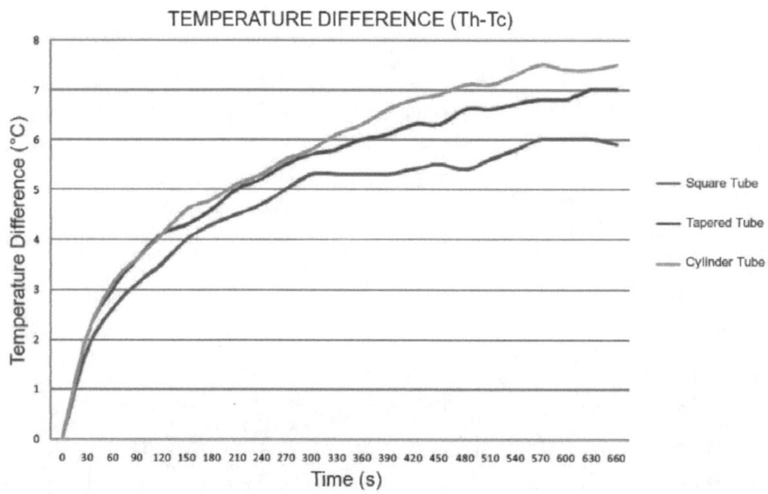

Figure 6. Comparison of Temperature difference , ΔT, for all resonators

Figure 6 is calculations of temperature difference that occurred at both ends of the stack, for three variations of resonator shape. It shows the comparison of ΔT for each resonator. It indicates that the difference in temperature at the ends of the stack of the cylinder tube resonator is larger than other resonator shapes.

4. Conclusion

Experiments have been conducted to observe the temperature difference that occurred at the ends of the stack of a thermoacoustic cooling device. Each of three different shapes of acrylic resonator which have the same volume is tested at constant ambient temperature and same experimental set up. The input resonance frequency for each of resonator is different because the resonance frequency is the

IOP Conf. Series: Materials Science and Engineering **539** (2019) 012041 doi:10.1088/1757-899X/539/1/012041

function of resonator's length and shape. The results showed that the cylindrical resonator has the largest temperature difference of 7.5°C, followed by truncated conical resonator with the temperature difference of 7°C. The cubical resonator produced the smallest temperature difference of 6.1°C. However, it is arguable that the resonator shape alone will have significant impact on temperature difference at stack ends of thermoacoustic cooling device.

References

[1]. Zink,F, Vipperman,J.S, Schaefer, L.A., 2009,. *Environmental motivation to switch to thermoacoustic refrigeration,*. Applied Thermal Engineering 30 (2010), 119-126.
[2]. Hartulistiyoso, E and Yulianto,M., 2013. *Potential Use of Bamboo as Resonator Tube of Thermoacoustis Refrigeration.* Jurnal Keteknikan Pertanian, 27. 143-150
[3]. Tijani, M.E.H., 2001. *Loudspeaker-driven thermo-acoustic refrigeration*. Tesis Ph.D. Technische Universiteit Eindhoven, Belanda.
[4]. Putra, N. and Agustina D, 2013. *Influence of stack plate thickness and voltage input on the performance of loudspeaker-driven thermoacoustic refrigerator.* Journal of Physics: Conference Series, Vol. 423
[5]. Swift, G. W. 1988. *Thermoacoustics Engines*. J. Accoust.soc.Am.84, 1145-1180.
[6]. Wetzel, Martin, Herman. C, 1997. *Design Optimization of Thermoacoustics Refrigeration.* Vol 20 (1997) 3 – 21.
[7]. Kinsler, L. E., Frey, A. R. Coppens, A.B. dan Sanders, J.V., 1999. *Fundamentals of Acoustics.* Wiley, New York. Edisi ke-4. Bab 8.

IOP Conf. Series: Materials Science and Engineering **539** (2019) 012042 doi:10.1088/1757-899X/539/1/012042

Optimization on strength spot welding joint trough finite element modelling indentation approach

I N Budiarsa[1], I M Astika[1], I N G Antara[1] and I M G Karohika[1]
Mechanical Engineering. Dept, Faculty of Engineering, Udayana University. Kampus Bukit Jimbaran Bali 80361. Bali

Email: nyoman.budiarsa@unud.ac.id

Abstract. Failure behaviour of spot welding are important in order to design the weld strength on dissimilar spot welded joint as a significant step to reduce vehicle weight in the development of urban transportation systems. Environmental friendly. Including in this case to take positive measures to reduce CO_2 emissions and reduce fuel consumption. The material failure behaviour, properties parameter of plastics materials and fracture of materials can be easily determined when the specimen material standards (i.e parent metal) but for the results of spot welding joint testing standards are not applicable to characterize the HAZ and spot welding nugget because it is constrained by a complex structure and a small sample size. One potential approach to be developed in the behaviour of material failure, parameter plastic properties of material is to combine the numerical-experimental method. Where ph-curve indentation experimental results combined with numerical approach through Finite Element (FE) modelling. Approach was established to be used for reverse (inverse) testing of the nature of constitutive material (i.e the yield stress (σy), the coefficient of strain hardening (n)) for nugget, HAZ and the parent material (base). In this approach the Load data of material deformation used as input data for finite element model (FE) which simulates the geometry and boundary conditions of the experiment. Finite element model (FE) Model-based approaches are developed, validated then used to predict the effect of nugget size and thickness of the sheet metal on the strength of the weld joint points with different materials (dissimilar material).

1. Introduction

One effective way is the vehicle weight reduction with dissimilar metal welding solutions. The application of dissimilar metal welding is not only to reduce the weight of the car body also improves protection against corrosion and improve crashworthiness [1], [2] Until now many dissimilar weld joint realized by mechanical assembly [3]. dissimilar welding joint is still intensively researched and developed, as interaction between the phenomenon of electrical, thermal, metallurgical and mechanical [4]. Most of the research done on the effect of the thickness and length of the welding point to the emphasis on physical and mechanical properties with the variation of the welding time. [5]. others to the problems nugget diameter size [6]. One area of research that is still active is the prediction of the dimensions and strength of the weld joint to simulate the welding process with finite element modeling [4], [7]. Another active area of research is the study of the microstructure [4]. Earlier work showed hardness value may be associated with a representative stress (σr) according to a representative strain (εr) which is the average value of the plastic strain generated in the indentation process [8], [9], [10] The concept was developed by modeling finite element (FE) has been successfully used in analyzing the sharp indenter (vickers) where a representative strain and stress are well defined using fixed indenter angle. In this case, the resulting relationship between the material parameters and P-h curve eventually be used to estimate the hardness of the material parameters. Simulations with validated models that have been done on various material properties (σy: 100-900 MPa, and n: 0.0-0.3) the relationship between the yield stress (σy), strain hardening coefficient (n), HV and HRB have been known. This will lead to the prediction results with input hardness value, then the properties of plastics (σy and n) on a complex material such as welds, can be predicted more accurately

2. Materials and Experimental

Two specimens with different materials and thicknesses (Stainless steel and mild steel) used in this study. As stated in Tabel.1.

Tabel 1. Chemical composition of material spot welding test

Concentration	C	Cr	Ni	Mn	Si	P	S
Stainless Steel G304	<0.08%	17.5-20%	8-11%	<2%	<1%	<0.045%	<0.03%
Mild steel	0.14%	0.01%	0.01%	0.32%	0.03%	0.2%	0.05%

Stainless steel used is stainless steel grade 304 with 25 mm width, 0.8 mm thickness. Other specimens are Mild steel with a width of 25mm and a thickness of 1.44 mm. The Welding on different materials combination was prepared and tested, for hardness testing samples of solid steel rod with a diameter elliptical Φ5 mm and a length of 90 mm and has a seat on the edge. Vickers hardness test is done by using Duramin-1 testing machine Struers Vickers hardness using direct loading method with a variety of loads of up 19.61 N to 490.3mN. Indenter has the right pyramidal shape with a square base and an angle of 136o to face the opposite side [11].

Welded joint on combinations of dissimilar material sheet metals are prepared and tested. Tensile shear tests performed using the Lloyd LR 30K Universal material testing machine to test tensile shear and compression. Machines with a maximum capacity of 30 kN with readings accurate to 0.5 % of the force. This machine is connected to a microcomputer which allows the graphics output of the test results and test data can be obtained and stored. Tensile shear test specimens with an initial load of about 50 N, clamped with two gaskets to avoid bending during testing. Tensile test performed on loading rate of 5 mm/min based on standard ASTM E 8-04.

3. Material Models

Based on Hooke's law and Von Mises criteria, then the true strain (ε) true stress (σ) is generally expressed as.

$$.\varepsilon = \begin{cases} \sigma/E & \text{for } \sigma \leq \sigma_y \\ \sigma_y/E\left(\frac{\sigma}{\sigma_y}\right)^{1/n} & \text{for } \sigma > \sigma_y \end{cases} \tag{1}$$

Where E is the Young's modulus, n is the strain hardening coefficient. In the region of the plastic, the strain could be described as the yield strain (σy) and true plastic strain. As for knowing the behavior of the failure and fracture properties of spot welded joint used Gurson Model approach. Gurson model is widely used in ductile fracture mechanics, where the fracture of materials is considered as a result of the growth of emptiness (void) in the volume of material. Gurson model most commonly used in the analysis of the failure behavior as described in the Gurson Tvergaard Needleman (GTN) model[12] assumed properties of plastics are produced from a resilient porous material, in which the yield surface is a function of the void (void) as follows

$$\Phi = \frac{3S_{jk}S_{jk}}{2\sigma_{ys}^2} + 2\,q_1\,\mathrm{f}\cosh\left(\frac{3\,q_2\,\sigma_m}{2\,\sigma_{ys}}\right) - (1 + q_3\mathrm{f}^2) = 0 \tag{2}$$

Where σy is the yield stress of the material, σm is the mean stress, f is the volume fraction of voids. f=0 means that the material is fully dense and conditions Gurson yield reduces von Mises; f = 1 means that the material is full void and has no stresses. S_{jk} is deviator components (j,k=1,2,3), is defined as $S_{jk}=\sigma_{jk}-\sigma_m\,\delta_{jk}$ and δ_{jk} is Kronecher delta $\delta_{jk} = 1$ if j=k and $\delta_{jk} = 0$ if j≠k. Some parameter values q_1,

q_2, q_3 are material constants, the values of the most encountered in most of the literature is as. $q_1 = 1.5$, $q_2 = 1$, $q_3 = 2.25$. In the power law of Elasto-plastic if representative strain (ε_r) is a plastic strain point of reference, the stress Representative strains (ε_r), represents the mean plastic strain defined by Tabor (1948). The techniques for estimating a point on the curve true stress-true strain is developed, in this case representative stress (σ_r) can be expressed as:

$$\sigma_r = \sigma_y \left(1 + \frac{E}{\sigma_y} \varepsilon_r\right)^n \quad (3)$$

In indenting occurred on Elasto plastic generally follows the Power law where the load P can be written as: [13]

$$P = P(h, E, v, Ei, vi, \sigma y, n) \quad (4)$$

Where E=Young's modulus indenter, and v= Poisson ratio. By combining the effects of the elasticity of an elastic and elasto indenter solid plastic can be written:

$$P = P(h, E^*, \sigma y, n) \quad (5)$$

Where E*

$$E^* = \left[\frac{1-v^2}{E} + \frac{1-{v_i}^2}{E_i}\right]^{-1} \quad (6)$$

In this research material systems under investigation is steel, so that the value of E remains at 200 GPa value E*= (~187 GPa with E indenter = 1220 GPa and E steel = 200 GPa) to avoid uncertainty in the value of E* from different sources. By applying the theorem in dimensional analysis of equation (5). So loading curvature C can be written as

$$C = \frac{P}{h^2} = \sigma_y \prod 1 \left(\frac{E^*}{\sigma_y}, \frac{\sigma_r}{\sigma_y}\right) \quad (7)$$

When the dimensionless function given in equation (5), the normalization is required in connection with approach against σy or σr. By simulating know the relationship between Cv Vs normalized relationship between the properties of materials and material properties normalization Vs Cs (strain hardening exponent (n) and yield stress (σy)). Curve P-h for both Vickers and Spherical indentation which has the following relationship

$$C = P/h^2 \quad (8)$$

Where P is load and (h) indentation depth at each load curve. Cv is the coefficient curvature of Vickers indenting and the ball indenter designated as Cs and Curvature is a function of the relationship between the yield stress (σy) and work hardening coefficient (n). This will give potential relationship allows prediction of material parameters of the test continuous indentation

4. Prediction Plastic Properties Spot weld

Methods to determine the nature of the plastic material with dual indenter is developed through the conclusion inverse prediction using input hardness values of materials are known to identify the constitutive material properties (σy, n). In the first phase, systematically developed FE models in a simulation room includes a wide range of potential material properties. In the next stage, the P-h curve developed are used in space limit simulation. Comparative approaches have been developed to predict

IOP Conf. Series: Materials Science and Engineering **539** (2019) 012042 doi:10.1088/1757-899X/539/1/012042

material set has actually indented curvature. Through the force data (F) vs indentation depth associated with a hardness value HV and HRB. Simulations with validated models that have been done on various material properties (σy: 100-900 MPa, dan n:0.0-0.5) the relationship between the yield stress (σy), strain hardening coefficient (n), HV and HRB can be known (Figure 1)

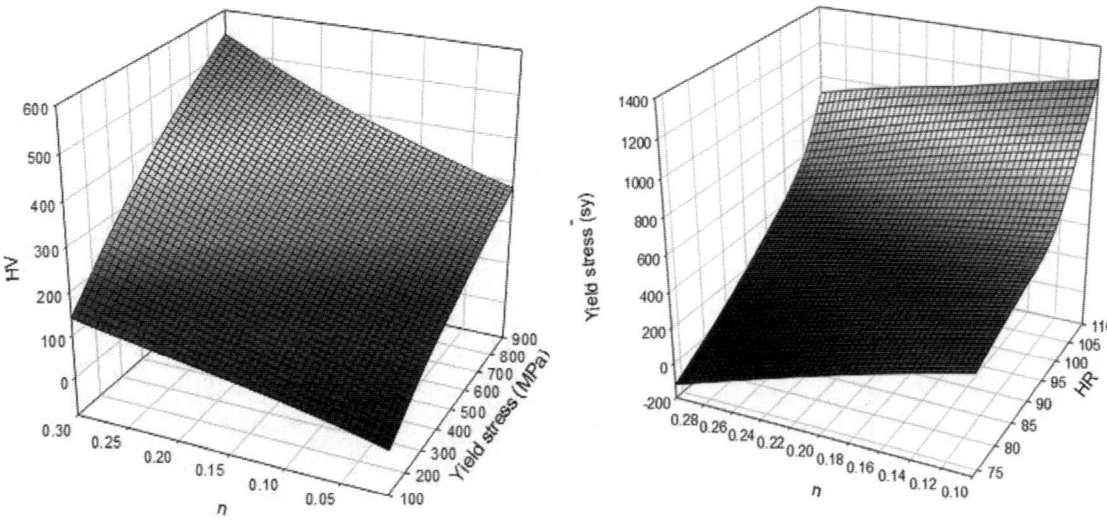

Figure 1. Typical Surface 3D plot as the model simulation results FE Model Indentation

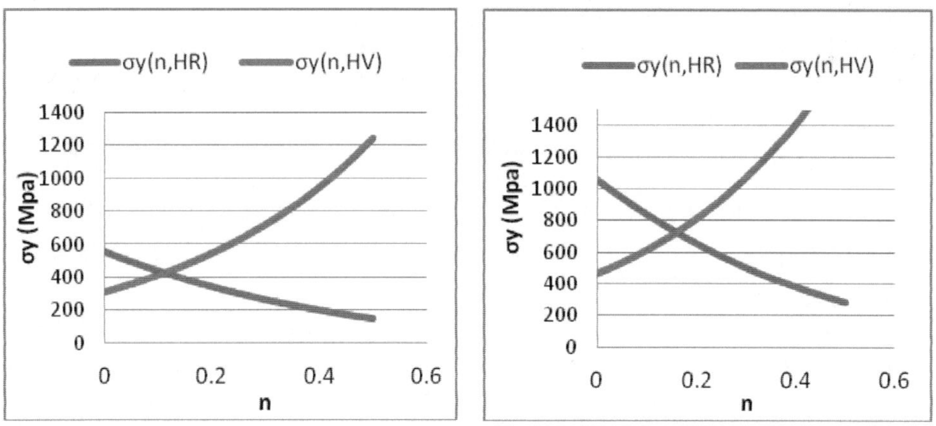

Figure 2 typical intersection curve σy (n, HRB) and σy (n, HV) in the FE Modeling approach for the prediction of parameters constitutive material (σy, n) for the parent metal (base)

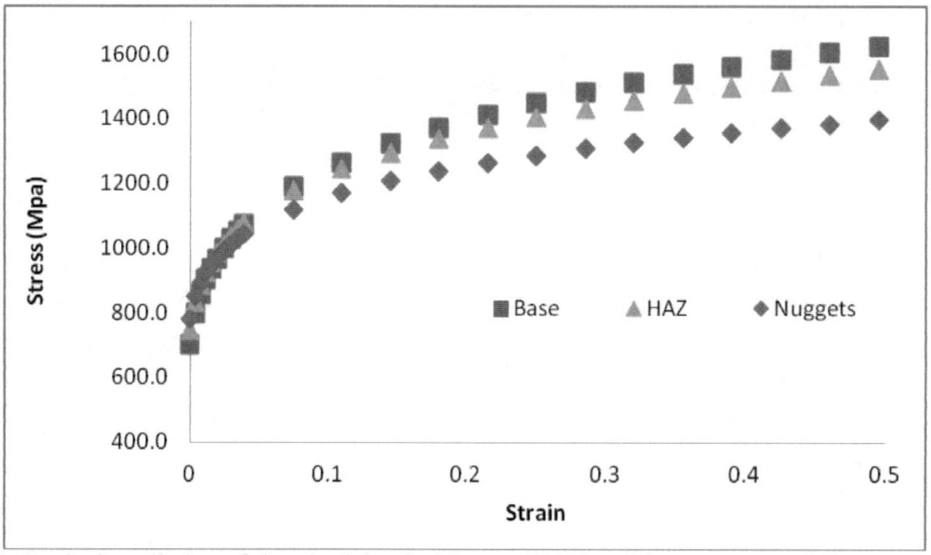

Figure 3. Typical prediction of the plasticity for Nugget, HAZ and the parent metal (base metal) as an application of FE Modeling constitutive parameters in the prediction of material (σy, n)

5. Conclusion

Methods to determine the nature of the plastic material with dual indenter is developed through the conclusion inverse prediction Inverse method using a numerical approach to the value of the constitutive material (σy, n) based on static indentation has been developed and validated. Evaluation is based on the experimental data showed similarities accurate numerical approaches indented continuous curve. The developed approach has been successfully used to characterize the plasticity of different zones at welds spot welding. (Nuggets; σy = 780 MPa, n= 0.12, HAZ: σy = 745 MPa, n=0.15, Base: σy =700 MPa, n=0.17). This plastic material parameters used in finite element modeling for tensile shear deformation spot welding and welded joints showed good correlation with experimental results. Validated FE models are then used to predict the effect of nugget size and thickness of the sheet metal on the strength of the weld joint points with different materials (dissimilar material)

Acknowledgements

This research/article's publication is supported by the United States Agency for International Development (USAID) through the Sustainable Higher Education Research Alliance (SHERA) Program for Universitas Indonesia's Scientific Modeling, Application, Research and Training for City-centered Innovation and Technology (SMART CITY) Project, Grant #AID-497-A-1600004, Sub Grant #IIE-00000078-UI-1

References

[1] Todd Litman, Comprehensive Evaluation of Energy Conservation and Emission Reduction Policies,"*Transportation Research A*, Vol. 47, January (2013), pp. 153-166
[2] Anaraki A.G., et al., Experimental and numerical analysis of low cycle fatigue of spot welded joints under peel-tension loading, *Mechanical Computational*, vol 221, (2002) pp.786-804.
[3] Irfan Dost, et all, Mechanical Evaluation of Joining Methodologies in Multi Material Car Body, International Journal of Advances in Engineering & Technology, ISSN: 2231-1963 Vol. 5, Issue 1, pp. 259-268 Nov. 2012.

IOP Conf. Series: Materials Science and Engineering **539** (2019) 012042 doi:10.1088/1757-899X/539/1/012042

[4] Rahman M.M.,et al.,.An Investigation into the effects of spot diameter and sheets thickness on fatigue life of spot welded structure based on FEA, *Research Journal of Applied Science*, 3(1):10-15, 2008
[5] Nachimani C., An Experimental Investigation on Spot Weld Growth on Dissimilar Joints of 304L Austenitic Stainless Steel and Medium Carbon Steel (Part1), International Journal of Advances in Applied Sciences (IJAAS) Vol. 2, No. 1, (2013), pp. 25~32
[6] Aslanlar S.,The effect of nucleus size on mechanical properties in electrical resistance spot welding of sheets used in automotive industry, *Materials and Design*, 27.125-131. 2006
[7] Hou Z., et al., Finite element analysis for the mechanical features of RSW process, *Journal of materials and Processing Technology*, 185, 1-3, (2007)160-165
[8] I. N. Budiarsa, M. Jamal, "P-h Curves and Hardness Value Prediction for Spherical Indentation Based on the Representative Stress Approach", Applied Mechanics and Materials, Vol. 493, pp. 628-633, 2014
[9] N. Budiarsa, "Indentation Size Effect (ISE) of Vickers Hardness in Steels: Correlation with H/E", Applied Mechanics and Materials, Vol. 391, pp. 23-28, 2013
[10] N. Budiarsa, A. Norbury, X. X. Su, G. Bradley, X. J. Ren, "Analysis of Indentation Size Effect of Vickers Hardness Tests of Steels", Advanced Materials Research, Vols. 652-654, pp. 1307-1310, 2013
[11] Bucaille.J.L.,et al., Determination of plastic properties of metals by instrumented indentation using instrumented sharp indentation. *Acta Materialia*, Vol.49, (2003) pp. 3899-3918
[12] G. Cricrì, et al., A consistent use of the Gurson-Tvergaard-Needleman damage model for the R-curve calculation 24 (2013) 161-174; DOI: 10.3221/IGF-ESIS.24.17
[13] Dao M., Chollacoop N., Van Vliet K. J., Venkatesh T. A. and Suresh S., Computational modelling of the forward and reverse problems in instrumented sharp indentation, Acta Materialia, Vol. 49, 2001, pp. 3899–391

IOP Conf. Series: Materials Science and Engineering **539** (2019) 012043 doi:10.1088/1757-899X/539/1/012043

THE EFFECT OF COMPOSITION OF PLASTIC WASTE LOW DENSITY POLYETHYLENE (LDPE) WITH SAND TO PRESSURE STRENGTH AND DENSITY OF SAND/LDPE COMPOSITES

I M Susila[1], N P G Suardana[1], C I P K Kencanawati[1], I N A Thanaya[2] and I W B Adnyana[1]

[1]Mechanical Engineering Udayana University

[2]Civil Engineering Udayana University

Email: npgsuardana@unud.ac.id

Abstract. Plastic waste especially plastic bags have not been widely reused or recycled so that the amount increases polluting the environment every day. One way to overcome this problem is by utilizing plastic waste into paving blocks as a substitute for cement. The problem is how the composition of plastic and sand. The purpose of this study is to determine the effect of using LDPE waste into a mixture of sand with ratios of 1:3, 1:5, and 1:7 with heating temperature of 200°C to the compressive strength and density of sand/LDPE composites. Based on the results of the study, the highest average compressive strength was resulted by the sample which has a ratio of 1:3 plastic and sand with 3mm sand grain and a heating temperature of 200°C which is equal to 32.7MPa. While the lowest average compressive strength was obtained by the sample which has a ratio of 1:7 plastic and sand, 3mm granules and temperature 200°C with a compressive strength of 12.0 (MPa). Increasing the mixture of plastic and sand 1:3, 1:5, 1:7 in which of the followed by increasing of density of the sample.

1. Introduction

Plastic has become an important component in this life. Almost all people use plastic material in their daily activities. Moreover, plastic material could change the woods and metal because of its characteristics such as light and strong, anti-corrosive, transparent and easy to be colored as well as having a good isolative characteristic. The use of plastic can be said not eco-friendly since it is hardly degraded by the environment without any limit. If this problem is not seriously handled so it will be a serious issue for our environment or the earth. Therefore, there should be an alternative of reusing plastic waste. The applicable alternatives that can be done are through cutting, melting, and mixing the plastic waste with the brace of sand to create one of new materials i.e. paving block from the plastic waste and reduce environmental pollution caused by plastic. Plastic waste functions to reduce the use of cement in making paving. This study aimed to preduce paving high compressive strength and good material density by adding LDPE (*Low Density Polyethylene*) plastic matrix into mixture of paving forming material (Arumi D, et.al, 2012).

2. Research Method

The sand was washed to omit the dust and other dirt and then dried naturally under the sunshine for 24 hours. Cleaning up the plastic waste, and then dried up under the sunshine. Cut the plastic with plastic cutting machine. Melt the plastic waste on a stove with temperature variation of 200°C on the stove and mix with sand with some composition of plastic: sand (1:3, 1:5, and 1:7) and sand grains 3mm (6 Mesh). Cut the test specimen with standard ASTM D695 for compressive test and density test by using ASTM D 792-08.

3. Result and Discussion

Data collection was done since the beginning of material mixing process i.e. material of the LDPE plastic and sand with ratios of 1:3,1:5,1:7 of 3 mm sand grains in 200°C, continued to specimen molding process until testing process such as compressive test, density test and SEM. The following is the research findings:

3.1. Data of Compressive Test

Collection of compressive test data was done by using mortar compression machine to find out the paving compressive strength. The following table is the calculation result of paving compressive strength by using mortar compression machine.

Table 1. The Calculation Result of Compressive Strength by Using Mortar Compression Machine

Comparison of Paving Mixture	Sample	Press Load (KN)	Cross section (Cm^2)	Compressive Strength (Mpa)	Average (MPa)
Mixture. 1:3,	Specimen 1	115	36	31.9	
Grain 3mm,	Specimen 2	120	36	33.3	32.7
Temperature 200°C	Specimen 3	118	36	32.8	
Mixture 1:5,	Specimen 1	70	36	19.4	
Grain 3mm,	Specimen 2	95	36	26.4	20.8
Temperature 200°C	Specimen 3	60	36	16.7	
Mixture 1:7,	Specimen 1	40	36	11.1	
Grain 3mm,	Specimen 2	50	36	13.9	12.0
Temperature 200°C	Specimen 3	40	36	11.1	

3.2. Discussion of Compressive Test

Based on the result of specimen molding and testing, the measurement of compressive strength with mixture variation of LDPE plastic and sand that has been made resulted data of average compressive strength from every samples as described by Fig. 1 which showed that different mixture of plastic and sand influences the paving compressive strength. In the testing process by placing the specimen in the middle of testing device and then turning the machine on as well as arranging the pressure to the maximum and finally turning the machine off to record the applied force. On the specimen of the mixture component of plastic and sand with ratio of 1:3, 3mm sand grain in temperature 200°C recorded the compressive strength of 32.7 (Mpa). On the specimen of the mixture component of plastic and sand with ratio of 1:5, 3mm sand grain in temperature 200°C recorded the compressive strength of 20.8 (Mpa). On the specimen of the mixture component of plastic and sand with ratio of 1:7, 3mm sand grain in temperature 200°C recorded the compressive strength of 12.0 (Mpa). Based on the result of measurement, the highest average compressive strength was resulted by the sample which has a ratio of 1:3 plastic and sand with 3mm sand grain and a heating temperature of 200°C which is equal to 32.7 (MPa). Meanwhile the lowest average compressive strength was obtained by the sample which has a ratio of 1:7 plastic and sand, 3mm granules and temperature 200°C with a compressive strength of 12.0 (MPa). It is because the composite in sample with ratio of 1:3 got all of the sand were bounded by matrix (as shown by SEM Fig. 3) so the compressive strength was better than specimen with ratio of 1:7, in which there were many cavities found as resulted by the lack of matrix to bound the entire surface of the sand as shown by SEM Fig. 4.

Afterward, the data were plotted into bar graph to describe the average of compressive strength of the specimens.

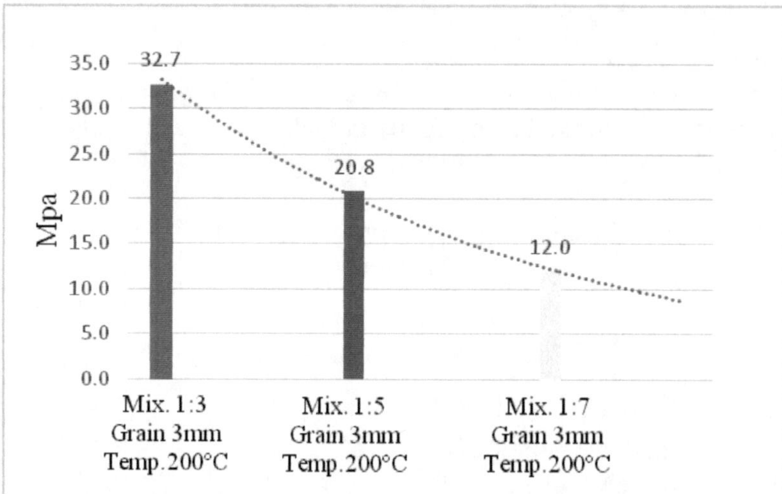

Figure 1. Graph of Compressive Strength Average of the Specimen with Plastic and Sand Mixture Ratios of 1:3,1:5,1:7 of 3 mm Sand Grains in 200°C.

3.3. Data of Density Test

Data of the density test was collected by comparing samples mass by using digital scale and recorded as dry mass (ms). Then, the sample was soaked in water and the mass was recorded as wet mass (mb) and the sample that was hanged by wire in water (mg).

Table 2. Data of Density Test

Variation	Sample	mk (gr)	ms (gr)	mg (gr)	mb (gr)	ρ (gram/cm³	Average
Mix.1:3,	1	2.21	19.04	9.12	20.08	1.443	
Grain 3mm,	2	2.21	19.13	9.17	20.11	1.452	1.455
Temp.200°C	3	2.21	19.45	9.21	20.20	1.471	
Mix.1:5,	1	2.21	19.89	9.43	20.40	1.506	
Grain 3mm,	2	2.21	20.35	10.11	20.96	1.537	1.545
Temp.200°C	3	2.21	20.43	10.59	21.33	1.574	
Mix.1:7,	1	2.21	21.41	13.11	23.61	1.680	
Grain 3mm,	2	2.21	21.55	13.59	24.11	1.688	1.765
Temp. 200°C	3	2.21	22.65	14.35	25.05	1.749	

Afterward, the data were plotted into bar graph to describe the average of density of the specimens.

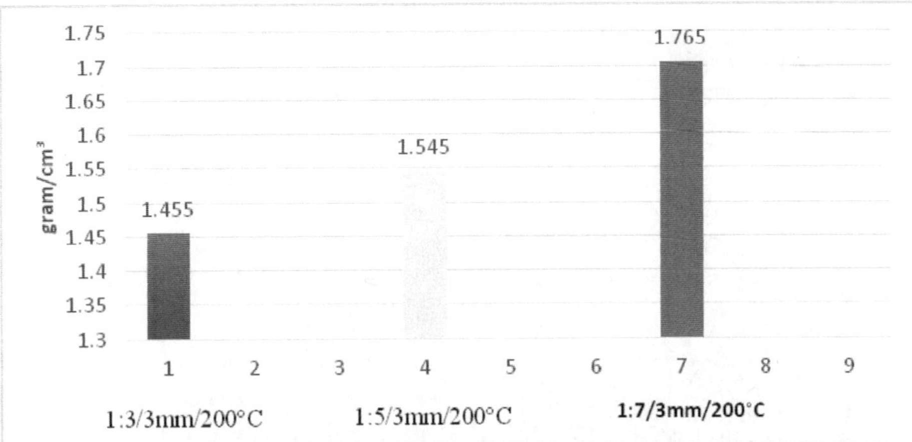

Figure 2. Graph of Density Average of the Specimen with Plastic and Sand Mixture Ratios of 1:3,1:5,1:7 with 3 mm Sand Grains in 200°C.

3.4. Discussion of Density Test

The measurement of density with mixture variation of plastic and sand that has been made resulted data of average density from every samples as described by Fig. 3. Fig. 3, shows that the highest density was gained from mixture variation of plastic and sand of 1:7 with 3mm sand grain in temperature 200°C while the lowest average was gained by sample of 1:3 with 3mm sand grain in temperature 200°C. It is caused by the composition of plastic on ratio 1:7 was fewer than ratio 1:3 in the same volume in which plastic density is lighter than the sand so it influences the composite density.

3.5. The Result of Material Characteristic by Using SEM (Scanning Electron Microscope) Type Phenom G2 Pro.

The following is the result of SEM observation upon the paving block specimen test with ratio variation of 1:3, 1:5 and 1:7 by using 3 mm sand grain in temperature of 200°C.

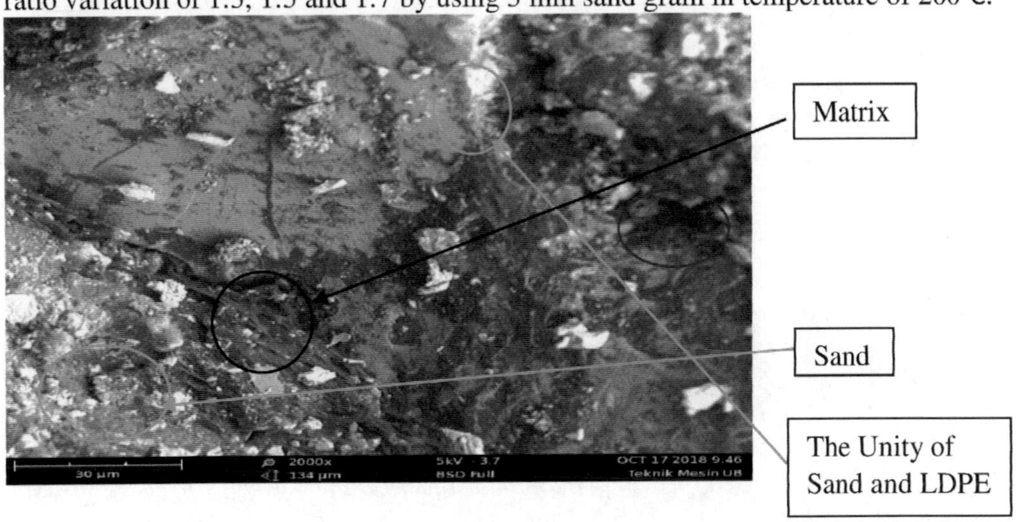

Figure 3. The Result of Micro Structure from the Specimen with Ratio of 1:3, with 3 mm Sand Grains in 200°C with 2000x Magnification.

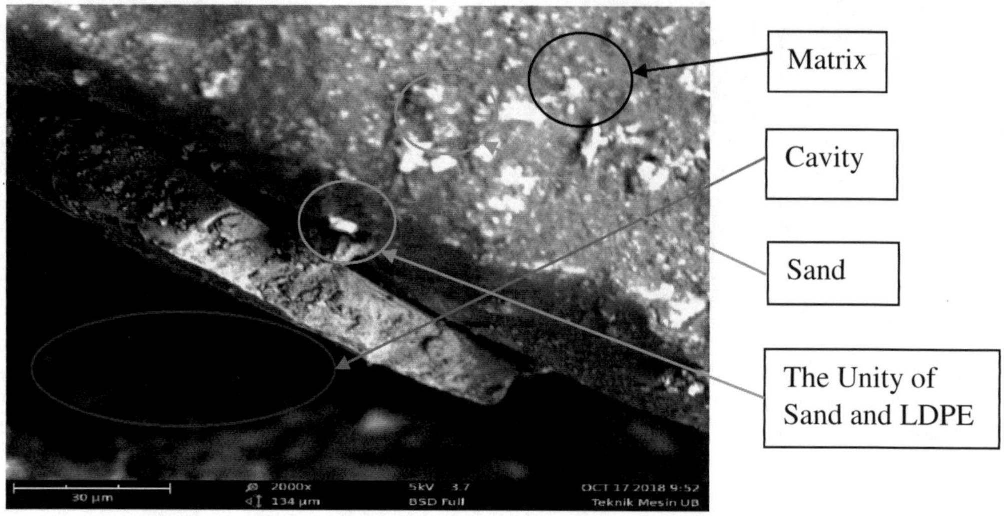

Figure 4. The Result of Micro Structure from the Specimen with Ratio of 1:7, with 3 mm Sand Grains in 200°C with 2000x Magnification.

4. Conclusion

Based on the research result and discussion on previous chapters, some conclusion can be drawn as follows:

1. The higher the ratio of plastic and sand, the lower the compressive strength of the composite.
2. The higher the ratio of plastic and sand, the higher the density of the composite.

Acknowledgment

This work was supported by the Directorate of Research and Community Services, Directorate General of Research and Development Strengthening, Ministry of Research, Technology, and Higher Education of Republic Indonesia, with agreement letter No. 171.120/UN14.4.A/LT/2018

References

[1] Aditya Bhardwaj, 2017.et.al. Int.Journal of Engineering Research and Application www.ijera.com ISSN : 2248-9622, Vol. 7, Issue 4, (Part -4), pp.79-81

[2] AR Hakim, 2012. *Pemanfaatan Limbah Padat (plastik) Sebagai Bahan Baku Pembuatan Bahan Bakar Pengganti Bensin.* Jurnal Jurusan Teknik Lingkungan FTSP UPN "Veteran" Jatim.

[3] Arumi, D, dkk, 2012. *Ecoplafie Paving (Economic Plastic Fiber)SebagaiProduk Perkerasan Jalan Berkonsep Ramah Lingkungan..* JurnalTeknik Sipil Universitas Sebelas Maret.

[4] Astm American Society for Testing And Materials, Copyright © 2004, West Conshohocken, PA. All rights reserved.

[5] Kumar S., Panda, A.K., dan Singh, R.K., 2011, A Review on Tertiary Recycling of High-Density Polyethylene to Fuel, Resources, Conservation and Recycling Vol. 55 893– 910

[6] Nasrun, Eddy Kurniawan, Inggit Sari 2016 *Pengolahan Kantong Plastik Jenis Kresek Menjadi Bahan Bakar Menggunakan Proses Pirolisis.* Jurusan Teknik Kimia, Fakultas Teknik, Universitas Malilkussaleh Kampus Bukit Indah, Muara Satu, Lhokseumawe, Aceh 24352

[7] Ngakan Putu Gede Suardana, Min Seuck Ku, Jae Kyoo Lim, Effects of diammonium phosphate on the flammability and mechanical properties of bio-composites. Bahans and Design 32 (2011) 1990–1999.

[8] Ravindra N. Patil, Shubham D. Kothawade, Hitesh A. Shinde, Rahul G. Katore, Prerak Jha, Himani P. Rane, 2017. *International Journal of Innovative Research in Science, Engineering and Technology* (An ISO 3297: 2007 Certified Organization) Website: www.ijirset.com Vol. 6

[9] Sherliana, 2016. *Studi Kuat Tekan Paving Block Dengan Campuran Tanah, Semen, Dan Abu Sekam Padi Menggunakan Alat Pemadat Modivikasi.*

[10] Surono, Untoro Budi, 2013. *Berbagai Metode Konversi Sampah Plastik Menjadi Bahan Bakar Minyak.* Jurnal Teknik VOL. 3 NO.1, hal. 32-40

[11] Syafitri, C. 2001, *Analisis Aspek Sosial Ekonomi Pemanfaatan Limbah Plastik*, Program Pascasarjana Institut Pertanian Bogor.

[12] R. Manikandran, D. S. Nirmala, G. Dhinakaran 2015. *Effect Of Ldpe Raw Material On Strength, Corrosion And Sorptivity Of Concrete,* Journal of Engineering Science and Technology Vol. 10, No. 4 (2015) 485 - 495 © School of Engineering, Taylor's University